Topics in Applied Physics Volume 71

Topics in Applied Physics Founded by Helmut K. V. Lotsch

Volumes 1–56 are listed on the back inside cover

The Monte Carlo Method in Condensed Matter Physics

Edited by K. Binder

With Contributions by
A. Baumgärtner K. Binder A. N. Burkitt
D. M. Ceperley H. De Raedt A. M. Ferrenberg
D. W. Heermann H. J. Herrmann D. P. Landau
D. Levesque W. von der Linden J. D. Reger
K. E. Schmidt W. Selke D. Stauffer
R. H. Swendsen J.-S. Wang J. J. Weis A. P. Young

With 83 Figures

Springer-Verlag
Berlin Heidelberg New York
London Paris Tokyo
Hong Kong Barcelona Budapest

Professor Dr. Kurt Binder

Institut für Physik
Johannes-Gutenberg-Universität Mainz
Staudingerweg 7
D-6500 Mainz,
Fed. Rep. of Germany

ISBN 3-540-54369-4 Springer-Verlag Berlin Heidelberg NewYork
ISBN 0-387-54369-4 Springer-Verlag NewYork Berlin Heidelberg

Library of Congress Cataloging-in-Publication Data.

The Monte Carlo method in condensed matter physics / edited by K. Binder; with contributions by A. Baumgärtner ... [et al.]. p. cm. – (Topics in applied physics ; v. 71) Includes bibliographical references and index. ISBN 0-387-54369-4 (alk. paper). – ISBN 3-540-54369-4 (alk. paper) 1. Monte Carlo method. 2. Statistical physics. 3. Condensed matter. I. Binder, K. (Kurt), 1944–. II. Baumgärtner, A. (Artur) III. Series. QC 174.85.M64M65 1992 530.4'1'01519282–dc20 91-42398 CIP

Printed in Germany

Typesetting: Thomson Press (India) Ltd., New Delhi;
Offsetprinting: Color-Druck Dorfi GmbH, Berlin; Binding: Lüderitz & Bauer, Berlin
54/3150-5 4 3 2 1 0 – Printed on acid-free paper

Preface

The Monte Carlo method is now widely used and commonly accepted as an important and useful tool in solid state physics and related fields. It is broadly recognized that the technique of "computer simulation" is complementary to both analytical theory and experiment, and can significantly contribute to advancing the understanding of various scientific problems.

Widespread applications of the Monte Carlo method to various fields of the statistical mechanics of condensed matter physics have already been reviewed in two previously published books, namely *Monte Carlo Methods in Statistical Physics* (Topics Curr. Phys., Vol. 7, 1st edn. 1979, 2nd edn. 1986) and *Applications of the Monte Carlo Method in Statistical Physics* (Topics Curr. Phys., Vol. 36, 1st edn. 1984, 2nd edn. 1987). Meanwhile the field has continued its rapid growth and expansion, and applications to new fields have appeared that were not treated at all in the above two books (e.g. studies of irreversible growth phenomena, cellular automata, interfaces, and quantum problems on lattices). Also, new methodic aspects have emerged, such as aspects of efficient use of vector computers or parallel computers, more efficient analysis of simulated systems configurations, and methods to reduce critical slowing down at phase transitions. Taken together with the extensive activity in certain traditional areas of research (simulation of classical and quantum fluids, of macromolecular materials, of spin glasses and quadrupolar glasses, etc.), there is clearly a need for a new book complementing the previous volumes. Thus, in the present book we present selected state-of-the-art reviews of those fields which have seen dramatic progress during the last couple of years and provide the reader with an up-to-date guide to the explosively growing original literature. Although the present book contains *several thousand references*, it must be realized that by now it has become impossible to describe all the research that uses Monte Carlo methods in condensed matter physics! We apologize to all colleagues whose work has been only briefly mentioned, or even not quoted at all, for the need to make a selection where certain topics had to be emphasized more than others. We hope that nevertheless the present book constitutes a useful landmark in a rapidly evolving field. Again, it is a great pleasure to thank the team of expert authors that contributed to the present book for their coherent and constant efforts and their fruitful collaboration.

Mainz, December 1991 *Kurt Binder*

Contents

Contributors

Baumgärtner, Artur

Institut für Festkörperforschung, Forschungszentrum Jülich, W-5170 Jülich, Fed. Rep. of Germany

Binder, Kurt

Institut für Physik, Johannes-Gutenberg-Universität Mainz, Staudingerweg 7, W-6500 Mainz, Fed. Rep. of Germany

Burkitt, Anthony N.

Fachbereich 8: Physik, Bergische Universität – GHS, Gaußstraße 20, W-5600 Wuppertal 1, Fed. Rep. of Germany

Ceperley, David M.

National Center for Supercomputing Applications, and Department of Physics, University of Illinois, Urbana, IL, 61801, USA

De Raedt, Hans

Institute for Theoretical Physics, University of Groningen, P.O. Box 800, NL-9700 AV Groningen, The Netherlands

Ferrenberg, Alan M.

Center for Simulational Physics, and Department of Physics and Astronomy, University of Georgia, Athens, GA 30602, USA

Heermann, Dieter W.

Institut für Theoretische Physik, Universität Heidelberg, Philosophenweg 19, W-6900 Heidelberg, Fed. Rep. of Germany

Herrmann, Hans J.

Höchstleistungsrechenzentrum, Kernforschungsanlage Jülich, Postfach 1913, W-5170 Jülich, Fed. Rep. of Germany

Landau, David P.

Center for Simulational Physics, University of Georgia, Athens, GA 30602, USA

Levesque, Dominique

Laboratoire de Physique Théorique et Hautes Energies, Bât. 211, Université de Paris-Sud, F-91405 Orsay Cedex, France

Linden, Wolfgang von der

Department of Mathematics, Imperial College, Huxley Building, Queen's Gate, London SW7 2BZ, UK
Permanent address: Max-Planck-Institut für Plasmaphysik, W-8046 Garching bei München, Fed. Rep. of Germany

Reger, Joseph D.

Institut für Physik, Johannes-Gutenberg-Universität Mainz, Staudingerweg 7, W-6500 Mainz, Fed. Rep. of Germany

Schmidt, Kevin E.

Department of Physics and Astronomy, Arizona State University, Tempe, AZ 85287, USA

Selke, Walter

Institut für Festkörperforschung, Forschungszentrum Jülich, W-5170 Jülich, Fed. Rep. of Germany

Stauffer, Dietrich

Institut für Theoretische Physik, Universität zu Köln, Zülpicher Str. 77, W-5000 Köln 41, Fed. Rep. of Germany

Swendsen, Robert H.

Department of Physics, Carnegie-Mellon University, Pittsburgh, PA 15213, USA

Wang, Jian-Sheng

Department of Physics, Hong Kong Baptist College, 224 Waterloo Road, Kowloon, Hong Kong

Weis, Jean-Jacques

Laboratoire de Physique Théorique et Hautes Energies, Bât. 211, Université de Paris-Sud, F-91405 Orsay Cedex, France

Young, Allan P.

Department of Physics, University of California, Santa Cruz, CA 95064, USA

1. Introduction

Kurt Binder

With 9 Figures

This chapter begins with some remarks on the scope of this book. Then it focuses attention on some technical developments on the analysis of Monte Carlo simulations which are of widespread applicability, such as progress obtained with finite size scaling at phase transitions, and the analysis of statistical errors.

1.1 General Remarks

"Computer simulation" now is recognized as an important tool in science, complementing both analytical theory and experiment (Fig. 1.1). In physics, the comparison of theoretical concepts with experimental facts has a long tradition, and has proven to be a primary driving force for continuous improvement of our understanding of the phenomena in question. However, there are many cases where comparing results of an analytical theory to experimental data leaves very important questions open: The analytical theory starts out with some model and in addition involves some mathematical approximations, which often are not justified from first principles. For example, in the theoretical description of an alloy phase diagram one may start out from an Ising-like Hamiltonian with certain assumptions about parameters describing the interactions between the various kinds of atoms [1.1]. The statistical mechanics of this *model* then is treated by an *approximate* analytical technique, e.g. the so-called "cluster variation" method [1.2]. Clearly, there will be some discrepancies between the outcome of such a theoretical treatment and experiment—and in many cases progress is hampered since it is not clear whether the problem is due to an inadequate model, an inaccurate approximation, or both. Sometimes such problems are obscured because the analytical theory contains some adjustable parameters: it may be that a fit of an inaccurate theory to experimental data is still possible, but the data resulting from such a fit are systematically in error.

Under such circumstances, simulations can be very helpful: the computer simulation can deal with a model without further approximations, apart from statistical errors and *controllable* systematic errors (such as finite size effects, for instance) the statistical mechanics of the problem is dealt with exactly. Thus it can be much better clarified whether a particular model is an adequate description of a condensed-matter system or not. Conversely, the simulation may also

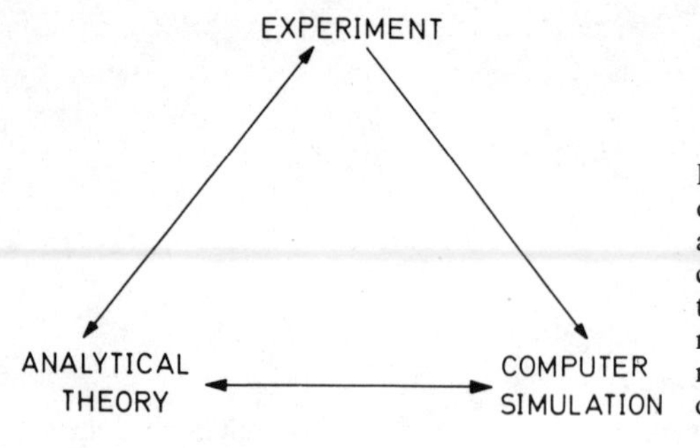

Fig. 1.1. Schematic illustration of the role of analytical theory, numerical simulation and experiment: the computer simulation can either be compared directly with a theory (to test the accuracy of mathematical approximations) or with experiment (to test whether a model faithfully describes a real system)

be used to simulate exactly that model that forms the basis of an approximate theoretical treatment to check the accuracy of the approximations made. In cases where a satisfactory theory has not yet been formulated, the insight into the physics gained from simulations may yield very valuable hints towards the development of a new theory. After all, the simulation can yield very detailed information on a model system, on microscopic scales of length as well as of time. This information may give qualitative insight into the validity of physical concepts and general assumptions about what the relevant processes going on are. The later chapters of this book will contain many illustrative examples of all these points.

In view of these distinct advantages of simulations and the fact that high-speed computers have become much more readily available and easier to access in recent years, it should not come as a surprise that the field has seen a rapid growth of applications. In fact, a good measure of this flourishing expansion is the relatively large number of books that have been published in the last couple of years: some of these are intended as a first introduction to the newcomer in the field [1.3–7], while some of them are monographs on specialized aspects of computer simulation. Quantum Monte Carlo methods are discussed in [1.8–10], applications of molecular dynamics methods are described in [1.7, 11–13]; computer simulation of lattice gauge theory has become a very active discipline [1.14] as well as the simulation of irreversible growth phenomena [1.15]. Also technical aspects, such as special purpose computers for simulations [1.16] or algorithmic implementation issues such as simulations on parallel computers [1.17], have received increasing attention.

While the field of "simulational condensed matter physics" is covered by the series of annual workshops at the Center for Simulational Physics at the University of Georgia, proceedings of which are available [1.18–20], the short papers contained therein describe studies of particular cases, similar in spirit to [1.21], rather than presenting comprehensive reviews of the respective fields. While two books describing the applications of the Monte Carlo method to the statistical thermodynamics of condensed matter have been edited by the present author [1.22, 23], this material is now somewhat outdated: extensive simulations have been carried out since then on all the subjects reviewed there, and also some fields which have not been covered there in detail at all (e.g., irreversible growth phenomena, quantum Monte Carlo studies of lattice models, etc.) have seen a lot of activity. Thus, in spite of the fact that this introductory chapter already refers to about 20 other books on computer simulation, there is still an urgent need for the present book!

One exciting feature of Monte Carlo simulation is the interdisciplinary character of this technique: methods developed for the analysis of simulation studies of the Ising model of magnetism [1.24] have become fruitful for the Monte Carlo study of polymer mixtures [1.25] as well as for the Potts glass model of orientational glasses [1.26]; concepts developed in the context of percolation clusters [1.27] have become fruitful for the Monte Carlo study of hot nuclei where a fragmentation analogous to a liquid–gas transition occurs [1.28], as well as for the description of the "poisoning" of catalytic surfaces by chemical reactions [1.29], and for the study of the initial stages of phase separation in off-critical binary alloys [1.30]. Thus progress obtained for one subject may well trigger better understanding of a rather different problem, throughout the whole range of condensed matter physics, and even neighbouring fields.

Hence in the present book, we aim at giving a representative overview of applications in the whole range of condensed matter physics, without attempting full completeness, which would be very hard to achieve because of the rapid expansion of the field. Thus the coverage of subjects in this book is somewhat selective: we want to complement two older books on the same subject, by adding chapters on subjects which were treated not at all extensively in the previous books [1.22, 23], such as quantum Monte Carlo methods for lattice models (Chap. 8), simulation of interfaces (Chap. 11), of irreversible growth phenomena (Chap. 5) and of cellular automata (Chap. 10). Aspects of program implementation and efficiency both on vectorising computers (Chap. 2) and on parallel computers (Chap. 3) are also treated, as well as a discussion of methods for avoiding "critical slowing down" and of the most efficient "data analysis" (Chap. 4). In addition, for several subjects already treated previously [1.22, 23] more recent high precision Monte Carlo studies have added so much progress, that a new description of the "state of the art" seems warranted: in this spirit, the present book contains chapters on classical fluids (Chap. 6) and on quantum fluids (Chap. 7), on spin glasses (Chap. 12), on percolation (Chap. 10) and on polymer simulation (Chap. 9). Clearly, there is recent simulation work on phase transitions in systems far from equilibrium [1.31], on the kinetics of the formation of ordered domains out of disordered initial configurations [1.32] and of phase separation [1.30, 33] and other diffusion phenomena [1.34], on the calculation of phase diagrams of alloys [1.1, 35] and adsorbed monolayers on surfaces [1.36], and numerous problems about critical phenomena [1.37]. Also new methods which potentially may become quite powerful have been developed, such as the combination of the Coherent Anomaly Method (CAM) [1.38] with Monte Carlo techniques (MC–CAM) [1.39]. In order to save space and to keep the effort in collecting material for this book manageable, all these subjects will not be discussed in detail here.

In the remainder of the present chapter, a discussion of the progress in the understanding of finite-size effects is given (Sect. 1.2) and some recent work on statistical and systematic errors in Monte Carlo sampling is reviewed (Sect. 1.3). Some final remarks are given in Sect. 1.4.

1.2 Progress in the Understanding of Finite Size Effects at Phase Transitions

Computer simulations necessarily deal with systems of finite size. In systems undergoing phase transitions of first or second order, the finite size of the system wipes out the singular behaviour occurring in the thermodynamic limit. For any quantitative studies of phase transitions with simulations, such finite size effects must be understood—therefore they are emphasised in the present article. For a more complete discussion, we refer to a recent book [1.40].

1.2.1 Asymmetric First-Order Phase Transition

By a "symmetric" first-order transition we mean a situation such as the transition of the Ising ferromagnet from positive to negative magnetisation at zero field, driven by changing the field H from positive to negative values. This simple case is well understood [1.40–44]. In particular, we note that the spin up–spin down symmetry of the problem implies that the finite size causes a rounding of the transition but no shift (i.e., it still occurs at $H=0$ also in a finite system).

By "asymmetric" first-order transitions we mean transitions where no such particular symmetry relating both phases exists [1.45]. In particular, transitions driven by temperature are normally asymmetric; e.g. in the q-state Potts model [1.46] with $q>4$ states in $d=2$ dimensions, a first-order transition occurs and its location is known exactly [1.47]. Clearly, there is no symmetry between the q-fold degenerate ordered phase and the (nondegenerate) disordered phase. We then also expect a shift of the "effective" transition temperature $T_c(L)$, where the rounded peak of the specific heat representing the smeared delta function of the latent heat has its maximum $C_L^{\max}$, relative to the true transition temperature $T_c(\infty)$,

$$T_c(L)-T_c(\infty)\propto L^{-\lambda}, \quad C_L^{\max}\propto L^{\alpha_m}, \quad \delta T\propto L^{-\theta}. \tag{1.1}$$

Here we have defined three exponents λ, α_m and Θ for the shift, the height of the peak and the temperature interval δT over which the singularity is smeared out.

A quantitative description of this smearing was first attempted in a pioneering paper by *Challa* et al. [1.48], who proposed a phenomenological theory in terms of a superposition of Gaussians representing each competing phase, and tested it by Monte Carlo simulation for the case $q=10$. Although this classic description has been widely quoted in the literature [1.4, 40, 45], it rests on a Landau mean-field-type assumption for the free energy of the model, which implies that for $T_c(\infty)$ all $(q+1)$ coexisting phases are described by Gaussian distributions of exactly the same *height*. This assumption, of course, was not obvious, and a recent analysis applying rigorous methods of statistical mechanics [1.49, 50] has shown that it is even incorrect: at $T_c(\infty)$ all $(q+1)$ coexisting phases should be described by Gaussian distributions of exactly the same *weight* (i.e., area underneath the peaks) [1.51]. It turns out that with this modification the theory is in even better agreement with the available numerical data for $q=10$ [1.48, 52]. However, care in the interpretation of Monte Carlo data is needed for transitions which are only weakly first order [1.52, 53]: already for $q=8$, prohibitively large lattice sizes [1.52] would be needed to reach the scaling regime where the theory [1.48, 50] holds, and for $q=5, 6$ even the qualitative distinction of the first-order character of the transition is difficult [1.53].

We now summarise the main statements of the theory [1.48–50]. The probability distribution of the internal energy E per lattice site is written as, for a d-dimensional hypercube with linear dimensions L,

$$P_L(E) \sim \frac{a_+}{\sqrt{C_+}} \exp\left(\frac{(E-E_+-C_+\Delta T)^2 L^d}{2k_B T^2 C_+}\right) + \frac{a_-}{\sqrt{C_-}} \cdot \exp\left(\frac{(E-E_--C_-\Delta T)^2 L^d}{2k_B T^2 C_-}\right), \tag{1.2}$$

where $\Delta T = T - T_c(\infty)$ and the specific heat for $L \to \infty$ near $T_c \equiv T_c(\infty)$ behaves as

$$\lim_{T\to T_c^-} C(T) = C_-, \quad \lim_{T\to T_c^+} C(T) = C_+, \tag{1.3}$$

and the weights a_+, a_- are related to the internal energies E_+, E_- of the disordered (ordered) branch by

$$a_+ = \exp\left(\frac{\Delta T(E_+ - E_-)L^d}{2k_B T T_c}\right), \quad a_- = q\exp\left(-\frac{\Delta T(E_+ - E_-)L^d}{2k_B T T_c}\right). \tag{1.4}$$

From (1.2) it is straightforward to obtain the energy $\langle E\rangle_L$ and specific heat $C(T,L)$ as

$$\langle E\rangle_L = \frac{a_+E_+ + a_-E_-}{a_+ + a_-} + \Delta T \frac{a_+C_+ + a_-C_-}{a_+ + a_-}, \tag{1.5}$$

$$C(T,L) = \frac{\delta\langle E\rangle_L}{\delta T} = \frac{a_+C_+ + a_-C_-}{a_+ + a_-} + \frac{a_+a_-L^d}{k_B T^2} \frac{[(E_+ - E_-) + (C_+ - C_-)\Delta T]^2}{(a_+ + a_-)^2}. \tag{1.6}$$

From (1.6) it follows that the specific heat maximum occurs at

$$\frac{T_c(L) - T_c}{T_c} = \frac{k_B T_c \ln q}{E_+ - E_-} 1/L^d, \tag{1.7}$$

and its height is

$$C_L^{\max} \approx \frac{(E_+ - E_-)^2}{4k_B T_c^2} L^d + \frac{C_+ + C_-}{2}. \tag{1.8}$$

Since the temperature region δT over which rounding of the delta function peak occurs is given just by taking the argument of the exponential functions in (1.4) of order unity,

$$\delta T \approx \frac{2k_B T_c^2}{(E_+ - E_-)L^d}, \tag{1.9}$$

one concludes that the exponents λ, α_m and Θ defined in (1.2) are all equal to the dimensionality

$$\lambda = \alpha_m = \Theta = d. \tag{1.10}$$

A quantity proposed by *Challa* et al. [1.48] as an indicator for the order of the transition,

$$V_L = 1 - \frac{\langle E^4 \rangle_L}{3\langle E^2 \rangle_L^2}, \tag{1.11}$$

can be shown [1.50] to exhibit a minimum at

$$\frac{T_V(L) - T_c}{T_c} = \frac{k_B T_c}{E_+ - E_-} \ln\left(q \frac{E_-^2}{E_+^2}\right) \frac{1}{L^d}, \tag{1.12}$$

and the value of this minimum is [1.50]

$$V_L^{\min} = \frac{2}{3} - \frac{1}{3}\left(\frac{E_+^2 - E_-^2}{2E_+ E_-}\right)^2 + \mathrm{O}(L^{-d}), \tag{1.13}$$

while at a second-order transition, $V_L(T_c) \to 2/3$ for $L \to \infty$. Finally, *Borgs* et al. [1.50] show that a third criterion for locating the transition temperature from the intersection of $E(T, L) \equiv \langle E \rangle_L$ and $E(T, 2L)$,

$$E(T_i, L) = E(T_i, 2L), \tag{1.14}$$

yields an estimate T_i which no longer differs from T_c by a term of order L^{-d}, but only by exponentially small corrections.

For $q = 10$ in $d = 2$, using the exactly known values for T_c, E_+ and E_- [1.47] in (1.7, 8, 12, 13) yield (where J is the coupling of the Potts Hamiltonian),

$$\mathscr{H}_{\text{Potts}} = -J \sum_{\langle i,j \rangle} \delta_{S_i S_j}, \quad S_i = [1, 2, \ldots, q],$$

$$\frac{k_B T_c(L)}{J} = 0.7012 + 1.63\, L^{-2}, \tag{1.15}$$

$$C_L^{\max} \approx 0.2464\, L^2 + \frac{C_+ + C_-}{2}, \tag{1.16}$$

$$\frac{k_B T_V(L)}{J} = 0.7012 + 2.39\,L^{-2}, \tag{1.17}$$

$$V_L^{\min} = 0.559 + O(L^{-2}). \tag{1.18}$$

Figures 1.2–4 show that there is very good agreement between these predictions and the numerical results of *Challa* et al. [1.48]. Note that *Challa* et al. [1.48] could not yet explain the observed distinction in the coefficient of L^{-2} in the relations for $T_c(L)$ and $T_V(L)$, unlike (1.15) and (1.17). In addition, it is gratifying that an independent new Monte Carlo study of the same model [1.52] is also in excellent agreement with (1.15–18). Thus it is fair to conclude that finite size effects at asymmetric first-order transitions are now rather well understood, although it is clearly difficult to reach the asymptotic regime (i.e. large enough L) when this description holds, for weakly first-order transitions [1.52, 53]. In this context, we mention encouraging results which have been found for the three-dimensional three-state Potts model [1.54].

To conclude this subsection, we emphasise that the description presented here refers strictly to the *canonical ensemble.* In the Monte Carlo simulation one is free to choose whichever ensemble of statistical mechanics that turns out to be convenient for the problem at hand. *Challa* and *Hetherington* [1.55] proposed as an interpolation between the canonical and the microcanonical ensemble, a scheme termed the "Gaussian ensemble". In this ensemble, as well as in the microcanonical ensemble, a first-order transition may show up as a loop in the $E(T, L)$ vs T curve. They [1.55] suggest that the use of this new ensemble may be more economical (with regard to the necessary use of computer time) than the use of the canonical ensemble. However, a careful assessment of

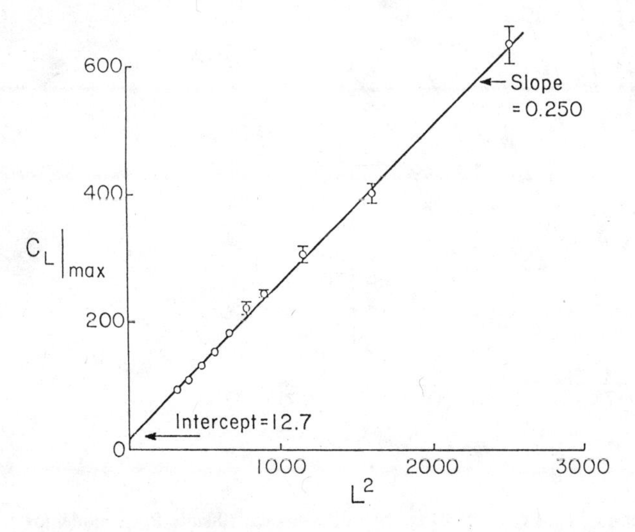

Fig. 1.2. Variation of the specific heat maxima of the two-dimensional 10-state Potts model with L^2. The slope is within 1.6% of its theoretical value, Eq. (1.16). (From [1.48])

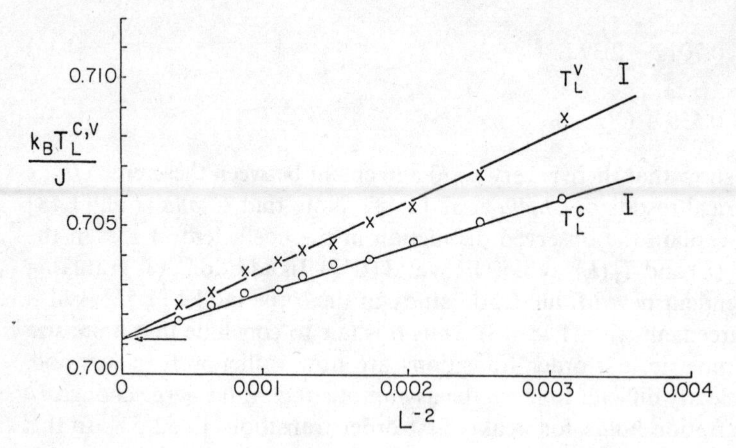

Fig. 1.3. Extrapolation of the characteristic temperature of the two-dimensional 10-state Potts model versus L^{-2}. T_c^L is the location of $C_L|_{max}$ and T_L^v is the location of $V_L|_{min}$. The arrow shows the exact result for the infinite lattice. The maximum values of the respective error bars are also indicated. (From [1.48])

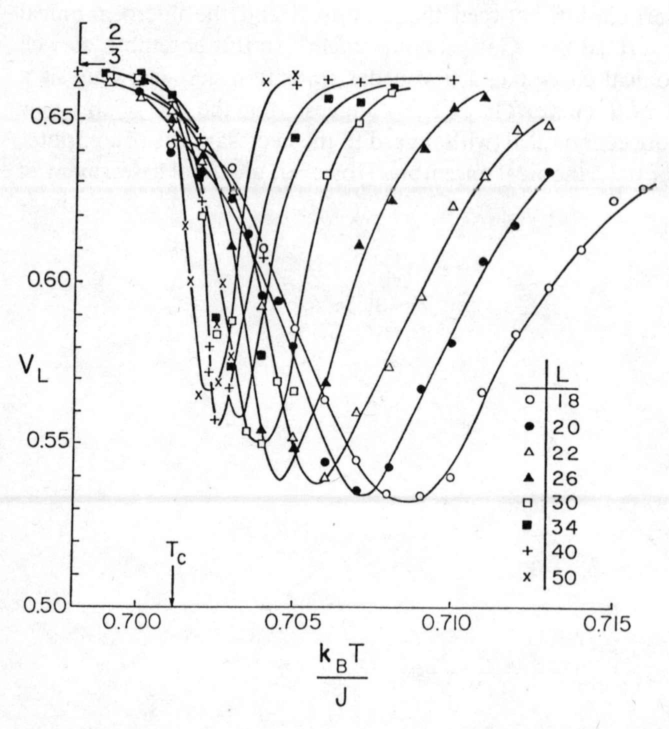

Fig. 1.4. Temperature variation of V_L, Eq. (1.11), of the two-dimensional 10-state Potts model for various lattice sizes. The transition temperature T_c and the trivial limit (2/3) for $V_L \to \infty$ are indicated. The nontrivial limit [$V_L|_{min} \to 0.559$, Eq. (1.18)] is consistent with the data. (From [1.48])

the finite size effects in this ensemble remains to be given, and thus this topic is left out of consideration here.

1.2.2 Coexisting Phases

As a second variation on our theme, we consider the liquid–gas transition occurring when we lower the temperature in a fluid, carrying out simulations in the *NVT* ensemble. Recently, a phenomenological description of size effects on this problem has been suggested [1.56] and corresponding simulation results on a two-dimensional fluid with (truncated) Lennard–Jones potential were presented [1.56, 57].

When the temperature is low enough that the considered density $\rho = N/V$ falls in the two-phase coexistence region, $\rho_{\text{gas}}(T) < \rho < \rho_{\text{liq}}(T)$, the system in equilibrium must separate into a (dense) liquid [of density $\rho_{\text{liq}}(T)$] surrounded by (less dense) gas [of density $\rho_{\text{gas}}(T)$], or vice versa. We want to estimate the coexisting densities from the simulation, estimate the corresponding compressibilities K_T^{gas}, K_T^{liq}, and locate the critical point $(\rho_{\text{crit}}, T_c)$ where this liquid–gas transition becomes continuous. For this purpose, understanding of the finite size effects involved is again clearly crucial.

Rovere et al. [1.56] suggested that a convenient method to study this problem can be based on an analysis of the density distribution $P_L(\rho)$ in subsystems of the total system, where the subsystem linear dimension L is much smaller than the linear dimension $V^{1/d}$ of the total system. Assuming $L \gg \xi$, the correlation length of density fluctuations, a double Gaussian approximation in full analogy to (1.2) is made [1.56]:

$$P_L(\rho) \sim \frac{\rho_{\text{liq}} - \langle\rho\rangle}{\rho_{\text{liq}} - \rho_{\text{gas}}} \frac{1}{\rho_{\text{gas}}(K_T^{\text{gas}})^{1/2}} \exp\left(-\frac{(\rho - \rho_{\text{gas}})^2 L^d}{2\rho_{\text{gas}}^2 k_B T K_T^{\text{gas}}}\right) + \frac{\langle\rho\rangle - \rho_{\text{gas}}}{\rho_{\text{liq}} - \rho_{\text{gas}}} \frac{1}{\rho_{\text{liq}}(K_T^{\text{liq}})^{1/2}} \exp\left(-\frac{(\rho - \rho_{\text{liq}})^2 L^d}{2\rho_{\text{liq}}^2 k_B T K_T^{\text{liq}}}\right); \tag{1.19}$$

here $\langle\rho\rangle$ is the average density, $\rho_{\text{gas}}, \rho_{\text{liq}}$ are the densities of the coexisting gas and liquid phases, and K_T^{gas}, K_T^{liq} the corresponding compressibilities at the coexistence curve.

For the reduced cumulant defined as

$$U_L = 1 - \frac{\langle(\Delta\rho)^4\rangle_L}{[3\langle(\Delta\rho)^2\rangle_L^2]}, \tag{1.20}$$

where $\Delta\rho \equiv \rho - \langle\rho\rangle$, one then derives from (1.19) an interesting behaviour in the limit $L \to \infty$, namely [$x \equiv (\langle\rho\rangle - \rho_{\text{gas}})/(\rho_{\text{liq}} - \rho_{\text{gas}})$ being the resulting fraction of the liquid phase at the considered state in the two-phase coexistence region in this limit]

$$U_\infty = -\frac{6x^2 - 6x + 1}{3x(1 - x)}. \tag{1.21}$$

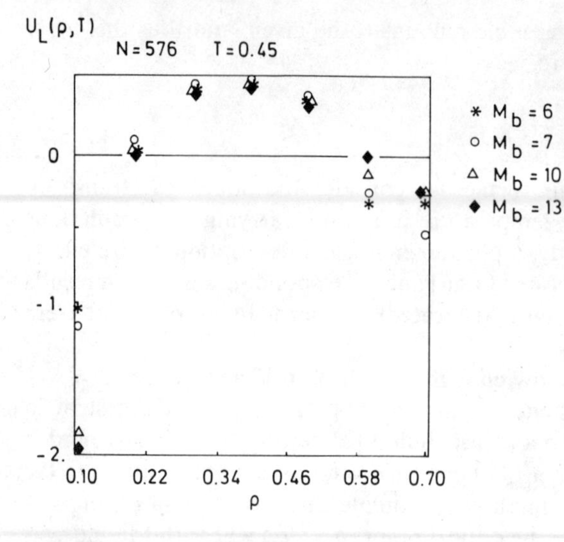

Fig. 1.5. Variation of cumulant $U_L(\rho, T)$ with density ρ for subsystems whose dimensions are a factor of M_b smaller than the total size of the system, which always contained $N=256$ atoms interacting with a Lennard–Jones 6–12 potential truncated at a distance $r=2.5\sigma$, $v(r)=4\varepsilon[(\sigma/r)^{12}-(\sigma/r)^6]$. The density is measured in units of σ^2, the temperature in units of ε/k_B. Note that the critical temperature is estimated from other data as $T_c=0.50\pm0.02$, while at $T=0.45$ the coexisting densities were estimated as $\rho_{gas}=0.021$, $\rho_{liq}=0.717$. (From [1.56])

This result shows that U_∞ tends to $-\infty$ at the coexistence curve ($x\to0$ or $x\to1$, respectively). The maximum value of U_∞, $U_\infty^{max}=2/3$, which occurs in the grand-canonical ensemble for $T<T_c$, is only reached at the density corresponding to the "rectilinear diameter", $\langle\rho\rangle=(\rho_{gas}+\rho_{liq})/2$. Simulations of subsystems of a two-dimensional Lennard–Jones fluid with $N=576$ atoms have given qualitative evidence for this behaviour (Fig. 1.5) [1.56]. Since the linear dimensions in this simulation were rather small, a quantitative test of (1.21) was not possible. Nevertheless, the calculation of the subsystem probability $P_L(\rho)$ presented in [1.56], showed that it is possible to apply finite size scaling concepts, which are so successful for various lattice models, to off-lattice models of fluids as well. Numerical data for $P_L(\rho)$ such as that shown in Fig. 1.6 [1.56] have been used successfully to estimate the compressibility of the system and densities of coexisting phases, and to locate the critical temperature T_c of the gas–liquid transition. This is evidence that the concept of analysing data from subsystems of a large system [1.58] is also useful for off-lattice models [1.59].

1.2.3 Critical Phenomena Studies in the Microcanonical Ensemble

The use of microcanonical simulation algorithms [1.60], e.g. the Creutz "demon algorithm" [1.61–63] for the Ising spin model, may be advantageous for various reasons: a Monte Carlo algorithm avoids the use of random number generators during the main part of the simulation, and hence can be executed particularly fast; molecular dynamics algorithms, desirable for many problems because they yield dynamical correlations on the basis of Newton's equations of motion, conserve the total energy of the system and thus also provide a microcanonical sampling of states. The basic independent variable is then no longer the temperature, of course, but rather the enthalpy density e. Now near a critical point we

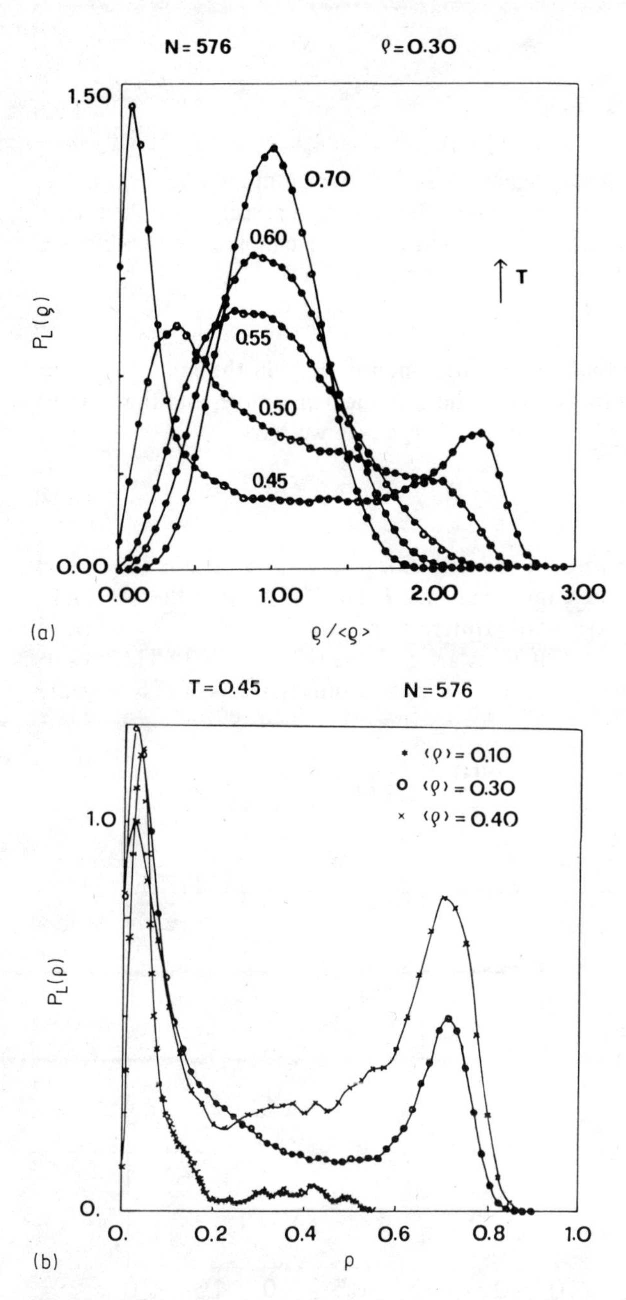

Fig. 1.6. (**a**) Subsystem density distribution function $P_L(\rho)$ for the two-dimensional Lennard–Jones system (Fig. 1.5) for $\langle\rho\rangle = 0.30$, $N = 576$, $M_b = 6$ and various temperatures. (**b**) $P_L(\rho)$ plotted vs ρ for $N = 576$, $T = 0.45$ and three densities $\langle\rho\rangle$, as indicated. (From [1.56])

have with $t = (T - T_c)/T_c$,

$$e^* \equiv e - e_c \approx A \frac{t^{1-\alpha}}{1-\alpha} + A't + \cdots, \tag{1.22}$$

where α is the specific heat exponent, A and A' are amplitude factors, and e_c is the enthalpy density at the critical point. Now finite size scaling for the microcanonical ensemble can still be formulated as in the canonical ensemble, e.g. the order parameter m of the transition scales as [1.60]

$$m = L^{-\beta/\nu} \tilde{m}_{\mathrm{mce}}(L/\xi); \tag{1.23}$$

the only distinctions are that the scaling function $\tilde{m}_{\mathrm{mce}}$ in the microcanonical ensemble (mce) is distinct from that in the canonical ensemble, and that ξ must be expressed in terms of e^* instead of t. For $\alpha > 0$, we have

$$\frac{L}{\xi} \propto L(e^*)^{\tilde{\nu}}, \quad \tilde{\nu} = \frac{\nu}{(1-\alpha)}, \tag{1.24}$$

where ν is the correlation length exponent. For $\alpha < 0$, however, it is the second term in (1.22) that dominates, implying that $L/\xi \propto L(e^*)^{\nu}$, since then e^* and t are linearly related. A particularly interesting case occurs for $\alpha = 0$, i.e. a logarithmic term due to the $t \ln t$ term arises from $t^{1-\alpha}$ for $\alpha = 0$. This occurs for the two-dimensional Ising model. No simple scaling in terms of e^* is possible. However, since [1.60]

$$\tilde{e}(T) = \frac{e(L,T) - e_c(L)}{t} = \frac{4}{\pi} \ln L + \frac{A(\tau)}{\tau}, \quad \tau \equiv Lt, \tag{1.25}$$

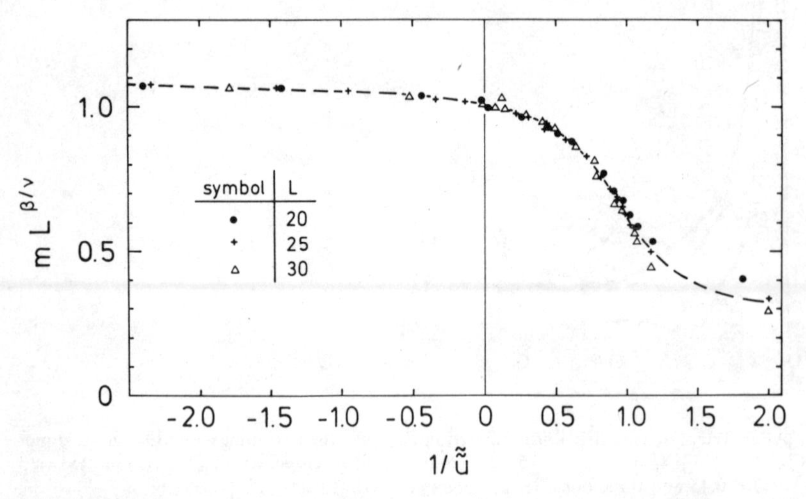

Fig. 1.7. The scaled magnetisation $mL^{\beta/\gamma}$ for finite $L \times L$ Ising square lattices with nearest-neighbour interactions plotted vs $l/\tilde{u}$, where $\tilde{u}$ is the scaling variable defined in the text, for $T_0 = 2.817\,\mathrm{J}/k_B$. (From [1.60])

where $A(\tau)$ is of the form $\tau \ln|\tau|$ as $\tau \to 0$, one can eliminate the $\ln L$ term by forming the difference $\tilde{u} = [\tilde{e}(T) - \tilde{e}(T_0)]/T_c$, where T_0 is some temperature far from T_c. Figure 1.7 shows that by studying $mL^{\beta/\nu}$ as function of $\tilde{u}$ one can in fact obtain a satisfactory scaling. A check of the scaling according to (1.24), which would be appropriate for the two-dimensional 3-state and 4-state Potts model universality classes, for instance, still remains to be done.

1.2.4 Anisotropy Effects in Finite Size Scaling

Recently finite size effects have been considered both for systems of "anisotropic shape" (e.g. geometries $L_\perp^{d-1} \times L_\parallel$, with linear dimension $L_\parallel$ in one direction much larger than the linear dimension $L_\perp$ in the remaining $d-1$ transverse directions) [1.64, 65] and for anisotropic critical phenomena in the bulk [1.64] and at interfaces [1.66]. While in the former case one simply has to take into account an additional scaling variable $L_\parallel/L_\perp$, characterising the shape of the system, in the case of anisotropic critical behaviour the situation is considerably more involved: we assume that two different correlation lengths $\xi_\parallel, \xi_\perp$ exist, with

$$\xi_\parallel \propto t^{-\nu_\parallel}, \quad \xi_\perp \propto t^{-\nu_\perp}, \tag{1.26}$$

with the critical exponents $\nu_\parallel, \nu_\perp$ being different in the two directions. While for bulk equilibrium critical phenomena this case is rather rare (it occurs, for instance, for uniaxial Lifshitz multicritical points [1.67]), it occurs in driven systems far from equilibrium [1.31] and for wetting transitions at surfaces [1.68]. If the generalised hyperscaling relation holds [$(d-1)\nu_\perp + \nu_\parallel = 2 - \alpha$], an additional scaling variable appears as $L_\parallel/L_\perp^{\nu_\parallel/\nu_\perp}$. Clearly, this fact makes a finite size scaling analysis very complicated, since $\nu_\parallel/\nu_\perp$ is not known a priori, and when the variation with one scaling variable (e.g. $tL^{1/\nu_\perp}$) is studied the other variable $L_\parallel/L_\perp^{\nu_\parallel/\nu_\perp}$ needs to be kept fixed. Thus it is no surprise that numerical studies of such complicated problems are scarce; only the simpler case of system shape effects for the two-dimensional Ising model (with isotropic critical behaviour!) has been tested extensively, and the dependence on the scaling variable $L_\parallel/L_\perp$ indeed works [1.64, 65]. On the other hand, for diverse systems such as the charged lattice gas in an electric field, hyperscaling is supposed to fail, and then the situation is very complicated [1.64]. We do not give any details on these classes of problems here, but refer to another recent review [1.69].

1.3 Statistical Errors

When planning a computer simulation study of a statistical mechanical model within a fixed computational budget, one is faced with a difficult choice between performing long simulations of small systems or shorter simulations of larger systems. This problem has been recently addressed by *Ferrenberg* et al. [1.70]. In order to use the available computer time as efficiently as possible it is important to know the sources of errors, both statistical and systematic ones, and how they depend on the size of the system and the number of updates performed.

Statistical errors occur because one can make a finite number n of statistically independent "measurements" in the course of a Monte Carlo simulation. As is well known [1.3–6, 22, 23, 71], successive system configurations generated in a Metropolis importance sampling method are not independent of each other. Suppose we perform in total N "measurements" taken at "time" intervals δt (measured in units of Monte Carlo steps per site for a lattice model). Then we have approximately [1.71]

$$n \approx \frac{N}{1 + 2\tau/\delta t}, \tag{1.27}$$

where τ is the integrated correlation time. The statistical uncertainty in the measured value of a quantity then is proportional to $n^{-1/2}$ for $n \gg 1$. For thermodynamic quantities the proportionality constant between the statistical uncertainty and $n^{-1/2}$, which is simply the standard deviation of the measured quantity, depends on the thermodynamic parameters (e.g. temperature and magnetic field, if we consider an Ising magnet) as well as on the system size. This size dependence depends on the *degree of self-averaging of the considered quantity* [1.72]. For self-averaging quantities, the proportionality constant decreases as the system size increases, and then the statistical errors decrease with increasing system size. An example of this is the magnetisation m of an Ising magnet; the square of the error $\overline{(\delta m)^2}$ is [1.71]

$$\overline{(\delta m)^2} = \frac{k_\mathrm{B} T \chi}{L^d} \frac{1}{n} \cong \frac{k_\mathrm{B} T \chi}{L^d} \frac{1 + 2\tau/\delta t}{N}, \tag{1.28}$$

where χ is the susceptibility and we consider a d-dimensional lattice of linear dimensions L. Outside of the critical region, χ tends to a finite value independent of L, and then $\overline{(\delta m)^2}$ decreases inversely proportional to $L^d N$. On a serial computer, a given amount of computer time corresponds to $L^d N = \text{const}$, if we disregard the time needed to equilibrate the system: then the same error results if we make fewer observations at larger systems or more observations at smaller systems, for a given amount of computer time. On a vectorising computer, the number of updates per second is often a decreasing function when L increases and then a better accuracy is reached for larger systems. This is no longer true at the critical point T_c, of course, where both χ and τ diverge as $L \to \infty$, $\chi \sim L^{\gamma/\nu}$, $\tau \sim L^z$, z being the "dynamic exponent" and γ the susceptibility exponent [1.73]. Then $\overline{(\delta m)^2} \sim L^{\gamma/\nu - d}/n \sim L^{\gamma/\nu + z - d}/N$. In this case of "weak self-averaging" the error strongly increases with increasing system size, for a given amount of computer time.

For the sampling of response functions from fluctuation relations, one has a complete lack of self-averaging [1.72]. Then the statistical error increases as $L^{d/2}$ away from T_c and as $L^{(d+z)/2}$ at T_c, for fixed computational effort (on a serial computer). An example of this behaviour is the susceptibility χ sampled from magnetisation fluctuations, for which

$$\frac{(\overline{\chi^2} - \bar{\chi}^2)}{\bar{\chi}^2} = 2 - 3U_L, \tag{1.29}$$

U being the fourth-order cumulant of the magnetization distribution, $U_L = 1 - \langle m^4 \rangle_L / [3\langle m^2 \rangle_L^2]$. Above T_c, $U_L \xrightarrow[L\to\infty]{} 0$, while at T_c, U_L tends to a universal constant U^*. Equation (1.29) implies that the relative error $[\overline{(\delta\chi)^2}]^{1/2}/\bar{\chi}$ of χ is independent of L, it decreases with $n^{-1/2}$ as $n \to \infty$, the proportionality constant being given by (1.29). In addition, one has to be careful because there is a *systematic error* in the MC sampling of response functions due to the finiteness of n: the response function is systematically underestimated [1.70]. This effect comes simply from the basic result of elementary probability theory [1.74], that estimating the variance s^2 of a probability distribution using n independent samples, the expectation value of the variance thus obtained $E(s^2)$ is systematically lower than the true variance of the distribution σ^2, by a factor $1 - 1/n$, $E(s^2) = \sigma^2(1 - 1/n)$. Applying this result to the susceptibility, we have from (1.27),

$$\chi_N = \chi_\infty \left(1 - \frac{1 + 2\tau/\delta t}{N}\right), \tag{1.30}$$

where χ_N denotes the expectation value of the susceptibility when N measurements are used at time intervals δt, and χ_∞ is the true value of the susceptibility. This effect becomes important particularly at T_c, where one uses the values of χ from different system sizes to estimate the critical exponent γ/ν. Because the correlation time depends on the system size, the systematic error resulting from (1.30) will be different for different system sizes.

To demonstrate these effects, *Ferrenberg* et al. [1.70] performed calculations for the nearest neighbour ferromagnetic Ising model right at the critical temperature T_c of the infinite lattice, known from other work ($T_c^{-1} = 0.221654$ [1.75]) on systems ranging from $L = 16$ to $L = 96$ using a fast multispin coding algorithm for the Cyber 205 at the University of Georgia [1.76]. Several million Monte Carlo steps per site (MCS) were performed, taking data at $\delta t = 10$, dividing the total number of observations N_{tot} into M bins of "bin length" N, $N_{\text{tot}} = MN$, and calculating χ_N from the fluctuation relation (of course, the result is averaged over all M bins). Figure 1.8 shows the expected strong dependence of χ_N on both N and L: while for $L = 16$ the data have settled down to an N-independent plateau value for $N \gtrsim 10^3$, for $L = 48$ the point even for $N = 10^4$ falls still slightly below the plateau, and for $L = 96$ the plateau is only reached for $N \gtrsim 10^5$. Thus if one uses a constant number N as large as 10^4 in this example for a finite size scaling analysis, the true finite-system susceptibility for large L is systematically underestimated, and so is γ/ν. However, if one *measures* τ for the different values of L and uses (1.30), it is possible to correct for this effect. In the present example, the appropriate correlation time for the susceptibility is [1.70] $\tau = 395$, 1640, 3745, 6935 and 15480, for $L = 16$, 32, 48, 64 and 96, respectively (note that a

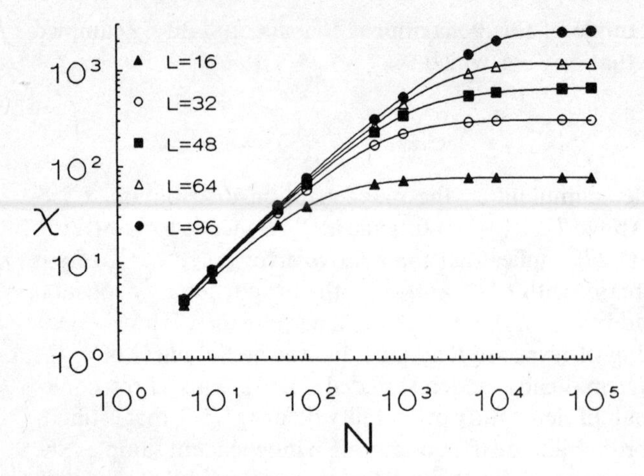

Fig. 1.8. Variation of χ_N for the "susceptibility" of $L \times L \times L$ ferromagnetic nearest-neighbour Ising lattices at the critical temperature as a function of "bin length" N. Different symbols indicate various values of L, as indicated in the figure. (From [1.70])

smaller time τ_c applies for the sampling of the specific heat from energy fluctuations!). Using these values one can rewrite (1.30) as $\chi_N = \chi_\infty(1 - 1/n)$, computing n from (1.27). Figure 1.9 shows that when $\chi_N L^{-\gamma/\nu}$ is plotted versus n, all the data collapse on a universal function, which for $n \gtrsim 5$ is compatible with the simple function $1 - 1/n$. *Ferrenberg* et al. [1.70] suggest that for any attempt to obtain high-precision estimates of critical exponents from the finite size scaling analysis of any model a scaling analysis of the data as a function of scaled bin length, as shown in Figs. 1.8 and 1.9, should be carried out.

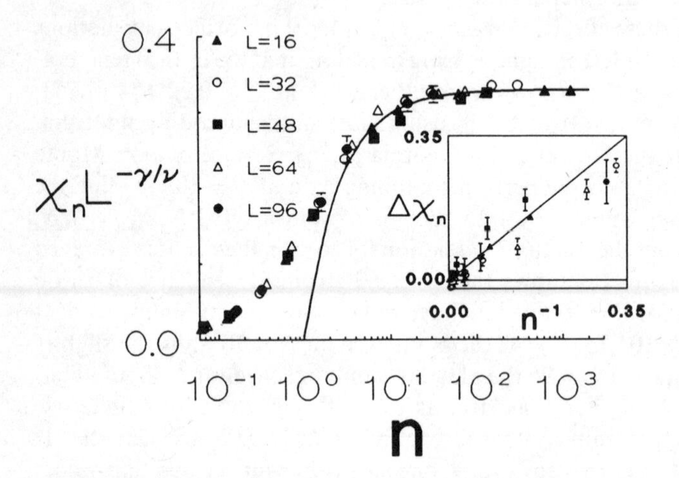

Fig. 1.9. Scaled susceptibility plotted vs scaled bin length n. The solid line is the function $\chi_\infty L^{-\gamma/\nu}(1 - 1/n)$, using the accepted value $\gamma/\nu = 1.97$ [1.75]. In the inset the reduced systematic error, $\Delta\chi_n = (\chi_\infty - \chi_n)/\chi_\infty$ is plotted vs n^{-1} to highlight the large bin length behaviour [the solid line, with slope = 1, is predicted by (1.30)]. (From [1.70])

A somewhat related but different discussion of error estimates of correlated data is given by *Flyvbjerg* and *Petersen* [1.77]. Although their method of successive "blocking" of data in the one-dimensional space of simulation time, similar to real-space renormalisation group ideas, has already been used for a long time (e.g. [1.78]), [1.77] gives a very pedagogic account which we briefly summarise here. Suppose we wish to calculate the error of the magnetisation $\bar{m}$ of an Ising model, obtained from a set of correlated data $m_1, m_2, \ldots, m_N$, without following the above approach, which requires the estimation of the relaxation time τ. The "blocking" is now defined as a transformation of the set of data $\{m_i\}$ into half as large a set $m'_1, \ldots, m'_{N'}$

$$m'_i = (m_{2i-1} + m_{2i})/2, \quad N' = N/2. \tag{1.31}$$

Now it is straightforward to show that both the mean value and the variance are invariant under this transformation,

$$\bar{m} = \bar{m}', \quad \sigma^2(m) = \sigma^2(m'). \tag{1.32}$$

Defining a correlation function γ_t as

$$\gamma_t \equiv \gamma_{i,j} \quad \text{for} \quad t \geqq |i-j| \quad \text{with} \quad \gamma_{i,j} \equiv \langle m_i m_j \rangle - \langle m_i \rangle \langle m_j \rangle, \tag{1.33}$$

one can show from (1.31) that for the correlation function of the primed variables we have

$$\gamma'_t = \begin{cases} 1/2\gamma_0 + 1/2\gamma_1, & \text{for } t = 0, \\ 1/4\gamma_{2t-1} + 1/2\gamma_{2t} + 1/4\gamma_{2t+1}, & \text{for } t > 0. \end{cases} \tag{1.34}$$

Since we can express the variance $\sigma^2(m)$ as [1.71, 77]

$$\sigma^2(m) = \frac{1}{N^2} \sum_{i,j=1}^{N} \gamma_{i,j} = \frac{1}{N}\left[\gamma_0 + 2\sum_{t=1}^{n-1}\left(1 - \frac{t}{n}\right)\gamma_t\right], \tag{1.35}$$

we have $\sigma^2(m) \geqq \gamma_0/N$. Now by repeated application of the blocking transformation, the true value of the variance emerges from the set of values γ_0/n, γ'_0/n', $\gamma''_0/n''\ldots$; every time the blocking transformation is applied, $2\gamma'_0 = \gamma_0 + \gamma_1$ increases unless γ_1 is already zero, because the correlation has already decayed during the "blocking time interval". Thus, from there onwards the ratio γ_0/n stays invariant under further blockings, it would reach a "fixed point" if one could let the number of blockings go to infinity. At this "fixed point", the blocked variables m'_i are independent Gaussian variables by the central limit theorem, of course.

Now one has only a finite number N of original observations available, and thus only a finite number k of blockings is possible, since we must have $N/2^k \gg 1$, otherwise one has errors that are too large in the estimate for γ_0. To correct for the bias due to the finite number of observations, it is now useful to consider the sequence of values

$$\frac{\bar{c}_0}{N-1}, \quad \frac{\bar{c}'_0}{N'-1}, \quad \frac{\bar{c}''_0}{N''-1}, \ldots, \tag{1.36}$$

where

$$\bar{c}_0 \equiv \frac{1}{N} \sum_{k=1}^{N} (m_k - \bar{m})^2, \tag{1.37}$$

remembering that c_0 is a biased estimator of γ_0, its expectation value being (for uncorrelated data) $\langle c_0 \rangle = \gamma_0 - \sigma^2(m)$ [1.74]. The strategy is then to look for a plateau in a plot of $c_0^{(k)}/(N^{(k)} - 1)$, where k is the number of blockings applied. Of course, at the end of the plateau (where $N/2^k$ is no longer very large) the error in $\bar{c}_0^{(k)}$ becomes so large that huge fluctuations appear. *Flyvbjerg* and *Petersen* [1.77] demonstrate the practical usefulness of this method by examples given for the two-dimensional Ising model at $J/k_B T = 0.30$, where for a 20×20 lattice 2^{17} measurements were made (at $\delta t = 1$). Since in this example the relaxation time was quite short ($\tau \approx 5.1$), it was easy to reach a convincing plateau. Further demonstrations of the practical usefulness of this method in critical phenomena studies would also be desirable.

Another very important problem is the error in estimates of relaxation times. Since in general the decay law with time t for a correlation function such as γ_t in (1.33) is not known, this problem has significantly hampered the study of critical slowing for Ising and Potts models [1.79] and dynamic simulations of self-avoiding random walks [1.80]. We refer to the original papers [1.79, 80] for a discussion.

1.4 Final Remarks

While in the early days of computer simulation Monte Carlo studies of model systems had mainly a qualitative exploratory character, there are now more and more quantitatively reliable numbers that one wishes to extract from the simulations. This aim makes it extremely important to control all systematic errors as reliably as possible, to extract the maximum information from the statistical data that are at hand, and to optimise the performance of the algorithm. Chapters 2 and 3 will take up this latter question, both for vectorising computers and for parallel computers, while Chap. 4 should be consulted for a discussion of optimum data analysis. In the present chapter, we have considered only two sources of systematic errors—finite size effects and finite observation time effects. However, one must also pay attention to possible errors due to correlations among pseudorandom numbers, artificial spatial correlations introduced in studies of dynamic phenomena such as domain growth when one updates sites regularly rather than randomly for the sake of efficient vectorisation or parallelisation [1.81], and to sort out in dynamic studies phenomena depending on the local transition probability used [1.82]. We have not reviewed these problems in detail here, since we feel more work is desirable to fully clarify these issues. Notwithstanding the great progress that has already been obtained with many applications, there is still a need for further improvements of some methodological details. Thus, there are many reasons why Monte Carlo simulation will remain an active and lively discipline in the future.

Acknowledgements. The author is greatly indebted to M.S.S. Challa, R.C. Desai, A.M. Ferrenberg, D.W. Heermann and D.P. Landau for a fruitful and stimulating collaboration on the problems which have been described in detail here. His understanding of finite size effects at first-order transitions has greatly benefited from interactions with V. Privman and R. Kotecky. This article has also profited from the numerous preprints on various subjects that have reached the author; it is a pleasure to thank all the authors of this very interesting work, and a duty to apologise to all these colleagues whose work was not mentioned appropriately in the present article, due to the necessity to limit both the length of this chapter and the effort in preparing it. Finally, partial support from the "Deutsche Forschungsgemeinschaft" under Grant N. SFB 262 is gratefully acknowledged.

References

1.1 K. Binder: In *Festkörperprobleme* (*Advances in Solid State Physics*), Vol. 26. ed. by P. Grosse Vieweg, Braunschweig 1986) p. 133
1.2 For a review, see T. Mohri, J.M. Sanchez, D.De. Fontaine: Acta Metall. **33**, 1171 (1985)
1.3 D.W. Heermann: *Computer Simulation Methods in Theoretical Physics*, 2nd edn. (Springer, Berlin, Heidelberg 1990)
1.4 K. Binder D.W. Heermann: *Monte Carlo Simulation in Statistical Physics*, Springer Ser. Solid-State Sci., Vol. 80 (Springer, Berlin, Heidelberg 1988)
1.5 M.H. Kalos, P.A. Whitlock: *Monte Carlo Methods*, Vol. 1. (Wiley, New York 1986)
1.6 H. Gould, J. Tobochnik: *An Introduction to Computer Simulation Methods—Application to Physical Systems* (Addision-Wesley, Reading, MA 1988)
1.7 W.G. Hoover: *Molecular Dynamics*, Lecture Notes Phys., Vol. 258. (Springer, Berlin, Heidelberg 1986)
1.8 M.H. Kalos (ed): *Monte Carlo Methods in Quantum Problems.* (Reidel, Dordrecht 1984)
1.9 M. Suzuki (ed): *Quantum Monte Carlo Methods*, Springer Ser. Solid-State Sci., Vol. 74 (Berlin, Heidelberg 1987)
1.10 J.D. Doll, J.E. Gubernatis (eds.): *Quantum Simulations* (World Scientific, Singapore 1990)
1.11 M.P. Allen, D.J. Tildesley: *Computer Simulations of Liquids* (Clarendon Oxford 1987); G. Cicotti, D. Frenkel, I.R. McDonald (eds.): *Simulation of Liquids and Solids* (North-Holland, Amsterdam 1987)
1.12 G. Ciccotti, W.G. Hoover (eds.): *Molecular-Dynamics Simulation of Statistical Mechanical Systems* (North-Holland, Amsterdam 1986)
1.13 R.W. Hockney, J.W. Eastwood: *Computer Simulation Using Particles* (Adam Hilger, Bristol 1988)
1.14 B. Bunk, K.H. Mütter, K. Schilling (eds.): *Lattice Gauge Theory, A challenge in Large-Scale Computing* (Plenum, New York 1986)
1.15 T. Vicsek: *Fractal Growth Phenomena* (World Scientific, Singapore 1989); H.G. Herrmann: Phys. Reps. **136**, 153 (1986); P. Meakin: In *Phase Transitions and Critical Phenomena*, Vol. 12, ed. by C. Domb, J.L. Lebowitz (Academic, New York 1988) Chap. 3
1.16 B.J. Alder (ed.): *Special Purpose Computers* (Academic, New York 1988)
1.17 D.W. Heermann, A.N. Burkitt: *Parallel Algorithms in Computational Science*, Springer Ser, Info. Sci., Vol. 24 (Springer, Berlin, Heidelberg 1991)

1.18 D.P. Landau, K.K. Mon, H.B. Schüttler (eds.): *Computer Simulation Studies in Condensed Matter Physics I*, Springer Proc. Phys., Vol. 33 (Springer, Berlin, Heidelberg 1988)
1.19 D.P. Landau, K.K. Mon, H.B. Schüttler (eds.): *Computer Simulation Studies in Condensed Matter Physics II*, Springer Proc. Phys., Vol. 45 (Springer, Berlin, Heidelberg 1990)
1.20 D.P. Landau, K.K. Mon, H.B. Schüttler (eds.): *Computer Simulation Studies in Condensed Matter Physics III*, Springer Proc. Phys., Vol. 53 (Springer, Berlin, Heidlberg 1991)
1.21 O.G. Mouritsen: *Computer Studies of Phase Transitions and Critical Phenomena* (Springer, Berlin, Heidelberg 1984)
1.22 K. Binder (ed.): *Monte Carlo Methods in Statistical Physics* Topics, Curr. Phys., Vol. 7, 2nd edn. (Springer, Berlin, Heidelberg 1986)
1.23 K. Binder (ed.): *Applications of the Monte Carlo Method in Statistical Physics*, Topics Curr. Phys., Vol. 36, 2nd edn. (Springer, Berlin, Heidelberg 1987)
1.24 K. Binder: Phys. Rev. Lett. **47**, 693 (1981); Z. Physik B **43**, 119 (1981)
1.25 A. Sariban, K. Binder: J. Chem. Phys. **86**, 5859 (1987); Macromolecules **21**, 711 (1988)
1.26 M. Scheucher, J.D. Reger, K. Binder, A.P. Young: Phys. Rev. B **42**, 6881 (1990)
1.27 D. Stauffer: Phys. Reps. **54**, 1 (1979); *Introduction to Percolation Theory* (Taylor and Francis, London 1985)
1.28 D.H.E. Gross: Rep. Prog. Phys. **53**, 605 (1990); H.R. Jaquaman, G. Papp, D.H.E. Gross: Nucl. Phys. A **514**, 327 (1990)
1.29 R.M. Ziff, E. Gulari, Y. Barshad: Phys. Rev. Lett. **56**, 2553 (1986); I. Jensen, H.C. Fogedby, R. Dickman: Phys. Rev. A **41**, 3441 (1990)
1.30 G. Lironis, D.W. Heermann, K. Binder: J. Phys. A **23**, L329 (1990); S. Hayward, D.W. Heermann, K. Binder: J. Stat. Phys. **49**, 1053 (1987)
1.31 S. Katz, J.L. Lebowitz, H. Spohn: Phys. Rev. B **28**, 1655 (1983); J. Stat. Phys. **34**, 497 (1984); J. Marro, J.L. Lebowitz, H. Spohn, M.H. Kalos: J Stat. Phys. **38**, 725 (1985); J.L. Valles, J. Marro: J. Stat Phys. **43**, 441 (1986); ibid. **49**, 89 (1987); J. Marro, L.Valles, J. Stat. Phys. **49**, 121 (1987); M.Q. Zhang, J.-S. Wang, J.L. Lebowitz, J.L. Valles; J. Stat. Phys. **52**, 1461 (1988); J.-S. Wang, J.L. Lebowitz, K. Binder: J. Stat. Phys. **56**, 783 (1989); R. Dickman: Phys. Rev. B **40**, 7005 (1989); J.V. Andersen, O.G. Mouritsen: Phys. Rev. Lett. **65**, 440 (1990)
1.32 For an early review, see e.g. K. Binder: Ber. Bunsenges. Phys. Chem. **90**, 257 (1986); some later pertinent references are: J.V. Andersen, H. Bohr, O.G. Mouritsen: Phys. Rev. B **42**, 283 (1990); P.J. Shah, O.G. Mouritsen: Phys. Rev. B **41**, 7003 (1990); O.G. Mouritsen, E. Praestgaard: Phys. Rev. B **38**, 2703 (1988); C. Roland, M. Grant: Phys. Rev. Lett. **63**, 551 (1989); J.D. Gunton, E.T. Gawlinski, K. Kaski: In *Dynamics of Ordering Processes in Condensed Matter*, ed. by S. Komura, H. Furukawa (Plenum, New York 1988, p. 101; H.C. Kang, W.H. Weinberg: J. Chem. Phys. **90**, 2824 (1989); C. Roland, M. Grant: Phys. Rev. B **41**, 4646 (1990); J. Vinals, M. Grant: Phys. Rev. B **36**, 7036 (1987); H.F. Poulsen, N.H. Andersen, J.V. Andersen, H. Bohr, O.G. Mouritsen: Phys. Rev. Lett. **66**, 465 (1991)
1.33 See e.g. J. Amar, F. Sullivan, R.D. Mountain: Phys. Rev. B **37**, 196 (1988); A. Milchev, D. Heermann, K. Binder: Acta Metall. **36**, 377 (1988); T.M. Rogers, K.R. Elder, R.C. Desai: Phys. Rev. B **37**, 9638 (1988); S. Puri, Y. Oono: Phys. Rev. A **38**, 1542 (1988); E.T. Gawlinski, J.D. Gunton, J. Vinãls: Phys. Rev. B **39**, 7266 (1989); K. Yaldram, K. Binder: Acta Metall. **39**, 707 (1991); J. Stat. Phys. **62**, 161 (1991); M. Laradji, M. Grant, M.J. Zuckermann, W. Klein: Phys. Rev. B **41**, 4646 (1990); C. Roland, M. Grant: Phys. Rev. B **39**, 11971 (1989); Phys. Rev. Lett. **60**, 2657 (1988)
1.34 K.W. Kehr, K. Binder, S.M. Reulein: Phys. Rev. B **39**, 4891 (1989); R. Czech: J. Chem. Phys. **91**, 2498 (1989); J.W. Haus, K.W. Kehr: Phys. Reps. **150**, 263 (1987)
1.35 See e.g. B. Dünweg, K. Binder: Phys. Rev. B **36**, 6935 (1987); H.T. Diep, A. Ghazali, B. Berge, P. Lallemand: Europhys. Lett. **2**, 603 (1986); U. Gahn: J. Phys. Chem. Solids **47**, 1153 (1986)
1.36 For a recent review, see K. Binder, D.P. Landau: In *Molecule–Surface Interactions*, ed. by K.P. Lawley (Wiley, New York 1989) p. 91
1.37 Some of this work will be reviewed in later chapters of this book. See also P.-A. Lindgard, O.G. Mouritsen: Phys. Rev. B **41**, 688 (1990); A.A. Aligia, A.G. Rojo, B.R. Alascio: Phys. Rev. B **38**, 6604 (1988); Z.-X. Cai, S.D. Mahanti: Solid State Commun. **67**, 287 (1988); Phys. Rev. B **40**, 6558 (1989); C.P. Burmester, L.J. Wille: Phys. Rev. B **40**, 8795 (1989); N.C. Bartelt, T.L. Einstein, L.T. Wille: Phys. Rev. B **40**, 10759 (1989); T. Aukrust, M.A. Novotny, P.A. Rikvold, D.P. Landau: Phys. Rev. B **41**, 8772 (1990); A. Milchev, B, Dünweg, P.A. Rikvold: J. Chem. Phys. **94**, 3958 (1991); O.G. Mouritsen: Chem. Phys. Lipids, in press R.M. Nowotny, K. Binder: Z. Phys. B **77**, 287 (1989); J.H. Ipsen, O.G. Mouritsen, M.J. Zuckermann: J. Chem. Phys. **91**, 1855 (1989)

1.38 M. Suzuki: J. Phys. Soc. Jpn. **55**, 4205 (1986); M. Suzuki, M. Katori, X. Hu: J. Phys. Soc. Jpn. **56**, 3092 (1987); X. Hu, M. Katori, M. Suzuki: J. Phys. Soc. Jpn. **56**, 3865 (1987); X. Hu, M. Suzuki: J. Phys. Soc. Jpn. **57**, 791 (1988); M. Katori, M. Suzuki: J. Phys. Soc. Jpn. **57**, 807 (1988); M. Suzuki: Prog. Theor. Phys. Suppl. **87**, 1 (1986); Phys. Lett A **127**, 440 (1988)
1.39 M. Katori, M. Suzuki: J. Phys. Soc. Jpn. **56**, 3113 (1987); N. Ito, M. Suzuki: Phys. Rev. B **43**, 3483 (1991); A. Patrykiejew, P. Borowski: Phys. Rev. B **42**, 4670 (1990)
1.40 V. Privman (ed.): *Finite-Size Scaling and Numerical Simulations of Statistical Systems* (World Scientific, Singapore 1990)
1.41 K. Binder: [Ref. 1.23, Chap. 1]
1.42 V. Privman, M.E. Fisher: J. Stat. Phys. **33**, 385 (1983)
1.43 K. Binder, D.P. Landau: Phys. Rev. B **30**, 1477 (1984)
1.44 M.E. Fisher, V. Privman: Phys. Rev. B **32**, 447 (1985)
1.45 For a general review on first-order phase transitions see K. Binder: Reps. Prog. Phys. **50**, 783 (1987)
1.46 R.B. Potts: Proc. Cambridge Philos. Soc. **48**, 106 (1952); for a review, see F.Y. Wu: Revs. Mod. Phys. **54**, 235 (1982)
1.47 R.J. Baxter: J. Phys. C **6**, L445 (1973); T. Kihara, Y. Midzuno, T. Shizume: J. Phys. Soc. Jpn. **9**, 681 (1954)
1.48 M.S.S. Challa, D.P. Landau, K. Binder: Phys. Rev. B **34**, 1841 (1986)
1.49 C. Borgs, R. Kotecky: J. Stat. Phys. **61**, 79 (1990)
1.50 C. Borgs, R. Kotecky, S. Miracle-Sole: J. Stat. Phys. **62**, 529 (1990)
1.51 V. Privman, J. Rudnick: J. Stat. Phys. **60**, 551 (1990)
1.52 J. Lee, J.M. Kosterlitz: Phys. Rev. B **43**, 3265 (1991); 1268 (1991)
1.53 P. Peczak, D.P. Landau: Phys. Rev. B **39**, 11932 (1989)
1.54 J. Lee, J.M. Kosterlitz: Phys. Rev. Lett. **65**, 137 (1990); M. Fukugita, H. Mino, M. Okawa, A. Ukawa: J. Stat. Phys. **59**, 1397 (1990)
1.55 M.S.S. Challa, J.H. Hetherington: Phys. Rev. Lett. **60**, 77 (1988); Phys. Rev. A **38**, 6324 (1988)
1.56 M. Rovere, D.W. Heermann, K. Binder: J. Phys. Cond. Matter **2**, 7009 (1990)
1.57 M. Rovere, D.W. Heermann, K. Binder: Europhys. Lett. **6**, 585 (1988)
1.58 K. Binder: Z. Phys. B **43**, 119 (1981); K. Kaski, K. Binder, J.D. Gunton: Phys. Rev. B **29**, 3996 (1984)
1.59 K. Strandburg, J.A. Zellweg, G.V. Chester: Phys. Rev. B **30**, 2755 (1984)
1.60 R.C. Desai, D.W. Heermann, K. Binder: J. Stat. Phys. **53**, 795 (1988)
1.61 M. Creutz: Phys. Rev. Lett. **50**, 1411 (1983); D. Callaway, A. Rahman: Phys. Rev. Lett. **49**, 613 (1982)
1.62 R. Harris: Phys. Lett. **111A**, 299 (1985)
1.63 G. Bhanot, M. Creutz, H. Neuberger: Nucl. Phys. B **235** [FS 11], 417 (1984)
1.64 K. Binder, J.-S. Wang: J. Stat. Phys. **55**, 87 (1989)
1.65 E.V. Albano, K. Binder, D.W. Heermann, W. Paul: Z. Phys. B **77**, 445 (1989)
1.66 E.V. Albano, K. Binder, D.W. Heermann, W. Paul: J. Stat. Phys. **61**, 161 (1990)
1.67 R.M. Hornreich, M. Luban, S. Shtrikman: Phys. Rev. Lett. **35**, 1678 (1975)
1.68 For a review see S. Dietrich: In *Phase Transitions and Critical Phenomena*, Vol. 12, ed. by C. Domb, J.L. Lebowitz (Academic, New York 1988, p. 1
1.69 K. Binder: [Ref. 1.40, p. 173]
1.70 A.M. Ferrenberg, D.P. Landau, K. Binder: J. Stat. Phys. **63**, 867 (1991)
1.71 H. Müller-Krumbhaar, K. Binder: J. Stat. Phys. **8**, 1 (1973)
1.72 A. Milchev, K. Binder, D.W. Heermann: Z. Phys. B **63**, 521 (1986)
1.73 For reviews of critical phenomena, see: M.E. Fisher: Rev. Mod. Phys. **46**, 597 (1974); P.C. Hohenberg, B.I. Halperin: Rev. Mod. Phys. **49**, 435 (1977)
1.74 For example, see H.W. Alexander: *Elements of Mathematical Statistics* (Wiley, New York 1961)
1.75 G.S. Pawley, R.H. Swendsen, D.J. Wallace, K.G. Wilson: Phys. Rev. B **29**, 4030 (1984); H.W.J. Blöte, J. de Bruin, A. Compagner, J.H. Croockewit, Y.T.J.C. Fonk, J.R. Heringa, A. Hoogland, A.L. Willigen: Europhys. Lett. **10**, 105 (1989); H.W.J. Blöte, A. Compagner, J.H. Croockewit, Y.T.J.C. Fonk, J.R. Heringa, T.S. Smit, A.L. van Willigen: Physica A **161**, 1 (1989); A.J. Liu, M.E. Fisher: Physica A **156**, 35 (1989), and references therein
1.76 S. Wansleben: Comput. Phys. Commun. **43**, 315 (1987)
1.77 H. Flyvbjerg, H.G. Petersen: J. Chem. Phys. **91**, 461 (1989)
1.78 C. Whitmer: Phys. Rev. D **29**, 306 (1984); S. Gottlieb, P.B. Mackenzie, H.B. Thacker, D. Weingarten: Nucl. Phys. B **263**, 704 (1986)

1.79 For a recent review see D.P. Landau, S. Tang: J. de Phys. **49**, C8-1525 (1988), and Chap. 2 of this book
1.80 N. Madras, A. Sokal: J. Stat. Phys. **50**, 109 (1988), and references therein
1.81 W. Schleier, G. Besold, K. Heinz: J. Stat. Phys. (1991, in press)
1.82 H.C. Kang, W.H. Weinberg: J. Chem. Phys. **90**, 2824 (1989)

2. Vectorisation of Monte Carlo Programs for Lattice Models Using Supercomputers

David P. Landau

With 11 Figures

A description of the use of vectorisation in Monte Carlo simulations is presented. We provide an overview of algorithmic strategies and their implementation on modern vector computers followed by a description of some results which have been obtained using vector processors.

2.1 Introduction

After the classic paper of *Metropolis* et al. [2.1], on Monte Carlo simulations of hard disks, first exposed the Monte Carlo method as a useful numerical method in statistical mechanics, an overwhelming number of studies of a wide variety of lattice models have appeared. Many of these have been reviewed in earlier monographs on Monte Carlo methods in statistical physics [2.2] so they will not be reviewed here. Most of this work has used essentially some variant of the "Metropolis method", although there have been a few studies using event-driven algorithms [2.3]. Because of the relative simplicity of the lattice models which have been simulated, algorithmic optimisation and effective code were rapidly developed e.g. for Ising models and other classical systems. However tightly written these codes may have been, they were still severely constrained by scalar cycle times. As a consequence in 1963 *Yang* [2.4] reported only 3×10^3 μtrials/second on an IBM 7090 for an Ising square lattice; by 1976 scalar computers had increased substantially in speed and *Landau* [2.5] obtained 29×10^3 μtrials/second using an IBM 370 model 168. (As a unit of measure of the speed of an algorithm we shall use the number of single spin flip trials, or μtrials. Another commonly used time scale is the MCS/site, which is equivalent to a complete sweep throughout the entire lattice. The number of MCS/site per unit time thus depends upon the size of the lattice under consideration, even if there is no intrinsic dependence in the algorithm on lattice size, and is therefore not a suitable unit here.) Many of the problems for which the Monte Carlo method is most suitable, involve the study of phase transitions for which the number of configurations which must be generated on large systems is the limiting factor. For these simulations it is essential to have much higher performance than that quoted above, if the full potential of the method is to be achieved.

As the rate of increase in scalar speed with time began to level off, the appearance of vector machines offered the possibility of making another major jump in performance. Vector machines offer great increases in speed by operating on entire vectors of data instead of single elements alone. From the operational point of view this means that all elements of all vectors to be used in the vector operation must be present at the very start of the operation. The Monte Carlo method as first devised, cannot be simply vectorised and in the next section we will discuss the simple modifications which can be made to allow the method to be vectorised. Although many of our comments will be appropriate to either array processors or pipeline machines, we will concentrate on the latter throughout this chapter. We shall see that the combination of algorithm development and vectorisation have led to improvements of the order of 10^3 in "Metropolis-like" methods over the past 15 years. This increased performance has led to a dramatic increase in the system sizes which can be simulated, as well as in the number of configurations which can be generated for a particular system. Before vector supercomputers became available, typical systems sizes were 10^3–10^4 sites and 10^5 particles comprised a very large sample whereas today 10^6 (or more) sites are not uncommon. Similarly most runs contained data for 10^4–10^5 MCS/site but runs of 10^6–10^7 MCS/site have now been made. In addition to allowing better results to be obtained for some problems which had already been under consideration, this increase in performance has allowed the study of some phenomena which were previously so far beyond existing resources that simulations were not even considered. So far, recently developed cluster flipping algorithms [2.6] have not been vectorisable and are thus not discussed in this chapter.

In the next section we shall provide some necessary background on vectorisation to help the reader understand the development of some of the most important vectorised Monte Carlo algorithms which are presented in the three following sections. In the section thereafter we shall review some of the typical results which were obtained, in most cases relatively recently, using the methods described in the preceding sections or minor modifications of these techniques. Since the Monte Carlo literature is now quite substantial (in some cases papers do not clearly state if vectorised algorithms were used) and several other articles in this volume discuss particular areas of study in some detail, we cannot hope to be exhaustive in our review. We therefore apologise at the outset to those authors whose work is not mentioned in the following sections.

2.2 Technical Details

2.2.1 Basic Principles

The simple approach to Monte Carlo sampling of an d-dimensional lattice involves examining the spin at each site, either sequentially or randomly, and allowing it to "flip" before going on to the next site. The sites of this system, e.g. L^d in shape, can be identified using d indices or, alternatively, by numbering them sequentially and using a single index. This situation would seem to be ideal for vectorisation with the lattice index serving as the vector subscript. However, there are several complications which must be addressed before vectorisation can be implemented.

First, in each spin-flip trial the update of the spin depends upon the orientations of its surrounding, interacting neighbours *at the time of the spin-flip trial*! If all spins are being considered at once, the question of whether the "new" value of a neighbour should be taken or the "old" value leads to ambiguity. Even if there were no other difficulties with the vectorisation this problem prevents the simple consideration of each spin successively as part of a single vector. (*Landau* and *Stauffer* [2.7] simulated an Ising square lattice using parallel updating on a Cray XMP and showed that the 2nd and 4th moments of the magnetisation distribution are incorrect although the estimate for T_c is quite reasonable.)

Second, the treatment of boundary conditions is non-trivial. If periodic boundaries are used, the data may easily be interpreted using finite size scaling theory [2.8], but periodic boundary conditions are not trivial in a vectorised algorithm. The simplest boundary condition from the programming perspective is a skew boundary in which the last site in one row has a nearest neighbour in the first site in the next row. Skew boundaries introduce seams in the lattice and produce a correction to the simple finite size scaling.

Lastly, the ordinary algorithm is usually written in a form that the "inner loop" does not vectorise effectively. Usually the number of parallel nearest neighbours will be summed over and the sum is used as an index for a look-up table of transition probabilities. For most models, the number of interacting neighbours with the same coupling strength is quite small so the sum over nearest neighbours is not efficiently vectorisable. This means that some reorganisation of the algorithm is needed; although this is not usually difficult it does mean that some effort must be expended to find the most intelligent way to do so.

The first efforts at creating algorithms which are in principle vectorisable were actually devised for scalar machines. In the earliest one of which we are aware, *Friedberg* and *Cameron* [2.9] simulated 4×4 Ising models on an IBM 7094 computer with 36 bit words. The lattice is divided into two interpenetrating sublattices such that no pair of neighbouring sites belongs to the same sublattice. Thus in principle, each sublattice can be treated in vector fashion without any danger of altering the spins on either sublattice incorrectly. (Unfortunately, the particular algorithm which they used did not satisfy ergodicity.)

Generally speaking a vector operation consumes an initial "start-up" time which is independent of the vector length, plus an execution time which is proportional to the number of elements in the vector. For the machines of interest here, the vectorisation is carried out by the computer through the use of pipelining. By this we mean that each functional unit in the computer is segmented so that instructions can be carried out in steps as in an automotive assembly line; after the pipeline is full, results appear at the rate of one per clock cycle regardless of the number of cycles needed to carry out a single operation! The maximum speedup is therefore equal to the number of steps needed to perform a particular operation. Of course, the ultimate limitation in performance for a program is due to the scalar portions of the program and can be calculated using Amdahl's law [2.10]. The performance ratio P, the speedup relative to the scalar processing rate, for a machine with a ratio of vector to scalar speed of E is

$$P = \frac{E}{(1-f)E + f}, \tag{2.1}$$

where f is the fraction of the code which is scalar. Therefore, even if the vector pipe is infinitely fast, for a code which is 80% vectorised the performance ratio is only 5!

The "start-up" time, or time needed to fill the pipeline, depends on whether the machine uses register-to-register (e.g. Cray Y-MP) or memory-to-memory (e.g. Cyber 205) architecture and differs substantially between different vector processors. To be effective, an algorithm must then produce vectors which are sufficiently long that the "start-up" time is an insignificant fraction of the total execution time. On a Cyber 205, for example, the vector pipeline is 65 536 elements long and scalar computing is actually faster for vectors with less than about 100 elements. Most other vector processors are more efficient with short vectors, e.g. Cray supercomputers have vector registers of length 64 and the IBM 3090 series has length 128. For these machines, however, the speed as a function of vector length looks like a sawtooth; maximum speed is obtained when the vector fills a pipe completely (a multiple of 64 or 128 respectively) and if the vector is a single element longer another start-up time is used to treat the extra element. Thus, available computer architecture has influenced the details of vector algorithms.

2.2.2 Some "Do"s and "Don't"s of Vectorisation

We first consider a simple DO-loop involving three vectors of length N:

$$\text{DO } 1\ I = 1, N$$

$$1\ \ C(I) = A(I) + B(I)$$

As mentioned earlier, a vector operation can be efficient only if the vector length is "long", therefore N must be sufficiently large that the "start-up" time is an insignificant fraction of the total execution time. This DO-loop should thus be vectorised only for large enough values of N and any algorithm must be organized so that vectors are sufficiently long.

Consider another simple DO-loop involving a recursive relation:

$$\text{DO } 1\ I = 1, N$$

$$1\ C(I) = A(I) + C(I - M)$$

where M is some predetermined integer. All of the elements in each vector should be available when the operation begins so recursion is not "legal". In some instances it may be possible to force vectorisation for this loop, but the results may then be correct only if M is sufficiently large that the newly calculated value of $C(I-M)$ is available when it is needed in the DO-loop.

Another impediment to vectorisation is a dependency, e.g.

$$\text{DO } 1\ I = 2, N$$

$$B(I) = A(I-1)$$

$$1\ A(I) = C(I)$$

so that the values of A are needed before they are calculated. Here too, the algorithm must be organised to avoid such structures.

Note that other complications to vectorisation can come from I/O or subroutine calls within a vector loop; in order to write very efficient programs one must thus become familiar with the shortcomings of the machine to be used, with regard to these statements.

2.3 Simple Vectorisation Algorithms

The first actual implementation of a vectorised, spin-flip Monte Carlo algorithm of which we are aware was done by *Oed* [2.11] who considered an Ising square lattice with nearest neighbour interactions. He introduced periodic boundary conditions by copying the rows and columns on the edges of the system to their respective opposite sides (virtual spins). After each pass through the lattice these rows and columns are then renewed to match to any changes which might have occured on the boundaries. The entire system is then divided into two interpenetrating sublattices in which each sublattice consists of sites which are part of a next-nearest-neighbour network. The two different sublattices are sometimes named red and black after the squares of a checkerboard, and this approach is sometimes called the "checkerboard algorithm". (This "checkerboard" idea is basically the same as used by *Friedberg* and *Cameron* [2.9] but the remainder of the algorithm is different.) Two separate spin vectors (composed of contiguous elements) are used to represent the two different sublattices. The sum of the nearest neighbours of each spin is then carried out by shifting one vector with respect to the other and summing the elements of the resulting vectors. The energy associated with flipping is then computed by multiplying each sum by the reference spin, also a vector operation. (Note that in the scalar algorithm, the "outer loop" is usually over the different sites in the lattice and the "inner loop" is over the neighbours, whereas in the vectorised algorithm the order of the loops is inverted.) When run on a Floating Point Systems AP-190L, roughly a factor of five increase in speed of the total program was obtained, including those parts which would not vectorise. This program can be made portable and executed on any vector machine. If a model contains more distant neighbour interactions, the method may still be implemented by the subdivision of the lattice into more sublattices such that no two spins on the same sublattice interact. As the number of sublattices grows, however, the number of sites per sublattice decreases for any constant lattice size and hence the lengths of the spin vectors become smaller. At some point the vector length becomes so small that vectorisation is no longer profitable for that lattice size. The procedure may still be used effectively but only with larger systems.

An alternative method to determining nearest neighbour products is to use vector "gather" statements to create vectors for the different nearest neighbours which can then be combined directly with the vector for the sublattice to be flipped. This procedure is in some ways more transparent, but it is generally also slower.

Spin-exchange Monte Carlo algorithms were not vectorised until quite recently. More sophisticated approaches will be described in the next sections, but we shall briefly describe the simple, "checkerboard" approach here. *Desalvo* et al. [2.12] present such an implementation for the square, simple cubic, and diamond lattices. Each checkerboard sublattice is further divided into additional sublattices, four in the case of the square lattice, and it is each of these "sparse sublattices" which form the vectors upon which the entire exchange operation is applied in a single vector operation, in a manner which is rather similar to the spin-flip program. The site with which each spin in the sublattice will attempt an exchange is chosen randomly and virtual spins are placed around the edges to provide periodic boundary conditions. On a Cray Y-MP the resultant code ran in excess of 10^6 spin-exchange trials/second for all three lattices.

2.4 Vectorised Multispin Coding Algorithms

One technique which had been proposed for scalar computers was to pack multiple spin values into a single word and to "flip" all those spins in a single operation. This procedure, known as multispin coding [2.13], offers the advantage of reduced storage needs and increased speed. When vectorised properly, multispin coding algorithms offer tremendous improvements in speed, although the resultant complexity makes them quite difficult to modify and they are seldom portable from one machine to another.

A truly high speed, mature, multispin coding algorithm was described by *Wansleben* [2.14] who used a different random number for each spin-flip trial and reached a speed of 38×10^6 spin-flip trials/second on a two-pipe Cyber 205. (If random numbers are reused, as in the method to be described in the next section, Wansleben estimates that a speed of 130×10^6 updates per second will result.) This approach was built on earlier developments for scalar Cyber machines [2.13] and a previous implementation [2.15] on the Cyber 205. The algorithm is suitable for variable transition probabilities and does not use a

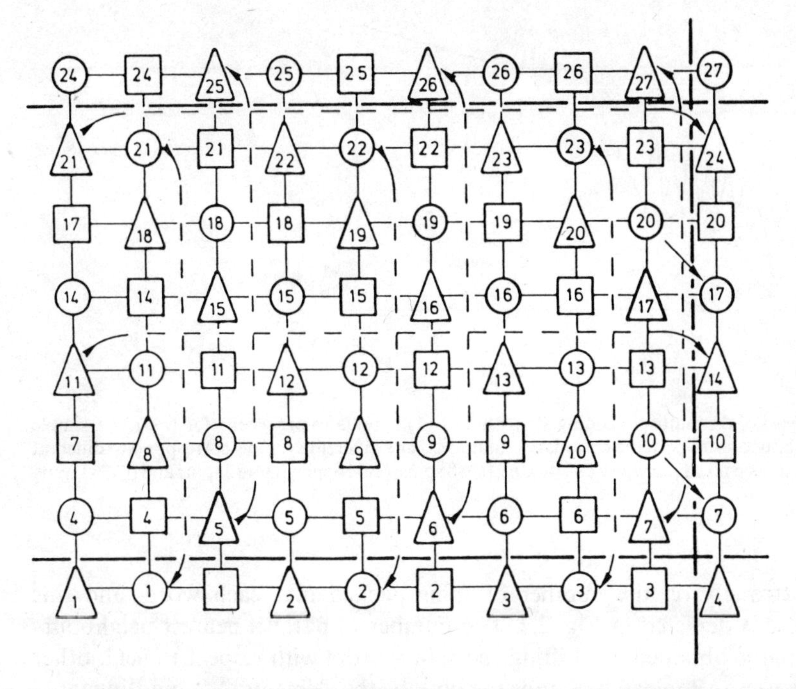

Fig. 2.1. Example for the implementation of periodic boundary conditions within a single plane for the simple cubic Ising model in combination with vectorisation. The different geometric shapes represent different sublattices. The numbers within the symbols indicate the position within the vector for each spin. Before the treatment of the "squared" sublattice, for example, spins are copied following the arrows onto "virtual spins" which lie outside the heavy lines. [2.14]

look-up table. The lattice is broken up into three interpenetrating lattices and periodic boundary conditions are implemented by adding "virtual spins" to the lattice as shown in Fig. 2.1. The sublattices are labelled 1, 2 and 3 and are denoted by different symbols. Note that none of the spins on a given sublattice are nearest neighbours of each other so they are mutually non-interacting. The rows of the lattice in the 1-direction are stored using the multispin storage technique. The number of spins packed into each machine word is "mspc". The number of words needed to store one row of the lattice is therefore L/mspc and L must be a multiple of mspc. The first spin of a row is stored in the first bit of the first multispin coding word, the second spin in the first bit of the second multispin coding word and so forth until all of the multispin coding words have been treated. The next spin is then stored in the second bit of the first multispin coding word, the next spin in the second bit of the second multispin storage word, etc. All spins stored in a multispin storage word belong to the same sublattice. Because of the sublattice structure and the limits on vector length, the values of L for which the algorithm is applicable ranges from $L=3$ to $L=192$, and L must be a multiple of mspc. The speed of the method

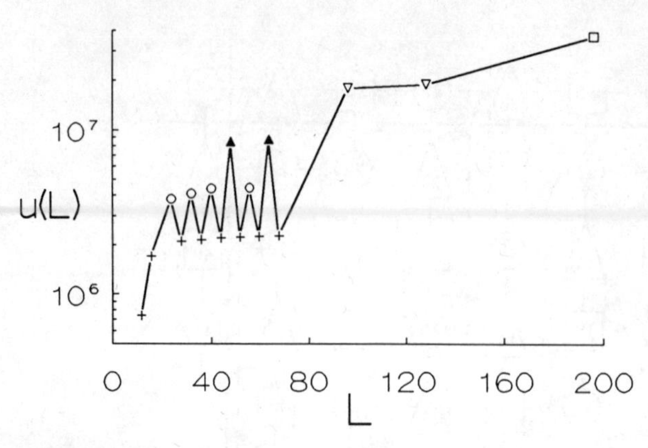

Fig. 2.2. Speed of the multispin coding algorithm [2.14] in updates per second for $L \times L \times L$ simple cubic Ising lattices with periodic boundary conditions. The different symbols correspond to different numbers of spins packed per word: (+) 4 spins; (○) 8 spins; (△) 16 spins; (▽) 32 spins and (□) 64 spins

depends strongly on the number of spins packed into each word, and this dependence is depicted in Fig. 2.2. The number of parallel nearest neighbours of each spin is obtained by shifting the spin vectors with respect to each other and using vector Boolean functions to combine the elements. A "spin-flip word" is then created by switching bits in an initial word according to the desired transition probabilities, and the spin-flips are accomplished by an "exclusive-or" operation. The logic of this procedure is shown schematically as follows [2.14]:

The multispin storage word is called "is", the transition probabilities for 4, 5, and 6 parallel nearest neighbours respectively are bolt4, bolt5, and bolt6, and rand is a random number between 0 and 1

```
c—compare random number with transition probability
   if(rand.gt.bolt6) then
      ifl = 1
   else
      ifl6 = 0
   endif
   if(rand.gt.bolt5) then
      ifl5 = 1
   else
      ifl5 = 0
   endif
   if(rand.gt.bolt4) then
      ifl4 = 1
   else
      ifl4 = 0
   endif
```

```
c
c—switch 1-bits into 0-bits within the flipword iscr1 according to transition
   probability
      iscr=AND(ie3,ie2)
      iscr=AND(iscr,ifl6)
      iscr1=ANDN(1,iscr)
c
      iscr=AND(ie3,ie1)
      iscr=AND(iscr,ifl5)
      iscr1=ANDN(iscr1,iscr)
c
      iscr=OR(ie1,ie2)
      iscr=ANDN(ie3,iscr)
      iscr=AND(iscr,ifl4)
c
c—perform spin flip
      is=XOR(iscr1,is)
```

This procedure is implemented using the special vector Fortran Q8 calls which are unique to the Cyber 205/ETA-10; the program is thus not portable. Each "if-block" in the above example becomes a single CALL Q8CMPGE statement; the program spends only 20 ns per spin generating the random number and comparing it with all the transition probabilities. All other steps together consume the remaining 7 ns per spin.

A previous study of random number generators on the Cyber 205 had shown the limitations of the standard congruential method. Instead a *Tausworthe* [2.16], or shift-register, generator was used. The algorithm for producing a new random number using the exclusive-or operation $\oplus$ is

$$x_k = x_{k-p} \oplus x_{k-q}, \tag{2.2}$$

where $p > q$ (p and q must be carefully chosen) and a table of p random numbers must first be generated by some other method. The values $p = 250$ and $q = 103$ have been the most popular values for shift-register methods, but there are many different pairs of q, p which produce excellent random number quality. In the following discussion we shall restrict ourselves to this particular case. *Kirkpatrick* and *Stoll* [2.17] had already provided a scalar version for use on IBM machines, but because of the high speed of the multispin coding Monte Carlo method, a high speed, vectorised version of this random number generator was also needed. The method which was developed [2.18] actually has recursive properties but, as described below, will still deliver correct answers when implemented properly. We begin by defining a vector IRAND which will be of length LENGTH + 251, where LENGTH must be greater than 251, and a vector IA of length LENGTH which is equivalent with the first LENGTH elements of IRAND. Another vector IB of length LENGTH is defined, but this vector actually overlaps with IA, i.e. the first element of IB is actually element 148 in

IA. IA and IB are then XOR'd together in a vector operation and the results stored in vector IRAND beginning with element 251. The newly produced random numbers are then stored back into the first 251 elements of IRAND. The statements needed to do this are listed below:

```
c—do the vector "exclusive-or"
      CALL Q8XORV(0,,IA,,IB,,IRANDD)
c—store new random numbers back into the "seed" vector
      CALL Q8VTOV(0,,IC,,0,,ISEEDD)
```

both a vector of random numbers as well as a new seed vector are thus generated with a speed of two numbers per cycle. In order to ascertain whether or not the recursive nature of this procedure is "dangerous", one must determine how many cycles the computer needs to store a result while performing the vector operation. A check showed that with the compiler then in use the recursion length must exceed 79 to guarantee the correct implementation of the algorithm. When this method was used on the ETA-10 it was found [2.19] that a longer recursion length was needed in order to provide the correct result. This is easily accomplished by using, e.g. $p = 1279$, $q = 1063$.

The vectorised multispin coding algorithm for "spin-flip" Monte Carlo was extended to "spin-exchange" by *Zhang* [2.20] on a Cyber 205. He divided the simple cubic lattice into 16 sublattices and allowed spin-exchange between spins on adjacent sublattices using an algorithm which was similar to Wansleben's. In this case the particle jump is carried out by a logical "exclusive-or" operation between a pair of multispin-storage words controlled by an "exchange word". Using logical operations, a fast technique was used to set the 1 or 0 bits in the exchange word according to the transition probability. All operations were, of course, carried out using vector Fortran, and a peak speed of 10.2×10^6 spin-exchange trials per sec was achieved. *Amar* et al. [2.21] constructed a multispin coding algorithm for the two-dimensional square lattice by dividing the system into 16 interpenetrating sublattices. They used "demon bits" to avoid a table look-up, and this procedure will be explained in the following section. Vector Fortran commands were used on the Cyber 205 and the resultant code ran at a speed of 15×10^6 spin-exchange trials/second.

2.5 Vectorised Multilattice Coding Algorithms

A different approach to multispin algorithms was introduced by *Bhanot* et al. [2.22], who packed spins from 64 different lattices into each 64 bit words. Since the lattices were distinct, the spins in each word were in principle independent of each other. The necessary information about each random number is coded into two bits and each bit pair is used exactly once for each of the 64 lattices. The flipping was accomplished without a look-up table using logical commands in the following way. Three bit variables, B3V, B2V and B1V were defined and interpreted as the bits in an octal integer (the "flip integer"); they are initialised to 0, 0 and 1, respectively. A spin product variable $x_{i,\mu}$ was determined for each nearest neighbour $\sigma_{i+\mu}$ of the reference spin σ_i by an exclusive-or operation

$$x_{i,\mu} = \mathrm{XOR}\,(\sigma_i, \sigma_{i+\mu}). \tag{2.3}$$

The six different values of $x_{i,\mu}$ are added to the flip integer with logical instructions. If the leftmost bit, B3V, is equal to 1 following the addition, the spin is flipped. If B3V is zero the spin should be flipped with an exponential probability $e^{-\beta\Delta S}$, where $\Delta S = 12 - \sum_{\mu} x_{i,\mu}$. The flipping is carried out using two additional variables, or "demons", called D2V and D1V (forming a "demon integer") which have the values (0, 1), (1, 0), and (1, 1) with probabilities p_{01}, p_{10}, and p_{11} given by

$$p_{01} = e^{-4\beta} - e^{-8\beta},$$
$$p_{10} = e^{-8\beta} - e^{-12\beta},$$
$$p_{11} = e^{-12\beta}, \tag{2.4}$$

where $\beta = 1/kT$. The "demon integer" and the "flip integer" are added and if B3V is 1 the flip is accepted. This procedure is carried out for each site in a checkerboard sublattice in vector fashion using Q8 calls. The distribution of random numbers is compared with the theoretically predicted form for a perfectly random distribution and the "effective temperature" of the run is determined and used to correct estimates for the bulk properties. On a two-pipe Cyber 205 a peak speed of 98×10^6 spin flip trials/second is achieved although, since the same random numbers are used for each lattice, the paths through configuration space are not truly independent. (Now that histogram techniques [2.23] have been developed as effective tools to study critical behaviour, "high quality" runs at a single temperature have gained in importance; the correlations introduced here by reuse of random numbers may give rise to an additional systematic error when a histogram analysis is used.) The program was later modified [2.24] to run Ising spin glasses with a minor reduction in speed to 80×10^6 updates/second on a two pipe Cyber 205.

Ito and *Kanada* [2.25] then wrote a version of this program for the HITAC S820/80. The program packs 32 different system spins into a single 32 bit word. Each system might be at a different temperature. They use the bubble sort method of *Williams* and *Kalos* [2.26] to decide if a spin is to be flipped. Their implementation on the HITAC S820/80 runs at a speed of 847×10^6 spin flip trials/second and a similar version generates 73.4×10^6 spin-flip trials/second on a single processor of the Cray XMP.

Heuer [2.27] later modified this algorithm by using self-consistent boundary conditions [2.28] and writing a program in Fortran 77 which was optimised to run on the Cray-YMP. The peak speed of this program was 305×10^6 spin-flip trials/second on a single processor. A modification to allow for site dilution reduced the performance to 205×10^6 spin-flip trials/second.

2.6 Vectorised Microcanonical Algorithms

A completely different set of algorithms exist in which the energy of the system never changes during the "spin-flip".

The first approach is the "over-relaxation" method [2.29], which may be used for simulating classical spins. For each spin the "effective field" due to its interacting near neighbours is calculated, and the spin is then allowed to precess about this field. The lattice is decomposed using some checkerboard decomposition, and then each sublattice is treated in vector fashion. As just stated, this algorithm is deterministic and not ergodic so "ordinary" Monte Carlo spin-flip trials must be added in some proportion to make the method generally applicable. This method cannot be used for Ising models.

A microcanonical method for the study of Ising spins on a simple cubic lattice has been vectorized by *Creutz* et al. [2.30] for the Cyber 205. In the original approach [2.31] a "demon" capable of carrying a small amount of energy moved through the lattice and as it touched each site it could give up or receive energy as needed for a spin to overturn. The demon energy could not exceed a predetermined value nor could it become negative. The temperature is determined from the distribution of demon energies. In the vectorised version multiple demons are used and Q8 calls are used extensively. A speed of 234×10^6 spin-flip trials/second was obtained on a 2-pipe Cyber 205 and a maximum possible speed of 2658×10^6 spin-flip trials/second was estimated for an eight processor ETA-10. A version of this method was formulated for *bcc* and *fcc* lattices by *Drouffe* and *Moriarty* [2.32].

A second microcanonical algorithm for Ising spins is really a cellular automata model in disguise [2.33]. A lattice is randomly occupied with 0's and 1's which are packed into 64 bit word (on a Cray or Cyber 205). Using an XOR operation each spin on a sublattice is flipped if, and only if, the spin has the same number of up and down nearest neighbours. (Periodic boundary conditions are introduced in the usual way described in Sect. 2.4.) If the four computer words I,J,K,M contain the four neighbours of a site (with 64 sites packed into each 64 bit word) the condition to flip a spin is

((I.XOR.J).AND.(K.XOR.M)).OR.((I.XOR.K).AND.(J.XOR.M)).

Note that the method is really deterministic and stochasticity is introduced only in the determination of the initial state using random numbers. A vectorised code for the Cray is described by *Herrmann* [2.34], and modifications were later introduced by *Zabolitsky* and *Herrmann* [2.35]. A speed of 234×10^6 updates/second were generated on a 4-pipe Cyber 205, 670×10^6 updates/second on the Cray X-MP, and 4300×10^6 updates/second using four processors on a Cray-2. By inverting the order of the basic double DO-loop described by *Herrmann* [2.34], *Zabolitsky* and *Herrmann* [2.35] were able to make more efficient use of memory access and speed up the program over the original version. Note that other microcanonical algorithms [2.36, 37] use "demons" which take up or give up energy locally to facilitate spin-flips and only conserve energy approximately whereas this algorithm has exact energy conservation. Although from the point of view of speed the Herrmann algorithm [2.34, 35] may seem advantageous, one must not forget that it has severe ergodicity problems at low temperatures and slow relaxation towards equilibrium occurs throughout the critical region.

2.7 Some Recent Results from Vectorised Algorithms

2.7.1 Ising Model Critical Behaviour

There have now been many Monte Carlo simulations of simple Ising models, but results obtained using scalar computers have been modest in resolution. High resolution calculations of critical behavior in the three-dimensional Ising model have been made using a special purpose computer [2.38], and these indicated a breakdown in finite size scaling theory with apparent kinks in the size dependence. Using the multilattice algorithm on the Cyber 205 described in Sect. 2.5, *Bhanot* et al. [2.39] simulated $L \times L \times L$ lattices with L varying between 8 and 44, retaining between 0.64×10^6 and 2.56×10^6 MCS/site for calculating averages. They performed simulations using different values of p, q in the Tausworthe random number generator, but found no difference in the data obtained for different versions of the random number generator. A scaling plot of the maximum in the magnetic susceptibility χ_m, shown in Fig. 2.3, revealed no such behaviour; the finite size behaviour is smooth for all lattice sizes. (This result was confirmed by other simulations and suggests that the special purpose

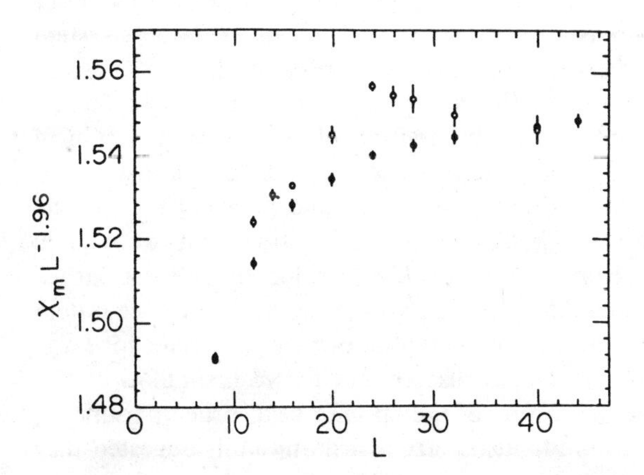

Fig. 2.3. Scaling behaviour of the maximum of the magnetisation fluctuation for the simple cubic Ising model. *Solid dots* are data obtained using the multilattice algorithm on a Cyber 205 and *open dots* were (questionable) data obtained using the Santa Barbara special purpose processor [2.39]

processor delivers faulty results.) From extrapolations to infinite lattice size they also estimate that $\gamma/\nu = 1.964(3)$.

Ferrenberg and *Landau* [2.40] used the multispin coding technique of Sect. 2.4 to simulate $L \times L \times L$ Ising lattices on a Cyber 205 with L as large as 96. Between 3×10^6 and 10×10^6 MCS/site were generated at the estimate for the critical temperature suggested by Monte Carlo renormalisation group calculations, and the data were reweighted to different temperatures using histogram methods [2.23]. Different thermodynamic derivatives were calculated and analysed using finite size scaling theory, including corrections to scaling, yielding extremely accurate values for critical exponents as well as for T_c; $J/k_B T_c = 0.2216595(27)$, $\nu = 0.6274(16)$. Even with such a detailed study, however, it is still not possible to calculate the corrections to scaling with much precision.

Alves et al. [2.41] used a modified multilattice coding method to simulate the three-dimensional Ising model on the Cyber 205 and on the ETA-10 E, ETA-10 Q, and ETA-10 G. The speed of the program varied with processor but on average about 2.1 clock cycles were needed for each update. They determined the density of states from which they calculated the zeros of the partition function. In addition, they calculated the correlation length and using finite size scaling extracted an estimate for ν. They concluded that $J/k_B T_c = 0.22157(3)$ and $\nu = 0.6285(19)$ or $0.6321(19)$ depending on the method used.

2.7.2 First-Order Transitions in Potts Models

Vectorised algorithms have also enabled large scale studies of systems undergoing first-order phase transitions. *Challa* et al. [2.42] used a vectorised, checkerboard algorithm on a Cyber 205 to simulate $L \times L$ $q = 10$ Potts models. This program, which was written for general q-value, produced about 0.5×10^6 spin-flip trials/second. In an infinite system the phase transition is first order and the transition temperature is known exactly [2.43]. For the largest system simulated, $L = 50$, 35×10^6 MCS/site were needed to produce data of sufficient quality. The finite size behaviour of the results was analysed using a simple theory which describes the probability distribution of energies as the sum of two different Gaussians, one for the ordered state and one for the disordered state [2.44]. This study was later extended by *Peczak* and *Landau* [2.45] to the $q = 5$ Potts model which has a very weak first-order transition. Runs were made for lattices ranging in size from $L = 30$ to $L = 240$; for the largest lattice fluctuations were quite slow and 5.5×10^6 MCS/site were needed to accurately produce the probability distributions for energy and order parameter for large systems. This finding removed an earlier discrepancy between simulation and theory [2.46]. The finite size behaviour of the specific heat, order parameter, specific heat, correlation length and cluster size distribution all suggested that the transition was second order. By comparing the correlation length for the $q = 5$ Potts model they were able to estimate that the correlation length at the transition was about 2000 lattice constants!

Billoire et al. [2.47] studied the ferromagnetic 3-state Potts model in three dimensions with small amounts of next-nearest-neighbour antiferromagnetic coupling J_{nnn}. They used a multilattice multispin coding algorithm, generating 3.6×10^6 spin-flip trials/second on a Cray X-MP. Approximately 10^6 MCS/site were generated for lattices as large as $48 \times 48 \times 48$, and a finite size scaling analysis was made of the bulk properties and the energy distributions. They conclude that with $J_{nnn}/J_{nn} = -0.2$ the transition is still first order with residual finite size corrections to the asymptotic form.

2.7.3 Dynamic Critical Behaviour

Dynamic critical behaviour is particularly difficult to study with Monte Carlo simulations because extremely long runs are needed to describe the time-displaced correlation function accurately. Vectorisation has allowed impressive improvements to be made in this direction. (A review of recent studies of dynamic critical phenomena, made using a variety of methods, is given in [2.48].) Using a Cyber 205 *Kalle* [2.49] combined a multispin coding simulation with a vectorised blocking method to determine the decay of the magnetisation from an ordered state in both the original as well as renormalised lattices. If two different block sizes b and b' are considered, the time at which the two different lattices have the same magnetisations should be given by

$$t_b/t'_b = (b/b')^z. \tag{2.5}$$

Kalle simulated two-dimensional Ising lattices as large as 8192×8192 and three-dimensional systems up to 512^3. Approximately 250 decays were followed for each lattice and the results were averaged together. The resulting estimates for the dynamic critical exponents are $z = 2.14 \pm 0.02$ for $d = 2$ and $z = 1.965 \pm 0.010$ for $d = 3$.

Wansleben and *Landau* [2.5] used the multispin coding algorithm described in Sect. 2.4 to simulate $L \times L \times L$ simple cubic Ising lattices with fully periodic boundary conditions and L as large as 96. The simulations were made at the infinite lattice critical temperature and a minimum of 3×10^6 MCS/site were generated for each lattice size. Time displaced correlation functions were calculated and the long time behaviour fitted to a multi-exponential decay. The resultant relaxation times were analysed using finite size scaling theory and the dynamic exponent z was estimated from Fig. 2.4 to be $z = 2.04 \pm 0.03$. This value removes the apparent contradiction between renormalisation group theory and earlier simulations. Because of the high statistics, sources of both systematic and statistical errors were explored and some light was shed on why other Monte Carlo work might have yielded other results.

Tang and *Landau* [2.51] used a vectorised, checkerboard algorithm on the Cyber 205 to study the time-dependent behaviour of two-dimensional q-state Potts models with $q = 2$, 3, and 4. Time-displaced correlation functions were determined and fitted to a two-exponential decay with the longest decay time being the correlation time τ. Since the exact values ot T_c were known it was

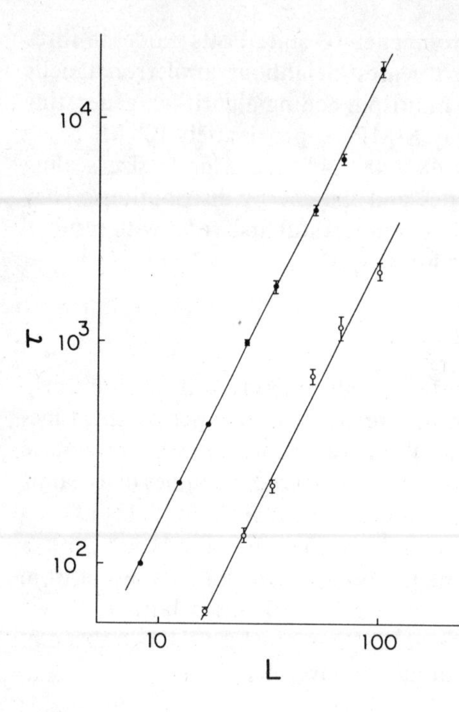

Fig. 2.4. Linear (equilibrium time-displaced) relaxation times for the magnetisation (●) and internal energy (○) of the simple cubic Ising model at the critical temperature. (After [2.50])

possible to perform an accurate finite size scaling analysis of τ, although z seems to increase very slightly with q, varying from 2.13 for $q=2$ to 2.19 for $q=4$. A quite similar result was found by *Bonfim* [2.52] who used multispin coding to study relaxation in large 1200×1200 q-state Potts models. From the finite size dependence of the correlation times he estimated $z=2.16 \pm 0.05$ for $q=2$, $z=2.16 \pm 0.04$ for $q=3$, and $z=2.18 \pm 0.03$ for $q=4$.

Mori and *Tsuda* [2.53] used a vectorised version of the multispin (superspin) coding algorithm on a HITAC S-810/20 supercomputer to study relaxation in enormously large, $L=10\,240$, Ising square lattices. At each temperature between 5 and 20 decays of the magnetisation were averaged together and the linear relaxation time τ was extracted from the exponential decay of the magnetisation as $M \to 0$. The non-linear relaxation time was obtained by integrating the decay. By fitting the divergence of the relaxation times as $T \to T_c$ they estimated that $z=2.076 \pm 0.005$, a value which is substantially below that which has been obtained from a number of other studies.

Zabolitsky and *Herrmann* [2.35] used a microcanonical method (Q2R cellular automaton) to study enormous two-dimensional systems, up to $123\,008 \times 123\,008$. They find that the time scale for decay of the magnetisation is much slower than for Metropolis Monte Carlo for temperatures just above the critical temperature. Just below T_c the magnetisation decays to the Ising value and at T_c they both decay with the same exponent. Results for $L=8320$ are shown in Fig. 2.5.

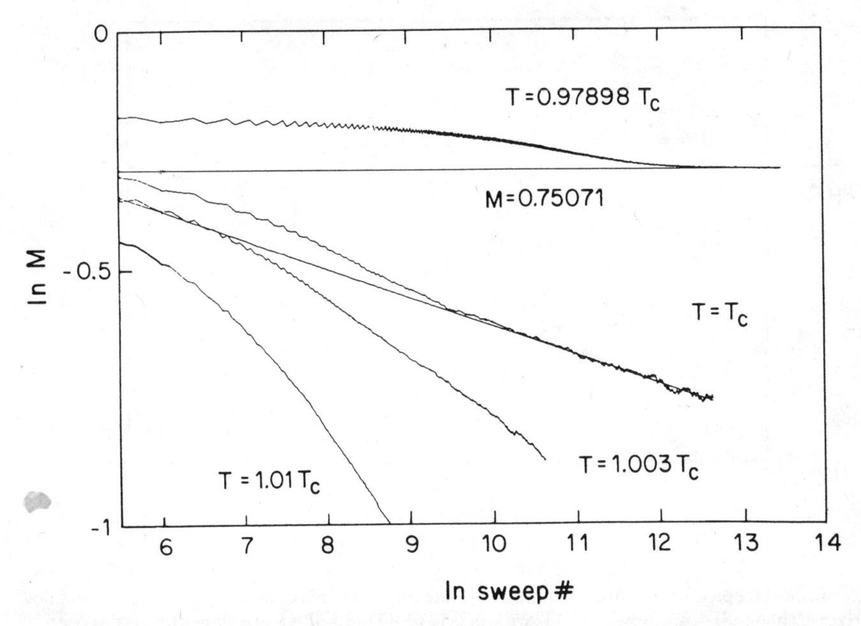

Fig. 2.5. Decay of the magnetisation for a 8320 × 8320 lattice using a microcanonical (Q2R cellular automaton) method. For the lowest temperature the line marked $M = 0.75071$ shows the exact Ising model value. (After [2.35])

2.7.4 Surface and Interface Phase Transitions

The connection between surface phase transitions in an Ising model in the presence of bulk and surface fields had been drawn by *Nakanishi* and *Fisher* [2.54]. There had been several studies [2.55] for $L \times L \times D$ Ising films with modified exchange in the surface layers and periodic boundary conditions in both directions parallel to the surface. *Kikuchi* and *Okabe* [2.56] used a multispin coding algorithm on the HITAC S810–20, storing 10 spins per word, and obtained a speed of 9.6×10^6 spin-flip trials/second. They calculated the surface layer magnetisation for systems of size $32 \times 32 \times 32$ and $32 \times 32 \times 64$ as a function of (bulk) magnetic field and temperature. From their data they calculated a surface equation of state, shown in Fig. 2.6 for different values of surface exchange. Attempts to sort out the global phase diagram had been complicated by the presence of capillary wave fluctuations which necessitated the use of large cross-sectional areas as well as thick samples. Using a multispin coding program with preferential sampling of layers near the surface, *Binder* et al. [2.57] produced accurate data for large lattices, typically $L = 128$, $D = 40$, although for some runs L as large as 512 or D as large as 160 were used. In some cases the use of such large lattices was crucial in reducing the "noise" sufficiently to allow reasonable conclusions to be extracted from the data. An example is shown in Fig. 2.7, where the surface-layer susceptibility data [2.58] for a $50 \times 50 \times 40$ system (i.e. 10^5 spins) gives no useful information about a

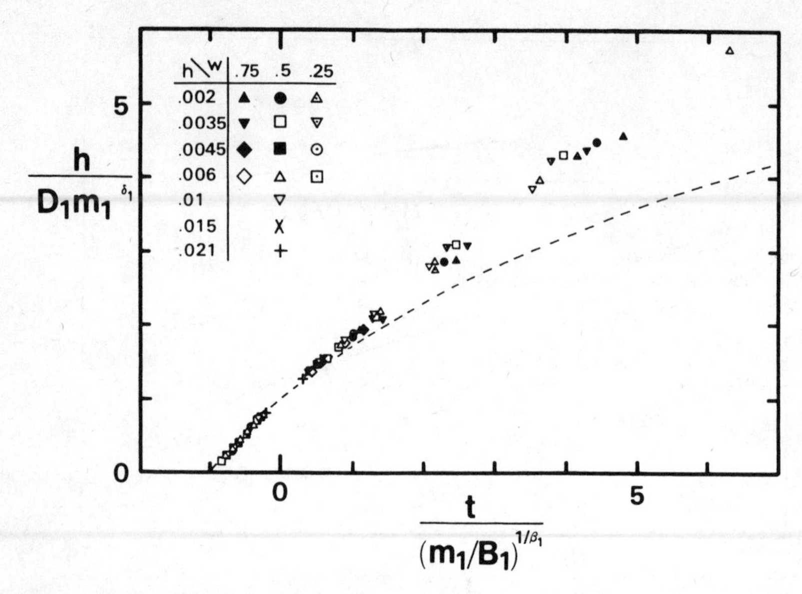

Fig. 2.6. Scaling plot for the surface equation of state for the simple cubic Ising model with modified surface exchange $J_s = wJ$. (See [2.56].) The mean field result (*dashed curve*) is shown for comparison

possible phase transition, but with $L = 128$ and $D = 80$ a peak can be clearly seen. Previous Monte Carlo studies [2.58] of critical wetting exponents for the three-dimensional Ising model in the presence of a surface field disagreed with the predictions of renormalisation group theory [2.59] based on a capillary wave Hamiltonian. Using a vectorised multispin coding program on the Cyber 205, *Mon* et al. [2.60] measured the anisotropic interfacial tension $\tau(\theta, T, L)$ associated with a tilted interface in the simple cubic Ising model for a wide range of temperature T, lattice size L, and tilt angle θ. By introducing an antiperiodic boundary condition perpendicular to the interface plane, they eliminated any finite size effects due to constraining a wandering interface between two surfaces. (The speed of the program was 30×10^6 updates/second.) The data were analysed by fitting small θ results to

$$\tau(\theta, T, L) = \tau(0, T, L) + \tau'(0, T, L)|\theta| + \tau''(0, T, L)\frac{\theta^2}{2}. \tag{2.6}$$

The data show that τ' scales as $1/L$ and is not zero for finite L; in fact it may be comparable to the quadratic term. The use of a capillary wave Hamiltonian, which assumes that $\tau' = 0$, is thus a questionable approximation, except near T_c, and the RG theory may simply not be relevant. Another interpretation of the findings of [2.57] is that the "interfacial stiffness constant" of the Ising model at the temperature used for the simulations is much smaller than was previously thought [2.61].

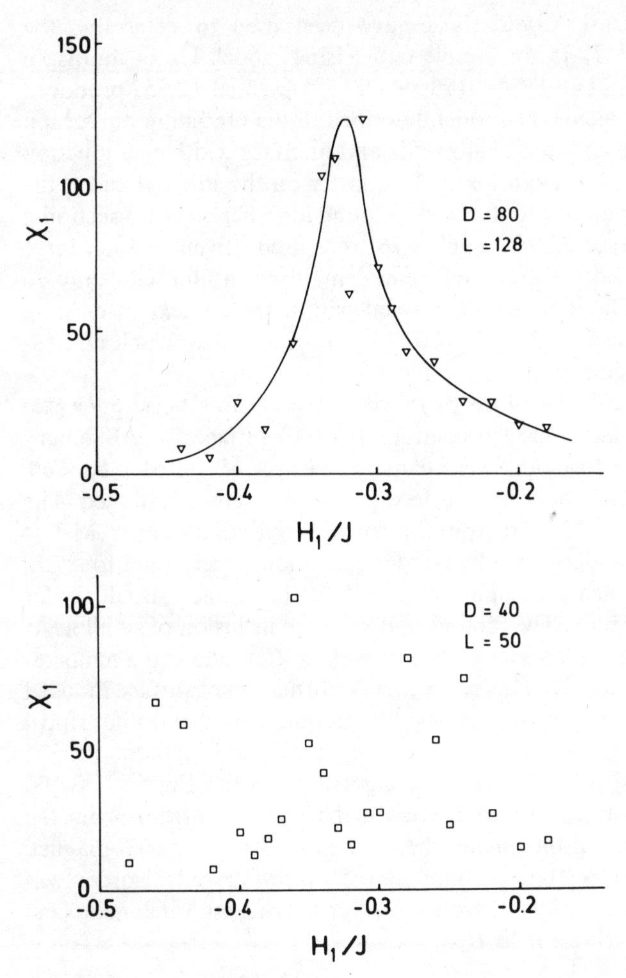

Fig. 2.7. Surface layer susceptibility χ_1 at $J/kT = 0.23$ and $J_s = J$ plotted vs surface field. The lower part shows data for a $50 \times 50 \times 40$ system; the upper part shows data for a $128 \times 128 \times 128$ system obtained using multispin coding. The curve in the upper part of the figure is only a guide to the eye. (See [2.57])

Landau and *Binder* [2.62] used a multispin coding algorithm on the Cyber 205 to study the "extraordinary transition" in $128 \times 128 \times 80$ simple cubic Ising films with enhanced surface exchange. In contrast to theoretical predictions only a free-energy-like singularity was found in the surface magnetisation.

Landau et al. [2.63] and *Peczak* and *Landau* [2.64] used a vectorised checkerboard algorithm on the Cyber 205 and the ETA-10 P108 to study surface behaviour in simple cubic XY-films as large as $50 \times 50 \times 59$ with modified surface exchange. Both the "ordinary" and "surface" phase boundaries were studied, and comparisons were drawn to the corresponding behavior of Ising films.

Extensive Monte Carlo simulations have been used to determine the roughening temperature T_R in the simple cubic Ising model. Using multispin coding on a Cyber 205 and an ETA-10 Piper Q216, *Mon* et al. [2.65] produced equilibrium configurations for the model described in the preceding paragraph with $\theta = 0$ and extracted an interfacial profile and interface width as a function of temperature. (The program ran about 10% faster on the Piper than on the Cyber 205.) Quite strong finite size effects were found for different cross-sectional areas and systems as large as $960 \times 960 \times 26$ were used. (Even in such large systems the interfacial width stayed sufficiently small that a film with only 26 layers was adequate.) The data were analysed within the context of existing theory and yielded the estimate $T_R/T_c = 0.542 \pm 0.005$, a value which convincingly excludes the possibility that $T_R = T_c^{2d}$.

Using a checkerboard algorithm (with very sparse sublattices) *Schweika* et al. [2.66] studied surface phase transitions in (100) surfaces of AB binary alloys on an fcc lattice. Most runs were made on lattices of size $61 \times 60 \times 60$, and profiles as well as both bulk and surface properties were determined. The program ran at a speed of 2.5×10^6 spin-flip trials/second on a Cray Y-MP. It is known that the bulk undergoes a first-order transition in zero field to a $L1_0$ structure, and for only nearest-neighbour coupling the surface should order simultaneously with the bulk. They found that with the inclusion of small next-nearest-neighbour coupling the surface orders in a $c(2 \times 2)$ structure at a temperature which is *above* the bulk T_c. This is the first confirmation of surface induced ordering [2.67], but the data are not yet of sufficient quality to determine critical exponents.

Helbing et al. [2.68] used a vectorised algorithm on the Fujitsu VP-100 and the Cray-XMP to study surface effects on the phase transition from the DO_3 ordered state to a paramagnetic phase in a bcc Ising antiferromagnet. Films as large as $20 \times 20 \times 91$ were simulated. Surface induced disorder was observed with the thickness of the disordered layer varying with magnetic field as $\ln(H - H_c)$ near the critical field H_c.

2.7.5 Bulk Critical Behaviour in Classical Spin Systems

The classical Heisenberg model described by the Hamiltonian

$$\mathscr{H} = -J \sum_{nn} \boldsymbol{S}_i \cdot \boldsymbol{S}_j, \tag{2.7}$$

where $\boldsymbol{S}_i$ are classical spins of unit length, has been studied using the simple checkerboard approach of Sect. 2.3. *Peczak* et al. [2.69] simulated the simple cubic Heisenberg model on lattices as large as $24 \times 24 \times 24$, using a vectorised checkerboard algorithm on the Cyber 205, and applied a histogram analysis to estimate T_c with high accuracy and to estimate critical exponents. They were able to show that values of T_c determined by other numerical means [2.70, 71] were too low and that using this estimate for T_c a finite size scaling analysis gives exponent estimates which are quite close to those predicted by ε-expansion renormalization group theory [2.72]. *Lau* and *Dasgupta* [2.73] used a 4-sublattice checkerboard algorithm on the Cray 2 to study the role of topological defects in the three-dimensional (3D) Heisenberg model.

Peczak and *Landau* [2.74] studied critical relaxation in the 3D Heisenberg model using both the vectorised Metropolis algorithm and one which mixes the Metropolis spin flips with a vectorised, over-relaxation method. Without over-relaxation the program generated about 0.22×10^6 spin-flip trials/second; additional over-relaxation moves required about one-third as much time. For the purely canonical simulations they found a dynamic exponent of $z = 1.96 \pm 0.06$, and as over-relaxation trials were added they observed a dramatic decrease in z all the way down to $z \approx 1.1$.

Minnhagen and *Nylen* [2.75] performed a simulation of the 2D XY-model on 63×64 lattices using a Cray-1 to study the temperature dependence of the helicity modulus. The data agreed well with the Coulomb gas scaling function.

2.7.6 Quantum Spin Systems

It is now well known that using the Trotter transformation [2.76] a d-dimensional quantum spin model can be mapped onto a $(d + 1)$-dimensional Ising model with modified interactions. As an example we consider 1D spin systems of length L with nearest neighbour interactions. This model maps onto an $L \times 2m$ square latice (where m is the Trotter index) with two spin couplings in the real lattice direction and four spin couplings on plaquettes which correspond to squares of a single colour on a checkerboard. Because of conservation laws emanating from the transformation, not all possible plaquette states are allowed [2.77]. *Okabe* and *Kikuchi* [2.78] assign a "plaquette spin" value to each state, see Fig. 2.8, which have the binary representations shown in the figure. Note that the only way to change a plaquette spin into some other allowed state is to flip two of the real spins on the plaquette. For the system as a whole, the spin flips are no longer single spin events but involve either "local" processes, in which the spins around an "empty" plaquette overturn, or "global" updates, in which all spins along a straight line path which circles the lattice (Fig. 2.8). Each "plaquette spin flip" is carried using an exclusive-or operation; for example, the 4-spin flip process is carried out by

$$p_{\text{new}} = \text{XOR}(p_{\text{old}}, 4).$$

Similar exclusive-or operations can be used to flip any two spins on a plaquette. The local and global updates may be vectorised by dividing the system into multiple sublattices, of different symmetry for the different processes, as shown in Fig. 2.9. The procedure is then analogous to that described in Sect. 2.3. A speed of 34×10^6 spin-flip trials/second was reached on a NEC SX-2.

Using this approach *Okabe* and *Kikuchi* [2.79] were able to reproduce the energy of the XY-chain to within 1 part in 10^4 by extrapolating values obtained for a 128 spin chain and Trotter indices between 8 and 32. They later applied

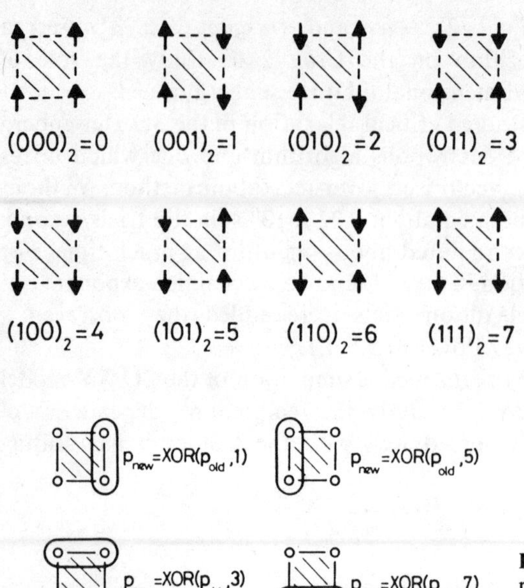

Fig. 2.8. Upper part, the "plaquette" spin representation of the four-spin configuration on "interaction squares" produced by the Trotter transformation of a 1D *xyz*-model. The lower part shows the different "exclusive-or" operations which will change one plaquette into another. (After [2.78])

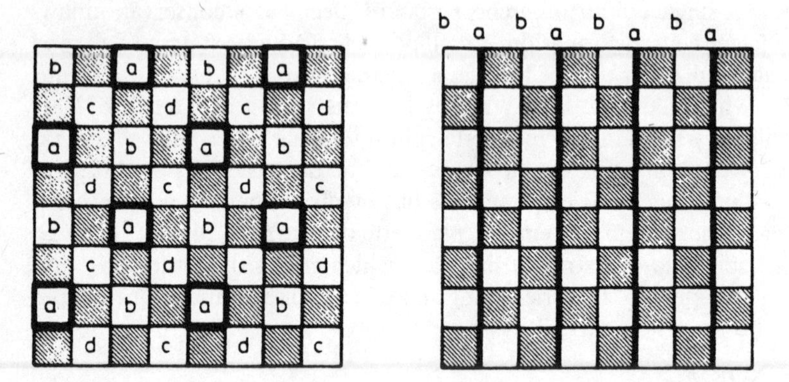

Fig. 2.9. Subgroup decomposition of the Monte Carlo trials for the local processes (**a**) and global process (**b**). (After [2.78])

this method to study [2.80] the 1D Heisenberg model using a cluster spin decomposition as well as a pair spin decomposition. The anisotropic $s = 1/2$ Heisenberg square lattice was simulated using a 3D version of this algorithm. Through extrapolations, first letting $M \to 0$ and then $L \to 0$, they extracted very low temperature estimates for the ground-state energy and order parameter.

Jacobs et al. [2.81] simulated an $L \times L$ array of Josephson junctions both with and without a transverse magnetic field. After the mapping was made, using the Feynman path integral formulation, to yield $L \times L \times L_\tau$ lattices, simulations were made on the Cray 2 and Cray X-MP using a discrete Z_N subgroup. The superfluid density, internal energy, and specific heat were studied for different lattice sizes and temperatures and phase diagrams were determined.

2.7.7 Spin Exchange and Diffusion

Kinetics of domain growth have also been studied using a Cyber 205 with a vectorised, multispin coding procedure for spin-exchange in the 2D Ising model. Using different measures of domain size $R(t)$, *Amar* et al. [2.21] found that a growth law of the form

$$R(t) = A + Bt^{1/3} \tag{2.8}$$

describes the long time data extremely well. Effective exponents at intermediate times are found to be substantially smaller but extrapolate rather accurately to 1/3 as shown in Fig. 2.10.

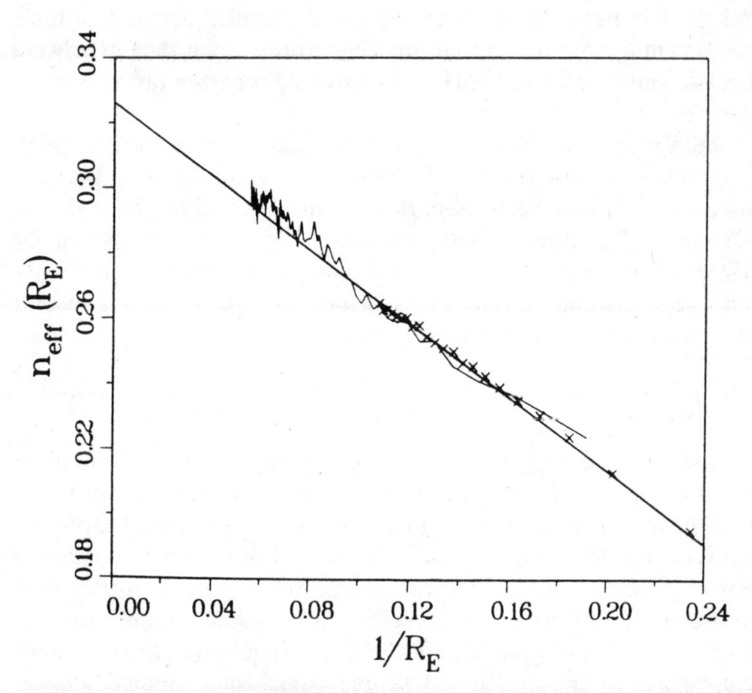

Fig. 2.10. Variation of the "effective exponent" for the growth of domains in the 2D Ising model by spin-exchange. Data are obtained from averages over multiple runs for 512 × 512 lattices; crosses are from averages over 100 runs of 100 000 MCS/site each, and the solid curve is from averaging over 10 runs of 10^6 MCS/site each. (After [2.21])

Diffusion without any double occupancy of sites has also been vectorised using a version of the simple, vectorised checkerboard algorithm [2.82] by *Paetzold*. The lattice is decomposed into 3^d sublattices, where d is the lattice dimension, so that particles on a given sublattice can move toward each other in the same time step and still not attempt to land on the same site. Sublattices are randomly chosen for consideration, and each particle on the sublattice attempts to jump to a nearest neighbour site, if that site is empty. This procedure is essentially a spin-exchange algorithm for a system at infinite temperature. On a Cray X-MP the speed was 2.05×10^6 updates/second whereas the standard approach on the IBM3090 yielded only 0.07×10^6 updates/second. On a 30^3 lattice *Paetzold* found [2.83] an exponent k for the rms displacement r, i.e. $r \propto t^k$ to be $k = 0.183 \pm 0.010$. The algorithm was also used to study collective and tracer diffusion [2.84].

Roman [2.85] performed extensive Monte Carlo simulations of the "ant in a labyrinth" problem on a Cray Y-MP. The diffusion of non-interacting particles on a percolation cluster in simple cubic lattices as large as 960^3 was studied. The exponent k for the rms displacement r is calculated to be $k = 0.190 \pm 0.003$ which is outside the error bars of the Alexander–Orbach conjecture value of $k = 0.201 \pm 0.001$.

Sahimi and *Stauffer* [2.86] simulated diffusion in random porous media in both two and three dimensions. They give a rather detailed account of their version of the "ant in a labyrinth" program. The lattices which they used were up to 1024^3 in size and on a Cray Y-MP the vectorized program ran at a speed of 4×10^6 steps/second.

Random walks is an area which is closely related to diffusion and for which Monte Carlo simulations have been used extensively, but for the most part the algorithms used have not been vectorised. *Berretti* and *Sokal* [2.87] have shown how to implement an algorithm for self-avoiding walks (SAWs) on a Cray. By rewriting and optimising the code and using a "sliding bit table" to store walk configurations, they were able to achieve a speed of 1.4×10^6 iterations/second and to simulate SAW's in lattices as large as $(2^{16})^3$.

2.7.8 Impurity Systems

Braun et al. [2.88] used a vectorised, multispin coding method to simulate ferromagnetic and antiferromagnetic simple cubic Ising models with site impurities. For the ferromagnet they found an increase in the critical exponent β but no abrupt change with impurity concentration. For the antiferromagnet they found a peak developed in the uniform susceptibility when impurities were added. *Faehnle* et al. [2.89] used a fully vectorised multispin coding program [2.88] on a Cray Y-MP to carry out Monte Carlo renormalisation group studies of the critical behaviour in site diluted simple cubic Ising models. Helical boundary conditions were used in one direction and periodic boundaries in the other two directions; lattices as large as 64^3 were considered with as many as 20% ($x = 0.2$) site impurities. They find critical exponents which apparently vary with dilution so that v changes from 0.63 with no impurities to $v = 0.69$ for $x = 0.2$. Using a multilattice Monte Carlo program [2.27] modified for the Cray from the HITAC program by *Ito* and *Kanada* [2.25], *Heuer* [2.90] obtained concentration dependent exponents which seemed to level off for $x = 0.5$. He also obtained an estimate for T_c for $x = 0.2$ which agreed well with that obtained from the MCRG calculation.

Landau and *Wansleben* [2.91] used a vectorised, multispin coding program to study the decay of magnetisation in a site dilute Ising square lattice at the percolation threshold. Lattices as large as 2048 × 2048 were used. The time dependence of the decay of the magnetisation was well described by a stretched exponential with a power which depended on temperature.

2.7.9 Other Studies

A simple vectorised, "checkerboard" algorithm was used to study the possible contributions of solitons to the bulk behaviour of classical spin chains in a symmetry breaking magnetic field [2.92, 93]. For the XY-ferromagnet, no indication was found for the predicted [2.94] contribution to the specific heat, but an examination of the configurations generated showed clear 2π-twists in the spin directions which formed solitons. (The Monte Carlo generated equilibrium configurations also served as starting points for numerical integration of the coupled equations of motion for the system.) When the sign of the interaction was changed, the system immediately went into a spin-flop state when a magnetic field was included. In this system "π-solitons" were formed by the exchange of spin orientation between sublattices. From an examination of

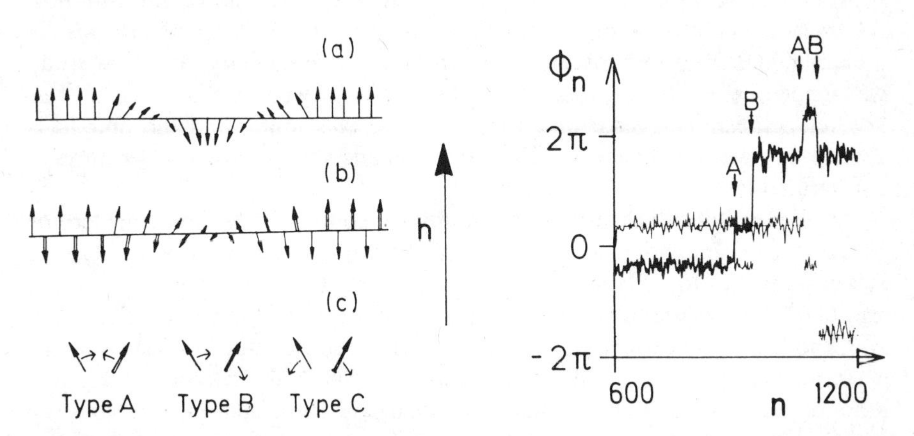

Fig. 2.11. (*Left*) Solitons in the XY-chain: (**a**) 2π-soliton in the ferromagnet, (**b**) π-soliton in the anisotropic antiferromagnet, (**c**) "π-like" solitons in the isotropic antiferromagnet. The spins are in a spin-flop state and each type of soliton is produced by allowing the sublattices to rotate in the XY-plane in the directions shown until the sublattices are interchanged. (*Right*) The phase of the spins as a function of spin number for $k_B T/J = 0.14$, $h/J = 1.5$ for a chain of 2000 spins. The solid and dashed lines represent the two different sublattices. Solitons are marked by arrows and labelled by type. (After [2.93])

Monte Carlo configurations a new "type B" soliton was discovered; the spin phase angle shown in Fig. 2.11 reveals the existence of two types of "π-solitons" in addition to a single 2π twist involving only spins on a single sublattice. Monte Carlo simulations have also been made for an anisotropic Heisenberg model spin chain [2.94].

Dawson et al. [2.95] simulated a 3D Ising model with competing interactions as a model for microemulsions. A vectorised, checkerboard algorithm was used on an IBM 3090-600E and in some cases six different planes were run in parallel on the different processors to further increase the performance. They studied $47 \times 47 \times 47$ lattices, and using information obtained from low temperature series and ε-expansions together with the simulations, they extracted a global phase diagram. *Stauffer* et al. [2.96] used a vectorised Monte Carlo method which generated 3.6×10^6 updates/second on the Cray-YMP to study the Widom model of microemulsions in a porous medium. Randomly diluted simple cubic lattices as large as 252^3 were simulated and a phase diagram was determined.

A vectorised checkerboard algorithm was used to study Blume–Capel and Blume–Emery–Griffiths models. *Kimel* et al. [2.97] simulated the antiferromagnetic Blume–Capel model in two dimensions using systems as large as 100×100 with up to 2×10^6 MCS/site. The full phase diagram was mapped out as a function of both uniform magnetic field and single-ion potential strength. They also conclude that the tricritical point does not decompose into a double critical point and critical endpoint as suggested by mean-field theory [2.98]; this conclusion was also reached for the next-nearest-neighbour antiferromagnet in two dimensions by *Landau* and *Swendsen* [2.99] using MCRG. A similar study was made of the 3D Blume–Capel model, but here a decomposition of the tricritical point was found [2.100]. Quantitatively the Monte Carlo results are quite different from those predicted from mean field theory. A similar study was made of the Blume–Emery–Griffiths model in two dimensions [2.101]. A low temperature staggered quadrupolar phase was found and the umbilicus formed between this phase and the ferromagnetic state at low temperatures is conjectured.

Kohring [2.102] has shown how multispin coding can be implemented for Monte Carlo simulations of neural network models. On a Cray Y–MP/832 systems with up to 10^9 couplings were treated and speeds between 400×10^6 and 6600×10^6 coupling evaluations per second were obtained.

Another area in which there has been much recent work has been damage spreading [2.103]. Two initial configurations which differ only by a small amount, e.g. a single spin, are simulated using the same spin flip rule and the identical random number sequence. The "difference", or Hamming distance Δ, between the resulting pair of configurations is then calculated. The temperature dependence of the resulting difference is studied and phase transitions between a frozen state $\Delta(t \to \infty) \to 0$ and a chaotic state where Δ approaches a constant as $t \to \infty$. Damage spreading is discussed in detail later in this volume so we shall not comment on it further here.

2.8 Conclusion

In this chapter we have attempted to describe some of the algorithms which have been used with great success on vector processors and to review some of the work which has been successfully completed using these techniques. Since published articles do not always indicate when vectorised algorithms have been used and the literature on Monte Carlo simulations has been growing rapidly, this chapter is certainly not exhaustive. We have, for example, neglected an entire area of lattice simulations, those dealing with lattice gauge theory; however, the interested reader can find a nice overview of algorithms which are in large part vectorisable in a review by *Toussaint* [2.104]. (An example of a vectorised SU(3) program for the Cyber 205 and ETA-10 is given by *Vohwinkel* et al. [2.105].) As we have shown in the previous section the vast majority of the highly efficient vectorised programs have been written for the Cyber 205. Since there is no "line of succession" for this machine, algorithms are now being developed to exploit the architecture of other supercomputers. As new non-vectorisable algorithms are developed there will clearly be better ways of carrying out some kinds of studies which were previously best undertaken using vectorised code. For many problems, however, a vectorised algorithm will still continue to provide the maximum computing power, although a combination of parallelism and vectorisation is likely to increase in importance. This is true on two different levels: some parallel machines pipeline instructions internally, and some machines like the Hypercube or the Cray Y–MP have "nodes" which have attached vector facilities. With the emergence of Fortran 90 with "vector syntax" it will become possible to write programs (not optimised) which will run on either vector machines or on SIMD machines, like the Connection Machine, without change. Vectorisation is thus likely to continue to play an important role in Monte Carlo simulations for some time into the future.

Acknowledgements. The author is indebted to K. Binder, A.M. Ferrenberg, and H.J. Herrmann for their interaction and helpful comments about this manuscript. The support of NSF grant #DMR-8715740 is gratefully acknowledged.

References

2.1 N. Metropolis, A.W. Rosenbluth, M.N. Rosenbluth, A.H. Teller, E. Teller: J. Chem. Phys. **21**, 1087 (1953)
2.2 K. Binder (ed.): *Monte Carlo Methods in Statistical Physics*, 2nd edn., Topics Curr. Phys., Vol. 7 (Springer, Berlin, Heidelberg 1986); K. Binder (ed.): *Applications of Monte Carlo Methods in Statistical Physics*, 2nd edn., Topics Curr. Phys., Vol. 36 (Springer, Berlin, Heidelberg 1987)
2.3 A. Bortz, M. Kalos, J. Lebowitz: J. Comput. Phys. **17**, 10 (1975)
2.4 C.P. Yang: Appl. Meth. **15**, 351 (1963)
2.5 D.P. Landau: Phys. Rev. B **13**, 2997 (1976)

2.6 R.H. Swendsen, J.-S. Wang: Phys. Rev. Lett. **58**, 86 (1987); U. Wolff: Phys. Rev. Lett. **62**, 361 (1989)
2.7 D.P. Landau, D. Stauffer: J. Phys. **50**, 509 (1989)
2.8 M.E. Fisher: In Proc. of the Intl. Summer School Enrico Fermi, Course 51, Varenna, 1970, ed. by M.S. Green (Academic, New York 1971); V. Privman (ed.): *Finite Size Scaling and Numerical Simulation of Statistical Systems* (World Scientific, Singapore 1990)
2.9 R. Friedberg, J.E. Cameron: J. Chem. Phys. **52**, 6049 (1970)
2.10 W. McIntyre: In *Supercomputers*, ed. by J.R. Kirkland, J.H. Poore (Prager, New York 1987)
2.11 W. Oed: Angew. Inf. **7/82**, 358 (1982)
2.12 A. Desalvo, G. Erbacci, R. Rosa: Comput. Phys. Commun. **60**, 305 (1990)
2.13 R. Zorn, H.J. Herrmann, C. Rebbi: Comput. Phys. Commun. **23**, 337 (1981)
2.14 S. Wansleben: Comput. Phys. Commun. **43**, 315 (1987)
2.15 S. Wansleben, J.G. Zabolitzky, C. Kalle: J. Stat. Phys. **37**, 271 (1984)
2.16 R.C. Tausworthe: Math. Comput. **19**, 201 (1965)
2.17 S. Kirkpatrick, E. Stoll: J. Comput. Phys. **40**, 517 (1981)
2.18 C. Kalle, S. Wansleben: Comput. Phys. Commun. **33**, 343 (1984)
2.19 P.A. Slotte, D.P. Landau: In *Computer Simulation Studies in Condensed Matter Physics*, ed. by D.P. Landau, K.K. Mon, H.-B. Schüttler, Springer Proc. Phys., Vol. 33 (Springer, Berlin, Heidelberg 1988)
2.20 M.Q. Zhang: J. Stat. Phys. **56**, 939 (1989)
2.21 J.G. Amar, F.E. Sullivan, R.D. Mountain: Phys. Rev. B **37**, 196 (1988)
2.22 G. Bhanot, D. Duke, R. Salvador: J. Stat. Phys. **44**, 985 (1986)
2.23 A.M. Ferrenberg, R.H. Swendsen: Phys. Rev. Lett. **63**, 1195 (1989); ibid. **61**, 2635 (1988)
2.24 G. Bhanot, R. Salvador, D. Duke, K.J.M. Moriarty: Comput. Phys. Commun. **49**, 465 (1988)
2.25 N. Ito, Y. Kanada: Supercomputer **25**, 31 (1988), and to be published
2.26 G. Williams, M. Kalos: J. Stat. Phys. **37**, 283 (1984)
2.27 H.-O. Heuer: Comput. Phys. Commun. **59**, 387 (1990)
2.28 H. Mueller-Krumbhaar, K. Binder: Z. Phys. **254**, 269 (1972)
2.29 M. Creutz: Phys. Rev. D **36**, 515 (1987); F.R. Brown, T.J. Woch: Phys. Rev. Lett. **58**, 2394 (1987)
2.30 M. Creutz, K.J.M. Moriarty, M. O'Brien: Comput. Phys. Commun. **42**, 191 (1986)
2.31 M. Creutz: Phys. Rev. Lett. **50**, 1411 (1983)
2.32 J.M. Drouffe, K.J.M. Moriarty: Comput. Phys. Commun. **52**, 249 (1989)
2.33 Y. Pomeau: J. Phys. A **17**, L415 (1984)
2.34 H.J. Herrmann: J. Stat. Phys. **45**, 145 (1986)
2.35 J. Zabolitsky, H.J. Herrmann: J. Comput. Phys. **76**, 426 (1988)
2.36 M. Creutz: Phys. Rev. Lett. **50**, 1411 (1983)
2.37 M. Creutz, P. Mitra, K.J.M. Moriarty: Comput. Phys. Commun. **33**, 361 (1984)
2.38 M.N. Barber, R.B. Pearson, D. Toussaint, J.L. Richardson: Phys. Rev. B **32**, 1720 (1985)
2.39 G. Bhanot, D. Duke, R. Salvador: Phys. Rev. B **33**, 7841 (1986)
2.40 A.M. Ferrenberg, D.P. Landau: Phys. Rev. B **44**, 5081 (1991); see also A.M. Ferrenberg, D.P. Landau, P.K. Peczak: J. Appl. Phys. **69**, 6153 (1991)
2.41 N.A. Alves, B.A. Berg, R. Villanova: Phys. Rev. B **41**, 383 (1990)
2.42 M.S.S. Challa, D.P. Landau, K. Binder: Phys. Rev. B **34**, 1841 (1986)
2.43 R.J. Baxter, J. Phys. A **15**, 3329 (1982)
2.44 This theory was an extension of that first introduced by K. Binder, D.P. Landau: Phys. Rev. B **30**, 1477 (1984)
2.45 P. Peczak, D.P. Landau: Phys. Rev. B **39**, 11932 (1989)
2.46 E. Katznelson, P.G. Lauwers: Phys. Lett. B **186**, 385 (1987); in *Nonperturbative Methods in Quantum Field Theory*, ed. by Z. Horvath, L. Palla, A. Patkos: (World Scientific, Singapore 1987)
2.47 A. Billoire, R. Lacaze, A. Morel: Nuclear Physics B **340**, 542 (1990)
2.48 D.P. Landau, S.Y. Tang, S. Wansleben: J. de Phys. **49**, Colloq. 8, 1525 (1989)
2.49 C. Kalle: J. Phys. A **17**, L801 (1984)
2.50 S. Wansleben, D.P. Landau: J. Appl. Phys. **61**, 4409 (1987); Phys. Rev. B **43**, 6006 (1991)
2.51 S.Y. Tang, D.P. Landau: Phys. Rev. B **36**, 567 (1987)
2.52 O.F. de Alcantara Bonfim: Europhys. Lett. **4**, 373 (1987)
2.53 M. Mori, Y. Tsuda: Phys. Rev. B **37**, 5444 (1988)
2.54 H. Nakanishi, M.E. Fisher: Phys. Rev. Lett. **49**, 1565 (1982)
2.55 K. Binder, D.P. Landau: Phys. Rev. B **37**, 1745 (1988)
2.56 M. Kikuchi, Y. Okabe: Prog. Theor. Phys. **74**, 458 (1985)

2.57 K. Binder, D.P. Landau, S. Wansleben: Phys. Rev. B **40**, 6971 (1989)
2.58 K. Binder, D.P. Landau, D.M. Kroll: Phys. Rev. Lett. **56**, 2276 (1986)
2.59 E. Brezin, B.I. Halperin, S. Leibler: Phys. Rev. Lett. **50**, 1387 (1983); R. Lipowsky, D.M. Kroll, R.K.P. Zia: Phys. Rev. B **27**, 1387 (1983)
2.60 K.K. Mon, S. Wansleben, D.P. Landau, K. Binder: Phys. Rev. Lett. **60**, 708 (1988); Phys. Rev. B **39**, 7089 (1989)
2.61 G. Gompper, D.M. Kroll, R. Lipowsky: Phys. Rev. B **42**, 961 (1990)
2.62 D.P. Landau, K. Binder: Phys. Rev. B **41**, 4786 (1990)
2.63 D.P. Landau, R. Pandey, K. Binder: Phys. Rev. B **39**, 12302 (1989)
2.64 P.K. Peczak, D.P. Landau: Phys. Rev. B **43**, 1048 (1991)
2.65 K.K. Mon, D.P. Landau, D. Stauffer: Phys. Rev. B **42**, 545 (1990)
2.66 W. Schweika, K. Binder, D.P. Landau: Phys. Rev. Lett. **65**, 3321 (1990)
2.67 See R. Lipowsky: J. Appl. Phys. **55**, 2485 (1984); Ferroelectrics **73**, 69 (1987)
2.68 W. Helbing, B. Duenweg, K. Binder, D.P. Landau: Z. Phys. B **80**, 401 (1990)
2.69 P. Peczak, A.M. Ferrenberg, D.P. Landau: Phys. Rev. B **43**, 6087 (1991)
2.70 D.S. Ritchie, M.E. Fisher: Phys. Rev. B **5**, 2668 (1972)
2.71 M.P. Nightingale, H.W.J. Bloete: Phys. Rev. Lett. **60**, 1562 (1988)
2.72 J.C. LeGuillou, J. Zinn-Justin: Phys. Rev. Lett. **39**, 95 (1977); Phys. Rev. B **21**, 3976 (1980); K.G. Wilson, M.E. Fisher: Phys. Rev. Lett. **28**, 240 (1972); J.C. LeGuillou: J. Phys. Lett. **46**, L137 (1985)
2.73 M.-H. Lau, C. Dasgupta: Phys. Rev. B **39**, 7212 (1989)
2.74 P. Peczak, D.P. Landau: J. Appl. Phys. **67**, 5427 (1990)
2.75 P. Minnhagen, M. Nylen: Phys. Rev. B **31**, 1693 (1985)
2.76 H.F. Trotter: Proc. Am. Math. Soc. **10**, 545 (1959)
2.77 M. Suzuki: Prog. Theor. Phys. **56**, 1454 (1976)
2.78 Y. Okabe, M. Kikuchi: Phys. Rev. B **34**, 7896 (1986)
2.79 Y. Okabe, M. Kikuchi: J. Phys. Soc. Jpn. **56**, 1963 (1987)
2.80 Y. Okabe, M. Kikuchi: J. Phys. Soc. Jpn. **57**, 4351 (1988)
2.81 L. Jacobs, J.V. Jose, M.A. Novotny, A.M. Goldman: Phys. Rev. B **38**, 4562 (1988)
2.82 O. Paetzold: Comput. Phys. Commun., in press
2.83 O. Paetzold: J. Stat. Phys. **61**, 495 (1990)
2.84 O. Paetzold, K.W. Kehr: Phys. Lett. **146**, 397 (1990)
2.85 H.E. Roman: J. Stat. Phys. **58**, 375 (1990)
2.86 M. Sahimi, D. Stauffer: Chem. Eng. Sci. **46**, 2225 (1991)
2.87 A. Berretti, A.D. Sokal: Comput. Phys. Commun. **58**, 1 (1990)
2.88 P. Braun, U. Staaden, T. Holey, M. Faehnle: Int. J. Mod. Phys. B **3**, 1343 (1989)
2.89 M. Faehnle, T. Holey, U. Staaden, P. Braun: Festköperprobleme **30**, 425 (1990)
2.90 H.O. Heuer: Europhys. Lett. **12**, 551 (1990)
2.91 D.P. Landau, S. Wansleben: J. Appl. Phys. **63**, 3039 (1988)
2.92 R.W. Gerling, D.P. Landau: Phys. Rev. B **37**, 6092 (1988)
2.93 M. Staudinger, R.W. Gerling, C.S.S. Murty, D.P. Landau: J. Appl. Phys. **57**, 3335 (1985)
2.94 O.G. Mouritsen, J. Jensen, H.C. Fogedby: Phys. Rev. B **30**, 498 (1984)
2.95 K.A. Dawson, B.L. Walker, A. Berera: Physica A **165**, 320 (1990)
2.96 D. Stauffer, J.S. Ho, M. Sahimi: J. Chem. Phys. **94**, 1385 (1991)
2.97 J.D. Kimel, S. Black, P. Carter, Y.-L. Wang: Phys. Rev. B **35**, 3347 (1987)
2.98 M. Kincaid, E.G.D. Cohen: Phys. Lett. A **50**, 317 (1974)
2.99 D.P. Landau, R.H. Swendsen: Phys. Rev. Lett. **46**, 1437 (1981)
2.100 Y.-L. Wang, J.D. Kimel: J. Appl. Phys., in press
2.101 Y.-L. Wang, F. Lee, J.D. Kimel: Phys. Rev. B **36**, 8945 (1987)
2.102 G.A. Kohring: Int. J. Mod. Phys. C **1**, 259 (1990)
2.103 H.E. Stanley, D. Stauffer, J. Kertesz, H.J. Herrmann: Phys. Rev. Lett. **59**, 2326 (1987)
2.104 D. Toussaint: Comput. Phys. Commun. **56**, 69 (1989)
2.105 C. Vohwinkel, B.A. Berg, A. Devoto: Comput. Phys. Commun. **51**, 331 (1988)

3. Parallel Algorithms for Statistical Physics Problems

Dieter W. Heermann and Anthony N. Burkitt

With 8 Figures

The rapid development of computer simulation methods [3.1–5] over the last couple of years has been matched only by the equally rapid advances that have taken place in the field of computer technology. The many-fold increases in the speed, memory size and the flexibility of computers has opened up a vast number of new possibilities for studying science and engineering problems, and enabled both new insights and new lines of enquiry. However, the complexity of the problems studied and the accuracy of the results required are such that the computing resources available are hardly able to keep up with the demand.

The last decade has been the birth of a very different type of computer architecture that provides a *qualitative* improvement in machine performance, namely the development of parallel computers. There are an astonishing diversity of such parallel machines, and they have been used to study an equally wide variety of problems. The performance of the present generation of parallel machines is now overtaking that of the best scalar and vector machines, particularly on problems for which there are suitable parallel algorithms.

In a chapter like this it is clearly not possible to provide a comprehensive presentation of all the methods and algorithms that are used, nor all the machines that are available, to simulate models of interest in physics. Insofar as such reviews [3.6] and articles are available, we provide appropriate references to the literature. We do seek to provide, however, a flavour of the central features that are involved in computer simulations on parallel computers, and we seek to illustrate these with examples. Some of the various paradigms for parallel computing are introduced as a means of distinguishing both different types of algorithms and different types of machines for carrying out simulations. We then present a number of such algorithms that are appropriate for Single Instruction–Multiple Data (SIMD) machines, followed by algorithms that are appropriate for Multiple Instruction–Multiple Data (MIMD) machines.

But before proceeding to look at various paradigms of parallel computing, we need to make some remarks about the importance of parallelism for computational science. The computer simulation of a physical system is in essence a *numerical experiment* that differs from its laboratory counterpart in a number of crucial ways. In a computer simulation we have the freedom to choose both the type of model we wish to study and the conditions (i.e., temperature, interparticle couplings, etc.) under which we wish to study it, often in ways that are not possible in a laboratory experiment. This gives us an enormous range of phenomena that can be investigated and a tremendous flexibility in investigating even the basic assumptions of our understanding of such systems, although we must recognise also that limitations will be encountered. These limitations, which arise from the finite simulation time, finite system size and so forth, are reflected in the accuracy of the results that we obtain.

The most obvious advantage offered by parallel computing is the promise of vast increases in computer resources, either existing or to become available in the near future. Possibilities still exist for increasing the performance of single-processor machines, but it is now apparent that such increases are clearly limited. By using a large number of processors to solve parts of a problem simultaneously, however, we have the possibility of almost unlimited computing power. The extent to which it is feasible or even possible to achieve this in practice is a question that we will examine here, both in relation to the particular architecture of the parallel computer and the type of algorithm that is implemented. The *speed-up* that it is possible to gain by using a multi-processor machine is a central concept, and we will look more closely at how best to define this quantity and what factors play a role in determining the speed-up that a parallel computer is actually able to deliver in regard to a specific problem.

In addition to this raw increase in computer power, however, the study of parallelism can bring a fresh view of physical processes with it. The inherent parallelism of many processes in nature becomes more transparent and is capable of being reflected in the models and methods we use to understand these physical processes. This new perspective often allows us to see another facet of the problem that was not apparent through serial methods of analysis and solution.

3.1 Paradigms of Parallel Computing

Given the enormous wealth of both architectures and algorithms for parallel computers, it is necessary to provide some general framework for discussing them [3.7, 8]. Rather than discussing all the different schemes that have been devised for such a classification, we will instead look at the two essentially different viewpoints that it is possible to adopt. We have called these the *physics-based description* and the *machine-based description* for reasons that will become apparent shortly.

3.1.1 Physics-Based Description

The physics-based strategy for describing parallel computing begins by looking at the type of parallelism that is contained in the physical processes that are being studied, or the algorithm that is used for their solution. This is essentially the model devised by the Southampton group [3.9], and it is basically independent of the hardware on which one may ultimately want to run the algorithm. The three types of parallelism that have been identified are *event parallelism*, *geometric parallelism* and *algorithmic parallelism*. The machine-based description on the other hand, starts by looking at the architecture of the computer and then describing algorithms in terms of their relationship to the architecture.

These concepts are illustrated by considering the simulation of the Ising model

$$\mathscr{H}_{\text{Ising}}(s) = -J \sum_{\langle ij \rangle} s_i s_j - H \sum_i s_i, \quad s_i = \pm 1, \tag{3.1}$$

where $\langle ij \rangle$ are nearest-neighbour pairs of lattice sites, H is the external magnetic field, and the exchange coupling J is restricted to be positive, and the N-particle system has a Hamilton function

$$\mathscr{H} = \frac{1}{2} \sum_i m v_i^2 + \prod_{i<j} V(r_{ij}), \tag{3.2}$$

where r_{ij} is the distance between particles i and j, and $V(r_{ij})$ is the interparticle potential, such as the Lennard–Jones potential or the electrostatic potential. In a Metropolis update [3.10] of a spin system, it is necessary to calculate the energy between the original spin configuration and a new trial configuration, usually in which one spin has a different orientation. On a serial computer this calculation and the decision to accept or reject the trial orientation are carried out for one spin after another, by moving through the lattice either sequentially or randomly. However, if we allow the possibility of spins being updated simultaneously then we need to give the condition of *detailed balance* some closer consideration. In a system with nearest-neighbour coupling, for example, it is important that no two neighbouring spins are updated simultaneously. In molecular dynamics simulations, on the other hand, the equations of motion are iterated step-wise forward in time and therefore this problem does not arise.

(a) Event Parallelism. This is perhaps the most straightforward type of parallelism, since it involves tasks that can be carried out independently of each other. The only interprocessor communication that is necessary is that of allocating the tasks to particular processors and collecting the results when the tasks are complete. Many of the problems that we encounter in computational science require us to investigate a set of independent parameters $p_1, \ldots, p_m$, which might, for example, consist of a set of temperatures and couplings to scan a phase diagram. Since the parameters are independent they can be investigated simultaneously on a set of processing elements. Alternatively, we may want to gain better statistics for a simulation by replicating the program on the available processors and assigning each processor a different initial configuration. Although this is the simplest form of parallelism it is also extremely efficient, since each processing element is continuously engaged in productive work. It is further possible to divide this category into those cases where each task is exactly identical (i.e., the data may be different but the computation is line-for-line identical) and those where the amount of computation, and indeed even the task itself, is data-dependent and may therefore be quite different on each processor.

(b) Geometric Parallelism. For a system that involves only spatially limited interactions it is natural to divide the volume into equal-sized portions that are assigned to the various processing elements. *Geometric parallelism*, also called *domain decomposition*, can be achieved by partitioning the space in a number of ways. This is also sometimes referred to as *data parallelism*, but this term has been used to describe algorithms that are either geometric *or* algorithmic in nature. It is therefore clearer to reserve the term *geometric parallelism* for those algorithms that involve the partitioning of the physical space in which the simulation takes place.

This is the most widely used type of parallelism and we will discuss various particular algorithms shortly. We shall also address the question of the efficiency in terms of the ratio of the computation that each processor carries out in comparison to the amount of communication. For simulations of lattice structures, such as the Ising model, this is a particularly effective and elegant way of achieving a good *load balancing*, i.e., ensuring that each of the processors has the same computational load. There are, of course, a number of ways of implementing this type of parallelism, depending upon the type of parallel computer available. For fine-grained machines it is possible to assign one spin to each processor, as illustrated on the left of Fig. 3.1. The white and black sites indicate the two interpenetrating sublattices on which the spins can be updated simultaneously using a local-update algorithm without violating detailed balance. This checker-board pattern has also been widely used to perform Monte Carlo simulations on vector machines [3.11]. For coarse-grained parallel computers, on the other hand, each processor is assigned a part of the lattice, such as is shown on the right of Fig. 3.1, where a two-dimensional (2D) lattice is partitioned

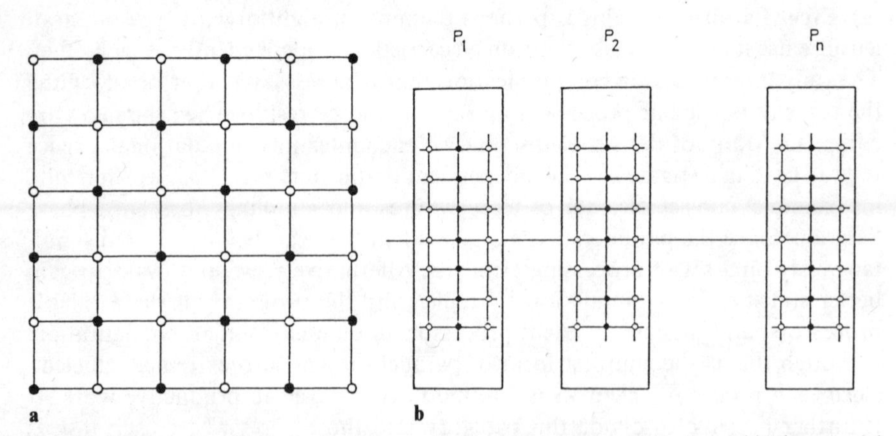

Fig. 3.1. The decomposition of a simple square lattice into two interpenetrating sublattices as used for fine-grained parallel computers or pipeline machines (**a**). Also shown is the decomposition into sublattices as used for coarse-grained parallel processors with local memory and message passing (**b**)

into strips. Again, special consideration must be given to the update of the spins on the boundaries of the strips to ensure that detailed balance is satisfied there. The scheme requires some communication structure between the processors which reflects the type and range of the interactions of the entities being simulated.

A geometric parallelisation is also appropriate for molecular dynamics systems with short-range interactions, i.e., with a cutoff well below the linear extension on the box. In such a case we can partition the volume into *cells* such that only neighbouring cells interact. An algorithm which performs such a partitioning will be discussed shortly.

(c) Algorithmic Parallelism. In algorithmic parallelism it is the algorithm itself which is divided between the processors, much like a production line in a factory. Each processor carries out one part of the computation on the data and then passes its results to the next processor in the line. The speed of such a scheme is determined by its slowest component, which means that considerable effort is required to ensure a good load balancing. This type of parallelisation is sometimes employed in conjunction with one of the other types. A simple example of such a combined algorithm is a Monte Carlo simulation with a geometrical parallelisation, but where each segment of the lattice is assigned *two* processors—one for the update part of the alogrithm and one for generating random numbers [3.12]. This scheme can be extended to situations where there are a set of processors, a *supernode*, associated with each spatial domain of the system. Each processor within this supernode may then have a different function [3.13].

3.1.2 Machine-Based Description

An alternative model for describing parallel computing is based on the design of the machine hardware. The most fundamental distinction between various parallel computers is whether there is a single instruction flow to all the processors or whether each processor has its own independent instruction stream. Machines are further distinguished by the nature of their data storage, the computational performance of each individual processor and the structure of the communication network by which the processors are connected.

(a) SIMD Architecture. In a single instruction–multiple data machine there is only one stream of instructions that is carried out by all the processors synchronously. Such an architecture, which is illustrated diagramatically on the left of Fig. 3.2, is particularly appropriate for problems involving a homogeneous data structure, such as lattice systems. For the Monte Carlo update of a nearest-neighbour spin system we can assign one spin variable to each processor. Detailed balance can be satisfied by updating the *even* and *odd* lattice sites alternately, as shown by the white and black sites in Fig. 3.1.

Existing SIMD machines are very *fine-grained*, i.e., they contain a very large number of processors which are each relatively simple. For example, the ICL

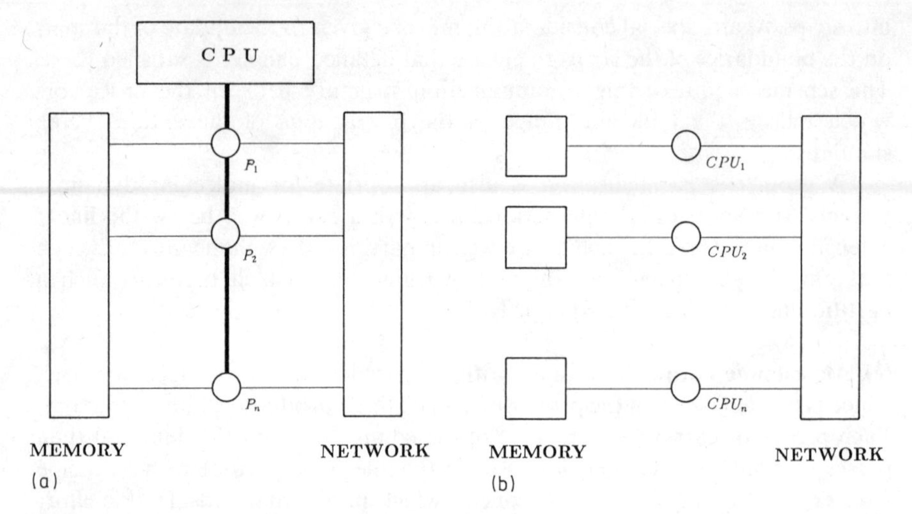

Fig. 3.2. Generic SIMD (**a**) and MIMD machines (**b**)

DAP [3.14, 15] contains 4096 bit-processors connected as a 64 × 64 cyclic 2D grid and the Connection Machine [3.16] contains up to 65 536 bit-serial processing elements which have the topology of a 12D hypercube.

(b) MIMD Architecture. A multiple instruction–multiple data machine allows a greater flexibility in the types of problems of algorithms that can be handled. Each processor can behave as a separate computer in its own right and carry out instructions independently, as illustrated on the right of Fig. 3.2.

MIMD architectures typically allow a wider spectrum of algorithms than SIMD architectures since they have a greater flexibility. Such machines tend to be *coarse-grained*, i.e., they consist of a smaller number of powerful processors connected by high-speed communication network. Such machines can achieve virtually 100% efficiency with replication algorithms, such as ray-tracing [3.17]. For geometrically parallelised algorithms each processor is assigned a portion of the volume, as illustrated on the right of Fig. 3.1. The effectiveness of such a partitioning will depend mainly upon the amount of computation that each processor carries out in comparison to the time spent communicating information to its neighbours. For homogeneous systems, such as lattice–spin systems, this ratio is proportional to the ratio of the volume contained by each processor (which is a measure of the amount of computation done) to the surface area (which is a measure of the amount of communication that must be carried out).

(c) The Connectivity. The traditional way to exchange messages is via a shared memory. However, in a parallel computer there is an alternative option of having the memory distributed over the processors. This solves the problem of two processors simultaneously reading or writing to the same memory location, but at the same time it presents a new problem of how best to connect the processors together to enable them to communicate information with each other.

There are a vast number of possibilities for such communication networks. Solutions range from the use of fast buses, to flat arrays, butterfly networks, hypercubes and even topologies that can be varied. A discussion of all of the possibilities is clearly outside the scope of this article, but in each machine the connectivity and the band-width between the processors will be a determining factor upon the effectiveness of any parallel algorithm.

Another useful concept is that of *granularity* [3.19], which is the amount of computation that a processor can do before it is necessary to communicate with other processors. This concept is clearly dependent upon both the machine and the algorithm.

The above type of classification in terms of the *synchronicity*, *connectivity* and *granularity* can also be applied to physical systems (and their methods of solution) directly [3.18]. For example, the connectivity of a lattice spin model is just its coordination number, whereas a system of charged particles with electrostatic interactions will have a global connectivity since each particle interacts with every other particle. We can likewise define the granularity of an algorithm so that, for example, every spin on a lattice is a grain or each ray in a ray-tracing algorithm is a grain. These grains can be made coarser, for example by grouping together spins into a block as we have seen in geometrically parallelised algorithms.

It will be clear from the above discussion and examples that algorithms and the machines on which they are run are not two different subjects, but rather two facets of the one topic. In order to achieve optimal performance on any parallel machine it is usually necessary to tailor an algorithm to the specific characteristics of the machine. Indeed, many special purpose machines have hardware architectural features that are designed to perform a specific algorithm optimally.

(d) Measurements of Machine Performance. The question of quantifying the performance of a parallel computer is an important but elusive one. It is immediately clear that we need to consider more than just the maximum available computing power (i.e., simply putting together the performances of the individual processors). Other factors, such as the speed of the communication between the processors, play a crucial role.

The *speed-up* is defined [3.20] as

$$S(p) = \frac{\text{time complexity on one processor}}{\text{time complexity on } p \text{ processors}}. \tag{3.3}$$

The optimum that we aim for in designing a parallel algorithm is to achieve a *linear speed-up*. In Fig. 3.3 the results are shown for the speed-up of two algorithms for the simulation of the 2D Ising model [3.21]. The results were obtained on a transputer-based computer by a straightforward partitioning of the lattice into strips in the way illustrated on the right of Fig. 3.1. The figure

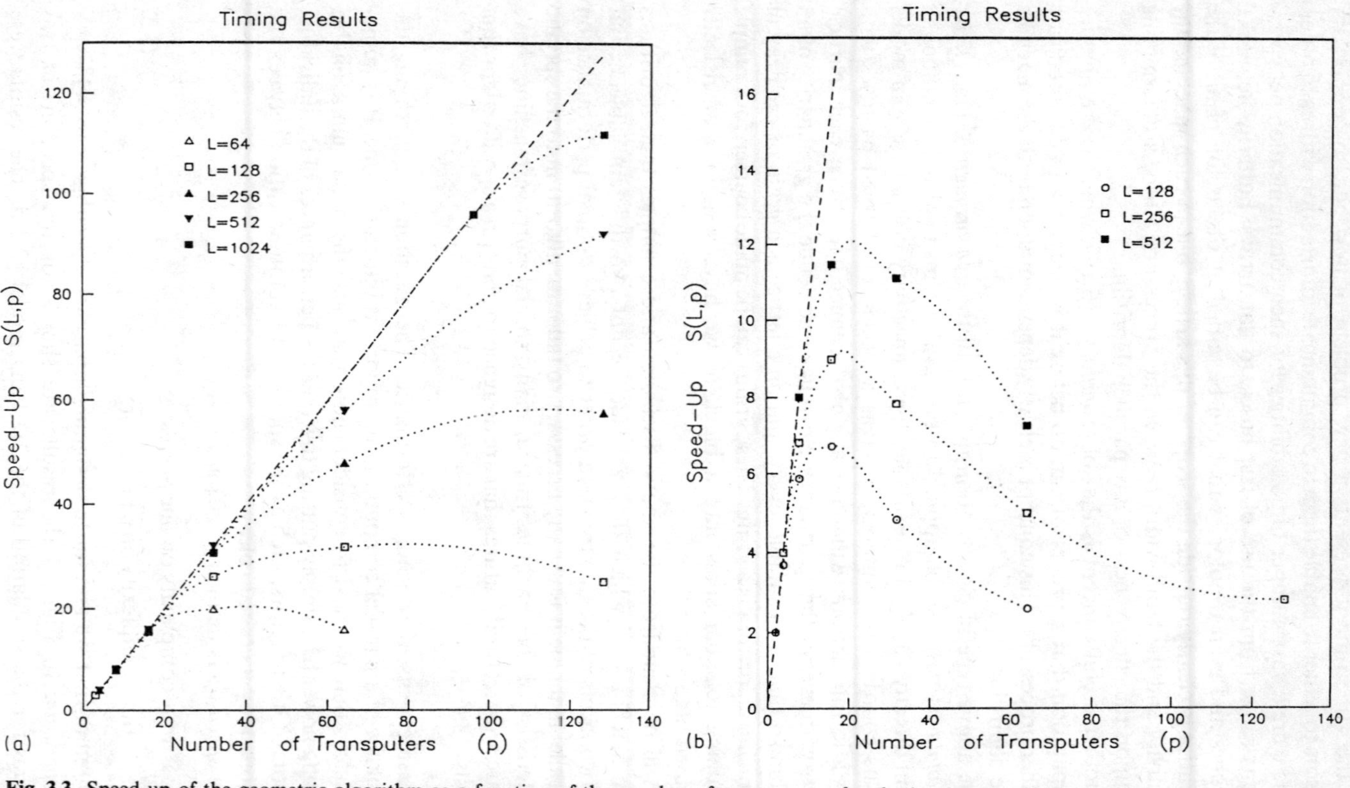

Fig. 3.3. Speed-up of the geometric algorithm as a function of the number of processors p for the local-update Ising problem **(a)**. Different curves denote different lattice sizes as indicated in the legend. The results for the speed-up of the non-local Ising model are also shown **(b)**

on the left shows the results for the local-update Metropolis algorithm and the figure on the right shows the results for a cluster algorithm for the Ising model, which will be discussed shortly.

This notion of speed-up, however, is not without problems. First and foremost, an algorithm designed to run optimally on a single processor will almost certainly look different from an algorithm designed to run on more than one processor. Moreover the communication costs, which play such an important role in practical applications, are disregarded in this definition. Neither the start-up times and latency for communication nor the problem-size dependence are taken into account, although these play an important role in the actual performance. In practice a more pragmatic approach is usually adopted, which depends upon the nature of the algorithm being investigated, and thus needs to be examined on a case-by-case basis. Ultimately it is, after all, the amount of wall-clock time taken that is important in judging the performance of an algorithm-machine pair. This problem of measuring the performance of parallel computers is extensively discussed in the literature [3.7, 8, 22–24].

The question of how to achieve the best possible performance on a parallel computer is even more vexed. In order to make the most effective use of the parallelism in a given situation we ideally require that each individual processing element is continually engaged in carrying out *useful* computation, i.e., both the time spent in communicating between processors and the administrative overhead involved in distributing the simulation over a number of processors should be kept negligible. In this situation we seek to achieve a good load-balancing of the individual processing elements, whereby each of the processors has the same amount of computation to do between the interaction points (i.e., the points in the program where they exchange messages), and therefore no processor has to wait for another processor to complete its part of the computation before proceeding. The need to minimise the time spent in communication between processing elements is clear, since this represents time during which the computation idles, although some types of processors can carry out computation and communication independently. The question of the administrative overheads associated with the parallelisation is rather more subtle, since there are a number of points that need consideration. It is clearly inefficient to duplicate some parts of the calculation on a number of processors, although it is often unavoidable. Likewise it appears inefficient to let any processor sit idle, but there may be efficient algorithms where this is actually necessary. This illustrates the point that such concepts as load balancing are not so much rules to be followed as guidelines to be considered, and every situation requires thinking afresh about the possibilities that it contains. A parallel algorithm will almost inevitably involve more steps (i.e., executed lines of code) than the equivalent scalar algorithm. It is important that this parallelisation overhead, which depends upon both the type of algorithm and the degree of granularity, should not grow excessively as we increase the number of processors.

3.2 Applications on Fine-Grained SIMD Machines

Massively parallel machines consisting of many thousands of processors, such as the ICL (now AMT) Distributed Array Processor (DAP) [3.14, 15] and the Connection Machine [3.16], have been extensively used to study diverse problems in computational science including image-processing, neural networks, cellular automata, quantum field theory (lattice gauge theory), spin systems and molecular dynamics (see [3.25, 26]). But before presenting examples of the last two applications we firstly give a brief outline of the principle hardware features of the two above mentioned machines.

The ICL DAP [3.14, 15] consists of a two-dimensional cyclic grid of 64×64 bit-serial processing elements with nearest-neighbour connections and with row and column broadcasting highways. The DAP is connected to a host machine which is responsible only for initialization and the input/output. Each bit-processor has 4 kbits of memory and has a clock rate of 5 MHz. The AMT DAP, which consists of an array of 32×32 elements, is based on VLSI technology and has a clock rate of 10 MHz. The machines are programmed in *DAP-Fortran*, which includes constructs such as scalars, vectors and matrices of logical (bit), integer and real variables.

The Connection machine [3.16] has a mixed architecture consisting of both 4 bit-processors and vector floating-point Weitek chips (one for each 32 processors). Each processor has 256 kbits of memory and the full version of the machine contains 65 536 processing elements which are connected as a 2^{12} hypercube. The machine can be programmed in parallel dialects of *Lisp*, *C* and *Fortran*. There are two types of communication networks available—a very fast grid network and a router (which enables any processor to communicate to any other in a way transparent to the user).

SIMD machines are particularly well suited for the simulation of lattice problems, and we will discuss two such applications here—spin systems and molecular dynamics—that have been carried out on the DAP and Connection Machine.

3.2.1 Spin Systems

In order to satisfy the condition of detailed balance it is necessary to ensure that only spins on *even* or *odd* sublattices are updated simultaneously, as illustrated earlier Fig. 3.1 (left). This apparently reduces the efficiency of the calculation by 50%, since only half the spins may be updated at any one step. In the simulation of the 2D ϕ^4-model on the Connection Machine [3.27] this problem was solved by explicitly putting one white and one black site on each processor. The virtual-processor facility of the Connection Machine enables the simulation of various lattice sizes without the need for making changes to the program, i.e., it is not necessary to have exactly the same number of lattice sites as processors.

The 2D grid structure of the DAP, together with its bit-processing capability, has made it an ideal machine for simulating the Ising model [3.28]. Systems of size $64 \times 64 \times 64$ have been studied by using one bit-plane of the DAP to represent one 64×64 section of the system, and the periodic boundary conditions are then automatically implemented by the cyclic nature of the DAP. The third dimension is achieved by simply introducing 64 such bit-planes into the machine, and implementing the periodic boundary conditions in this direction is then simply a matter of serial computing.

The problem of ensuring detailed balance without any loss in efficiency is solved by interleaving, or "puckering", adjacent planes, since even-sites on one bit-plane correspond to odd-sites on the neighbouring bit-plane. The efficiency is further increased by the fact that DAP-Fortran allows bit-operations.

An extremely accurate measurement of the critical coupling for the three-dimensional Ising model has been made, $K^c = 0.22654(6)$ [3.28], using the above updating scheme together with a real-space renormalisation group calculation. Rather than describing the method of the renormalisation group or the results obtained, which are given in [3.28], we instead discuss some of the issues relating to the parallelisation of the algorithm. The method proceeds by *blocking* a 64^3 configuration to a 32^3 configuration. This is done by replacing the 8 spins in a 2^3 block by a single spin, which is chosen to be parallel to the majority in the block (or choosen randomly if there are an equal number of up- and down-spins). This blocking process can be repeated further for lattices of size 16^3, 8^3 and so forth. A comparison of the spin–spin correlation functions between the different size configurations enables a very accurate determination of the critical coupling to be made. In blocking from a 64^3 lattice to a 32^3 lattice it would appear that only one in four processors would be used, which would represent a drop in efficiency of 3/4. This can be avoided by continuing the generation of three more 64^3 configurations for blocking and then interleaving these with the first at the 32^3 level. This same procedure can be continued at the further levels of blocking to 16^3 and 8^3 configurations, thus ensuring that the overall efficiency of the algorithm is extremely high.

3.2.2 Molecular Dynamics

An example of a molecular dynamics simulation carried out on a SIMD machine is an early study of the plastic-to-crystalline phase transition in sulphur hexafluoride, SF_6, carried out on the DAP [3.29]. In this study the molecules do not change their neighbours during the course of the simulation and they can be regarded as oscillating around some mean position on a regular lattice. This makes the problem ideal for simulating on a SIMD machine, since the molecules can simply be distributed over the processors in the same way as is done for a lattice–spin system. The octahedral-shaped SF_6 molecules undergo a phase transition upon cooling from the plastic-crystalline phase to the true crystalline phase with a body-centred cubic lattice. During the simulation the molecules undergo orientational displacements which can lead to reorientations. These studies have thrown some light upon our understanding of the plastic phase and the results have been supported by experiment.

3.3 Applications on Coarse-Grained MIMD Machines

The types of algorithms that are suitable for coarse-grained machines are much more varied. It is, of course, possible to implement virtually all the above applications, with appropriate modifications, on coarse-grain machines. However there are also fundamentally different types of algorithms that have been very effectively implemented.

3.3.1 Molecular Dynamics

In molecular dynamics problems [3.3, 30–33] one considers the integration of the equations of motion of monatomic particles or collections of polymeric aggregates. These N particle systems, which for convenience we assume at the moment to be monatomic particles, are usually confined inside a volume V with periodic boundary conditions. The type of parallelisation that it is appropriate to adopt depends upon the range of the interaction between the particles. In the case of a system with long-range interactions each step of the simulation involves the calculation of all $N(N-1)/2$ forces, and an appropriate algorithm for such a situation will be discussed shortly when we consider data-parallel algorithms. After the forces have been calculated for each particle we can then iterate the equations of motion one time-step forward. Here we consider the

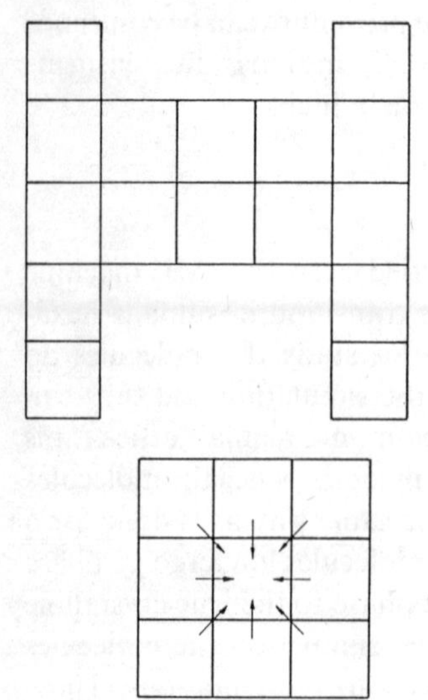

Fig. 3.4. Decomposition of the MD cell into strips or squares

case where the interaction is short ranged, in the sense that the potential range is less than the linear size L of the volume. This enables us to implement a geometric parallelisation.

Each processor is assigned responsibility for a spatial subregion of the volume. If, for purposes of illustration, we consider here a 2D simulation, then there are at least two possible ways to decompose the volume into segments—namely a decomposition into strips or a decomposition into squares, as shown in Fig. 3.4. This can, of course, be straightforwardly generalised to higher dimensions.

The block width, whether for a strip decomposition or a square decomposition, is naturally required to be greater than the range of the interaction between the particles, in order to ensure that only nearest-neighbour blocks need to exchange information. Moreover, if blocks are too thin then there is the possibility that particles can jump between non-adjacent blocks on consecutive time steps. This puts a natural bound on the number of processors that can usefully be employed in any given simulation. The number of processors must,

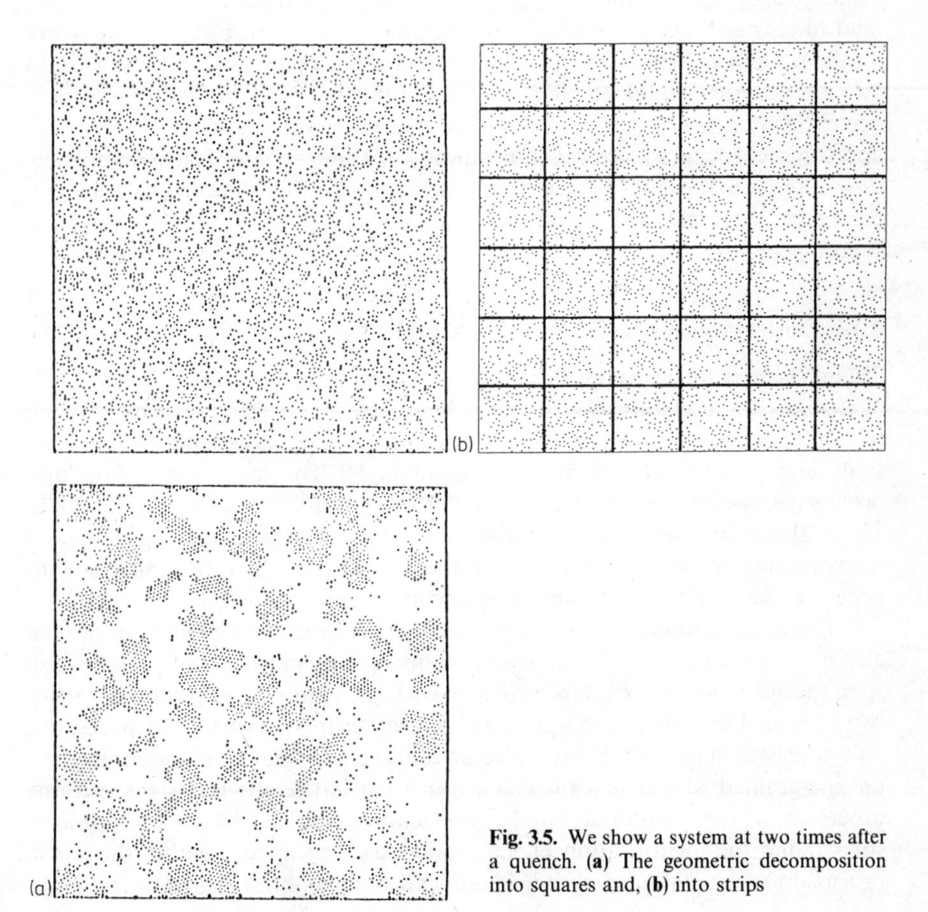

Fig. 3.5. We show a system at two times after a quench. (**a**) The geometric decomposition into squares and, (**b**) into strips

as always, correspond to the system size in such a way that the time spent communicating is negligible or small in comparison with the time spent doing useful computation, i.e., the surface-to-volume ratio must be small.

An efficient parallel algorithm for molecular dynamics based on the above geometric decomposition has been given by *Rapaport* [3.34]. In this algorithm, the most computationally intensive step, namely the calculation of the interactions, is carried out in parallel on all the processors. The information concerning particles near the boundary of a block, and thus within the range of interaction of particles on the neighbouring block, is communicated to the neighbour. The forces acting upon all the particles within each of the blocks can then be calculated and the particle coordinates and velocities iterated one time-step forward. The coordinates of particles which in this time-step move *out* of the block are then given to the appropriate neighbour.

Dynamical load balancing is a very important issue, but one that has hardly begun to be addressed. Consider the situation where the particles are not homogeneously distributed over the volume, as often arises in non-equilibrium situations. In this situation the distribution of the particles is such that they tend to cluster together, as shown in the time sequence in Fig. 3.5. The figure on the left shows how the strips would be filled with particles, and on the right how the squares would be filled. It is apparent that while initially the square decomposition may be best, in the later stages the strip geometry is the more appropriate. In the square decomposition some squares are almost completely unoccupied in the later stages, whereas others are fully occupied. The strip geometry distributes the workload in a much more balanced way than the square geometry.

3.3.2 Cluster Algorithms for the Ising Model

There has recently been considerable interest and progress in developing cluster algorithms for updating spin models. The correspondence between the Ising model and percolation [3.35, 36] has been used by *Swendsen* and *Wang* [3.37] to develop a cluster algorithm for the Ising model. This algorithm has recently been generalised by *Wolff* [3.38] to enable cluster updates of $O(n)$-spin models. These algorithms substantially reduce the effect of critical slowing down for temperatures in the neighbourhood of the critical point, and thus allows us to generate independent spin configurations much more quickly.

The computationally difficult part of this problem is to identify clusters in a spin system that is distributed over a number of processors. Such clusters can span the entire lattice and thus be distributed over many processors. The lattice is partitioned into strips (or squares) and each strip is assigned to a processor. (We note here in passing that it is also possible to carry out cluster identification on fine-grained SIMD machines in which each lattice site is assigned to one processor, as has been done for the connection machine [3.39].) Each processor handles the identification of the clusters as they appear within the single processor, using standard cluster-identification techniques (e.g., Appendix A.3

of [3.40]). In the next step, each processor communicates its result to a neighbour and joins together those clusters which are distributed over two processors. These pairs of processors then communicate their result to the neighbour pair and link their common clusters, and so the algorithm proceeds until we have all the clusters over the entire lattice. On the root level of this binary-tree identification procedure, all clusters are linked together and the identification of the clusters is complete. For a more complete description of the parallelisation of cluster algorithms see [3.41].

As an application of these algorithms we very briefly describe two simulations concerning the problem of the dynamics of the Swendsen–Wang method. These simulations were carried out on two transputer-based parallel computers—the Meiko Computing Surface [3.42] of the condensed matter theory group at the University of Mainz and the Parsytec Supercluster [3.43] at the GMD (Bonn).

We start by considering the quench of a system from a homogeneous equilibrium state to a state underneath the coexistence curve [3.44]. Let $l(t)$ denote the average size of the domains which appear after such a quench. The relaxation into a stable equilibrium state is characterised in most models by the growth law for the average domain size [3.45–7],

$$l(t) \propto t^n . \tag{3.4}$$

The relaxation dynamics are completely different for a dynamics which is induced by the cluster updating procedure. In the local-rule based updating schemes (e.g., based on single-spin *Metropolis* [3.10] or *Glauber* [3.48] update probabilities) domains can only grow by a diffusional motion of the interface. This constraint is removed when one allows arbitrarily large sized portions of the system to change, as for the Swendsen–Wang Ising model [3.37].

The simulations of the quench from a high temperature state to a state below the coexistence curve were carried out for the 2D Ising model on various system sizes ($N = L^2$, $L = 12, 24, 48, 96, 128, 256$) and temperatures ($T/T_c = 0.9$, 0.95, 0.98).

In contrast to the results from the simulations with a local-update rule, the data shows an exponential relaxation

$$l(t) \propto e^{ct} , \tag{3.5}$$

instead of an algebraic relaxation towards the stable equilibrium state, as can be seen in Fig. 3.6 for $T/T_c = 0.95$.

As a second example of an application of parallelised cluster methods we look at the critical slowing down at the critical point of the 2D Ising model [3.44]. We define the normalised autocorrelation function ρ of a quantity A as

$$\rho(t) = \frac{\langle A(0)A(t)\rangle - \langle A\rangle^2}{\langle A^2\rangle - \langle A\rangle^2} , \tag{3.6}$$

where t is the time, which is measured in Monte Carlo simulations [3.2, 3, 5] by the number of sweeps through the lattice. The exponential autocorrelation

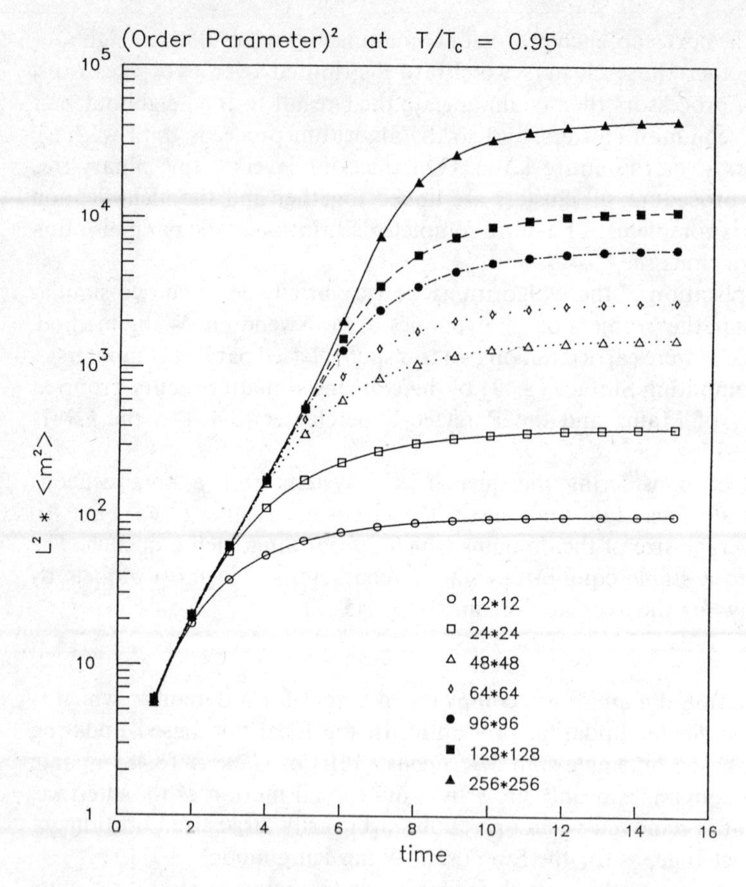

Fig. 3.6. A semi-log plot of the results of the average domain size R and time t as measured in Monte Carlo sweeps. The data presented here is for a temperature $T/T_c = 0.95$. Different symbols denote different system sizes as indicated in the legend

time is defined by

$$\tau_{\exp,A} = \limsup_{t\to\infty} \frac{t}{-\log \rho(t)}, \tag{3.7}$$

and the integrated autocorrelation time by

$$\tau_{\text{int},A} = \frac{1}{2} \sum_{t=-\infty}^{\infty} \rho(t). \tag{3.8}$$

Although these two correlation times are usually the same, there are some models, such as the self-avoiding random walk [3.49], where this is not the case.

At the critical point T_c, one finds for models such as the Glauber Ising [3.48] and Q2R [3.50, 51] (i.e., those with a local dynamics) that the autocorrelation

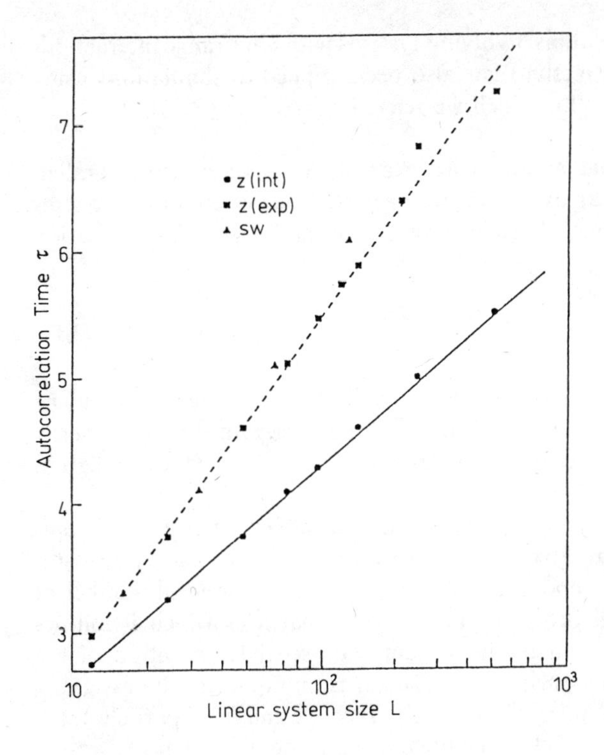

Fig. 3.7. Finite size dependence of the autocorrelation time for the 2D Swendsen–Wang Ising model at T_c. Shown is a semi-log plot of the autocorrelation time versus the linear system size L. The plot includes the data for the exponential and integrated autocorrelation time as well as data from Wang

time varies with the linear system size L as

$$\tau \propto L^z. \tag{3.9}$$

Data for the autocorrelation times are shown in Fig. 3.7. The figure covers two decades in the linear system size, over which it has a linear behaviour on a semi-log plot, indicating that the correlation between successive configurations is governed by a logarithmic law

$$\tau \propto \ln L. \tag{3.10}$$

3.3.3 Data Parallel Algorithms

There are some situations where data parallelisation is to be preferred over geometric parallelisation. A well-known example is a system with long-range interactions, such as an atomic system with electrostatic interactions. In such a system all atoms or particles inside the computational cell interact with each other and all parts of the partitioning must therefore communicate with each other. A geometrical partitioning would neither spread the workload evenly among the processors nor reduce the complexity.

In data parallel algorithms, the system is typically divided into units of groups of particles that have no spatial relationship to each other. We present here two types of such algorithms involving systems with long-range interactions and polymer systems. The method has also been applied to simulations using hybrid molecular dynamics, for which we refer the reader to [3.52].

(a) Long-Range Interactions. In the earlier example of a geometrically parallelised algorithm for molecular dynamics, we restricted our attention to systems with short-range interactions. Consider now a molecular dynamics simulation of a two-body gravitational or electrostatic force

$$F_{ij} = G\frac{m_i m_j}{r_{ij}^2}, \tag{3.11}$$

in which for a system of N particles, there are $N(N-1)/2$ interactions to be calculated at every time step. There is clearly nothing to be gained by a geometric partitioning of the volume since this provides no simplification in the calculation of the interparticle forces.

An algorithm for implementing this problem effectively on a parallel computer has been proposed by *Fox* and *Otto* [3.18, 53] in which each processor, of which we have say P, is randomly assigned a subset of the total number of particles N such that each processor has $n = N/P$ particles which it follows throughout the time evolution of the system. The spatial separation of the particles associated with any particular processor is unimportant. By assigning each processor an equal number of particles (which we call *local* particles) it is straightforward to achieve a good load balancing of the network. The processors are connected in a ring topology and the algorithm proceeds in the following way. Each processor starts by sending the mass and spatial coordinates of one of its local particles to the next processor in the forward direction around the ring, and at the same time, it receives a particle from its neighbour which lies in the other direction. The forces between this *travelling* particle and all the local particles are then computed. The travelling particle is then sent on to the next processor in the ring in a cyclical fashion and the whole procedure is repeated until the complete ring of processors has been visited. This entire procedure must then be carried out for every one of the n particles on each processor and the forces acting on each particle due to all the other particles are accumulated in order to move the particles forward one time step in the usual fashion.

This algorithm will provide an efficient means of parallelising the problem so long as the time spent in communication of the particle masses and positions is small in comparison with the calculation of the forces. This can be arranged simply by ensuring that the granularity is not too fine, i.e., that there are a sufficient number n of local particles on each processor. In such a case this algorithm proves to be extremely effective.

(b) Polymers. The simulation of a single polymer chain on a lattice using a geometric parallelisation of the lattice into strips or squares suffers from the problem of an unbalanced workload among the processors which operate on the parts of

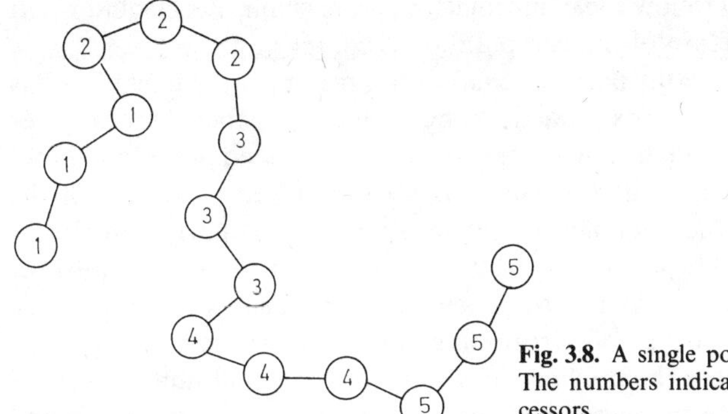

Fig. 3.8. A single polymer in data parallel form. The numbers indicate the mapping to the processors

the lattice. Each segment of the lattice (in this case strips) will contain, in general, an arbitrary number of monomer units. This problem can be overcome by using a data parallel algorithm in which the number of monomer units per processor is fixed.

In the data parallel algorithm, we divide the polymer chain into equal units of groups of monomers. Schematically this is shown in Fig. 3.8. Each processor is then responsible for N/P of the N monomer units, where "monomer units" are to be understood in a very broad sense. In some situations it may be better to use a different partitioning, for example into the repeat units of the polymer. In both cases there will be a fixed number of units per processor. For such a partitioning it is not relevant that the units are sitting on a lattice and we do not distinguish between lattice and off-lattice types of simulations.

The equal division of the data into parcels has the consequence that some of the neighbour relations are preserved here in the polymer problem, whereas for monatomic systems the neighbour relations are completely lost in a data parallel algorithm. Atoms which live on the same processor may be so far apart that they do not interact. In the polymer problem, the preservation of part of the neighbour relations yields a reduction in the amount of computation and communication necessary to generate the next conformation of the entire polymer chain.

The most straightforward configuration of the processors to handle the communication is the ring topology. Since there are possible connections between all data parcels, every processor needs to get data from all the other processors. However there are some polymer problems where the conformation is fairly rigid and this feature can be used to reduce the communication overhead, although this is not generally the case for monatomic systems. In such situations it would be desirable to have an adaptable topology for the processor connections, something which in principle is possible with transputer-based machines. If the conformation is slowly changing, or even fixed, and one only wants to compute some static properties of the system then the processor topology can be adapted to the neighbour relations of the particles, and the processor topology will follow the conformation of the chain. This would result in a considerable reduction of the computational complexity.

In the situation where the polymer chain is no longer simply linear but has side arms, there are several types of partitionings that are possible. If the polymer has a repeat unit where each unit comprises a side arm and some monomer units of the backbone, then it is possible to use this repeat structure of the chain. Each processor then handles a segment that includes part of the backbone and a side chain. The second possibility is to group together a number of monomers, irrespective of whether they belong to the backbone or the side chains. Although the first type of partitioning is usually the most appropriate, there are situations where the polymer may have only a small number of side arms in comparison with the number of monomer units in the entire chain and this type of partitioning would therefore not result in a reduction of the complexity of the computation. In this case, the second algorithm would be much more appropriate. The type of partitioning depends upon the granularity of the processor. If there are only a small number of very powerful processors then the first procedure would be more appropriate.

3.4 Prospects

The rapid development of parallel machines over the past decade has given rise to a vast variety of algorithms covering most branches of computational science, and the rate at which this field is expanding makes predictions about future developments hardly possible. Our aim here has been to present a brief overview of the types of methods that are employed and to give some specific examples. We have divided the presentation into those algorithms that are suitable for SIMD machines and those that are suitable for MIMD architectures. This division, although by no means rigid, reflects the two main streams of development in parallel computing; SIMD machines tend to consist of a very large number of fine-grained processors and they yield an extremely high performance for specialised problems, whereas MIMD machines typically consist of a smaller number of high-performance processors and can be used for a wider variety of algorithms. In order to achieve high performance from a parallel computer it is necessary to design tailored algorithms that are capable of fully utilising the parallel hardware.

Acknowledgments. We would like thank the Höchstleistungsrechenzentrum for their support during part of this work. The fruitful discussions with K. Binder and D. Stauffer, as well as the support from the Sonderforschungsbereich 262 and the Schwerpunkt Schi 257/2-2, are gratefully acknowledged.

References

3.1 K. Binder (ed.): *Monte Carlo Methods in Statistical Physics*, 2nd edn., Topics Curr. Phys. Vol. 7 (Springer, Berlin, Heidelberg 1986)
3.2 K. Binder, D.W. Heermann: *Monte Carlo Simulation in Statistical Physics.* Springer Ser. Solid-State Sci., Vol. 80 (Springer, Berlin, Heidelberg 1988)
3.3 D.W. Heermann: *Computer Simulation Methods in Theoretical Physics*, 2nd edn. (Springer, Berlin, Heidelberg 1990)
3.4 O.G. Mouritsen: *Computer Studies of Phase Transitions and Critical Phenomena*, Springer Ser. Comput. Phys. (Springer, Berlin, Heidelberg 1984)
3.5 M.H. Kalos: *Monte Carlo Methods* (Wiley, New York 1986)
3.6 D.W. Heermann, A.N. Burkitt: *Parallel Algorithms in Computational Science*, Springer Ser. Inf. Sci., Vol. 24 (Springer, Berlin, Heidelberg 1991)
3.7 R.W. Hockney, C.R. Jesshope: *Parallel Computers* (Adam Hilger, Bristol 1981)
3.8 R.W. Hockney, C.R. Josshope: *Parallel Computers 2* (Adam Hilger, Bristol 1989)
3.9 D.J. Pritchard, C.R. Askew, D.B. Carpenter, I. Glendinning, A.J.G. Hey, D.A. Nicole: In *PARLE—Parallel Architectures and Languages Europe*, Vol. I, ed. by J.W. de Bakker, L. Nijman, P.G. Treleaven, Lect. Notes Comput. Sci., Vol. 258 (Springer, Berlin, Heidelberg 1987) p. 278
3.10 N. Metropolis, A.W. Rosenbluth, M.N. Rosenbluth, A.H. Teller, E. Teller: J. Chem. Phys. **21**, 1087 (1953)
3.11 D. Barkai, K.M. Moriarty: Comput. Phys. Commun. **25**, 57 (1982)
3.12 C.R. Askew, D.B. Carpenter, J.T. Chalker, A.J.G. Hey, D.A. Nicole, D.J. Pritchard: Comput. Phys. Commun. **42**, 21 (1986)
3.13 C.R. Askew, D.B. Carpenter, J.T. Chalker, A.J.G. Hey, M. Moore, D.A. Nicole, D.J. Pritchard: Parallel Comput. **6**, 247 (1988)
3.14 S.F. Reddaway: *DAP-A Distributed Array Processor.* 1st Annu. Symp. on Comput. Arch. (IEEE/ACM), Florida, 1973
3.15 D. Parkinson: Comput. Phys. Commun. **28**, 325 (1983)
3.16 W. Hillis: The connection machine. Sci. Am. 108 (June 1987)
3.17 D.F. Rogers: *Procedural Elements for Computer Graphics* (McGraw-Hill, New York 1985)
3.18 G.C. Fox, M.A. Johnson, G.A. Lyzenga, S.W. Otto, J.K. Salmon, D.W. Walker: *Solving Problems on Concurrent Processors*, Vol. 1 (Prentice-Hall, Englewood Cliffs, NJ 1988)
3.19 H.T. Kung: In *Advances in Computers*, Vol. 19, ed. by M. Yovitts (Academic, New York 1980) p. 65–112
3.20 D. Nicol, F.R. Willard: J. Parallel Distrib. Comput. **5**; 404 (1988)
3.21 D.W. Heermann, A.N. Burkitt: Parallel Comput. **13**, 345 (1990)
3.22 I.S. Duff: Comput. Phys. Rep. **11**, 1 (1989)
3.23 A.J.G. Hey: Comput. Phys. Commun. **56**, 1 (1989)
3.24 A.J.G. Hey: In *PARLE '89—Parallel Architectures and Languages Europe*, Vol. II, ed. by E. Odijk, M. Rem, J.C. Syre, Lect. Notes Comput. Sci., Vol. 366 (Springer, Berlin, Heidelberg 1989)
3.25 G.S. Pawley, K. Bowler, R.D. Kenway, D.J. Wallace: Comput. Phys. Commun. **37**, 251 (1985)
3.26 H.D. Simon (ed.): *Scientific Applications of the Connection Machine* (World Scientific, Singapore 1989)
3.27 D.W. Heermann, A.N. Burkitt: In preparation (1990)
3.28 G.S. Pawley, R.H. Swendsen, D.J. Wallace, K.G. Wilson: Phys. Rev. B **29**, 4030 (1984)
3.29 G.S. Pawley, G.W. Thomas: Phys. Rev. Lett. **48**, 410 (1982)
3.30 D.J. Tildesley: In *Computational Physics*, ed. by R.D. Kenway, G.S. Pawley (SUSSP, Edinburgh 1987)
3.31 M.P. Allen, D.J. Tildesley: *Computer Simulation of Liquids* (Clarendon, Oxford 1987)
3.32 W.G. Hoover: *Molecular Dynamics*, Lect. Notes Phys. Vol. 258 (Springer, Berlin, Heidelberg 1986)
3.33 G. Ciccotti, W.G. Hoover (eds.): *Molecular Dynamics Simulation of Statistical Mechanics Systems* (North-Holland, Amsterdam 1986). Proc. of the Int. School of Physics "Enrico Fermi", Course XCVII, Varenna, 1985
3.34 D.C. Rapaport: Comput. Phys. Rep. **9**, 1 (1988)
3.35 P.W. Kasteleyn, C.M. Fortuin: J. Phys. Soc. Jpn. **26** (Suppl.), 11 (1969)
3.36 C.M. Fortuin, P.W. Kasteleyn: Physica (Utrecht) **57**, 536 (1972)

3.37 R.H. Swendsen, J.-S. Wang: Phys. Rev. Lett. **58**, 86 (1987)
3.38 U. Wolff: Phys. Rev. Lett. **62**, 361 (1989)
3.39 P. Tomayo, R. Brower: Boston University Preprint (1990)
3.40 D. Stauffer: *Introduction to Percolation Theory* (Taylor and Francis, London 1985)
3.41 A.N. Burkitt, D.W. Heermann: Comput. Phys. Commun. **54**, 201 (1989)
3.42 Computing Surface: Hardware Reference Manual (Meiko, Bristol 1987)
3.43 Supercluster: Hardware Reference Manual (Parsytec, Aachen 1988)
3.44 A.N. Burkitt, D.W. Heermann: Europhys. Lett. **10**, 207 (1989)
3.45 J.D. Gunton, M. san Miguel, P.S. Sahni: In *Phase Transitions and Critical Phenomena*, Vol. 8, ed. by C. Domb, J.L. Lebowitz (Academic, New York 1983)
3.46 K. Binder, D.W. Heermann: In *Scaling Phenomena in Disordered Systems*, ed. by R. Pynn, T. Skjeltrop (Plenum, New York 1985)
3.47 A. Sadiq, K. Binder: J. Stat. Phys. **35**, 517 (1984)
3.48 R.J. Glauber: J. Math. Phys. **4**, 294 (1963)
3.49 N. Madras, A.D. Sokal: J. Stat. Phys. **50**, 109 (1988)
3.50 G.Y. Vichniac: Physica **10D**, 96 (1984)
3.51 H.J. Herrmann: J. Stat. Phys. **45**, 145 (1986)
3.52 D.W. Heermann, P. Nielaba, M. Rovere: Comput. Phys. Commun. **60**, 311–318 (1990)
3.53 G.C. Fox, S.W. Otto: Phys. Today **37**(5), 50 (1984)

4. New Monte Carlo Methods for Improved Efficiency of Computer Simulations in Statistical Mechanics

Robert H. Swendsen, Jian-Sheng Wang and Alan M. Ferrenberg

In the past few years, new approaches to Monte Carlo simulations have produced substantial improvements in efficiency for both simulation and data analysis. This paper reviews recent developments in simulational methods to overcome problems due to the long relaxation times related to "critical slowing down", under the general category of "cluster methods". We also discuss general "histogram methods" for improving the efficiency of the data analysis. These methods are described along with their advantages, disadvantages and some new applications.

4.1 Overview

Monte Carlo (MC) computer simulations [4.1–5] have been very useful in statistical physics for a number of years, especially in the study of phase transitions. Computer simulations have also had a major impact in a number of other fields, notably lattice gauge theory calculations in high-energy physics, in which the mathematical structure has been similar, even though the underlying scientific problems have been quite different. Simulations are also used to study other systems where information away from a phase transition is important. In biology and chemistry [4.6, 7], simulations are providing new insight into the dynamics of proteins and other complex molecules.

Recently, several advances have been made in both the simulation techniques and the methods of analysis. This review will discuss some of the advances made in each aspect of MC simulation. We will present a particularly successful class of algorithms, known as "cluster" methods, which greatly reduce the difficulties associated with "critical slowing down", as well as discussing recent applications and advances in the use of histograms to increase the information obtained from simulation data.

"Critical slowing down" is probably the most important source of difficulties in the study of phase transitions, as well as being an intrinsically interesting physical phenomenon [4.8]. In the past few years, several approaches have been suggested for overcoming this problem: Fourier acceleration [4.9], multigrid Monte Carlo [4.10–14], algorithms based on renormalisation-group ideas [4.15], and over-relaxation [4.16].

This paper will present an approach [4.17] based on the Fortuin–Kasteleyn mapping [4.18] between Potts and percolation models [4.19]. This type of "cluster" algorithm has proven to be extremely effective in reducing critical slowing down and increasing efficiency [4.20–39].

Although histogram methods have been known for over three decades, their importance has recently been rediscovered with new applications to the study of phase transitions [4.40–71]. Their first great advantage is that they provide the best available techniques for determining the locations and magnitudes of peaks in various thermodynamic functions. This is vital, since finite-size-scaling methods allow accurate and reliable determinations of critical properties from this information, but direct MC data are not well-suited for providing information on narrow peaks. Normally, MC is used to provide estimates for some thermodynamic quantities at particular temperatures. To locate a peak, we then need many high-accuracy simulations, giving us discrete points, none of which is exactly at the maximum.

We will divide the discussion roughly into two parts, with cluster acceleration simulation methods in Sect. 4.2 and histogram methods in Sect. 4.3. To introduce the simulation methods, we will first discuss the physical origin of the phenomenon of critical slowing down. We then describe standard MC simulations, along with its advantages and disadvantages. The Fortuin–Kasteleyn (FK) transformation [4.18], is used to construct a new approach to MC simulations, and review recent developments and extensions and other related algorithms. We then describe the single-histogram technique and the multiple-histogram method, a new method for combining the results of several simulations for improved accuracy over a wider parameter range. We finish the discussion of histograms with a review of their history along with recent developments and applications.

4.2 Acceleration Algorithms

4.2.1 Critical Slowing Down and Standard Monte Carlo Method

To have a specific example, consider the Ising ferromagnet with the Hamiltonian

$$H = K_{\mathrm{I}} \sum_{\langle i,j \rangle} \sigma_i \sigma_j , \tag{4.1}$$

where factors of $-1/k_{\mathrm{B}}T$ have been absorbed into the coupling strength K_{I} $(=J/k_{\mathrm{B}}T)$. The spin on lattice site i is σ_i, which can take on the values $+1$ or -1.

As the temperature is lowered, the probability of neighbouring spins having the same sign increases, and patches of positive and negative spins are formed. At any given temperature all sizes are present up to a maximum size, given by the correlation length ξ. At the critical temperature, $\xi \to \infty$, and the critical properties of the system at the phase transition are closely related to the temperature dependence of the correlation length.

Dynamical behaviour can be measured using time-dependent correlation functions

$$f_E(t-t') = \frac{\langle E(t)E(t')\rangle - \langle E\rangle^2}{\langle E^2\rangle - \langle E\rangle^2}, \tag{4.2}$$

where E is the total energy (or some other property) of the system. The most useful definition of a correlation (or relaxation) time for our purposes is that of the integrated correlation time in units of lattice sweeps,

$$\tau_{\text{int}} = \sum_{t=1} f(t). \tag{4.3}$$

The sum is usually cut off at the first negative value of $f(t)$. The error in the average energy is higher by a factor of $\sqrt{1+2\tau_{\text{int}}}$ [4.72] over what would be obtained from independent estimates.

The correlation length ξ and the correlation time τ are related by a power law

$$\tau \propto \xi^z, \tag{4.4}$$

where z is the dynamical critical exponent. The value of z tends to be about 2, so that the correlation time increases rapidly as the critical temperature is approached. This increase in correlation time is responsible for inefficiency of standard MC simulations near a second-order phase transition.

In standard Monte Carlo, we update spins one at a time. The energy change ΔE necessary to "flip" a single spin is calculated, and the decision of whether or not to carry out the spin flip depends on this energy change.

Such algorithms are extremely efficient in equilibrating local fluctuations. However, the large clusters responsible for critical phenomena are not efficiently treated. Standard MC algorithms only have a significant effect on the boundary of a cluster. It is not too surprising that the diffusion of boundaries over given distance requires a time roughly proportional to the square of that distance giving $z \approx 2$.

For the finite systems used in computer simulations, the clusters are limited by the system size. Even at the critical temperature of the infinite system, the correlation length and correlation time remain finite. However, τ is still related to the size of the largest cluster, so that at T_c

$$\tau \propto L^z, \tag{4.5}$$

where L is the linear dimension. The time needed for a given accuracy grows as $1+2\tau$, so that the computer time grows as L^{d+z}. The factor L^d is not really a problem, since the larger system contains proportionally more information. However, the factor of L^z leads to inefficient simulations for large systems.

4.2.2 Fortuin–Kasteleyn Transformation

In contrast to thermal phase transitions, there is no critical slowing down in percolation transitions [4.19, 73, 74]. A typical (bond) percolation problem consists of a lattice of points with randomly placed bonds (with probability p) between neighbours. For small values of p, only small clusters will be formed. As p increases, the clusters grow. When p reaches a critical value p_c, one of the clusters will span the (infinite) lattice. For $p > p_c$, the largest cluster will occupy a non-zero fraction of the total lattice.

The most interesting feature of this problem from our point of view is that simulations of percolation models generate independent configurations with every sweep, so that the correlation time is exactly zero!

The clear similarity between these problems is due to a rigorous connection. The Fortuin–Kasteleyn (FK) transformation [4.18] maps a ferromagnetic Potts model to a corresponding percolation model.

The Potts model is described by the Hamiltonian

$$H = K \sum_{\langle i,j \rangle} (\delta_{\sigma_i,\sigma_j} - 1), \tag{4.6}$$

where the spin on lattice site i, σ_i, can take on the values $1, 2, \ldots, s$, so that s is the number of states. If $s = 2$, this is equivalent to the Ising model of (4.1). The interactions between Potts spins can have arbitrary range, and the transformation is valid in any number of dimensions on any lattice.

The transformation consists of replacing each interaction between the spins on two sites by either a "bond" (requiring the spins on these two sites to be identical) with a probability

$$p = 1 - \mathrm{e}^{-K}, \tag{4.7}$$

or with a probability $q = 1 - p$ for a bond between sites being absent.

This transformation is performed on all interactions leaving only configurations of bonds forming clusters. All spins on a given cluster must have the same value, but that value is independent of the spins on any other cluster. This problem, with both spins and bonds present, is known as "site-bond percolation", and was introduced through a related transformation by *Coniglio* and *Klein* [4.73].

The spin values can then be integrated out, giving a factor of s for each cluster. Denoting the total number of clusters (including clusters consisting of only a single site) by N_c, the number of bonds by b, and the number of missing bonds by m, we can write the Potts partition function in terms of a percolation problem as

$$Z = \sum_{\text{bonds}} p^b (1-p)^m s^{N_c}, \tag{4.8}$$

which is the main result originally derived by *Fortuin* and *Kasteleyn* [4.18].

4.2.3 Swendsen–Wang Algorithm

To construct a new MC algorithm, we apply the transformation directly to the spin configurations [4.17]. The algorithm consists of two underlying transformations: first, the spin configuration is replaced by a configuration of bonds based on the spins; and next, the bond configuration is used to construct a new configuration of spins. Both steps involve the use of random numbers.

We begin with a configuration of Potts spins. Bonds between nearest-neighbour sites with the same value of the spin are assigned with the probability $p = 1 - \exp(-K)$. No bonds are included between sites with different spin values. This gives a configuration of bonds that form clusters of spins, with all spins in each cluster having the same value.

We then assign new random values to the spins with the sole restriction that the same value be given to every site within a given cluster. The bonds have completed their purpose, and we have a new spin configuration.

The algorithm is clearly ergodic, since there is always a non-zero probability of going from any configuration to any other configuration in a single sweep.

The identification of the clusters is carried out with an algorithm due to *Hoshen* and *Kopelman* [4.75], which requires an amount of computer time proportional to the total number of spins. On a scalar computer, the total time to update the entire lattice once is about twice that for standard MC. On a vector or parallel machine, the problem of efficiently identifying clusters is still being worked on. Although progress has been made, this is still the weakest part of the algorithm [4.76–78].

Fortunately, even with the difficulties in vectorisation, the efficiency of the SW algorithm is so high, that for large lattices two or three orders of magnitude improvement have been achieved. The original calculations [4.17] indicated that the value of the dynamical critical exponent for the $d = 2$ Ising model was reduced from $z \approx 2.1$ to $z_{SW} \approx 0.35$. Later work with larger lattices and better statistics has shown that z is even lower, and *Heermann* and *Burkitt* have suggested that $\tau_{SW} \propto \ln L$, which would imply that $z_{SW} = 0$ [4.36].

For the $d = 3$ Ising model, the improvement is substantial, although not as good as in two dimensions. The dynamical critical exponent is reduced from a value near 2 to $z_{SW} \approx 0.55(3)$ [4.37, 38]. In four dimensions, it appears that $z_{SW} \approx 1$ [4.35].

The extension of the SW algorithm to include magnetic fields can be carried out in two ways [4.17]. In one approach, an extra "ghost" spin, which interacts with every spin in the system with a coupling constant equal to the magnetic field, can be introduced. Alternatively, each cluster can be treated as a single spin in a magnetic field equal to the true magnetic field times the number of sites in the cluster. The heat bath algorithm will then produce efficient relaxation to equilibrium.

Another direct extension of the method is to include negative (antiferromagnetic) interactions for the case of $s = 2$ (Ising model) [4.17]. This is actually an extension of the original FK transformation, noting that for $s = 2$, the condition that the spins are different is a unique specification. We can interpret antiparallel spins in terms of anti-bonds with a probability

$$p = 1 - e^{-|K|}. \tag{4.9}$$

Although this generalisation of the FK transformation is even valid for spin glasses, the corresponding algorithm is inefficient whenever frustration is introduced into the Hamiltonian. Consequently, this approach has been most useful for systems without frustration.

A side effect of the cluster algorithm is that equilibrium configurations of the bond configurations are also produced. This turns out to be extremely useful, since several properties of the thermal system can actually be evaluated more accurately in the percolation representation. This has been exploited by a number of people [4.20, 31, 79–81], beginning with *Sweeny* [4.82], who had developed a method for simulating 2D clusters directly.

4.2.4 Further Developments

Two particularly important advances have been made by *Wolff*. This first was the single-cluster algorithm [4.23]. The same FK transformation is used, but only a single cluster is formed by choosing a single site and creating bonds to that site as in (4.7) (for neighbours with the same spin). This process is continued with the more distant neighbours, until the cluster has reached its limits. This cluster is then flipped. "Time" is proportional to how many sites have been flipped.

In two dimensions, the single-cluster algorithm seems to have the same size dependence as the original SW algorithm, but the correlation time is reduced by a factor of 2. In three dimensions, the single-cluster algorithm has $z_{sc} \approx 0.4$ [4.39], which is better than SW. In four dimensions, the single-cluster algorithm seems to give $z_{sc} \approx 0$ [4.39].

Wolff then made an even more important extension of the cluster approach, which allows it to be applied to general $O(n)$ models [4.23, 25]. He introduced an embedding of an Ising variable into the $O(n)$ model, and simulated the resulting Ising model with a cluster algorithm. For the xy model, the embedding was done by randomly choosing a 2D unit vector and reflecting each spin through the plane described by this vector. For each spin, an Ising value of $+1$ can be assigned to the original direction and -1 to the reflected direction. The effective Hamiltonian has no frustration and can be efficiently simulated. The remarkable feature is that this approach completely eliminates critical slowing down in the low-temperature phase of the $d = 2\, xy$ model [4.25, 26], which is an even better result than has been achieved for the simpler Ising model. The efficiency of Wolff's method is also found for the $d = 3\, xy$ model, as shown recently by *Janke* [4.27].

A similar, although more restricted, embedding of Ising variables was developed independently by *Brower* and *Tamayo* to simulate the ϕ^4 model [4.28]. Two-dimensional embeddings have also been applied by *Wang* et al. for efficient MC simulations of general antiferromagnetic Potts models [4.29, 30]. Two spin values are picked at random and all spins that do not have one of these values are frozen. The unfrozen spins now form an Ising antiferromagnet on a diluted lattice, which is easily simulated with either the SW algorithm, or the single-cluster method. For large s, the spin values can be treated in pairs, reducing the problem at each step to non-interacting Ising antiferromagnets. This approach is very efficient in both two and three dimensions, especially when combined with a multiple-histogram analysis [4.42]. *Hasenbusch* and *Meyer* [4.33] have recently applied the Ising embedding idea to the roughening problem, to create an efficient method for simulating the interface of a 3D Ising model. Their approach has a dynamical critical exponent of only 0.44 ± 0.03, which is much more efficient than either standard Metropolis or a direct application of the SW algorithm.

Our understanding of why these non-local algorithms are so fast (and not faster) is still incomplete. *Klein* et al. [4.34] have suggested an explanation based on an effective updating size, which predicts a relationship with the dynamic critical exponent for Glauber dynamics,

$$z_{\text{sw}} = \frac{z_{\text{G}} - 2\gamma}{d_{\text{m}}\nu}, \tag{4.10}$$

where the mean fractal dimension of finite clusters satisfies $d - \beta/\nu \leqq d_{\text{m}} \leqq d$. However, recent values for z suggest that $z_{\text{sw}}(\text{2D}) = 0$ [4.36], so this equation is not obeyed.

Mean-field calculations for SW dynamics as well as Wolff's single cluster variation give $z_{\text{sw}} = 1$ [4.35] and $z_{\text{w}} = 0$ [4.39] for the Ising models above the upper critical dimension four.

A rigorous lower bound,

$$z_{\text{sw}} \geqq \frac{\alpha}{\nu}, \tag{4.11}$$

has been obtained by *Li* and *Sokal* [4.37], where α is the specific heat exponent.

The cluster algorithms have been applied to a number of problems for high precision calculations: xy model in two dimensions [4.25] and three dimensions [4.83]; dilute Ising models in two and three dimensions [4.84, 85]. The algorithms also stimulated several studies of clusters in the Ising models [4.80, 86–89]. Some nonequilibrium problems are also studied [4.90–93].

Fortran programs of the SW and Wolff algorithms for $d = 2$ Ising model are published in [4.94], and a program of the SW algorithm of the Potts model for arbitrary dimension d and Potts state s is in [4.95].

A number of directions are open for future developments. One of the most interesting is the combination of cluster methods with multigrid techniques. Progress has already been made which shows that significant reductions in correlation times can be achieved by such a combination [4.11, 12]. This approach involves a considerably greater challenge to implement the program, but it also introduces considerable flexibility in designing further improvements. Work is currently in progress in several areas to exploit current advances and develop the new approach further.

4.2.5 Replica Monte Carlo Method

One of the limitations of the cluster methods discussed above is that they are extremely inefficient for the study of phase transitions in systems which contain "frustration"; that is, systems with interactions that have conflicting effects on the alignment of the spins.

The best-known example of a frustrated system is that of the spin glass, in which the interactions between neighbouring spins are quenched and randomly positive and negative. This would be described by the Hamiltonian

$$H = K \sum_{\langle i,j \rangle} B_{ij} \sigma_i \sigma_j , \tag{4.12}$$

where σ_i takes on the values ± 1 and, as usual, the factor $-1/k_B T$ has been absorbed into the coupling constant K. The B_{ij} are dimensionless variables that describe the quenched, random interactions.

For frustrated systems, as discussed in Sect. 4.2.3, the Fortuin–Kasteleyn clusters are not the appropriate ones for developing an efficient algorithm. One approach toward generating efficient clusters is the replica Monte Carlo method [4.96], which has proven very efficient in two dimensions, and helpful in reducing correlation times in three dimensions.

The method simulates several independent "replicas" of the model in (4.12), each at a different inverse temperature, K. The replicas are regarded as forming a $(d+1)$-dimensional system with the Hamiltonian

$$H_R = \sum_{n}^{R} K^{(n)} \sum_{\langle i,j \rangle} B_{ij} \sigma_i^{(n)} \sigma_j^{(n)} . \tag{4.13}$$

Although each replica is statistically independent, we are allowed to make MC moves that involve more than one replica, as long as we satisfy the condition of detailed balance. Specifically, we will make moves based on "replica mixing". We rewrite the Hamiltonians of two neighbouring replicas by introducing new variables

$$\tau_i^{(n)} = \sigma_i^{(n)} \sigma_i^{(n+1)} , \tag{4.14}$$

to obtain

$$H_2 = \sum_{\langle i,j \rangle} B_{ij} [K^{(n)} + K^{(n+1)} \tau_i^{(n)} \tau_j^{(n)}] \sigma_i^{(n)} \sigma_j^{(n)} . \tag{4.15}$$

Changing $\sigma_i^{(n)}$ while holding $\tau_i^{(n)}$ fixed is equivalent to changing both $\sigma_i^{(n)}$ and $\sigma_i^{(n+1)}$. The effective interaction between $\sigma_i^{(n)}$ and $\sigma_j^{(n)}$ in (4.15) depends on the sign of the product $\tau_i^{(n)} \tau_j^{(n)}$. If the product is negative, the effective coupling is proportional to $K^{(n)} - K^{(n+1)}$, which can be small if the temperatures of the two replicas are close together.

We can now use the distribution of the τ's to determine an efficient choice of clusters to flip. The first definition of clusters that was used simply defined neighbouring spins to be in the same cluster if the product $\tau_i^{(n)} \tau_j^{(n)}$ was positive.

This produces clusters of sites for which the spins in neighbouring replicas have the same (or opposite) configuration.

Each cluster is then assigned a value $r_c = +1$, and a cluster Hamiltonian is constructed,

$$H_{cl} = \sum_{\langle b,c \rangle} k_{b,c} r_b r_c. \tag{4.16}$$

The effective interactions $k_{b,c}$ between clusters is proportional to $K^{(n)} - K^{(n+1)}$, since it comes entirely from the boundaries between clusters, for which the product $\tau_i^{(n)} \tau_j^{(n)}$ is negative. This interaction is further weakened by cancellations from positive and negative contributions of the B_{ij}'s.

An even more efficient algorithm can be obtained by applying the SW algorithm within a cluster of τ's with the same sign. This will break the system into still smaller clusters, without introducing more interactions.

Since replica MC satisfies detailed balance, it will not introduce any bias into the simulation, even though it mixes replicas which are statistically independent. However, the procedure is not ergodic, and must be used as part of a simulation that includes standard single-site MC updates. The single-site moves are efficient at high temperatures, and will ensure ergodicity in the total simulation.

Since many temperatures must be simulated at once in the replica MC method, a consideration for when the use of replica MC will be advantageous is whether simulations of that many temperatures are needed. In most cases, a wide range of temperatures is needed. However, if data is only required at a single temperature, the method does have the disadvantage that computer time will be increased proportional to the number of replicas used. The increase in efficiency may still be more than enough to make up for the extra CPU time, but the total gain will not be as great as the improvement in correlation times would indicate.

The method works extremely well in two dimensions. In three dimensions it reduces correlation, but does not eliminate the problem of critical slowing down. The difficulty in three dimensions can be traced to the percolation properties of the clusters formed during replica mixing. At low temperatures, only two large clusters form. This greatly reduces the independence of successive configurations, but still represents an improvement over standard methods.

The replica MC method has been applied to the $\pm J$ Ising spin glass in 2, 3, and 4 dimensions [4.96–98]. It has been used in conjunction with a Monte Carlo renormalisation-group analysis [4.99], which confirmed the absence of a phase transition in two dimensions, and confirmed the phase transition found by *Bhatt* and *Young* [4.100] and *Ogielski* and *Morgenstern* [4.101, 102].

4.2.6 Multigrid Monte Carlo Method

A particularly interesting development in the last two years has been the combination of multigrid methods with cluster Monte Carlo. The initial work was by *Kandel* et al. in 1988, who claimed to have completely eliminated critical slowing down [4.11, 12]. Recent work on the SW algorithm and the Wolff single cluster variation suggests that all of these algorithms have $z=0$, so that their claims are quite probably correct. However, in addition to having $z=0$, they also had very small coefficients, so that their correlation times were extremely short. The price, however, is a considerable increase in complexity of the algorithm, which would not be an important consideration for an extensive study, but would discourage its use for short investigations of new models.

A particularly successful application of multigrid ideas has been the application to the fully frustrated Ising model by *Kandel* with *Ben-Av*, and *Domany* [4.14]. Their method eliminated slowing down, even at $T=0$.

Another approach to implementing multigrid ideas for Monte Carlo simulations [4.10] involves a combination with a kind of inverse Monte Carlo renormalisation-group (MCRG) calculation. In one type of MCRG [4.15], a lattice of size 2^n (where n is some small integer) can be renormalised by defining some rule for creating block spins with a scale factor 2, yielding an new "renormalised" lattice of size 2^{n-1}. The main difficulty involves the accurate calculation of the effective renormalised couplings, which must be obtained to a very high accuracy if the critical properties are to be well simulated. This approach is very appealing, but has not yet produced the same efficiency and flexibility as other methods.

4.3 Histogram Methods

4.3.1 The Single-Histogram Method

Consider a MC simulation where σ represents a configuration or state of the system for the Hamiltonian

$$-\beta\mathscr{H}(\sigma)=K\hat{\mathscr{S}}(\sigma). \tag{4.17}$$

To simplify notation, we will use just a single parameter K, but the extension to more than one parameter is straightforward. The inverse temperature is $\beta=1/k_BT$, and K is the dimensionless coupling constant associated with the operator $\hat{\mathscr{S}}(\sigma)$.

A Monte Carlo simulation of N steps at an inverse temperature K_0 produces a histogram $H(S,K_0,N)$ for values S for the operator $\hat{\mathscr{S}}(\sigma)$. The normalized distribution function,

$$P_K(S)=\frac{H(S,K_0,N)\mathrm{e}^{(K-K_0)S}}{\sum_{\{S\}}H(S,K_0,N)\mathrm{e}^{(K-K_0)S}} \tag{4.18}$$

can then be used to compute the expectation value of any function of S in the neighbourhood of K_0,

$$\langle f(S)\rangle(K)=\sum_{\{S\}}f(S)P_K(S). \tag{4.19}$$

For a finite run, the histogram will be subject to statistical errors

$$[\delta H(S, K_0, N)]^2 = gH(S, K_0, N), \tag{4.20}$$

with $g = 1 + 2\tau$ [4.72] where τ is the correlation time. In particular, the relative uncertainty will become large in the wings of the distribution where the number of histogram entries is small. This is important because of the shift in the peak of the distribution when (4.18) is used. When $|K - K_0|$ becomes too large, the peak in the new distribution will occur for an S value in the wings of the measured distribution resulting in large errors.

For a single-peaked distribution, the width of the peak is related to the specific heat by

$$\delta S \propto [C(K_0)L^d]^{1/2}, \tag{4.21}$$

where L^d is the volume of the system. The shift of the peak position is related to the specific heat and ΔK by

$$\Delta S \propto [L^d C(K_0)]\Delta K. \tag{4.22}$$

The maximum ΔK that gives reliable results occurs when $\Delta S \propto \delta S$, so that

$$\Delta K_{\text{MAX}} \propto [C(K_0)L^d]^{-1/2}, \tag{4.23}$$

which agrees with the observation of *Chesnut* and *Salsburg* [4.41] that the range of reliable results for this method decreases as the system size increases. However, at a critical point another effect must be taken into account. The finite-size scaling region, which usually coincides with the region of the specific heat peak, also decreases as the system size increases. The rate of decrease, for a thermal operator, is given by L^{-y_T}. The specific heat at the peak increases as $L^{\alpha/\nu}$ so that ΔK_{MAX} decreases as $L^{-(\alpha/\nu + d)/2}$. Using the scaling result $\alpha = 2 - d\nu$, ΔK_{MAX} is seen to decrease as $L^{-1/\nu}$. Since $y_T = 1/\nu$, ΔK_{MAX} decreases at the same rate as the finite-size scaling region. This means that if a single simulation provides information over the entire finite-size scaling region for some value of L, similar simulations on larger lattices will also provide information over the full finite-size scaling region.

Away from the critical point, the specific heat is approximately independent of system size so that ΔK_{MAX} decreases as $L^{-d/2}$. (The width of the distribution increases as $L^{d/2}$.) Because most operators of interest are extensive, the number of simulations needed to provide information over the entire range of S increases as $L^{d/2}$.

4.3.2 The Multiple-Histogram Method

The multiple-histogram method optimises the combination of data from an arbitrary number of simulations to obtain information over a wide range of parameter values [4.42–44]. The method provides an optimised combination of data from different sources, produces results in the form of continuous functions for all values of interest, and can be applied to an arbitrary number of simulations. Errors can be calculated and provide a clear and simple guide to optimising the length and location of additional simulations to provide maximum accuracy.

Consider R MC simulations. The nth simulation, with N_n MC updates, is performed at K_n and provides a histogram $H_n(S)$, where S is one of the values in the spectrum of the operator $\hat{\mathcal{S}}$. Each of the R simulations performed will give a different estimate for the density of states, $W(S)$, and an improved estimate can be determined as a weighted sum. This is then optimised, *for each value of* S, by choosing the weights to minimise the error. The important idea of optimising separately for each value of S was first suggested by *Bennett* in connection with the related problem of calculating the difference in free energy between two temperatures when MC data at each of these temperatures is available [4.45].

The only errors in $W_n(S)$ come from statistical errors in the histogram $H_n(S)$ and are given by

$$[\delta H_n(S)]^2 = g_n \overline{H_n(S)}, \tag{4.24}$$

where a bar over an expression indicates the expectation value with respect to all MC simulations of length N_n, and $g_n = 1 + 2\tau_n$ where τ_n is the correlation time for the nth simulation [4.72].

Ferrenberg and *Swendsen* [4.42] made a self-consistent estimate for $\overline{H_n}(S)$, based on the full set of simulations. Another possibility would be to use only the information in the nth simulation and simply replace $\overline{H_n}(S)$ by $H_n(S)$, which was done by *Huang* et al. [4.46, 47]. We believe that the original self-consistent approximation provides a better approximation, although the differences should be fairly small.

The final multiple-histogram equations for the (unnormalised) probability distribution are

$$P(S,K) = \frac{\left[\sum_{n=1}^{R} g_n^{-1} H_n(S)\right] e^{KS}}{\sum_{m=1}^{R} n_m g_m^{-1} e^{K_m S - f_m}} \tag{4.25}$$

where

$$e^{f_n} = \sum_S P(S, K_n). \tag{4.26}$$

Equations (4.25) and (4.26) must be iterated to determine the f_n values self-consistently. As with other Monte Carlo techniques for calculating free energies, these equations determine the free energy to within an additive constant.

This method gives a particularly useful expression for the statistical errors in $W(S)$,

$$\delta W(S) = \left[\sum_n g_n^{-1} H_n(S)\right]^{-1/2} W(S). \tag{4.27}$$

This expression provides a clear guide for planning a series of simulations. The locations and heights of peaks in the relative error, plotted as a function of S, give direct quantitative indications of the optimum locations and lengths of additional MC simulations.

Once the f_n values are determined, (4.25) can be used to calculate the average value of any operator on S as a function of K:

$$\langle A\rangle(K) = \frac{\sum_{\{S\}} A(S)P(S,K)}{\sum_{\{S\}} P(S,K)}. \tag{4.28}$$

4.3.3 History and Applications

The application of histogram methods to MC simulations dates back to 1959, when *Salsburg* et al. [4.48] first wrote down the necessary single-histogram equations and explained how they could be used. This was developed by several people, including *Chesnut* and *Salsburg* [4.41], and was first implemented in a practical calculation by *McDonald* and *Singer* [4.49] in 1967.

In 1972, *Valleau* and *Card* [4.50] broadened the useful temperature range with the idea of *multistage sampling* (MSS) in which supplemental or *bridging* distributions are used to provide information in the wings of the original distribution. The only weak point was the implementation of the particular method for combining information from different simulations.

In 1976, *Bennett* [4.45] made the important contribution of optimising the estimates for the density of states *at each value of the energy*. However, since he was considering a more restricted problem, no general application to the multiple-histogram was made at that time.

In 1977, *Torrie* and *Valleau* [4.51] introduced the method of *umbrella sampling*, which was an attempt at generating wider probability distributions by simulating a different Hamiltonian than the one actually of interest. Although the idea has merit in certain situations, it generally leads to reduced efficiency, since the configurations generated are not those characteristic of the parameters of interest.

An important application of histograms for the study of phase transitions was made in 1982 by *Falcioni* et al. [4.52] and in 1984 by *Marinari* [4.53]. These workers applied the single-histogram method to calculate the zeroes of partition function. This turns out to be extremely useful in providing additional information about the critical behaviour at the transition [4.54]. Subsequently, *Bhanot* and co-workers [4.55–58] used a multistage sampling approach to produce plots of thermodynamic quantities over a wide range of parameters with these methods.

In 1988, *Ferrenberg* and *Swendsen* [4.40] pointed out the high efficiency of the single-histogram equations for determining the location and height of peaks in the neighbourhood of a phase transition. They also noted that the range of validity for single-histograms scales in exactly the right way to cover the same part of the finite-size scaling region for arbitrarily large systems.

In 1989, *Ferrenberg* and *Swendsen* introduced a multiple-histogram method [4.42] which combined the advantages of multistage sampling [4.50] with Bennett's method [4.45] for improved estimates of free-energy differences. Variations of this method have been suggested by *Alves* et al. [4.59].

Histogram methods have been used increasingly in the study of lattice gauge theories. Several groups have applied the approach to the problem of the phase transition in the SU(3) model, indicating that it is first-order [4.60, 61]. The multiple-histogram method has also been used to obtain more accurate information about phase transitions in the SU(2) model [4.62]. *Gottlieb* et al. used histogram techniques to study four-flavour QCD with intermediate-mass and light-mass quarks [4.63].

Histograms have also been applied to the 3D three-state Potts model by *Gavai* and *Karsch* [4.64], who confirmed the first-order nature of the transition even in the presence of antiferromagnetic second-neighbour interactions. Another study by *Fukugita* et al. [4.65] also used histograms to investigate the behaviour of this model.

Janke and *Kleinert* [4.66] used histogram techniques to investigate the Laplacian roughening model, and found evidence for a first-order transition, instead of two Kosterlitz–Thouless transitions.

Some very interesting work by *Caticha* et al. [4.67] used histogram techniques for a Monte Carlo renormalisation-group (MCRG) analysis of the 2D Ising model. They demonstrated that they could calculate RG trajectories both above and below criticality from a single MC simulation. The same group also applied histogram techniques, together with finite-size scaling, to calculate a non-universal critical exponent associated with a linear defect in the 2D Ising model [4.68].

Münger and *Novotny* have also studied the reweighting of Monte Carlo data in an MCRG calculation of an Ising antiferromagnet [4.69]. Their work also included a useful analysis of the systematic errors encountered when calculating specific heats and susceptibilities from fluctuations. As is well known, the fluctuations are biased estimators, but *Münger* and *Novotny* point out that the extent of the bias depends on the extent of the shift in temperature.

Mallezie simulated a quantum hard-rod system with MC methods and used histogram techniques to analyse the results and distinguish between first- and second-order phase transitions [4.70].

In recent studies of antiferromagnetic Potts models, *Wang* et al. [4.29, 30] used a new simulation procedure combined with the multiple-histogram method [4.42] to compute the zero-temperature entropy of the three-state model in both two and three dimensions.

Recent work by *Ferrenberg* and *Landau* [4.71] on a MC simulation of the 3D Ising model has actually duplicated the accuracy of the MCRG calculation of the critical temperature using histogram methods and points to the possibility of going beyond that accuracy in the next stage of the calculation.

4.4 Summary

In this paper, we have reviewed two complementary new approaches to improve the accuracy and efficiency of Monte Carlo computer simulations in the study of phase transitions. Substantial gains have already been achieved, and since these methods represent approaches to the problem of simulation that are still being developed, they promise further improvements in the future. In both cases, a key feature is the lifting of the restriction of thinking about computer simulations as simply "computer experiments," and recognising that there are certain aspects of simulations that can be exploited to greatly increase their value.

Acknowledgements. R.H.S. would like to acknowledge support by the National Science Foundation Grant No. DMR-9009475. J.-S.W. would like to acknowledge support by the Max-Planck Institut für Polymerforschung. A.M.F. would like to acknowledge support by National Science Foundation Grant No. DMR-8715740.

References

4.1 N. Metropolis, A.W. Rosenbluth. M.N. Rosenbluth, A.H. Teller, E. Teller: J. Chem. Phys. **21**, 1087 (1953)
4.2 K.G. Wilson: Phys. Rev. D **10**, 2445 (1974); and in *Recent Developments in Gauge Theories*, ed. by G. 't Hooft et al. (Plenum, New York 1980)
4.3 K. Binder: Phys. Rev. Lett. **47**, 693 (1981)
4.4 K. Binder: J. Comput. Phys. **59**, 1 (1985)
4.5 K. Binder (ed.): *Monte Carlo Methods in Statistical Physics*, Topics Curr. Phys. Vol. 7, 2nd edn. (Springer, Berlin, Heidelberg 1986); K. Binder (ed.): *Applications of the Monte Carlo Method in Statistical Physics*, Topics Curr. Phys., Vol. 36, 2nd edn. (Springer, Berlin, Heidelberg 1987)
4.6 J.A. McCammon, S.C. Harvey: *Dynamics of Proteins and Nucleic Acids* (Cambridge University Press, New York 1987), and references therein
4.7 C.H. Brooks III, M. Karplus, B.M. Pettitt: Proteins: A theoretical perspective of dynamics, structure and thermodynamics, in *Advances in Chemical Physics*, Vol. LXXI, ed. by I. Prigogine, S. Rice (Wiley, New York 1988)
4.8 P.C. Hohenberg, B.I. Halperin: Rev. Mod. Phys. **49**, 435 (1977)
4.9 G. Parisi: In *Progress in Gauge Field Theory*, ed. by G. 't Hooft et al. (Plenum, New York 1984) p. 531; G.G. Batrouni G.R. Katz, A.S. Kronfeld, G.P. Lepage, B. Svetitsky, K.G. Wilson: Phys. Rev. D **32**, 2736 (1985); E. Dagotto, J. Kogut: Phys. Rev. Lett. **58**, 299 (1987)
4.10 J. Goodman, A.D. Sokal: Phys. Rev. Lett. **56**, 1015 (1986); Phys. Rev. D **40**, 2035 (1989)
4.11 D. Kandel, E. Domany, D. Ron, A. Brandt, E. Loh: Phys. Rev. Lett. **60**, 1591 (1988)
4.12 D. Kandel, E. Domany, A. Brandt: Phys. Rev. B **40**, 330 (1988)
4.13 R. Ben-Av, D. Kandel, E. Katznelson, P.G. Lauwers, S. Solomon: J. Stat. Phys. **58**, 125 (1990)
4.14 D. Kandel, R. Ben-Av, E. Domany: Phys. Rev. Lett. **65**, 941 (1990)
4.15 K.E. Schmidt: Phys. Rev. Lett. **51**, 2175 (1983); M. Faas, H.J. Hilhorst: Physica A **135**, 571 (1986); H.H. Hahn, T.S.J. Streit: Physica A **154**, 108 (1988); E.P. Stoll: J. Phys. Condens. Matter **1**, 6959 (1989)
4.16 M. Creutz: Phys. Rev. D **36**, 515 (1987); F.R. Brown, T.J. Woch: Phys. Rev. Lett. **58**, 2394 (1987); S.L. Adler: Phys. Rev. D **38**, 1349 (1988)
4.17 R.H. Swendsen, J.-S. Wang: Phys. Rev. Lett. **58**, 86 (1987)
4.18 P.W. Kasteleyn, C.M. Fortuin: J. Phys. Soc. Jpn. Suppl. **26**s, 11 (1969); C.M. Fortuin, P.W. Kasteleyn: Physica (Utrecht) **57**, 536 (1972)

4.19 D. Stauffer: Phys. Rep. **54**, 1 (1979); J.W. Essam: Rep. Prog. Phys. **43**, 830 (1980)
4.20 U. Wolff: Phys. Rev. Lett. **60**, 1461 (1988); Nucl. Phys. B **300** [FS22], 501 (1988)
4.21 F. Niedermayer: Phys. Rev. Lett. **61**, 2026 (1988)
4.22 R.G. Edwards, A.D. Sokal: Phys. Rev. D **38**, 2009 (1988)
4.23 U. Wolff: Phys. Rev. Lett. **62**, 361 (1989)
4.24 U. Wolff: Phys. Lett. B **228**, 379 (1989)
4.25 U. Wolff: Nucl. Phys. B **322**, 759 (1989)
4.26 R.G. Edwards, A.D. Sokal: Phys. Rev. D **40**, 1374 (1989)
4.27 W. Janke: Phys. Lett. A **148**, 306 (1990)
4.28 R.C. Brower, P. Tamayo: Phys. Rev. Lett. **62**, 1087 (1989)
4.29 J.-S. Wang, R.H. Swendsen, R. Kotecký: Phys. Rev. Lett. **63**, 109 (1989)
4.30 J.-S. Wang, R.H. Swendsen, R. Kotecký: Phys. Rev. B **42**, 2465 (1990)
4.31 M. Hasenbusch: Nucl. Phys. B **333**, 581 (1990)
4.32 Ch. Frick, K. Jansen, P. Seuferling: Phys. Rev. Lett. **63**, 2613 (1989)
4.33 M. Hasenbusch, S. Meyer: Phys. Rev. Lett. **66**, 530 (1991)
4.34 W. Klein, T. Ray, P. Tamayo: Phys. Rev. Lett. **62**, 163 (1989)
4.35 T.S. Ray, P. Tamayo, W. Klein: Phys. Rev. A **39**, 5949 (1989)
4.36 D.W. Heermann, A.N. Burkitt: Physica A **162**, 210 (1990)
4.37 X.-J. Li, A.D. Sokal: Phys. Rev. Lett. **63**, 827 (1989); Phys. Rev. Lett. **67**, 1482 (1991)
4.38 J.-S. Wang: Physica A **164**, 240 (1990)
4.39 P. Tamayo, R.C. Brower, W. Klein: J. Stat. Phys. **58**, 1083 (1990)
4.40 A.M. Ferrenberg, R.H. Swendsen: Phys. Rev. Lett. **61**, 2635 (1988)
4.41 D.A. Chesnut, Z.W. Salsburg: J. Chem. Phys. **38**, 2861 (1963)
4.42 A.M. Ferrenberg, R.H. Swendsen: Phys. Rev. Lett. **63**, 1195 (1989)
4.43 A.M. Ferrenberg, R.H. Swendsen: Comput. in Phys. 101 (Sept./Oct. 1989)
4.44 R.H. Swendsen, A.M. Ferrenberg: In *Computer Simulation Studies in Condensed Matter Physics II*. ed. by D.P. Landau, K.K. Mon, H.-B. Schüttler, Springer Proc. Phys., Vol. 45 (Springer, Berlin, Heidelberg 1990)
4.45 C.H. Bennett: J. Comput. Phys. **22**, 245 (1976)
4.46 S. Huang, K.J.M. Moriarty, E. Myers, J. Potvin: Preprint (1990)
4.47 S. Huang, E. Myers, K.J.M. Moriarty, J. Potvin: Preprint (1990)
4.48 Z.W. Salsburg, J.D. Jackson, W. Fickett, W.W. Wood: J. Chem. Phys. **30**, 65 (1959)
4.49 I.R. McDonald, K. Singer: Discuss. Faraday Soc. **43**, 40 (1967)
4.50 J.P. Valleau, D.N. Card: J. Chem. Phys. **57**, 5457 (1972)
4.51 G. Torrie, J.P. Valleau: Chem. Phys. Lett. **28**, 578 (1974); J. Comput. Phys. **23**, 187 (1977)
4.52 M. Falcioni, E. Marinari, M.L. Paciello, G. Parisi, B. Taglienti: Phys. Lett. **108B**, 331 (1982)
4.53 E. Marinari: Nucl. Phys. B **235** [FS11], 123 (1984)
4.54 C.N. Yang, T.D. Lee: Phys. Rev. **87**, 404 (1952)
4.55 G. Bhanot, S. Black, P. Carter, R. Salvador: Phys. Lett. B **183**, 331 (1987)
4.56 G. Bhanot, K.M. Bitar, S. Black, P. Carter, R. Salvador: Phys. Lett. B **187**, 381 (1987)
4.57 G. Bhanot, K.M. Bitar, R. Salvador: Phys. Lett. B **188**, 246 (1987)
4.58 G. Bhanot, R. Salvador, S. Black, R. Toral: Phys. Rev. Lett. **59**, 803 (1987)
4.59 N.A. Alves, B.A. Berg, R. Villanova: Phys. Rev. B **41**, 383 (1990)
4.60 G.V. Bhanot, S. Sanielev: Phys. Rev. D **40**, 3454 (1989)
4.61 M. Fukugita, M. Okawa, A. Ukawa: Phys. Rev. Lett. **63**, 1768 (1989)
4.62 W. Bock, H.G. Evertz, J. Jersák, D.P. Landau, T. Neuhaus, J.L. Xu: Phys. Rev. D **41**, 2573 (1990)
4.63 S. Gottlieb, W. Liu, R.L. Renken, R.L. Sugar, D. Toussaint: Phys. Rev. D **40**, 2389 (1989)
4.64 R.V. Gavai, F. Karsch: Phys. Lett B **233**, 417 (1989)
4.65 M. Fukugita, H. Mine, M. Okawa, A. Ukawa: KEK preprint 89-74
4.66 W. Janke, H. Kleinert: Phys. Lett. A **140**, 513 (1989)
4.67 N. Caticha, J. Chahine, H.R. Drugowich de Félicio: Phys. Rev. A **40**, 7431 (1989)
4.68 J. Chahine, H.R. Drugowich de Félicio, N. Caticha: Phys. Lett. A **139**, 360 (1989)
4.69 E.P. Münger, M.A. Novotny: J. Phys. A, to be published
4.70 F. Mallezie: Phys. Rev. B **41**, 4475 (1990)
4.71 A.M. Ferrenberg, D.P. Landau: Preprint
4.72 H. Müller-Krumbhaar, K. Binder: J. Stat. Phys. **8**, 1 (1973)
4.73 A. Coniglio, W. Klein: J. Phys. A **13**, 2775 (1980)
4.74 C.-K. Hu: Phys. Rev. B **29**, 5103, 5109 (1984)
4.75 J. Hoshen, R. Kopelman: Phys. Rev. B **14**, 3438 (1976)

4.76 C.F. Baillie, P.D. Coddington: Concurrency: Practice and Experience **3**, 129 (1991)
4.77 H. Mino: Preprint
4.78 A.N. Burkitt, D.W. Heermann: Parallel Comput. **13**, 345 (1990)
4.79 C.-K. Hu, K.-S. Mak: Phys. Rev. B **40**, 5007 (1989)
4.80 M.D. De Meo, D.W. Heermann, K. Binder: J. Stat. Phys. **60**, 585 (1990)
4.81 F. Niedermayer: Phys. Lett. B **237**, 473 (1990)
4.82 M. Sweeny: Phys. Rev. B **27**, 4445 (1983)
4.83 M. Hasenbusch, S. Meyer: Phys. Lett. B **241**, 238 (1990)
4.84 J.-S. Wang, W. Selke, Vl.S. Dotsenko, V.B. Andreichenko, Europhys. Lett. **11**, 301 (1990); Physica A **164**, 221 (1990)
4.85 J.-S. Wang and D. Chowdhury: J. de Phys. **50**, 2905 (1989); J.-S. Wang, M. Wöhlert, H. Mühlenbein, D. Chowdhury: Physica A **166**, 173 (1990)
4.86 Z. Alexandrowicz: Physica A **160**, 310 (1989)
4.87 J.-S. Wang: Physica A **161**, 249 (1989)
4.88 A. Coniglio, F. Liberto, G. Monroy, F. Peruggi: J. Phys. A: Math. Gen. **22**, L837 (1989)
4.89 J.-S. Wang, D. Stauffer: Z. Phys. B **78**, 145 (1990)
4.90 A.N. Burkitt, D.W. Heermann: Europhys. Lett. **10**, 207 (1989)
4.91 D. Stauffer: Physica A **162**, 27 (1989)
4.92 T.S. Ray, P. Tamayo: J. Stat. Phys. **60**, 851 (1990)
4.93 T.S. Ray, J.-S. Wang: Physica A **167**, 580 (1990)
4.94 J.-S. Wang, R.H. Swendsen: Physica A **167**, 565 (1990)
4.95 L.-J. Chen, C.-K. Hu, K.-S. Mak: Comput. Phys. Commun. (1991), Preprint
4.96 R.H. Swendsen, J.-S. Wang: Phys. Rev. Lett. **57**, 2607 (1986)
4.97 J.-S. Wang, R.H. Swendsen: Phys. Rev. B **38**, 4840 (1988)
4.98 J.-S. Wang, R.H. Swendsen: Phys. Rev. B **38**, 9086 (1988)
4.99 J.-S. Wang, R.H. Swendsen: Phys. Rev. B **37**, 7745 (1988)
4.100 R.N. Bhatt, A.P. Young: Phys. Rev. Lett. **54**, 924 (1985)
4.101 A.T. Ogielski, I. Morgenstern: Phys. Rev. Lett. **54**, 928 (1985)
4.102 A.T. Ogielski: Phys. Rev. B **32**, 7384 (1985)

5. Simulation of Random Growth Processes

Hans J. Herrmann

With 17 Figures

Many phenomena, like the shape of a snowflake, the size distribution of asteroids or the roughness of a crack surface can only be understood as the end product of a dynamical process which has random and irreversible ingredients. In the last ten years much progress has been achieved in understanding this kind of phenomenon, by studying so-called "growth models" [5.1–3]. These models are very difficult to deal with analytically and for this reason most of the results obtained come from numerical simulations.

Reversible growth is also of interest, in particular when critical clusters are generated. Cellular automata with a probabilistic rule or random initial condition are, in general, used to achieve this. An example is the study of "damage spreading" in models as simple as the Ising model, in which case the probabilistic growth rule can be heat bath dynamics. One finds dynamical phase transitions that in some cases can be shown to coincide with critical points in equilibrium. At these transitions one finds fractal, dynamical clusters.

In this chapter we want to discuss some cases of growth clusters, present some of their simulation techniques and show results obtained in this way.

5.1 Irreversible Growth of Clusters

Growth models are defined by a growth rule which is applied at discrete time steps. At time $t = 0$ one sets off with a given initial configuration. After a time interval Δt, the rule is applied for the first time and applied again each time an interval Δt has passed. For a simulation on a computer, this is the most practical way of introducing time. The growth rule itself is also conveniently defined via FORTRAN instructions in order to avoid any ambiguity.

5.1.1 A Simple Example of Cluster Growth: The Eden Model

As the first example of a growth rule, let us consider the Eden model [5.4], which, because of its simplicity, is the "Ising model of growth models". It was originally formulated to describe the growth of tumours [5.4] and has also been applied to bacterial colonies [5.5]

Let us consider a regular lattice on which sites can be occupied or empty. The rule of the Eden model that one applies at each time step is to randomly

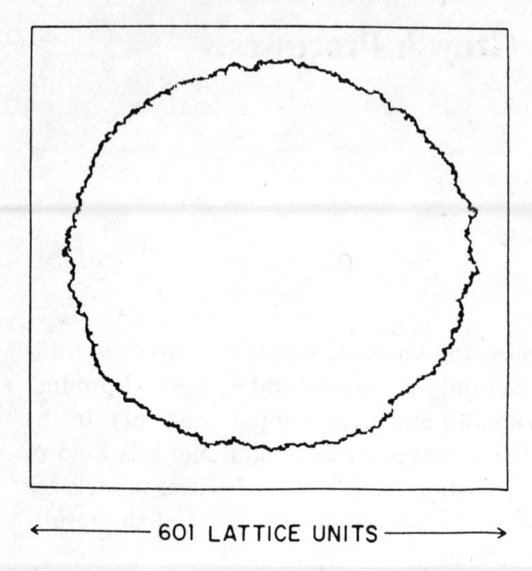

Fig. 5.1. Outer surface of a round Eden cluster of 2×10^5 sites, taken from [5.6]

choose one empty site that is a nearest neighbour of an occupied one, and to occupy it. If at $t = 0$ just one site was occupied one gets, on large scales and after very long times a round cluster as shown in Fig. 5.1, while if at $t = 0$ a whole line (plane in three dimensions) was occupied, one simulates surface growth and finds asymptotically flat surfaces.

More interesting is the scaling behaviour of the width w of the cluster surface defined as

$$w^2 = \sum_i (r_i^2 - \langle r \rangle^2), \tag{5.1}$$

where r_i is the shortest distance of a surface site i from the initial cluster at $t = 0$. In the flat geometry,

$$w = L^\zeta F(h/L^z) \tag{5.2}$$

has been proposed [5.7] where L is the length of the system, $h = Lt$ the average height of the growing surface and F a scaling function. In two dimensions the "roughening" exponent $\zeta = 1/2$ and the dynamic exponent $z = 3/2$ have been found analytically [5.8] while in higher dimensions only numerical estimates are known [5.9, 10].

The straightforward method for simulation of the Eden growth [5.11] is to store the information about which site has been touched (occupied or neighbour of occupied site) in a large lattice and to keep the coordinates of the p perimeter sites in a separate list. At every time step one randomly selects one element of the perimeter list, occupies the corresponding site and removes it from the list, stores the coordinates of the pth element where the newly occupied site was stored before and decreases p by one. Then, one checks for each nearest neighbour of the newly occupied site whether it had been touched before, and if not, one marks it in the lattice as touched, increases p by unity and adds the coordinates of the site to the perimeter list.

We see that the process is sequential because occupying one site depends upon previously occupied sites and therefore it cannot be easily vectorised. For this reason, the time needed to add one site is roughly the same on a scalar CDC-176 as it is on the Cray2 (about 6 μs in two dimensions [5.12]). This is in fact typical for most irreversible growth models. A modified vectorisable Eden growth algorithm has however also been developed on a Cray2 in [5.12]. Sixty-four different perimeter sites were selected simultaneously to be occupied. With a probability that decreases with increasing p, one can then make the error of twice selecting the same site to be occupied. The deviations from the usual Eden process vanish for infinite cluster sizes. If the inner loop is written in assembly language one effectively gains a factor of 4 in speed [5.12].

Computer memory can be saved when the cluster grows from a line by storing only the growth zone in the computer [5.13]. When the perimeter touches the upper boundary of the stored lattice, one determines the line (plane in three dimensions) on which the perimeter site with the smallest coordinate lies, and then shifts all coordinates such that this line becomes the first line. This shifting in computer memory vectorises automatically and can therefore be handled very efficiently. Typically the number of lines needed to contain the growth zone is about twice the w defined above. Another way of gaining memory is to store the information whether a site has been touched or not in a single bit, therefore packing many sites into one computer word. using this, clusters of 10^9 sites have been simulated [5.12]. The decoding of bit information, however, slows the program down (30% in [5.12]).

The results obtained for exponents of [5.2] are not always unambiguous. This is mainly due to slow and complicated convergence towards the asymptotic

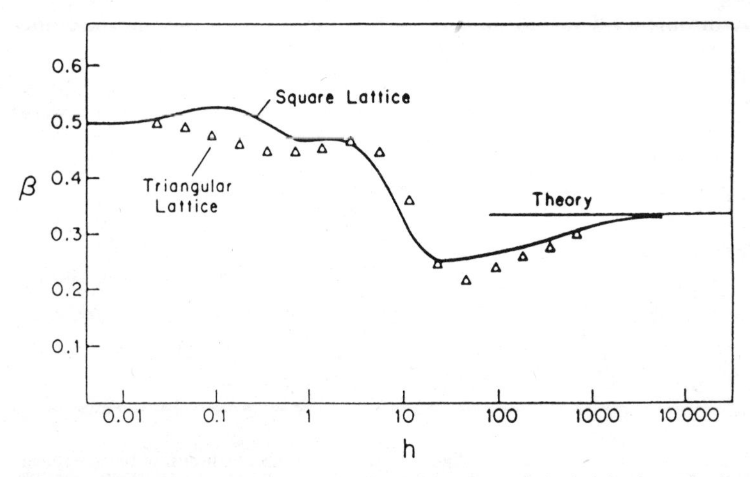

Fig. 5.2. Effective dynamical exponent β as a function of the height h as obtained in [5.12] for square and triangular lattices

behaviour. In two dimensions, controversy existed [5.14, 15] before the exact values were known [5.8]. As an example of slow and non-monotonic convergence, we show in Fig. 5.2 the behaviour of the effective exponent $\beta = d \ln w / d \ln h$ obtained for fixed L as a function of the height h. For large h one expects asymptotically $\beta = \zeta/z = 1/3$ in two dimensions.

It is possible to improve the convergence behaviour by considering different variants of the Eden model [5.10, 15], by explicitly making a correction to the scaling ansatz [5.16] or by using "noise reduction" [5.9, 17]. Nevertheless it has not been possible up to now to reliably confirm or disprove the conjecture $\zeta = 2/(2+d)$ [5.10] in higher dimensions despite much recent effort [5.18].

5.1.2 Laplacian Growth

(a) Moving Boundary Condition Problems. More complicated, but also more realistic, are growth models in which the growth probability on the surface of the cluster depends on some physical properties, which can be described by the solution of some equation valid outside the cluster. A famous example is the Stefan problem, i.e. the solidification in an undercooled melt [5.19]. In this case, the growth of the crystal is determined by the temperature field T which in equilibrium follows the Laplace equation

$$\Delta T = 0 \tag{5.3a}$$

On the external boundary (e.g. at infinity) a temperature T_0 below the melting temperature T_m is imposed (Fig. 5.3). On the interface the Gibbs–Thomson relation implies

$$T_I = T_m \left(1 - \frac{\gamma \kappa}{L}\right), \tag{5.3b}$$

where γ is the liquid–solid surface tension, κ the curvature of the interface and

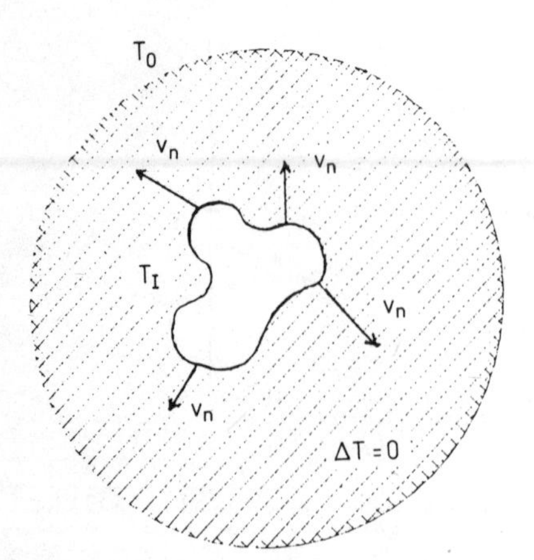

Fig. 5.3. Schematic representation of the solidification in an undercooled melt as a moving boundary problem

L the latent heat. The normal growth velocity v_n of the interface can be obtained through heat conservation, which, neglecting heat diffusion inside the solid, gives

$$v_n = \frac{D_T c_p}{L} \nabla_n T, \tag{5.3c}$$

where D_T is the thermal diffusion coefficient and c_p the specific heat.

Equations (5.3) define what is called a "moving boundary problem": At any instant one has a well-defined Dirichlet boundary condition (fixed values of T) of an elliptic (Laplace) equation and thus a unique solution for the temperature field outside the cluster. This solution determines [through (5.3c)] the motion of the internal boundary in a time interval Δt. The resulting change in the shape of the boundary after Δt defines a new Dirichlet condition and therefore another solution T. Strictly speaking, a given solution is only valid during and infinitesimally short time and the equations should be solved infinitely many times to see a finite growth of the cluster.

Not even the simplest moving boundary problem can be solved yet. The only analytical method that has been successfully applied is (linear) stability analysis of interface perturbations. Different types of instabilities have been found like in the Stefan problem the Mullins–Sekerka instability [5.10], which states that periodic excitations on a flat surface are unstable when their wavenumber is less than $k_s \propto L/\sqrt{D\gamma T_m c_p}$. These instabilities of the growth surface enhance any disorder present in the shape of the interface and are ultimately the origin for the appearance of the non-trivial, fractal clusters which we will discuss in the next section.

Many interesting growth patterns in nature can be formulated as moving boundary problems. Replacing temperature in (5.3) by pressure describes "viscous fingering", i.e. the penetration of a fluid into a much more viscous one (Saffman–Taylor instability). Dielectric breakdown (discharges) seem to follow (5.3) [slightly generalising (5.3c)] in the electrical potential for $\gamma = 0$, and for fracture one solves for the vector displacement field as we will discuss later.

(b) Numerical Simulation of Dielectric Breakdown and DLA. A direct numerical simulation of the moving boundary problem of (5.3) is possible by discretising the moving interface and using a Green's function formalism to solve for the gradient of the temperature field at each point. This method, which has been applied extensively [5.20], unfortunately only allows for resolutions of the boundary with roughly 500 points if one does not want to spend more than several hours on a Cray-XMP. Clusters obtained with such a poor resolution show various branches (about ten) due to the so-called "tip-splitting" instability [5.20] but are too small to be analysed for their fractal properties.

Larger clusters can be obtained by a numerical technique that was first formulated for the "dielectric breakdown model (DBM)" [5.21]:

$$\Delta\Phi = 0, \quad \Phi_I = 0 \qquad \text{and} \qquad v_n = (\nabla\Phi)^\eta, \tag{5.4}$$

where Φ is the electrical potential and η is a phenomenological parameter expressing the fact that no first-principles growth law is known. One considers a regular lattice and on each site a real variable Φ_i which is fixed to be zero on the "occupied" sites of the cluster and Φ_0 on the external boundary of the lattice. On all the other N sites, the discretised Laplace equation

$$\Phi_i = \frac{1}{z} \sum_{nn} \Phi_j \tag{5.5}$$

must be fulfilled, where the sum runs over the z nearest neighbours of site i. Equation (5.5) is a system of N coupled linear equations which can be solved in many ways [5.22]. Among all the (outer) perimeter sites of the cluster one is chosen randomly with a probability proportional to the value of Φ_i on the perimeter site and occupied. Then (5.5) must be solved again and the process is repeated until an occupied site touches the external boundary of the lattice.

By far the most CPU-time consuming part of the algorithm is the solution of (5.5). As a consequence, the growth of a cluster of typically 2000 sites (lattices with 2×10^4 sites) needs roughly one hour on a Cray-XMP. A straightforward but inefficient way of solving the equations is relaxation. One relaxation step consists in setting on each site the value of Φ_i given by (5.5). This can be done sequentially or simultaneously (Jacobi method). Although one needs many relaxation steps to reach equilibrium, in some cases one may consider that an approximate solution is good enough and thus only perform a small number of relaxation steps each time one site is grown. This is the case, for example, when one considers, as in [5.21], that in a finite system the noise from the probabilistic rule dominates over the details of the electric field. This approach can considerably reduce computer time (20000 site clusters in 1 hour of one Cray-XMP processor) but one must beware of the systematic errors which will show up particularly on large scales. If, however, one wants to know the solution of (5.5) with good precision, for instance to know the distribution of growth probabilities on the surface [5.23], the best method to use is the conjugate gradient technique [5.22]. This method finds the exact solution after N^3 steps but in practice one only needs about 50 steps to reach precisions of the order of 10^{-10} for the system sizes one can consider. For $N \geqq 10^4$ the CPU time can be decreased further using multigrid methods or Fourier acceleration [5.24]. We note, that all these methods vectorise very well [5.25].

The clusters obtained with the above method are big enough to assure that they are fractals, i.e. that their mass M (number of sites) grows with the radius R of the cluster as

$$M \propto R^{d_f}. \tag{5.6}$$

However, in order to reliably determine the value of the fractal dimension d_f with an accuracy of better than 20%, larger clusters are needed. This can be achieved through a very different algorithm using random walks which in the Laplacian version is known as "diffusion limited aggregation (DLA)" [5.26]. From a randomly selected position on a hyperspherical surface of radius R_m which just encloses the cluster, a particle is launched to start a random trajectory.

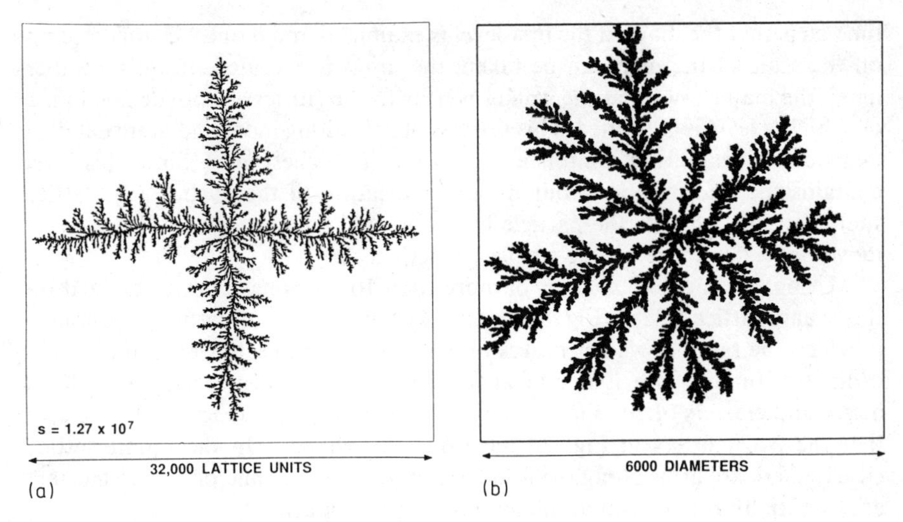

Fig. 5.4. DLA clusters: (**a**) 1.27×10^7 sites on the square lattice, (**b**) 10^7 sites off-lattice. Courtesy of Paul Meakin and Peter Ossadnik (Ref. 5.82)

If it touches the cluster it sticks irreversibly on its surface. After the particle has stuck or moved a distance R_f away from the cluster, a new particle is launched. In principle R_f should be infinite but even for $R_f = 2R_m$ no detectable distortions of the clusters have been observed as compared to $R_f = 100\,R_m$ in $d = 2$. The irreversible aggregation of particles forms fractal clusters (Fig. 5.4). The above described diffusion process is a discretisation of a Laplace equation in the density of diffusion particles and corresponds to the moving boundary problem of (5.4).

With the above algorithm it is possible to easily generate clusters of 50000 particles on the square lattice, although the algorithm is strictly sequential so that most of the data have been generated on scalar computers (e.g. on an IBM 3081 for Figs. 5.4 and 5.5). Over the years it has even been possible to improve the implementation of the algorithm on the computer, in particular in two dimensions [5.27, 28]. The most dramatic gain in speed is obtained by allowing the particles to take longer steps when they are in empty regions. If the particle is at the centre of an empty hypersphere then it will emerge at a random position on its surface after just one time step. In the case of a hypercube, the particle will emerge on the surface with a probability proportional to the Laplacian Green's function on the surface.

In order to efficiently keep track of the empty regions, especially the ones inside the cluster, a hierarchical map of regions was used in [5.29]. The system (lattice or continuum) is divided into k regions (maps) of equal size ($k = 4$ on the square lattice) which constitute the first level of the hierarchy. Each map is again subdivided into k maps (second level) and so on. On each level the random walker is located on a well defined map. To see how it can jump at the next time-step, first the map on the first level is examined and if one sees that a jump on the scale of the map can be taken, the jump is executed. If, on the other hand, the map shows that the walker is near to the cluster a more detailed map on a higher level is consulted. This process of consulting more and more detailed maps continues until a jump is executed on the highest level. Since this level contains the information about the exact location of the particle (e.g. lattice site) it reveals whether the particle has contacted the cluster or if it has to be moved one step more on the smallest length scale.

Using this method, clusters of more than 10^7 sites have been obtained on the square lattice [5.28] (Fig. 5.4a). One can also make off-lattice simulations in which the random walker moves on the continuum, i.e. has real number co-ordinates. In this case it is crucial to use the above-mentioned hierarchical maps, and clusters of 10^7 sites have been generated [5.82] (Fig. 5.4b). One sees that the two clusters in Fig. 5.4 have different shapes: On the square lattice clearly there are arms along the four axis of the lattice, while off-lattice clusters are essentially round with an increasing number of arms.

The calculation of the fractal dimension can be obtained by monitoring the dependence of the radius of gyration R_G of the cluster as a function of its number of sites s while it grows. The local slope β in a log-log plot is then the inverse fractal dimension. We show in Fig. 5.5 that β converges rather quickly to its asymptotic value for 2D off-lattice DLA, giving $d_f = 1.715 \pm 0.004$ [5.29].

Not all quantities converge as fast as the fractal dimension. For the variance ξ of the deposition radius, i.e. the distance that a sticking particle is from the centre of the cluster, one expects [5.14] $\xi \sim s^{\bar{\nu}}$. In Fig. 5.5 we see the convergence of the effective $\bar{\nu}$ as the cluster size increases obtained from the same clusters as the curve for β in this figure. Evidently $\bar{\nu}$ has not yet reached its asymptotic value for clusters of $s = 10^6$.

On the square lattice, clusters are cross-shaped on large scales (Fig. 5.4a), an effect that can be enhanced by using noise reduction techniques [5.30].

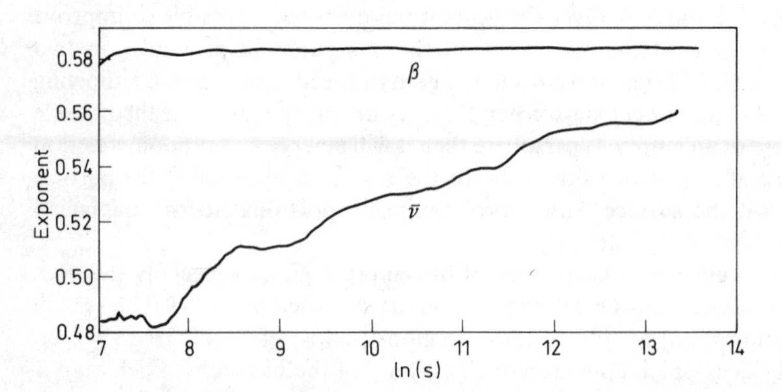

Fig. 5.5. Effective inverse fractal dimension β and width exponent $\bar{\nu}$ against the logarithm of the number s of sites considered within an off-lattice cluster of 10^6 sites as taken from [5.29]

Instead of defining a global fractal dimension for such clusters, it is better to distinguish between a direction along the arms and a direction perpendicular to the arms. The first is given by the exponent $\nu_{\parallel}$, which describes the growth of the length of the arm as function of its mass, and the second is given by $\nu_{\perp}$, which describes the growth of the width of the arm. $\nu_{\parallel} \approx 2/3$ and $\nu_{\perp} \approx 1/3$ have been found [5.28] for the square lattice.

In higher dimensions it is more difficult to grow big clusters so that, for instance, for off-lattice DLA only clusters of 350 000 sites in $d = 3$ and of 25 000 in $d = 7$ are available [5.29]. Correspondingly, the fractal dimensions are known with less accuracy except for $d = 3$, where $d_f = 2.485 \pm 0.005$ has been obtained [5.29].

Many variants of the DLA model have been investigated and also many other properties of DLA clusters have been studied in great detail, such as the distribution of growth probabilities on the surface, the anisotropy effects on lattices or the geometrical multifractality. We will not enter into these aspects here because they have been presented extensively in [5.3].

(c) Fracture. The breaking of solids due to an external load is a problem of evident technological importance and which has been studied very intensely [5.31]. Let us restrict our attention here to elastic solids, i.e. to what is called "brittle" fracture. In this case the medium is described on a mesoscopic level by the equations of motion of the displacement field **u**. In order to study how a void in the material becomes a crack and how this crack grows, it is useful to formulate a moving boundary problem in **u**. The equation of motion of the elastic solid is the Lamé equation

$$\nabla(\nabla \cdot \mathbf{u}) + (1 - 2\nu)\Delta \mathbf{u} = 0. \tag{5.7a}$$

which consists in fact of d coupled (elliptic) equations (ν is the "Poisson ratio"). On the external boundary, the condition is given by the imposed load. On the internal boundary, i.e. the crack surface, the condition is that the stress $\sigma_{\perp}$ (force) perpendicular to the surface be zero. Using Hooke's law this condition can be written in terms of displacements as

$$\partial_{\perp} u_{\parallel} + \partial_{\parallel} u_{\perp} = 0, \quad \text{and} \quad (1 - \nu)\partial_{\perp} u_{\perp} + \nu \partial_{\parallel} u_{\parallel} = 0, \tag{5.7b}$$

where $u_{\perp(\parallel)}$ are the components of **u** and $\partial_{\perp(\parallel)}$ the directional derivatives perpendicular (parallel) to the crack surface. The normal growth velocity v_n will be a function of the difference between the stress $\sigma_{\parallel}$ parallel to the crack surface and a material dependent cohesion strength σ_c. Since no first-principles law is known, one assumes the general behaviour

$$v_n = c(\sigma_{\parallel} - \sigma_c)^{\eta}, \tag{5.7c}$$

where c and η are constants.

The above moving boundary problem again has to be discretised for numerical treatment. A good way to do it is using the beam model [5.33]. Let us consider a square lattice. On each site i, one places three continuous degrees of

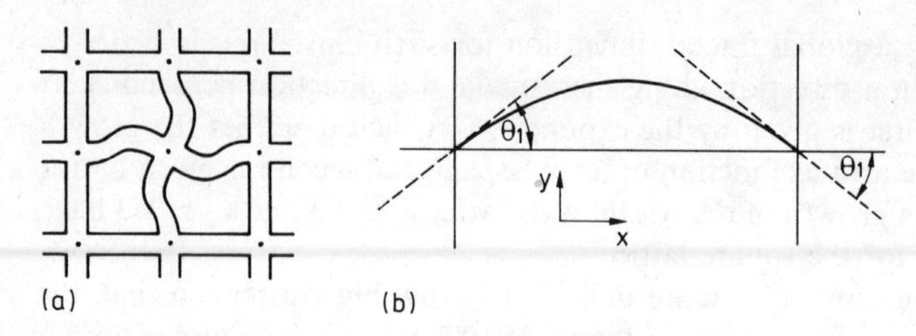

Fig. 5.6. Schematic representation of the beam model. (**a**) Rotation of one site at the centre of the lattice. (**b**) Beam flexed due to rotations at its extremities

freedom: the two displacement co-ordinates x_i and y_i and $z_i = l \cdot \varphi_i$, where $\varphi_i \in [-\pi, \pi]$ is a rotation angle and the lattice spacing l is usually set to unity. Nearest neighbouring sites are connected through a "beam" in such a way that it joins site i forming an angle φ_i with respect to the underlying undistorted lattice. In Fig. 5.6a we see the form of the beams when just the site in the middle is rotated. Each beam is assumed to have a finite cross-section A so that besides a traction force f along its axis one also has a shear s and moments m_i and m_j at its two ends i and j. These can be calculated [5.34] in linear (elastic) approximation. For a horizontal beam one obtains in two dimensions;

$$f = a(x_i - x_j), \tag{5.8a}$$

$$s = b[y_i - y_j + (z_i + z_j)/2], \tag{5.8b}$$

$$m_i = cb(z_i - z_j) + \tfrac{1}{2}b(y_i - y_j + \tfrac{2}{3}z_i + \tfrac{1}{3}z_j), \tag{5.8c}$$

where $a = EA, b = [(GA)^{-1} + (l^2/12)(EI)^{-1}]^{-1}$ and $c = EI/(l^2GA)$ with a Young's modulus E, a shear modulus G and a moment of inertia I. Analogous expressions are found for a vertical beam. To obtain for each site three discretised equations, one imposes that the sum of horizontal and vertical force components and the sum of the moments be zero at this site.

For the growth rule one can use

$$p = [f^2 + r \cdot \max(|m_1|, |m_2|)]^\eta \tag{5.9}$$

as the quantity to which the breaking probability is proportional. Here r measures the susceptibility of the material to breaking through bending as compared to stretching.

On top and on bottom of the lattice an external shear is imposed. In the centre of the lattice one removes one beam which represents the initial microcrack. Next we consider the six nearest neighbour beams of this broken beam, i.e. the two beams that are parallel to the broken beam and the four perpendicular beams that touch a common site with the broken beam. The beam model can be solved by a conjugate gradient method [5.22] to very high precision (10^{-20}) and the p's of (5.9) are calculated for each of the nearest neighbour beams.

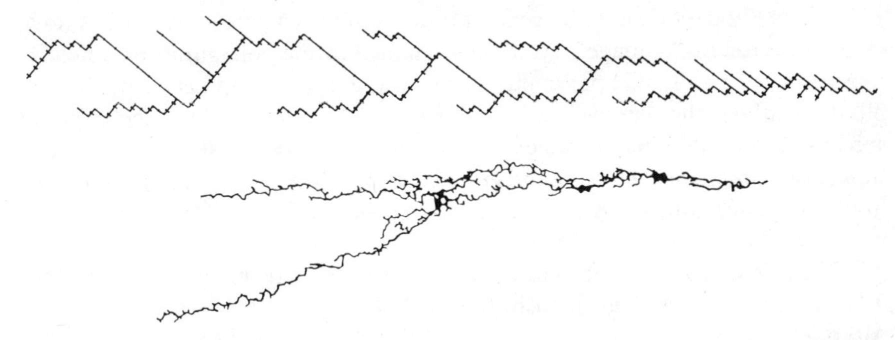

Fig. 5.7. Numerical and experimental cracks. The upper shape is the upper half of a crack grown in a 118×118 system under external shear with $r = 0$ and a vertical initially broken beam using $f = 1$ and $\eta = 0.2$. The lower picture shows the morphology of cracking in Ti -11.5 Mo -6 Zr -4.5 Sn aged 100 hrs at 750° K, and tested in 0.6 M LiCl in methanol at 500 mV under increasing stress intensity (taken from [5.39])

With this method, clusters of 2000 sites can be produced using several Cray-XMP hours. The clusters obtained in this way with the beam model and also using other discretisations of (5.7), are fractal [5.35]. Their fractal dimension is lower than that found for DLA, in agreement with experiments that can produce both problems ((5.3) and (5.7)) in one setup [5.36]. It is, however, not possible to give precise numbers because the clusters are too small. A stochastic algorithm analogous to the diffusion process of DLA is in principle possible [5.37], but has not yet been implemented.

It is also possible to grow deterministic cracks [5.38] by breaking the beam which has the largest value of

$$\bar{p}(t) = p(t) + f \cdot p(t-1), \tag{5.10}$$

where $p(t)$ is the value of (5.9) at time-step t and f is a memory factor which models the action of corroding agents. Stress corrosion cracking is determining for the breaking of glass and the ageing of planes. Only in the presence of memory ($f \neq 0$) do the cracks become fractal [5.38]. In Fig. 5.7, we see an example of such a deterministic crack compared to an experimental crack. The fractal dimension depends on η: $d_f = 1.3$ for $\eta = 1.0$ and $d_f = 1.1$ for $\eta = 0.2$.

5.2 Reversible Probabilistic Growth

5.2.1 Cellular Automata

Cellular automata are dynamical systems with many discrete degrees of freedom. Each cell that contains one discrete variable has a rule which defines its state at the next time step as a function of the values of the neighbouring cells and the cell itself. Cellular automata produce patterns in space and time that can be moving fractal clusters if the dynamics is chaotic.

To see if the dynamics of a cellular automaton is chaotic, one watches how easily an error or "damage", i.e. a small change in the configuration, spreads. In models with a tunable parameter one can find a transition between a frozen phase, in which the damage heals and a chaotic phase, in which the damage spreads to infinity. One example is the Kauffman model [5.40, 41], where the transition can be located using the gradient method [5.42] by letting a damage front spread in a gradient. Another example are thermal Monte Carlo simulations which will be discussed extensively later.

The rule of a cellular automaton can be deterministic or probabilistic. The first case has been studied intensively [5.43, 44] for binary states, e.g. 0 and 1. Although deterministic they can be disordered, either in the choice of the rule, like in the Kauffman model [5.40, 41], or in the initial configuration like in Q2R [5.45]. In all cases binary, deterministic automata can be coded in the computer as one-site one-bit ("multispin-coding") and then as many sites treated in parallel as there are bits in a computer word using logical bit-by-bit operations. In addition, the inner loops of the programs can be easily vectorised so that very high performances in speed can be achieved (1.1 GHz and 3.2 GHz for Q2R [5.46] on a Cray-2 processor and a full CM-2 respectively and 1.7 GHz for Q2R and 200 MHz for the Kauffman model [5.47] on one processor of a Cray-YMP). Very big systems can also be simulated due to the multi-spin-coding (1.5×10^{10} sites for Q2R [5.46]). More details on these programming techniques applied to the Kauffman model can be found in this book in Chap. 10.

Probabilistic automata use random numbers at each time-step. Examples are the Hamiltonian formulation of the Kauffman model [5.48] and annealed models [5.41, 49]. But the most prominent example is thermal Monte Carlo dynamics to which the rest of this chapter is devoted.

5.2.2 Damage Spreading in the Monte Carlo Method

Monte Carlo can be viewed as a dynamical process in phase space, i.e. starting from a given initial configuration one follows a trajectory in phase space, applying the Monte Carlo procedure. This trajectory will of course depend on the specific type of Monte Carlo (heat bath, Glauber, Metropolis, Kawasaki, etc.) but will also depend on the sequence of random numbers. The trajectories will always go towards equilibrium which means that the configurations will be visited by the trajectory with a probability proportional to the Boltzmann factors. Once in equilibrium, the trajectories will stay there as assured by the detailed balance condition.

We want to ask whether such dynamics is chaotic. More precisely, suppose we make a small perturbation in the initial condition. Will the new trajectory be just slightly different or will it be totally different? The second case in which two initially close trajectories will quickly become very different is generically called chaotic. The detailed definition of chaos may in some cases also include the speed with which the trajectories separate, but in our context we do not want to make this distinction.

In order that the concept of closeness of trajectories be meaningful one must have a definition of distance in phase space. If we consider a system of Ising variables, then a useful metric can be given by the "Hamming distance" or "damage"

$$\Delta(t) = \frac{1}{2N} \sum_i |\sigma_i(t) - \rho_i(t)|, \tag{5.11}$$

where $\{\sigma_i(t)\}$ and $\{\rho_i(t)\}$ are the two (time-dependent) configurations in phase space and N is the number of sites, labelled by i. This definition is certainly arbitrary and the results that we present in the following depend on this definition. Physically it just measures the fraction of sites for which the two configurations are different. The definition of (5.11) can easily be generalised to continuous variables [5.50] or to variables of q-states [5.51].

According to the discussion above, one would then call a dynamical behaviour chaotic when in the thermodynamic limit $\Delta(t)$ goes to a finite value for large times if $\Delta(0) \to 0$. The opposite of chaotic is called frozen, i.e. when $\Delta(\infty) = 0$ if $\Delta(0) \to 0$. If one wants to get the limit $\Delta(0) \to 0$ properly in a numerical simulation one can do the following [5.52]: Consider three configurations $\{\sigma_A\}$, $\{\sigma_B\}$ and $\{\sigma_C\}$ with $\Delta_{AB}(0) = \Delta_{BC}(0) = \frac{1}{2}\Delta_{AC}(0) = s$, where s is a fixed small number and Δ_{AB} denotes the distance between $\{\sigma_A\}$ and $\{\sigma_B\}$. Then

$$\Delta(t) = \Delta_{AB}(t) + \Delta_{BA}(t) - \Delta_{AC}(t) \tag{5.12}$$

is a very good extrapolation to $\Delta(0) \to 0$.

The crucial idea in order to study damage spreading in Monte Carlo is therefore to apply, on two configurations, the *same sequence of random numbers* in the Monte Carlo algorithm. In this way one is taking the same dynamics for both configurations. To get statistically meaningful results one must then average over many initial configurations of equal initial distance and over many different sequences of random numbers. Also note that, as in deterministic dynamical systems, once the two configurations become identical they will always stay identical.

5.2.3 Numerical Results for the Ising Model

In order to see if the Monte Carlo method can generate chaotic behaviour in the sense outlined above, Glauber dynamics was applied in [5.52] on the 2D Ising model. Using (5.12), the distance of initially close equilibrium configurations was calculated after 10^4 updates per site, i.e. after a long time. The result is shown in Fig. 5.8 as a function of temperature. Let us remark that since the calculation was performed in a finite system of size L, the final distance would have vanished after a time of the order of $\exp(L^2)$, because eventually two uncorrelated trajectories in a finite phase space will always meet. The times considered in Fig. 5.8 are, however, much smaller than these "Poincaré" times. We see from Fig. 5.8 that above a certain temperature T_s, the Glauber dynamics is chaotic while below T_s it is frozen. The order parameter for this transition between

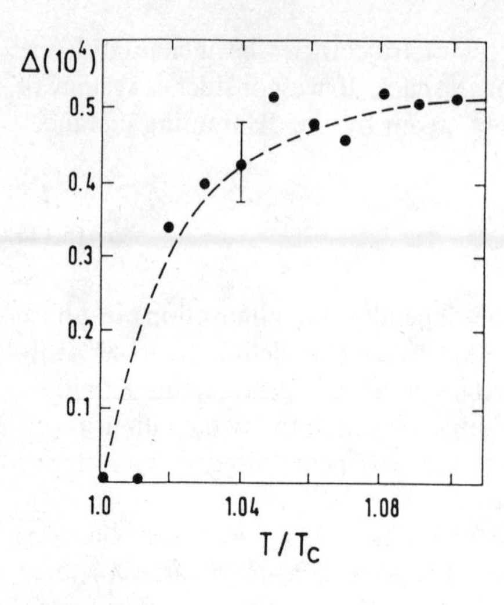

Fig. 5.8. Hamming distance 10^4 time step for $\Delta(0) \to 0$ using Glauber dynamics as a function of T/T_c for the 2D Ising model [5.52]

chaotic and frozen is the quantity plotted in Fig. 5.8. The data indicate that T_s is very close if not identical to the critical temperature T_c of the Ising model. It has not been possible up to now to determine the critical exponent β of this transition because the statistical fluctuations are very strong.

In 3 dimensions using Glauber dynamics qualitatively the same picture emerges as in 2 dimensions, only the temperature T_s seems to be 4% below T_c as found in [5.53]. Particularly interesting is the fact that one finds a transition line even in a homogeneous field [5.54], as shown in Fig. 5.9. This clearly indicates that the dynamic transition between chaotic and frozen is not identical to the transition between the ferromagnetic and the paramagnetic phase. The transition line of Fig. 5.9b neither agrees with the percolation transition of minority spins [5.54]. It is thus an open question whether this dynamical transition is related to any known property of the Ising model or is a novel phenomenon.

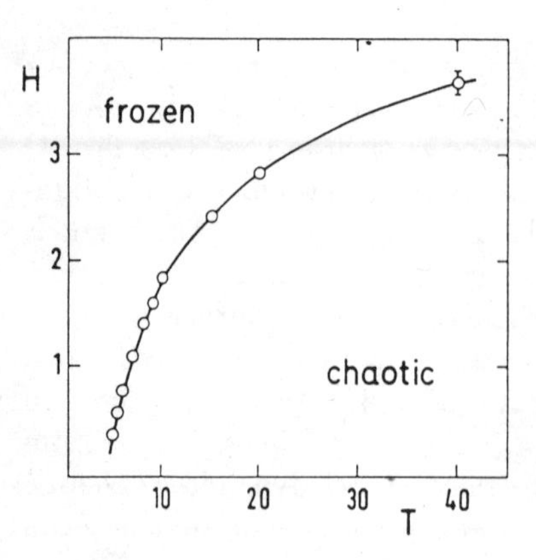

Fig. 5.9. Phase diagram in the field—temperature plane between the chaotic and the frozen phase for the 3D Ising model obtained by *Le Caër* [5.54] for Glauber dynamics in a 10^3 lattice

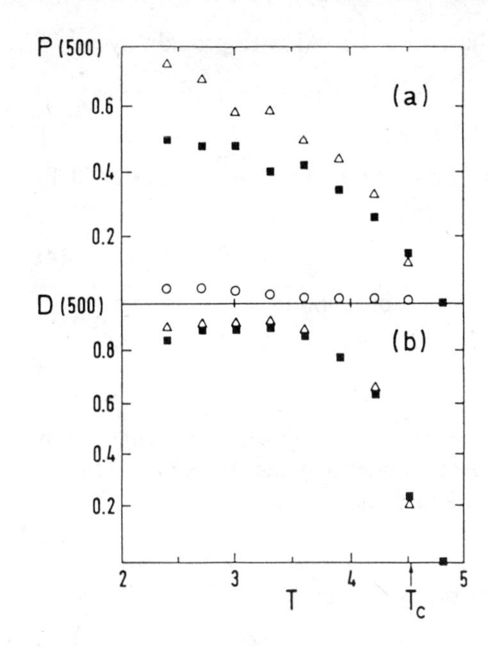

Fig. 5.10. Damage after 500 time steps in the 3D Ising model, as a function of temperature for $L = 12$ from [5.55] for various initial damages: $\Delta(0) \to 0$ (○), $\Delta(0) = \frac{1}{2}$(□) and $\Delta(0) = 1$(△); (**a**) fraction $P(t)$ of non-identical configurations and (**b**) distance $D(t)$ between two configurations provided they are still different

The same questions have been addressed using a slightly different dynamics, namely heat bath dynamics [5.55]. The fraction $P(t)$ of the pairs of configurations that had not yet become identical after a time t, and the average distance $D(t)$ between only those not yet identical configurations, were calculated, ($\Delta(t) = P(t) \cdot D(t)$). In Fig. 5.10 we show the result with not thermalised configurations and looking at different values of initial damage. After a time of 500 updates per site, one sees that the survival of damage depends on the initial damage (Fig. 5.10a) while when the configurations are different, their distance does not depend on the initial distance (Fig. 5.10b). We also see that if the initial distance goes to zero, the final distance is also going to vanish because of $P(t)$. This is in agreement with the result found for Glauber dynamics in the ferromagnetic phase. In the paramagnetic phase, however, the result for heat bath is strikingly different from the Glauber dynamics because, as seen from Fig. 5.10, heat bath does not show chaotic behaviour but on the contrary is in a frozen phase in which the final distance vanishes even if the initial distance was large. The opposite behaviour of the two dynamics is very surprising because normally it is thought that heat bath and Glauber dynamics are identical for the Ising model. We will investigate this question in the next section.

5.2.4 Heat Bath Versus Glauber Dynamics in the Ising Model

Let us consider variables $\sigma_i = \pm 1$ and define as $h_i = \sum_{nn} \sigma_j$ the local field acting on σ_i that comes from its nearest neighbours. Then one update in heat bath is given by setting the new value σ_i' of the spin to be $+1$ with probability p_i:

$$p_i = \frac{e^{2\beta h_i}}{1 + e^{2\beta h_i}}. \tag{5.13}$$

On the computer this is implemented by choosing a random number $z \in [0, 1]$ and setting

$$\sigma_i' = \operatorname{sign}(p_i - z). \tag{5.14}$$

In Glauber dynamics a spin is flipped with a probability

$$p(\text{flip}) = \frac{e^{-\beta \Delta E}}{1 + e^{-\beta \Delta E}}, \tag{5.15}$$

where ΔE is the difference between the energy of the would-be new configuration and of the old configuration. For the Ising model one has $\Delta E = 2\sigma_i h_i$. On the computer, one implements Glauber dynamics via

$$\sigma_i' = -\sigma_i \operatorname{sign}[p(\text{flip}) - z]. \tag{5.16}$$

Using (5.13) and (5.15), p(flip) can be expressed in terms of p_i by

$$p(\text{flip}) = \begin{cases} 1 - p_i & \text{if } \sigma_i = +1 \\ p_i & \text{if } \sigma_i = -1, \end{cases}$$

and (5.17)

$$p(\text{not flip}) = \begin{cases} p_i & \text{if } \sigma_i = +1 \\ 1 - p_i & \text{if } \sigma_i = -1. \end{cases}$$

One sees from (5.17) that for the Glauber dynamics, σ_i' is set $+1$ with probability p_i, and -1 with probability $1 - p_i$, just as in heat bath so that both dynamics have exactly the same probabilities.

Inserting (5.17) into (5.16) one finds how Glauber dynamics is implemented on the computer;

$$\begin{aligned} \sigma_i' &= \operatorname{sign}(z - (1 - p_i)) \quad \text{if } \sigma_i = +1 \\ \sigma_i' &= \operatorname{sign}(p_i - z) \qquad\quad \text{if } \sigma_i = -1. \end{aligned} \tag{5.18}$$

This means that depending on the value of σ_i, one uses the random number differently. So, if in one configuration the site σ_i was $+1$ and in the other configuration it was -1, but the nearest neighbours in both configurations are the same, the damage at site i will probably survive in Glauber dynamics while it will certainly heal in heat bath. This gives rise to the different behaviour in damage spreading between the two methods.

5.2.5 Relationship Between Damage and Thermodynamic Properties

Damage for heat bath can be related to correlation functions as has been shown in [5.56]. In the following we will briefly report on these relationships. We will consider Ising variables $\sigma_i = \pm 1$ which can also be expressed, as usual, by $\pi_i = \frac{1}{2}(1-\sigma_i) = 0, 1$.

In order to produce a small damage between configurations $\{\sigma_A\}$ and $\{\sigma_B\}$, we will fix the spin at the origin of configuration $\{\sigma_B\}$ to be always

$$\sigma_0^B = -1. \tag{5.19}$$

Thus this represents a source of damage. In principle two types of damage can be imagined at site i, damage $+-$ where $\sigma_i^A = +1$ and $\sigma_i^B = -1$ or damage $-+$ where $\sigma_i^A = -1$ and $\sigma_i^B = +1$. The probabilities of finding a certain type of damage at site i can then be expressed as

$$d_i^{+-} = \langle\!\langle (1-\pi_i^A)\pi_i^B \rangle\!\rangle \quad \text{and} \quad d_i^{-+} = \langle\!\langle \pi_i^A(1-\pi_i^B) \rangle\!\rangle, \tag{5.20}$$

where $\langle\!\langle \cdots \rangle\!\rangle$ denotes a time average. Let us define the difference between the damage and use (5.20);

$$\Gamma_i = d_i^{+-} - d_i^{-+} = \langle\!\langle \pi_i^A \rangle\!\rangle - \langle\!\langle \pi_i^B \rangle\!\rangle, \tag{5.21}$$

where we have also used (5.10).

We want to express the damage through thermodynamic quantities defined on an unconstrainted system $\{\pi_i\}$ with averages taken over many configurations. So we translate condition (5.19) by a conditional probability and use ergodicity to go from time averages to thermal averages. Thus

$$\langle\!\langle \pi_i^A \rangle\!\rangle = \langle \pi_i \rangle \quad \text{and} \quad \langle\!\langle \pi_i^B \rangle\!\rangle = \frac{\langle \pi_i(1-\pi_0) \rangle}{\langle 1-\pi_0 \rangle}, \tag{5.22}$$

where $\langle \cdots \rangle$ denotes a thermal average. Inserting this into (5.21) one finds

$$\Gamma_i = \frac{\langle \pi_i \pi_0 \rangle - \langle \pi_i \rangle \langle \pi_0 \rangle}{1 - \langle \pi_0 \rangle} = \frac{C_{0i}}{2(1-m)}, \tag{5.23}$$

where

$$C_{0i} = \langle \sigma_i \sigma_0 \rangle - \langle \sigma_0 \rangle^2 \quad \text{and} \quad m = \langle \sigma_0 \rangle \tag{5.24}$$

are just the correlation function and the magnetisation. Equation (5.23) relates a certain combination of the damages with thermodynamic functions. In its derivation we have used ergodicity but have not made any assumptions about the dynamics, the random numbers or the type of interaction and it is therefore of a very general validity.

If one had chosen another form of fixed damage the result would have changed slightly. For instance fixing $\sigma_0^B = -1$ and $\sigma_0^A = +1$ one would find $\Gamma_i = C_{0i}/(1-m^2)$.

If one considers variables with more degrees of freedom than Ising variables things can become more complicated but are still in principle feasible. As an example let us look at the Ashkin–Teller model [5.57] where on each site one has two binary variables $\sigma_i = \pm 1$ and $\tau_i = \pm 1$ and a Hamiltonian per site

$$\mathcal{H}_{ij} = -K(\sigma_i \sigma_j + \tau_i \tau_j) - 2L\sigma_i \sigma_j \tau_i \tau_j. \tag{5.25}$$

In this case there are twelve possible damages per site which we label such that left means configuration A, right configuration B, top means σ and bottom means τ; for example $\left(\begin{smallmatrix}+&-\\-&+\end{smallmatrix}\right)$ is $\sigma_i^A = +1$, $\tau_i^A = -1$, $\sigma_i^B = -1$ and $\tau_i^B = +1$. It can be shown [5.51] that if one fixes $\sigma_0^B = -1$ and $\tau_0^B = +1$ or $\sigma_0^B = +1$ and $\tau_0^B = -1$ then

$$\Gamma_i = \frac{\langle \sigma_i \tau_i \sigma_0 \tau_0 \rangle - \langle \sigma_i \tau_i \rangle \langle \sigma_0 \tau_0 \rangle}{2(1 - \langle \sigma_0 \tau_0 \rangle)} \tag{5.26}$$

with

$$\Gamma_i = \left(d_i^{\left(\begin{smallmatrix}+&+\\+&-\end{smallmatrix}\right)} + d_i^{\left(\begin{smallmatrix}+&-\\+&+\end{smallmatrix}\right)} + d_i^{\left(\begin{smallmatrix}-&-\\-&+\end{smallmatrix}\right)} + d_i^{\left(\begin{smallmatrix}-&+\\-&-\end{smallmatrix}\right)} \right) - \left(d_i^{\left(\begin{smallmatrix}+&+\\-&+\end{smallmatrix}\right)} + d_i^{\left(\begin{smallmatrix}+&-\\-&-\end{smallmatrix}\right)} + d_i^{\left(\begin{smallmatrix}-&+\\+&+\end{smallmatrix}\right)} + d_i^{\left(\begin{smallmatrix}-&-\\+&-\end{smallmatrix}\right)} \right), \tag{5.27}$$

and if one fixes $\sigma_0^B = -1$ then

$$\tilde{\Gamma}_i = \frac{\langle \sigma_i \sigma_0 \rangle - \langle \sigma_i \rangle \langle \sigma_0 \rangle}{2(1 - \langle \sigma_0 \rangle)} \tag{5.28}$$

with

$$\tilde{\Gamma}_i = \left(d_i^{\left(\begin{smallmatrix}+&-\\+&-\end{smallmatrix}\right)} + d_i^{\left(\begin{smallmatrix}+&-\\+&+\end{smallmatrix}\right)} + d_i^{\left(\begin{smallmatrix}+&-\\-&+\end{smallmatrix}\right)} + d_i^{\left(\begin{smallmatrix}+&-\\-&-\end{smallmatrix}\right)} \right) - \left(d_i^{\left(\begin{smallmatrix}-&+\\+&+\end{smallmatrix}\right)} + d_i^{\left(\begin{smallmatrix}-&+\\+&-\end{smallmatrix}\right)} + d_i^{\left(\begin{smallmatrix}-&+\\-&+\end{smallmatrix}\right)} + d_i^{\left(\begin{smallmatrix}-&+\\-&-\end{smallmatrix}\right)} \right). \tag{5.29}$$

Thus both types of correlation functions that one has in the Ashkin–Teller model can be expressed as a combination of damages.

The damage for which we have presented numerical data in the preceding section was not the quantity Γ_i but the sum of all the damages

$$\Delta = \sum_i \Delta_i \quad \text{with} \quad \Delta_i = d_i^{+-} + d_i^{-+}. \tag{5.30}$$

In order to express these quantities in terms of thermodynamic quantities it is necessary to make some assumptions. Let us therefore now restrict ourselves to ferromagnetic interactions, heat bath dynamics and the use of the same random numbers for both configurations. We again consider only Ising variables and fix the damage as in (5.19), i.e. $\sigma_0^B = -1$. Since at $t = 0$ the damage is of type $+-$ at the origin, we have

$$p_i^A \geqq p_i^B \quad \text{for all } i, \tag{5.31}$$

where p_i^A is the value defined in (5.13) for configuration A. Suppose one were to try to create a damage of type $-+$ at site i. Then one would need, in order to produce $\sigma_i^A = -1$, a random number z which fulfills $z \geqq p_i^A$ according to heat bath. Since one is using the same random number for configuration B this means using (5.31) that $z \geqq p_i^B$ and therefore $\sigma_i^B = -1$. It is therefore impossible to create a damage of type $-+$ and thus (5.31) will also be valid at the next time-step. By induction one can now conclude that (5.31) will always be valid

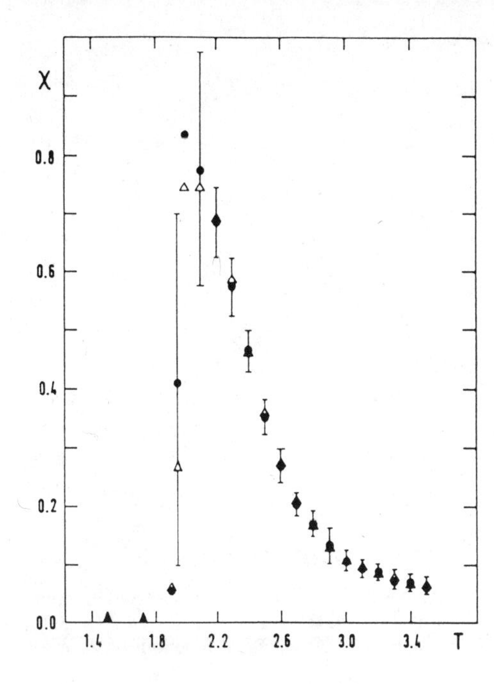

Fig. 5.11. Susceptibility χ(●) and $2\Delta(1-m)$ (△) as function of temperature from 30 systems of size 10×10 (from [5.56]) for the 2D Ising model

and that a damage of type $-+$ cannot be created under the conditions that we had imposed. Consequently we have proved that $d^{-+} = 0$ and it follows for the damage

$$\Delta_i = \frac{C_{0i}}{2(1-m)} \quad \text{and} \quad \Delta = \frac{\chi}{2(1-m)}, \tag{5.32}$$

where $\chi = \sum_i C_{0i}$ is the susceptibility.

In Figs. 5.11 and 5.12 we see how the two exact relations of (5.32) are realised numerically for the 2D Ising model. In Fig. 5.11 we see the susceptibility obtained from (5.32) on the one hand and from the fluctuations of the magnetisation on the other, in a small system. The data agree very nicely. In Fig. 5.12 we see the correlation function obtained via (5.32) (circles) and in the usual way (triangles) using for both methods roughly the same computational effort. One sees that in the usual method once the values are less than about 10^{-3}, the statistical noise takes over and the curve flattens. On the other hand, using (5.32) one gets to substantially smaller values without feeling the statistical noise. The reason why the use of damage is superior numerically stems from the fact that this method just monitors the difference between two configurations subjected to the same thermal fluctuations so that this noise is effectively cancelled. A similar fact was already pointed out for continuous systems by *Parisi* [5.58].

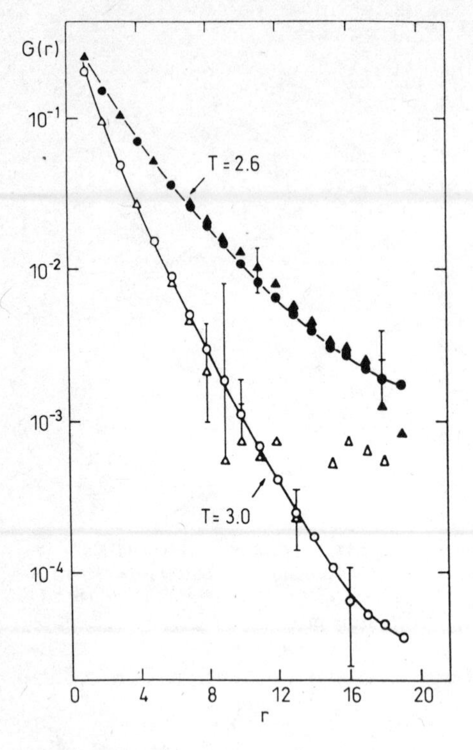

Fig. 5.12. Correlation function $G(r) = \sum_{|i|=r} C_{0i}$ (▲ for $T = 2.6$ and △ for $T = 3.0$) and $2\bar{\Delta}(r)(1-m)$ (● for $T = 2.6$ and ○ for $T = 3.0$) where $\bar{\Delta} = \sum_{|i|=r} \Delta_i$ as a function of r in a semi-log plot. The data come from 10 systems of size 40×40 and were taken from [5.56] for a 2D Ising model

5.2.6 Damage Clusters

The damage that was fixed at the origin at time $t = 0$ acts as a source from which damage constantly spreads. At a given instant, one can look at all the sites that are damaged and one will find a cloud or cluster of sites around the origin. These clusters fluctuate in size and shape and are not necessarily connected. In Figs. 5.13a and b we see two of these clusters for the 2D Ising model at T_c, which are just 38 time steps apart. Since heat bath was used, these clusters represent according to (5.32), the fluctuation of the magnetisation due to the application of a local field at the origin. In Fig. 5.13c we see what happens if a damage $\sigma_i^B = -1$ is fixed all along the boundary of the system. Then clusters of the type shown in Figs. 5.13a and b overlap and one finds fluctuations that seem to be of all sizes. Using the equivalence of the Ising model to a lattice gas, one can interpret these fluctuations as the fluctuations in local density which have actually been measured in a recent experiment [5.59].

Using (5.32) it is actually possible to calculate the dependence between radius and number of sites for the damage clusters. One finds that at the critical point the clusters are fractal, that means that if L is the size of the smallest box into which the cluster fits, and s is the number of sites in the cluster (averaging over all clusters), one has a relation

$$s \propto L^{d_f}. \tag{5.33}$$

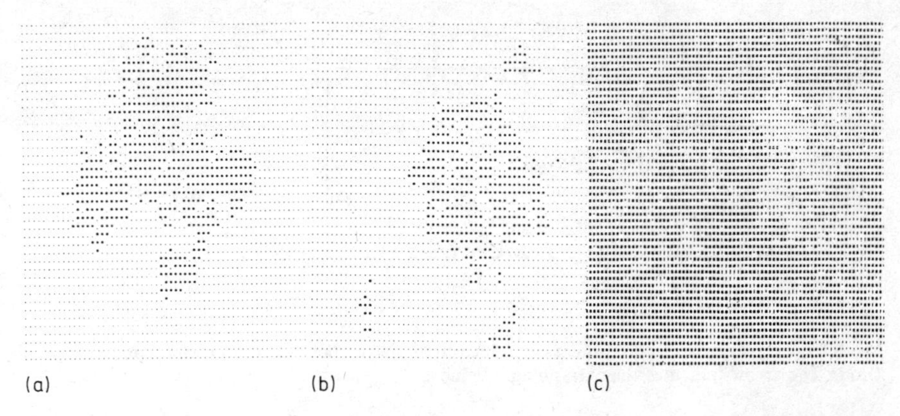

Fig. 5.13. Damage clusters of the 2D Ising model at T_c. In (**a**) and (**b**) the damage is fixed to be $\sigma_0^B = -1$ at the origin, and the size of the system is $L = 60$. In (**c**) the same kind of damage is fixed all along the boundary and $L = 100$. Heat bath and the same random number sequence were applied on both configurations (from [5.56])

Using (5.32) and the fact that at T_c one has $G(r) \propto r^{2-d-\eta}$, one obtains $d_f = d - \beta/\nu$ where β and ν are the critical exponents for the magnetisation and for the correlation length.

Since one does not expect several length scales in the problem one can also replace L in (5.33) by the radius of gyration R. In Fig. 5.14 we see results for

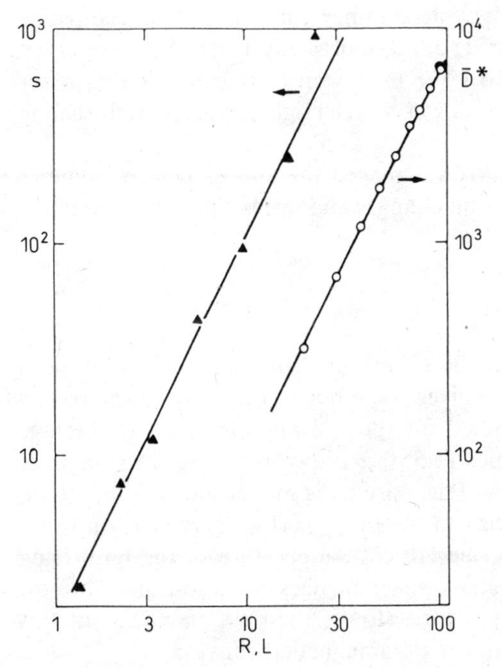

Fig. 5.14. Log–log plot of the number s of damaged sites against the radius R of the cluster (▲), for the 2D Ising model at T_c with $\sigma_0^B = -1$ fixed. We also show the total number $\bar{D}^*$ of damaged sites if the damage is fixed on the boundary of a system of size L as a function of L (○) (from [5.56])

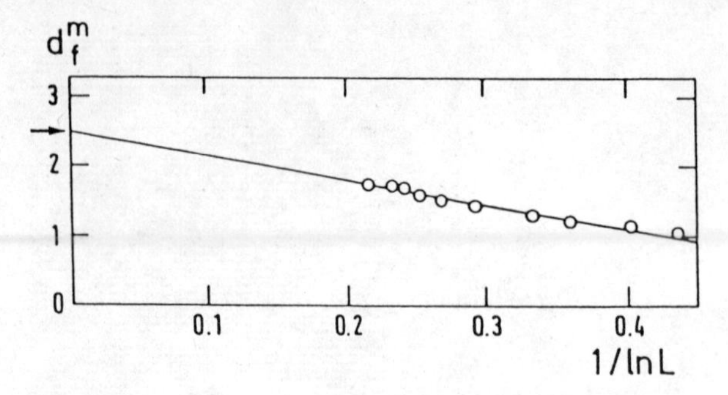

Fig. 5.15. The effective fractal dimension d_f measured via the box counting method as function of $1/\ln L$. The arrow indicates the extrapolated value $d_f = d - \beta/\nu \cong 2.5$ (from [5.60])

the numerical determination of d_f for the two dimensional Ising model at T_c and the agreement with $d_f = \frac{7}{4}$ is reasonably good.

It is also possible to make similar arguments for the case when the damage is fixed on the boundary as shown in Fig. 5.13c. There the density of damaged sites in the centre of the box decreases as a power law with the size of the box, and consequently one finds again a fractal dimension as shown in Fig. 5.14.

The fractal dimension of the clusters can also be measured by the touching method, which has been widely used in cellular automata. One lets the damage spread until it touches the boundary of the system. The number of sites S of the touching cluster that are within a box of size ρ around the origin, scales like $S \propto \rho^{d_{\text{eff}}}$, where d_{eff} is the effective fractal dimension which should converge with the system size L, to the d_f of (5.33). Numerically it has been observed [5.60] that this convergence goes like $1/\ln L$, as seen in Fig. 5.15. Using other methods the convergence can be even slower, which might be due to multiscaling [5.60].

Let us note that if one does not fix a damaged site and if one uses Glauber dynamics instead of heat bath, the result changes and one seems to find compact clusters [5.52].

5.2.7 Damage in Spin Glasses

Numerical work on spin glasses has been both challenging and frustrating in the past. Thus any new method that gives some hope of improving the results should be tested. The results obtained for the $\pm$ Ising spin glass of damage spreading, using heat bath dynamics [5.55], are shown in Fig. 5.16. In their spirit they are analogous to the data of the pure Ising model shown in Fig. 5.10. In three dimensions one believes that, at about $T_{SG} = 1.2$, there is a transition to a spin glass phase. At $T_G = 4.5$, the critical temperature of the pure Ising model, it is thought that the paramagnetic phase changes into a so-called Griffiths phase, in which correlations decay a little slower than exponential; but this phase is very difficult to detect from the paramagnetic behaviour.

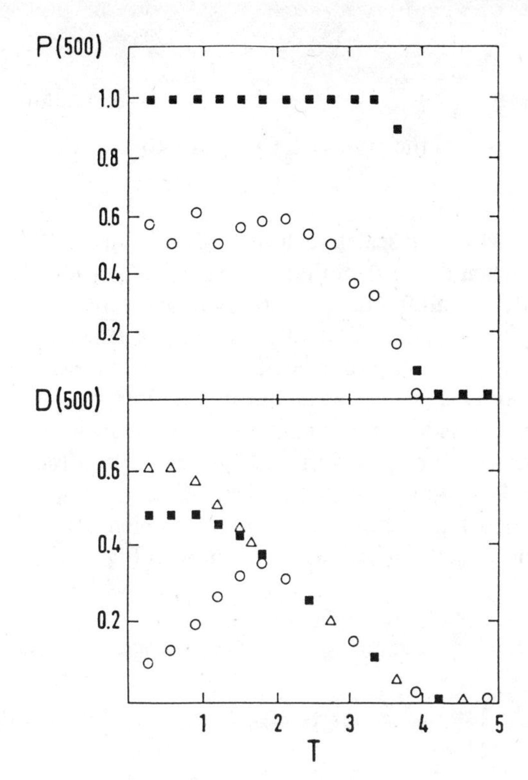

Fig. 5.16. Damage in the 3D $\pm$ spin glass after 500 time steps as a function of temperature for $L = 12$, taken from [5.55], for various initial damages: $\Delta(0) \to 0$ (○), $\Delta(0) = \frac{1}{2}$ (■) and $\Delta(0) = 1$ (△); fraction $P(t)$ of non-identical configurations and distance $D(t)$ between two configurations provided they are still different

From Fig. 5.16 one indeed sees two characteristic temperatures not too far from T_{SG} and T_G, which could be interpreted as separating three phases. Unlike the ferromagnetic phase of Fig. 5.10, both these low temperature phases are chaotic. This is because there is a non-zero probability that two configurations remain different after some time that is large, but much shorter than the Poincaré time. In the low temperature phase the final distance does depend on the initial distance, while in the intermediate phase the final distance is independent of the initial distance. One might argue that this behaviour reflects the phase-space structure of the spin-glass phase and of the Griffiths phase. But qualitatively, one finds the same picture in two dimensions [5.61] as the one seen in Fig. 5.16, although it is generally accepted that no spin glass phase exists in two dimensions.

However, there does exist a way to extract T_G for the 3D spin glass. During a certain interval of time a site will be damaged and healed again several times. For each site i, one can look at f_i, which is the number of times that it is damaged again during a fixed, long time interval. Next, one can look at the distribution $P(f)$, i.e. the number of sites that have a certain frequency f and analyse the moments of this distribution,

$$M_q = \sum_f f^q P(f) \quad q = 0, \pm 1, \ldots \tag{5.34}$$

and normalise them,

$$m_q = (M_q/M_0)^{1/q}. \tag{5.35}$$

These moments will scale with the size of the system L according to

$$m_q \propto L^{x_q}. \tag{5.36}$$

Usually the distribution is self-averaging and scales such that all x_q (with $q \neq 0$) are the same. But in specific cases, when the distribution has a particularly long tail, it can show what has been called "multifractality" [5.62] which means that the x_q vary with q and one has therefore an infinity of different scaling exponents.

In [5.63] the moments of (5.35) have been calculated for various models. For the 3D spin glass it has been found that the distribution is multifractal at T_G, as can be seen from Fig. 5.17a. This is in marked contrast to what has been found at T_{SG} for the 3D spin glass, at T_G or T_{SG} for the 2D spin glass or at the critical temperature T_c of the 3D Ising model, because in all these cases the lines for the different moments in the log–log plot are parallel to each other, which means that x_q is the same for all q. As an example, we show in Fig. 5.17b the data for the 3D Ising model.

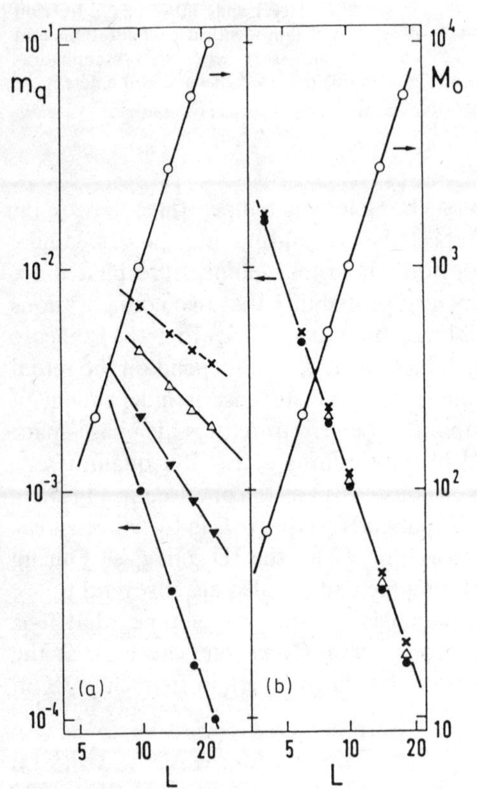

Fig. 5.17. Log–log plot of the moments of the distribution $P(f)$ of the redamaging of sites as a function of the system size L for **(a)** the 3D ± spin glass at T_G **(b)** the standard 3D Ising model at T_c. The moments are: $M_0(\bigcirc)$, $m_1(\bullet)$, $m_2(\blacktriangledown)$, $m_3(\triangle)$ and $m_4(\times)$ (from [5.63])

We can conclude that the temperature $T_G = 4.5$ of the 3D $\pm$ spin glass is special because it corresponds to a distribution of the redamaging frequency that is multifractal, while in the usual case, particularly at T_G in the 2D $\pm$ spin glass, one has a simple one-exponent scaling behaviour. This feature might be an empirical way to single out models that show spin glass behaviour. The numerical evidence that T_{SG} is indeed the transition to the spin glass phase can be strengthened by making a scaling ansatz for the relaxation times [5.64].

The same technique has also been applied to the 3D $\pm$ Ising spin glass in a homogeneous field [5.65]. A line is found on which the redamaging frequency is multifractal which indicates the existence of an Almeida–Thouless line, also in 3D.

The transition at T_G might *not* be related to the critical point of the pure Ising model. In fact a precise determination of this onset of spreading, using the gradient method [5.66], clearly shows for the 2D $\pm$ Ising spin glass, a value of T_G that is 6% below the exactly known value of the pure Ising model. Numerical and theoretical arguments indicate that the onset of damage is in fact related to a percolation phenomenon [5.67]. The use of Glauber dynamics for the damage spreading in the 3D $\pm$ Ising spin glass has also been investigated [5.68], and no transition at T_G was observed.

5.2.8 More About Damage Spreading

Apparently, damage spreading introduces a new type of dynamical phase transition which depends on the details of the algorithm used. Besides for heat bath and Glauber also Metropolis [5.69] and Swendsen–Wang [5.70] dynamics have been investigated. It is also possible to define a dynamics with a tunable parameter f that interpolates between two of them (for instance between heat bath and Glauber), and then phase transitions at a value f_c can appear between two different types of behaviour [5.71]. In mean field, the difference between parallel and sequential dynamics has also been studied [5.72].

Damage spreading has also been studied in other models. If in the Ising model the range of the interaction is increased, one finds qualitatively the same picture as in the simple Ising model except for an eventual shift in the onset of spreading temperature [5.73]. In the ANNNI model it seems that damage spreading constitutes a way to observe the floating phase in two dimensions [5.74]. Continuous degrees of freedom have also been considered, and both for the XY-model [5.50], as well as for the Heisenberg model [5.75], unexpected phase transitions have been found.

5.3 Conclusion

We have discussed various algorithms for growing clusters. Irreversible far-from-equilibrium growth which is responsible for many patterns found in nature is essentially a sequential process. However, if in a moving boundary problem one has to solve for the discretised equations, this most CPU-time consuming part can be vectorised.

Reversible growth can be simulated by a thermal Monte Carlo method as illustrated for the case of damage spreading in the Ising model. Multi-spin-coding and vectorisation can speed up these algorithms to a certain extent. At the dynamical phase transition, critical clusters spread and under certain conditions their density can be related to the correlation function in equilibrium. This gives an algorithm that suppresses the statistical noise in the correlation functions very efficiently.

The clusters can be analysed for various properties, including their fractal dimension and their surface roughness. In some cases, convergence is very slow and the limits of today's computational power are reached.

Obviously, in this chapter we have selected only some of the random cluster growth models that have been studied in the past. The variety of models that has been investigated is overwhelming. Irreversible growth, which is often also called random growth in the literature, is used to model deposition, corrosion, fragmentation, gelation, etc. Deposition, for instance, can occur ballistically on a surface [5.76] or under the condition that the particles are not allowed to pile up, as in the random sequential adsorption and the car parking model [5.77]. Adsorbed particles can then still undergo irreversible surface reactions and surface poisoning [5.78]. Reversible growth of clusters has been studied extensively as phase separation in thermal far-from-equilibrium models [5.79], in reaction-diffusion models for autocatalytic chemical reactions [5.80] and as the motion of interfaces in SOS models [5.81].

Acknowledgements. I thank P. Meakin for several figures and D. Stauffer for critically reading the manuscript.

References

5.1 T. Vicsek: *Fractal Growth Phenomena* (World Scientific, Singapore 1989)
5.2 H.J. Herrmann: Phys. Rep. **136**, 153 (1986)
5.3 P. Meakin: In *Phase Transitions and Critical Phenomena*, Vol. 12, ed. by C. Domb, J.L. Lebowitz (Academic, New York 1988)
5.4 M. Eden: In Proc. of the Fourth Berkeley Symp. on Mathematical Statistics and Probability, Vol. IV, ed. by F. Neyman (University of California, Berkeley 1961); in Symposium on Information Theory in Biology, ed. by H.P. Hockey (Pergamon, New York 1959) p. 359
5.5 T. Vicsek, M. Cserzö, V.K. Horváth: Physica A **167**, 315 (1990)
5.6 P. Meakin: J. Theor. Biol. **118**, 101 (1986)
5.7 F. Family, T. Vicsek: J. Phys. A **18**, L75 (1985)
5.8 M. Kardar, G. Parisi, Y.C. Zhang: Phys. Rev. Lett. **56**, 889 (1986)
5.9 D.E. Wolf, J. Kertész: Europhys. Lett. **4**, 651 (1987); J. Phys. A **20**, L257 (1987)
5.10 J.M. Kim, J.M. Kosterlitz: Phys. Rev. Lett. **62**, 2289 (1989)
5.11 P. Freche, D. Stauffer, H.E. Stanley: J. Phys. A **18**, L1163 (1985)
5.12 J.G. Zabolitzky, D. Stauffer: Phys. Rev. A **34**, 1523 (1986)
5.13 R. Hirsch, D.E. Wolf: J. Phys. A **19**, L251 (1986)
5.14 M. Plischke, Z. Rácz: Phys. Rev. Lett. **53**, 415 (1984); ibid. **54**, 2056 (1985)

5.15 P. Meakin, R. Jullien, R. Botet: Europhys. Lett. **1**, 609 (1986); R. Jullien, R. Botet: Phys. Rev. Lett. **54**, 2055 (1985); J. Phys. A **18**, 2279 (1985)
5.16 J. Kertész, D. Wolf: J. Phys. A **21**, 747 (1988)
5.17 J. Szép, J. Cserti, J. Kertész: J. Phys. A **18**, L413 (1985)
5.18 D. Liu, M. Plischke: Phys. Rev. B **38**, 4781 (1988); P. Devillard, H.E. Stanley: Physica A **160**, 298 (1989); L. Tang, B. Forrest: Phys. Rev. Lett. **64**, 1405 (1990); W. Renz: Private communication
5.19 J.S. Langer: Rev. Mod. Phys. **52**, 1 (1980)
5.20 D.A. Kessler, J. Koplik, H. Levine: Adv. Phys. **37**, 255 (1988)
5.21 L. Niemeyer, L. Pietronero, H.J. Wiesmann: Phys. Rev. Lett. **52**, 1033 (1984)
5.22 W.H. Press, B.P. Flannery, S.A. Teukolsky, W.T. Vetterling: *Numerical Recipes, The Art of Scientific Computing* (Cambridge University Press, Cambridge 1986)
5.23 T.C. Halsey, P. Meakin, I. Procaccia: Phys. Rev. Lett. **56**, 854 (1986)
5.24 G.G. Batrouni, A. Hansen: J. Stat. Phys. **52**, 747 (1988)
5.25 C. Evertsz, E. Hafkenscheid, M. Harmsma, L. Pietronero: In *Science and Engineering on Cray Supercomputers*, ed. by E.J. Petcher (Cray Research Inc., Minneapolis, 1988) p. 85
5.26 T.A. Witten, L.M. Sander: Phys. Rev. Lett. **47**, 1400 (1981)
5.27 R.C. Ball, R.M. Brady: J. Phys. A **18**, L809 (1985); P. Meakin: J. Phys. A **18**, L661 (1985)
5.28 P. Meakin, R.C. Ball, P. Ramanlal, L.M. Sander: Phys. Rev. A **35**, 5233 (1987)
5.29 S. Tolman, P. Meakin: Phys. Rev. A **40**, 428 (1989)
5.30 J. Kertész, T. Vicsek: J. Phys. A **19**, L257 (1986); J. Nittmann, H.E. Stanley: Nature **321**, 663 (1986)
5.31 H. Liebowitz (ed.): *Fracture*, Vols. I–VII (Academic, New York 1986)
5.32 H.J. Herrmann, S. Roux (eds.): *Statistical Models for the Fracture of Disordered Media* (Elsevier, Amsterdam 1990)
5.33 S. Roux, E. Guyon: J. de Phys. Lett. **46**, L999 (1985)
5.34 R.J. Roark, W.C. Young: *Formulas for Stress and Strain* (McGraw-Hill, Tokyo 1975) p. 89
5.35 E. Louis, F. Guinea, F. Flores: In *Fractals in Physics*, ed. by L. Pietronero, E. Tossatti (Elsevier, Amsterdam 1986); E. Louis, F. Guinea: Europhys. Lett. **3**, 871 (1987); P. Meakin, G. Li, L.M. Sander, E. Louis, F. Guinea: J. Phys. A **22**, 1393 (1989); E.L. Hinrichsen, A. Hansen, S. Roux: Europhys. Lett. **8**, 1 (1989)
5.36 H. Van Damme, E. Alsac, C. Laroche: C.R. Acad. Sci., Ser. II **309**, 11 (1988); H. Van Damme: In *The Fractal Approach to Heterogeneous Chemistry*, ed. by D. Avnir (Wiley, New York 1989) p. 199
5.37 S. Roux: J. Stat. Phys. **48**, 201 (1987)
5.38 H.J. Herrmann, J. Kertész, L. de Arcangelis: Europhys. Lett. **10**, 147 (1989)
5.39 M.J. Blackburn, W.H. Smyrl, J.A. Feeney: In *Stress Corrosion in High Strength Steels and in Titanium and Aluminum Alloys*, ed. by B.F. Brown (Naval Res. Lab., Washington, DC 1972) p. 344
5.40 S.A. Kauffman: J. Theor. Biol. **22**, 437 (1969)
5.41 B. Derrida, Y. Pomeau: Europhys. Lett. **1**, 45 (1986); B. Derrida, D. Stauffer: Europhys. Lett. **2**, 739 (1986)
5.42 L.R. da Silva, H.J. Herrmann: J. Stat. Phys. **52**, 463 (1988)
5.43 S. Wolfram: *Theory and Application of Cellular Automata* (World Scientific, Singapore 1986); E. Bienenstock, F. Fogelman-Soulié, G. Weisbuch (eds.): *Disordered Systems and Biological Organization*, NATO ASI Series, Ser. F (Springer, Berlin, Heidelberg 1986)
5.44 H.J. Herrmann: In *Nonlinear Phenomena in Complex Systems*, ed. by A.N. Proto (Elsevier, Amsterdam 1989)
5.45 Y. Pomeau: J. Phys. A **17**, L415 (1984); H.J. Herrmann: J. Stat. Phys. **45**, 145 (1986)
5.46 J.G. Zabolitzky, H.J. Herrmann: J. Comput. Phys. **76**, 426 (1988); S.C. Glotzer, D. Stauffer, S. Sastry: Physica A **164**, 1 (1990)
5.47 D. Stauffer: In *Computer Simulation Studies in Condensed Matter Physics II*, ed. by D.P. Landau, K.K. Mon, H.-B. Schüttler, Springer Proc. Phys., Vol. 45 (Springer, Berlin, Heidelberg 1990)
5.48 L. de Arcangelis, A. Coniglio: Europhys. Lett. **7**, 113 (1988)
5.49 H.J. Hilhorst, M. Nijmeijer: J. de Phys. **48**, 185 (1987)
5.50 O. Golinelli, B. Derrida: J. Phys. A **22**, L939 (1989)
5.51 A.M. Mariz: J. Phys. A **23**, 979 (1990)
5.52 H.E. Stanley, D. Stauffer, J. Kertész, H.J. Herrmann: Phys. Rev. Lett. **59**, 2326 (1987)
5.53 U. Costa: J. Phys. A **20**, L583 (1987)
5.54 G. Le Caër: J. Phys. A **22**, L647 (1989); Physica A **159**, 329 (1989)
5.55 B. Derrida, G. Weisbuch: Europhys. Lett. **4**, 657 (1987)

5.56 A. Coniglio, L. de Arcangelis, H.J. Herrmann, N. Jan: Europhys. Lett. **8**, 315 (1989)
5.57 J. Ashkin, E. Teller: Phys. Rev. **64**, 178 (1943)
5.58 G. Parisi: Nucl. Phys. B **180**, 378 (1981)
5.59 P. Guenoun, F. Perrot, D. Beysens: Phys. Rev. Lett. **63**, 1152 (1989)
5.60 L. de Arcangelis, H.J. Herrmann, A. Coniglio: J. Phys. A **23**, L265 (1990)
5.61 A.U. Neumann, B. Derrida: J. de Phys. **49**, 1647 (1988)
5.62 G. Paladin, A. Vulpiani: Phys. Rep. **156**, 147 (1987)
5.63 L. de Arcangelis, A. Coniglio, H.J. Herrmann: Europhys. Lett. **9**, 749 (1989)
5.64 I.A. Campbell, L. de Arcangelis: Europhys. Lett. **13**, 587 (1990)
5.65 L. de Arcangelis, H.J. Herrmann, A. Coniglio: J. Phys. A **22**, 4659 (1989)
5.66 N. Boissin, H.J. Herrmann: J. Phys. A **24**, L43 (1991)
5.67 L. de Arcangelis, A. Coniglio: Preprint; A. Coniglio: In *Correlations and Connectivity in Constrained Geometries*, ed. by H.E. Stanley, N. Ostrowsky (Kluwer, Dordrecht 1990)
5.68 H.R. da Cruz, U.M.S. Costa, E.M.F. Curado: J. Phys. A **22**, L651 (1989)
5.69 A.M. Mariz, H.J. Herrmann, L. de Arcangelis: J. Stat. Phys. **59**, 1043 (1990)
5.70 D. Stauffer: Physica A **162**, 27 (1990)
5.71 A.M. Mariz, H.J. Herrmann: J. Phys. A **22**, L1081 (1989); O. Golinelli: Physica A **167**, 736 (1990)
5.72 O. Golinelli, B. Derrida: J. de Phys. **49**, 1663 (1988)
5.73 S.S. Manna: J. de Phys. **51**, 1261 (1990)
5.74 M.N. Barber, B. Derrida: J. Stat. Phys. **51**, 877 (1988)
5.75 E.N. Miranda, N. Parga: J. Phys. A **24**, 1059 (1991)
5.76 D.N. Sutherland: J. Colloid Interface Sci. **25**, 373 (1967); P. Meakin: J. Colloid Interface Sci. **102**, 505 (1984)
5.77 E. Burgos, H. Bonadeo: J. Phys. A **20**, 1193 (1987); E. Hinrichsen, J. Feder, T. Jøssang: J. Stat. Phys. **44**, 793 (1986); P. Schaaf, J. Talbot: Phys. Rev. Lett. **62**, 175 (1989); R.D. Vipst, R.M. Ziff: J. Chem. Phys. **91**, 2599 (1989)
5.78 R.M. Ziff, E. Gulari, Y. Barshad: Phys. Rev. Lett. **56**, 2553 (1986); I. Jensen, H.C. Fogedby, R. Dickman: Phys. Rev. A **41**, 3411 (1990)
5.79 S. Hayward, D.W. Herrmann, K. Binder: J. Stat. Phys. **49**, 1053 (1987)
5.80 R. Dickman: Phys. Rev. B **40**, 7005 (1989)
5.81 D.B. Abraham, J. Heiniö, K. Kaski: Phys. Rev. B (1990)
5.82 P. Ossadnik: Physica A **176**, 454 (1991)

6. Recent Progress in the Simulation of Classical Fluids

Dominique Levesque and Jean Jarques Weis

This chapter is written with the same motivation as the two earlier reviews by *Hansen* and ourselves [6.1, 2], namely to present simulation work in the domain of classical fluids. The present review considers developments from mid 1984 onwards. In spite of the nearly exponential increase of number of publications in this field, we have attempted to give account of simulation work on the same topics previously considered in [6.1, 2], the most important being simple fluids, molecular fluids, solutions and interface phenomena. It was obviously not possible to describe in detail the enormous amount of work on complex molecules and for some specific cases we had to limit ourselves to just mentioning the references. Also, several topics, lying at the frontier of the field, as for instance adsorption phenomena or quantum processes in liquids have been omitted for reasons of space limitation. Several books dealing with simulation methods have been published recently (a complete list is given in Chap. 1). Here we mention the excellent book of *Allen* and *Tildesley* [6.3] which is closest to the matter considered in this chapter.

The chapter is divided into eight sections: the first four (written by D.L.) deal with simple monatomic systems and the last four (written by J.J.W.) with molecular (anisotropic) systems. Section 6.1 describes some new improvements of the MC methods. Simple monatomic fluids and Coulomb systems are considered in Sects. 6.2, 3, respectively, and inhomogeneous simple fluids in Sect. 6.4. Section 6.5 is devoted to molecular model systems and Sect. 6.6 to molecular systems interacting with realistic potentials. The simulations of solutions, with special emphasis on aqueous solutions, are reviewed in Sect. 6.7 and molecular interfaces are treated in Sect. 6.8.

6.1 Improvements of the Monte Carlo Method

This section is devoted to a discussion of recent technical improvements of the Monte Carlo (MC) method applied to the simulation of condensed fluid phases.

6.1.1 Metropolis Algorithm

A comprehensive review of the Metropolis algorithm has been published by *Bhanot* [6.4], the ergodicity and the convergence of the algorithm are presented from a mathematical point of view.

In MC simulations, a quasi-ergodic behaviour can lead to large systematic errors on the estimates of the thermodynamic quantities; it results from potential barriers between two or more domains of probable configurations in the phase space. Clearly, the difficulty is to generate, by the Metropolis algorithm, often very improbable configurations, which realise in phase space, the displacement of the system from a domain to another. *Frantz* et al. [6.5] propose a way to make this displacement; the method is based on the use of the configurations of the system at two different temperatures, the largest being high enough to allow to the Metropolis random walk to overcome the potential barriers. The simulations are done in two steps. First at high temperature a set of configurations of the system is generated and stored. Then, at low temperature, in order to avoid that the standard Metropolis random walk gives configurations staying near only one of the minima of energy, attempts will be made to jump from a low temperature configuration to a high temperature configuration. The jump corresponds to a displacement of all the particles of the system in one step towards a new configuration, which is one of the high temperature stored configurations. If the jump is successful, the new configuration will be probably in the vicinity of a different energy minimum. The standard Metropolis algorithm is continued at low temperature until the next jump attempt. In [6.5], the method is shown to be successful for unidimensional (1D) systems and also for small systems in three dimensions (3D); its efficiency for 3D systems of $\sim 10^3$ particles remains to be demonstrated.

The work [6.5] is in many aspects similar to the work of *Voter* [6.6], which very clearly defines the characteristics of a MC algorithm able to determine the free energy difference between states of a system which correspond to configurations localised near energy minima widely separated in phase space. In [6.6] the localisations of the energy minima are supposed known, and the displacement from a configuration near a minimum to a configuration near another minimum is obviously possible. Generally, the localisations of minima are unknown and the method of *Frantz* et al. gives a way to effect moves between domains of probable configurations without knowing, a priori, their positions in phase space.

In the "umbrella" sampling method, which was devised for the calculation of free energy differences between two thermodynamic states, the choice of the bias functions, needed to connect the distant domains of the phase space corresponding to these states, is not easy. In [6.7] *Mezei* proposes a systematic procedure for the derivation of these bias functions. *Harvey* and *Prabhakaran* [6.8] discuss and describe ways to check whether the choice of the bias functions is adequate for a specific MC calculation.

The convergence of the averages estimated in the course of the MC simulations has been attempted to be accelerated by various methods (cf. [6.1, 2]). One of these methods, the so-called force-biased sampling, is examined by *Goldman* [6.9]. He determines the optimal value of the parameter of the force-biased function. *Chapman* and *Quirke* [6.10] try to show that multiparticle moves in the Metropolis algorithm do not affect the convergence in MC simulation;

but the tests are done on very small systems. The optimisation of the a priori trial moves of a particle in the Metropolis algorithm is considered in detail by *Kolafa* [6.11]; he proposes different choices for the simulations of atomic and molecular fluids.

The estimate of statistical errors in MC computations is made difficult by the correlations between the successive configurations generated by the Metropolis algorithm. The statistical errors are not proportional to the inverse of the square root of the number n of trial moves but decrease more slowly. More reliable error estimates can be obtained if the number n^* of trial moves needed to obtain uncorrelated configurations is known; the number of truly independent configurations can be estimated equal to $n' = n/n^*$ and the decrease of statistical error is then proportional to $\sim (n')^{-1/2}$. The computation of the autocorrelation functions of relevant quantities (displacements of the molecules, potential energy, etc.) allows an evaluation of n^*. These questions are discussed by *Kolafa* [6.12], *Bishop* and *Frinks* [6.13].

6.1.2 Monte Carlo Simulations and Statistical Ensembles

(a) Canonical, Grand Canonical and Semi-grand Ensembles. The localisation of the phase transitions is one of the major challenges of MC simulations; one of the possible procedures is to calculate the low temperature isotherms of the considered system and to look for van der Waals loops. These loops occur only if the volume and the boundary conditions used in the simulations preclude the phase separations and allow to observe homogeneous metastable states in the two-phase region. However, van der Waals loops have been obtained only in few cases in MC or MD simulations. Two methods which seem able to generate van der Waals loops, are proposed by *Challa* and *Hetherington* [6.14] and *Caliri* et al. [6.15]. The first method, the so-called Gaussian ensemble, has been shown useful for the study of phase transitions of lattice systems, but its efficiency for the fluids is not established. Similarly the technique of [6.15] is to maintain a homogeneous density in the canonical (NVT) ensemble by introducing a chemical potential μ as an extra extensive variable in the ensemble average. The chemical potential governs the exchange of particles between the subvolumes of the system. As a test, the technique is shown able to generate the van der Waals loop of the liquid–solid transition of hard disks. But in this method, there is no way to choose a priori the appropriate value of μ, thus the authors must compare their results to the data of previous simulations in order to estimate an acceptable value of μ. This shortcoming is a strong limitation for a general use. A method of localisation of the liquid–solid transition for 2D systems is proposed by *Fraser* et al. [6.16] and tested on the system of hard disks; it is based on the analysis of the edge-length distribution of Voronoï polygons which is very sensitive to the structural changes in the system. But the method does not seem useful for an exact localisation of the transition which must be established from the values of the free energy.

The configurations generated by MC simulation in the canonical ensemble for a given thermodynamic state, can be "reweighted" for computing the thermodynamic functions of another state, possibly for a system with a different hamiltonian. The "weights" are the ratios of the Boltzmann factors of the two states; they are easily computed if the two states differ only by their temperatures. An efficient "reweighting" procedure is developed by *Shaw* [6.17], which associates "reweighting" and careful curve fitting of the configurational density of states. Shaw is able to calculate the equation of state of the Lennard–Jones (LJ) system over a large domain of densities and temperatures by using the configurational density of the atomic systems interacting by the r^{-12} potential. Clearly, a general use of this procedure is limited by phase transitions. For instance in [6.17], it seems that a bias on the calculated pressures and internal energies can appear in the vicinity of the critical point where, obviously, the configurations of the LJ atoms will be very different from the always homogeneous configurations of the r^{-12} atoms.

The choice of too small simulation cells distorts the sampling of the configurations in the grand canonical (GC) simulations; *Parsonage* [6.18] investigates this problem on simple 1D models.

In the simulations of polydisperse fluids, one of the aims can be the calculation of the composition, i.e. the relative concentration of the different particle species. A well adapted ensemble for this computation is the semi-grand canonical ensemble, which is defined by N, the total number of particles, V, the volume, T, the temperature and $\mu(c)$, the chemical potential which is a function of the composition c of the system. The sampling of the configurations and the compositions is done by using the Boltzmann factor $\exp\{-\beta[U_N + \mu(c)]\}$, U_N being the internal energy of the N particles and $\beta = 1/(kT)$. The sampling is possible if the dependence of μ on c is known. Assuming that in the vicinity of a homogeneous composition c_0 (system of identical particles) $\mu(c) - \mu(c_0)$ has a gaussian form, *Kofke* and *Glandt* [6.19] carry out simulations on weakly polydisperse fluids.

(b) Gibbs Ensemble. The most interesting advance in the domain of simulations of classical fluids is the simulation procedure proposed by *Panagiotopoulos* [6.20], the so-called Gibbs ensemble. This procedure is especially well suited for the computation of phase equilibria of pure fluids or mixtures. In the canonical ensemble, phase equilibria are difficult to simulate because of the fact that the interfaces fill a large amount of the total volume, the limited extension of the bulk phases can bias the localisation of the equilibrium. In the Gibbs ensemble, the interface region disappears and the equilibrium is obtained between two bulk systems. These systems are enclosed in volumes V_1 and V_2 with periodic boundary conditions and have respectively N_1 and N_2 particles. The volumes V_1 and V_2 can vary, the total volume $V = V_1 + V_2$ being kept constant. The systems can exchange also particles with the condition $N = N_1 + N_2$ constant. The particle moves in V_1 or V_2, the exchange of particles and the variations of the volumes are sampled according to generalised Boltzmann factors $\exp(-\beta W)$.

These factors are such that for a particle move in V_1 or V_2, W is equal to ΔU_1 or ΔU_2, the variation of the internal energy due to the move. For the exchange of a particle between V_1 and V_2, W is given by

$$\Delta U_1 + \Delta U_2 + \frac{1}{\beta}\log\frac{(V-V_1)(N_1+1)}{V(N-N_1)} \tag{6.1}$$

if the particle goes from V_1 to V_2; for a variation ΔV of the volume V_1, W has the form

$$\Delta U_1 + \Delta U_2 - \frac{N_1}{\beta}\log\frac{(V+\Delta V)}{V_1} - \frac{(N-N_1)}{\beta}\log\frac{(V-V_1-\Delta V)}{(V-V_1)}. \tag{6.2}$$

As shown in [6.20] and also in [6.21, 22], this sampling scheme guarantees that in the limit of a large number of trials, the pressures and the chemical potentials will be equal in V_1 and V_2. Then, a two-phase equilibrium can be generated if adequate values of V, N and T have been chosen. The method is very well suited for computing the gas–liquid coexistence line and for estimating the localisation of the critical point. For temperatures near the triple point, the method will be confronted with the usual difficulties of the GC simulations or the calculations of the chemical potential, due to the low probability to successfully insert particles in the dense systems. But the efficient sampling procedures devised for the computations of μ are easily applicable for the Gibbs ensemble simulations, allowing to obtain the equilibrium between phases with large differences in densities.

The extension of the Gibbs ensemble method to the cases of mixtures or membrane equilibria has been discussed by *Panagiotopoulos* et al. [6.23]. These equilibria are determined more easily in the Gibbs isothermal-isobaric (NPT) ensemble by the generalisation of the sampling procedure described above. The sampling procedure is identical with the sampling in the Gibbs canonical ensemble for the moves and the exchanges of particles; but for the variations of volume the generalized Boltzmann factor becomes

$$\exp\left(-\beta\left\{\frac{-N_1}{\beta}\log\frac{(V+\Delta V_1)}{V_1} - \frac{N_2}{\beta}\log\frac{(V+\Delta V_2)}{V_2}\right.\right.$$
$$\left.\left. + P(\Delta V_1 + \Delta V_2) + \Delta U_1 + \Delta U_2\right\}\right). \tag{6.3}$$

The total volume V of the two simulation cells is no more constant and its value is fixed by the value P of the pressure of the system. The variations ΔV_1 and ΔV_2 are independent. The Gibbs isothermal–isobaric ensemble is not adapted for the calculations of phase equilibria of pure systems, because obviously, if P and T do not correspond exactly to a phase equilibrium, the particles in the two volumes will be in identical thermodynamic states. For the mixtures, where the number of intensive thermodynamic variables is larger than in the pure systems, it will be possible to achieve equilibria between mixtures of different compositions.

The equilibrium between phases, separated by semi-permeable membranes, is obtained by performing exchanges between the two cells only for the species which can cross the membrane. The pressure P becomes the difference of pressure imposed by the membrane and the variations ΔV_1 and ΔV_2 are not independent $\Delta V_1 = -\Delta V_2$. The extension of the Gibbs ensemble technique to the semi-grand ensemble sampling is discussed by *Kofke* and *Glandt* [6.24].

The Gibbs ensemble is a promising and important advance for the study of phase equilibria. Due to the suppression of the interface region, it should allow an accurate determination of the bulk properties of phases in equilibrium and also facilitate the simulations of equilibria between the phases of ionic and dielectric systems.

(c) Monte Carlo Algorithm for "Adhesive" Particles. The modifications of the Metropolis algorithm entailed by the simulations of the systems of "sticky" (adhesive) hard disks or hard spheres are studied by *Seaton* and *Glandt* [6.25, 26]. The difficulty for an efficient sampling of the configurations of such systems is that, in the standard Metropolis algorithm, there is a zero probability to generate configurations with two disks or spheres exactly in contact and then to account for the effects of the "sticky" potential. For a given arrangement of the disks or spheres, the sampling of a change of the number of disks or spheres in contact must be included in the a priori trial move of the particles. This is made by computing the relative probabilities of the states, neglecting the possible core overlaps, where the displaced particle would be in contact with 0, 1, 2,... particles in the volume v_a which is accessible by a trial move. The choice between these states is made according to these probabilities; after the selection of any arrangement of the particle in the chosen state, the configuration is accepted or rejected following the occurrence of 0 or 1 (maybe more) overlap. The sampling must satisfy the detailed balance relation, this imposes an adequate choice of v_a. A choice of v_a well suited for the dense fluids is proposed by *Kranendonk* and *Frenkel* [6.27].

In conclusion of this section on new or improved MC algorithms, we mention that for MD simulations, new methods have been devised for the study of solid–solid phase transitions, melting, etc. The principle of these methods is the use of generalised Lagrangians which make possible simulations at constant pressure or temperature with a variable shape of the simulation cell. Clearly, MC simulations having the same motivations are achievable with the Hamiltonians associated to these Lagrangians. General discussions of these methods can be found in [6.28–30].

6.1.3 Monte Carlo Computation of the Chemical Potential and the Free Energy

(a) Chemical Potential. The chemical potential μ for large finite systems is the free energy difference between systems with N and $N-1$ particles,

$$\mu = F(N, V, T) - F(N-1, V, T). \tag{6.4}$$

This relation is equivalent to writing

$$e^{\beta\mu} = \frac{Q_{N-1}}{Q_N}, \tag{6.5}$$

where Q_N is the configurational part of the partition function

$$Q_N = \int d\boldsymbol{r}_1 \cdots d\boldsymbol{r}_N \, e^{-\beta U_N}. \tag{6.6}$$

Equation (6.5) is the basis for the MC calculations of chemical potential, it can be written as

$$e^{-\beta\mu} = \langle e^{-\beta \Delta U_N} \rangle_{N-1}, \tag{6.7}$$

or

$$e^{\beta\mu} = \langle e^{\beta \Delta U_N} \rangle_N, \tag{6.8}$$

where $\langle \cdots \rangle_N$ and $\langle \cdots \rangle_{N-1}$ denote canonical averages of systems with N and $N-1$ particles and $\Delta U_N = U_N - U_{N-1}$. In principle these formulas are equivalent, but their efficiencies for practical computations of μ seem to differ. Each formula has obvious shortcomings. In (6.7) the difficulty arises from the sampling of the positions of the "virtual" Nth particle, contributing significantly to the ensemble average; in dense fluids, these positions are not efficiently localised by a random insertion which generally corresponds to a large positive energy of interaction between the "virtual" particle and the "real" particles. In (6.8), the configurations for which the Nth particle has a large positive energy, are not generated during a MC simulation with a limited number of steps and then a systematic bias will appear in the estimate of μ. This remark is also valid for the computations of μ by the MD method where the large positive values of ΔU_N can be forbidden by the energy conservation (cf. for instance [6.31, 32]). The efficiencies of (6.7) and (6.8) for a reliable estimate of μ have been compared by *Guillot* and *Guissani* [6.31] who conclude that (6.8) gives biased results.

Deitrick et al. [6.33] review the different methods for the calculation of μ. In this work, the authors propose a new method, the so-called "excluded volume map sampling" (EVMS) method. It consists in excluding, a priori, the insertion of the "virtual" Nth particle in a part of the volume of the simulation cell; this excluded volume is the union of sub-volumes surrounding each particle where the energy of a "virtual" particle inserted inside would be larger than a predetermined value.

A charging process is proposed by *Mon* and *Griffiths* [6.34] for the computation of μ. They use

$$\mu = \sum_{j=1,m-1} [F(N, V, T, \lambda_{j+1}) - F(N, V, T, \lambda_j)], \tag{6.9}$$

where $F(N, V, T, \lambda_j)$ is the free energy of a system in which the energy $U^N(\lambda_j)$ of the Nth particle is proportional to λ_j, with $\lambda_1 = 0$ and $\lambda_m = 1$. The internal energy has the form

$$U_N(\lambda_j) = U_{N-1} + U^N(\lambda_j) \tag{6.10}$$

for the jth step of the charging process. The gradual coupling of the Nth particle avoids, at high density, the problem of localising the empty cavities where the insertion of this particle gives a significant contribution to the ensemble average (6.7), but it increases the time of the computations. The test of this method performed in [6.34] is not very convincing.

Han et al. [6.35] have derived a method for the computation of μ which uses (6.8), and seems without systematic bias. The definition of $g(f)$, the density of probability such that the Nth particle has an internal energy f,

$$g(f)=\frac{\mathrm{e}^{-\beta f}\int d\boldsymbol{r}_1\cdots d\boldsymbol{r}_N\,\mathrm{e}^{-\beta U_{N-1}}\delta(\Delta U_N-f)}{\int\limits_{-\infty}^{\infty}\mathrm{e}^{-\beta f'}\int d\boldsymbol{r}_1\cdots d\boldsymbol{r}_N\,\mathrm{e}^{-\beta U_{N-1}}\delta(\Delta U_N-f')df'}, \tag{6.11}$$

allows us to write (6.8) as

$$\mathrm{e}^{\beta\mu}=\int\limits_{-\infty}^{\infty}\mathrm{e}^{\beta f}g(f)df. \tag{6.12}$$

But for large positive f, the precise estimate of the very small values of $g(f)$ which, multiplied by $\exp(\beta f)$, contribute to the integral on f in (6.12), is not possible by MC simulation. *Han* et al. propose to use the identity

$$\begin{aligned}\mathrm{e}^{\beta\mu}&=\frac{\int\limits_{-\infty}^{f_0}\Omega(f)df}{\int\limits_{-\infty}^{\infty}\mathrm{e}^{-\beta f}\Omega(f)df}\,\frac{\int\limits_{-\infty}^{\infty}\Omega(f)df}{\int\limits_{-\infty}^{f_0}\Omega(f)df}\\&=\frac{\int\limits_{-\infty}^{f_0}\mathrm{e}^{\beta f}g(f)df}{B},\end{aligned} \tag{6.13}$$

where $\Omega(f)=\int d\boldsymbol{r}_1\cdots d\boldsymbol{r}_N\,\mathrm{e}^{-\beta U_{N-1}}\delta(\Delta U_N-f)$.

B is evaluated from the ensemble average of the Heaveside function $\theta(f_0-\Delta U_N)$,

$$B=\langle\theta(f_0-\Delta U_N)\rangle_{N-1}. \tag{6.14}$$

The main interest of the formulas (6.13) and (6.14) is to show that μ is computable from configurations where the energy of the Nth particle is smaller than f_0. These configurations are easy to generate in a MC simulation if f_0 is not too large. The results obtained in [6.35] seem to prove that the "real" particle method in this formulation does not have the deficiency noticed in [6.31]. In the articles [6.31, 36], important size effects in the calculation of μ have been found. A formula for correcting these effects is derived by *Smit* and *Frenkel* [6.37]. Excluding the vicinity of the critical point, where the formula is not valid, these authors demonstrate that systems of ~ 100 particles are large enough to evaluate accurately μ. The articles of *Sloth* and *Sorensen* [6.38, 39] discuss the evaluation of μ in ionic systems, but the questionable use of the minimum image convention in the calculation of Coulomb interactions can seriously bias all the results obtained in these works. A similar remark applies to [6.40] on the same subject, also due to an approximate estimate of the long range Coulomb interactions in periodic systems. The formulas for the computation of μ in the Gibbs ensemble and the isobaric–isothermal ensemble are given, respectively, in [6.22] and [6.41].

(b) Free Energy. The evaluation of the free energy or the differences of free energy between two phases, is essential for the exact localisation of the first order phase transitions. The main methods for this computation are summarised by the relations

$$F(V_2,T)-F(V_1,T)=-\int_{V_1}^{V_2} P(V)dV,$$

$$F(\lambda)-F(0)=\int_0^{\lambda}\langle U_N(\lambda')\rangle_{\lambda'}d\lambda', \tag{6.15}$$

$$F_1-F_2=kT\log\frac{Q_1}{Q_2}, \tag{6.16}$$

where in (6.15) λ is a parameter the value of which modifies the expression of the potential energy and $\langle\cdots\rangle_\lambda$ is a canonical average for a given value of λ. These relations show that F can be evaluated by integration of the pressure along an isotherm, by a charging process or by the ratio of the partition functions of the same system (or different systems) in different thermodynamic states.

A charging process well suited for the calculation of the free energy of solids is proposed by *Frenkel* and *Ladd* [6.42]. Its validity is demonstrated in [6.42] and also by *Lutsko* et al. [6.43]. *Mon* [6.44] shows that the approximation

$$F(1)-F(\lambda)=\langle U(\lambda)-U(1)\rangle_\lambda \tag{6.17}$$

is useful if $U(1)$ is correctly chosen.

The evaluation of free energy differences in mixtures by a charging process supposes the choice of an adequate path in order to avoid too numerous intermediate steps. Proposals for the realisation of such evaluations are done and developed by *Jorgensen* et al. [6.45, 46] on specific examples. The charging process procedures in the isothermal-isobaric ensemble for the computation of Gibbs free energy differences in mixtures of simple or complex molecules are discussed, for instance, in [6.47–49]. These articles describe how the charging processes can be generalised to the transformation of the molecular species, allowing the calculation of Gibbs free energy differences between two aqueous solutions with different solute molecules, or between two conformations of macromolecules. *Tobias* and *Brooks* [6.50] discuss the same problem using the point of view of the potential of mean force between two molecules whose separation has a fixed value r. This potential is easily shown to allow the estimate of the free energy difference between states which differ only by the value of r. This result is used, for instance, by *van Eerden* et al. [6.51] to perform a calculation of the free energy of the association and of the decay of complexes formed between a molecule and an alkali-metal cation.

Chialvo [6.52] derives all the theoretical expressions useful for the practical evaluation of the infinite dilution activities, the Henry's constant ratios, etc. The same problem is considered by *Sindzingre* et al. [6.53] from the point of view of numerical simulations. An elegant method for the calculation of free energy differences between systems at different temperatures is derived by *Branka* and *Parrinello* [6.54]. For a given system, they consider the generalised Hamiltonian of *Nose* [6.28] devised for the MD simulations at constant T, and they remark that the averages of the moments of the variable which scales up and down the velocities during the time evolution, in order to keep T constant, are proportional to the ratio of the canonical partition functions of the system at two values of T. A method for estimating the entropy contribution of quasi-gaussian degrees of freedom is proposed by *Edholm* and *Berendsen* [6.55]; this method seems well adapted to this kind of computation for macromolecules.

In a phase equilibrium, the surface tension is the thermodynamic quantity characterising the interface; it is defined as the free energy derivative with respect to the interface area at constant volume and temperature. *Aloisi* et al. [6.56] take advantage of this definition for computing the surface tension from the free energy difference between diphasic systems enclosed in slabs of equal volumes but with slightly different sections.

The process of formation of an interface at constant volume and temperature, allows the calculation of the surface tension from the free energy difference between systems with and without an interface. By a "reweighting" of the configurations generated in this process, the surface tension for a slightly different temperature is estimated and the surface entropy, i.e. the temperature derivative of the surface tension, is obtained. *Wang* and *Stroud* [6.57] show the feasibility of such computations.

In the mixtures, the partial molar volumes, energies or enthalpies can be obtained from the derivatives of the thermodynamic potentials or from the computation of the averages of the fluctuation covariances of the concentrations, the pressure, the temperature, etc. *Debenedetti* [6.58] derives the formula for the use of this last method in simulation and proves its efficiency in [6.58, 59].

6.1.4 Algorithms for Coulombic and Dielectric Fluids

In the charged or dielectric systems, the range of the interactions is infinite. In the simulations of such systems with periodic boundary condition, the interactions between the particles inside the simulation cell and their periodically reproduced replicas cannot be neglected. The series of these interactions can be summed, but because the series is not absolutely convergent, a well defined mathematical procedure (Ewald summation method) must be used. This point is discussed by *de Leeuw* et al. [6.60, 61]. These authors give the following expression of the energy of N particles of charge q in a cubic cell of side L with periodic boundary condition, that we call the Ewald energy U_{Ew},

$$U_{\mathrm{Ew}} = \frac{1}{L}\left\{\sum_{\mathbf{n}}\sum_{ij}^{*} q_i q_j \operatorname{erfc}\frac{a|\mathbf{x}_{ij}+\mathbf{n}|}{|\mathbf{x}_{ij}+\mathbf{n}|} + \frac{1}{\pi}\sum_{\mathbf{k}>0}\left[\frac{1}{k^2}\exp\left(\frac{-\pi^2 k^2}{a^2}\right)\left|\sum_i q_i \exp(2\pi \mathrm{i}\mathbf{k}\cdot\mathbf{x}_i)\right|^2\right]\right\} + \frac{2}{3\pi L^3}\left(\sum_i q_i \mathbf{r}_i\right)^2 + C, \tag{6.18}$$

where $\mathbf{n}$ are vectors with integers components, $\mathbf{x}_{ij}$ is $\mathbf{r}_{ij}/L$, $\mathbf{k}$ are the reciprocals vectors of the cubic lattice of cells, a is a parameter, $\sum_{ij}^{*}$ indicates $i \neq j$ if $\mathbf{n}=0$, and C is a constant.

A different approach is developed by *Cichocki* et al. [6.62]. They solve the Poisson equation of the charges in the simulation cell, accounting for the periodicity by the condition that the normal component of the electric field is zero on the boundary of the cell. The internal energy is expressed in terms of the Wigner potential (potential of a charge q at the centre of the basic cell of a cubic lattice with a neutralizing background). This energy U_{W} is equivalent to U_{Ew}, but its mathematical expression is different. This approach has the advantage of making obvious the important points that, in U_{Ew} or U_{W}, the charge–charge interaction, being solution of the Poisson equation, satisfies the Gauss theorem and has all the physical properties of the Coulomb interaction (screening effects, Stillinger–Lovett rules, etc.), and that the interactions between electric multipoles are consistently derivable by repeated applications of the gradient operator on the charge–charge interaction. The previous remarks suggest that a bias can appear in simulations of Coulomb or dielectric systems where, for numerical convenience, the interaction between charges is not a solution of the Poisson equation, for instance, Coulomb potential and minimum image convention or truncated point dipole–dipole interaction and reaction field.

The numerical computation of U_{Ew} or U_{W} depends on their mathematical expressions. In U_{Ew}, with an adequate choice of a, the sum on $\mathbf{n}$ can be limited to the term $\mathbf{n}=0$ and the sum on $\mathbf{k}$ to ~ 100 terms. The Wigner potential is the sum of a Coulomb potential, a harmonic potential and a series of spherical harmonics. A precise estimate of this series is derived in [6.62] which can be used for a fast numerical evaluation. The expression of U_{W} for 2D systems is given in [6.63].

It is not simple to get an expression of the charge–charge interaction in 3D systems that are infinite and periodic in only two directions, and thus useful for numerical simulations. The derivation of U_{Ew} following the method of [6.60] leads to an expression where the double sum on the vectors $\mathbf{k}$ and the particles is replaced by a double sum on the vectors $\mathbf{k}$ and the pairs of particles, making

U_{Ew} useless for numerical simulations. In [6.64–67] different approaches are proposed to solve the problem. *Rhee* et al. [6.64] show that in U_{Ew} it is possible to retain a double sum on the **k** vectors and particles by using in the summation procedure (cf. [6.60]) an adequate factor of convergence. But with this factor, the sum on the **n** vectors in U_{Ew} is not exponentially convergent and to accurately compute U_{Ew}, at least eight terms with $\mathbf{n} \neq 0$ must be included. *Kuwajima* and *Warshel* [6.65] propose to calculate the sum on the **k** vectors by using an expansion on the basis of the harmonic polynomials. However the proof that this expansion is useful for simulations of charged inhomogeneous fluids is not established. In the articles [6.66, 67], Smith, and Cichocki and Felderhof discuss the extensions of their methods [6.60, 62] for the computation of U_{Ew} (or U_{W}) to the evaluation of the energy of a 2D lattice of cells of charged particles between parallel planes. Smith considers cells which are sandwiched by two semi-infinite identical dielectric continua and derives a numerical procedure which avoids the double summation on **k** vectors and the pairs of particles. But in practice, the use of this procedure does not seem easy. Cichocki and Felderhof derive the form of U_{W} for a 2D lattice of cells limited by two parallel planes along the third direction and give the numerical coefficients of the expansion of the Wigner potential on the set of spherical harmonics.

The representation of pair interactions by the sum of a potential computed in r-space for the short-range part, and of a Fourier series computed in k-space for the long-range part, is demonstrated useful if the range of the interactions span a dimension comparable to the size of the system by *Plischke* and *Abraham* [6.68].

6.2 Pure Phases and Mixtures of Simple Fluids

The recent numerical works on pure phases of simple fluids, hard disks, hard spheres, or LJ fluids, have been performed to obtain their equations of state with an improved accuracy compared to previous calculations. For the mixtures, the localisation of the phase transitions and the determination of excess mixing thermodynamic properties are the main motivations of the simulations.

6.2.1 Two-Dimensional Simple Fluids

The simulations of the hard disk system carried out by *Schreiner* and *Kratky* [6.69] allow an analysis of the finite size effects due to the shape of the area enclosing the disks and the value of N, the number of the disks. These effects are shown to be in agreement with theoretical predictions. The results of the most extensive computations on the hard disk systems since 1954 are presented in [6.70] by Erpenbeck and Luban. The size of the systems goes from $N = 1512$ to $N = 5822$ and up to 10^8 collisions are generated in a simulation run. By using the formulas for the extrapolation to the infinite system size derived in

[6.69] Erpenbeck and Luban obtain 10 values of the compressibility factor with an error of 10^{-4} in the range of density $1/30 - 1/1.4\,\rho_0$ (ρ_0 is the density of close packing). An analytic representation of these data by a Levin approximant leads to an extremely accurate equation of state of the fluid phase of the hard disks.

The adhesive or "sticky" hard disks are a simple system of strongly interacting particles which can undergo crosslinking to long chain molecules. The simulations [6.26, 27] demonstrate the feasibility of MC computations for such particles with singular pair interactions.

A complete determination of the equation of state of the 2D LJ fluid for the temperature range 0.45–5 and density range 0.01–0.8 (LJ reduced units), was performed by *Reddy* and *O'Shea* [6.71] on the basis of the data of 300 simulations and of the virial series of the pressure. The data are fitted by a polynomial equation of state depending on 33 parameters. The critical point is localised at $T_c^* = 0.54$ and $\rho_c^* = 0.36$. From the results of simulations for systems having from 256 to 10864 particles, along the isotherm $T^* = 1$, *Bruin* et al. [6.72] do not detect significant size effect in the 2D LJ fluid.

A computation of the gas–liquid equilibrium for the 2D LJ system is done by *Singh* et al. [6.73] in the Gibbs ensemble. These authors estimate that the critical point is located at $T_c^* = 0.47$ and $\rho_c^* \sim 0.33 \pm 0.2$; this value of T_c^* disagrees with the value reported in [6.71] and is not very convincing because at $T^* = 0.468$, *Singh* et al. obtain ρ (gas) $= 0.042$ and ρ (liquid) $= 0.603$, densities which are rather far from the estimated critical density. This doubt on the possibility of a precise determination of the critical point in 2D LJ fluid, is reinforced by the two works of *Rovere* et al. [6.74, 75] who attempt to apply to fluid, the finite-size scaling procedure used successfully for the lattice systems. No definitive conclusion can be attained for the reasons which are summarised in the final section of [6.75]. The most important point is that the finite-size scaling procedure in fluids demands considerable computational resources due to the fact that the update of one spin in a lattice system is ~ 100 faster to compute than the acceptance of a trial move of a particle in a fluid. The final estimate of *Rovere* et al. for the critical constants is $T_c^* = 0.50 \pm 0.02$ and $\rho_c^* \sim 0.36$.

The nature of the 2D fluid–solid transition which was believed first order, has been questioned by *Nelson* and *Halperin* [6.76] who predict an intermediate phase between the "true" solid and fluid, the so-called "hexatic phase" formed by an anisotropic liquid. A review of the theoretical aspects of this question is done in [6.77] by Strandburg. Extensive computer simulations have been devoted to the search for the "hexatic phase" in the 2D hard disks and LJ systems. The articles [6.78–81] are examples of these works. The unambiguous manifestation of the hexatic phase is the algebraic decrease of the two-body orientational correlation function which describes the orientational ordering of the first shell of the six neighbours of a disk or a LJ atom. Clearly, such an asymptotic behaviour can be strongly affected by size effects. Finally, no firm conclusion about the occurrence of the hexatic phase has been obtained, cf. the following quotations of [6.78]: "... our study of the orientational correlation function therefore clearly points towards a first-order melting" and of [6.81]: "We are therefore left with a puzzle. The thermodynamics of the melting transition in a 2D LJ system suggests a first order transition. But the 'two phases region' is highly unusual. We find no evidence for coexistence between bulk crystal and fluid."

The percolation of 2D LJ or square-well atoms in a fluid phase is studied by *Heyes* and *Melrose* [6.82]. For the LJ fluid, two atoms are considered connected if their distance is smaller than a definite value s_c; for the other fluid, the range of the well defines the distance of connection. The percolation thresholds are computed in the square-well fluid and in the LJ fluid for different choices of s_c.

6.2.2 Three-Dimensional Monatomic Fluids

Very extensive MD and MC simulations of the 3D hard sphere system are reported by *Erpenbeck* and *Wood* [6.83]. The simulations were performed for system sizes ranging from 108 to 4000 spheres and for each density the evolution of the system was calculated for 10^7 or 10^8 collisions. The extrapolation of the pressure towards the thermodynamic limit was made by using the known N dependence of the NPT ensemble; the values of the relative errors on the pressure for $N=\infty$ were estimated as 10^{-4}. Combining the known values or the MC estimates of the virial coefficients with the simulation results, a Padé approximant is determined which summarises the simulation data within their statistical error and is, today, the most accurate equation of state of the 3D hard sphere fluid phase. The radial distribution functions (rdf), $g(r)$, of the hard sphere system in the vicinity of the freezing transition are computed with an excellent precision by *Groot* et al. [6.84] and by *Labik* and *Malijevsky* [6.85]. From these simulation data, they are able to calculate the direct correlation functions $c(r)$ and to compare them to empirical analytic formulas with adjustable parameters. In spite of a very impressive agreement inside the hard core between these formulas and the MC estimates, uncertainties remain on the amplitude of the small oscillatory tail of $c(r)$ outside the hard core diameter.

The three body correlation functions $g_3(\boldsymbol{r},\boldsymbol{r}')$ and the three body direct correlation functions $c_3(\boldsymbol{r},\boldsymbol{r}')$ of the hard spheres are computed by *Haffmans* et al. [6.86] and *Rosenfeld* et al. [6.87]. The motivation of the simulations in [6.86] is a theoretical computation of the viscosity and diffusivity of noble gases. The $g_3(\boldsymbol{r},\boldsymbol{r}')$ functions are determined in Fourier space and only for the geometrical arrangements where two of the three spheres are in contact. A comparison is done with two approximate expressions of $g_3(\boldsymbol{r},\boldsymbol{r}')$ which are found to be in almost quantitative agreement with the MD data. In [6.87], the purpose of the MC calculation is to obtain a reliable numerical estimate of $c_3(\boldsymbol{r},\boldsymbol{r}')$, useful for testing the validity of the density functional theory (DFT) of melting; as in [6.86], the computation is made in Fourier space, but 10^9 configurations were needed to achieve an accuracy of $\sim 10\%$ on the numerical results. A theoretical form of $c_3(\boldsymbol{r},\boldsymbol{r}')$ is shown to be in quasi-quantitative agreement with the MC data.

The metastable amorphous state between the melting density and the random packing is studied by *Tobochnik* and *Chapin* [6.88] for 2D and 3D hard spheres. These authors use simulation cells which are the surfaces of the ordinary sphere and of the 4D hypersphere. This choice suits the study of the amorphous state by avoiding the nucleation of the system to the solid phase, because there is no regular periodic arrangement of the disks or of the spheres on such surfaces. The random closest packings are found for 85% (2D) or 68% (3D) of the close packing density and the rdf show characteristics typical of an amorphous phase.

The equations of state of the hard spheres of dimensionality higher than 3 have been computed by MC simulations for D equal to 4 and 5 by *Michels* and *Trappeniers* [6.89] and exactly for $D = \infty$ by *Frisch* et al. [6.90]. The simulations show that the fluid–solid transition occurs at lower density, relative to close packing density for the increasing value of D. At $D = \infty$, there is no transition. The equations of state from 2D to 5D are summarised in a unique analytical form by *Luban* and *Michels* [6.91].

The formulas needed for MD simulations of the hard sphere system in generalised ensembles [6.28–30] are derived by *de Smedt* et al. [6.92]. The quantum corrections for the hard sphere fluid at high temperature are estimated by *Yoon* and *Scheraga* [6.93]; they use a classical canonical partition function which is a product of two-body Slater sums for two hard spheres. The onset of the momentum condensation has been found.

The 3D adhesive ("sticky") hard sphere system is a prototype of the colloidal suspensions which have a tendency to form gels or colloidal crystals. The most interesting physical properties of this fluid are a gas–liquid transition and the dependence on the temperature of its percolation threshold. Both properties are studied by MC simulations by *Seaton* and *Glandt* [6.26, 94] and by *Kranendonk* and *Frenkel* [6.27]. The equation of state at low density, the rdf and the critical constants are in close agreement with their determination made by the Percus–Yevick (PY) integral equation theory. For higher densities, the system attains its percolation threshold and obvious large disagreements appear because of the unphysical predictions of the PY theory. The simulations reported in [6.26, 94] are made in the NVT ensemble and in [6.27] in the NPT ensemble. The NPT ensemble avoids the pressure estimate by a difficult extrapolation of the rdf contact values which relies on the hypothesis of the continuity of the function $\exp[\beta u(r)]g(r)$ (here $\beta u(r)$ is a formal notation for the hard core and the "sticky" potentials).

To test the feasibility of GCMC simulations in the dense fluids, *Mezei* [6.95] calculates the thermodynamic properties of the r^{-12} particles (soft spheres). For the same system, *Barrat* et al. [6.96] report the calculation of $c_3(\boldsymbol{r}, \boldsymbol{r}')$, validating an approximate theoretical form. *Balucani* and *Vallauri* [6.97] study the angular distribution of the directions of the vector distances between a particle and its nearest neighbours. This distribution gives informations on $g_3(\boldsymbol{r}, \boldsymbol{r}')$ in the liquids, the supercooled liquids or the glasses. The distribution calculated from the superposition approximation for $g_3(\boldsymbol{r}, \boldsymbol{r}')$ is in qualitative agreement with simulation data.

Extensive MD and MC simulations are reported by *Hayes* and *Woodcock* [6.98] for the systems of particles interacting by a hard core potential with attractive inverse power tails. They derive for these systems an analytic equation of state of van der Waals type.

6.2.3 Lennard–Jones Fluids and Similar Systems

The structural and thermodynamic properties of the LJ fluid are studied by *Vogelsang* and *Hoheisel* [6.99–101]. In the first article, the simulations are performed over wide ranges of density and temperature; the computations are used for checking the physical assumptions of the perturbation theories of the liquid state. In [6.100], the supercritical LJ fluid is investigated, the rdf and the three-body correlation functions calculated numerically allow the test of the validity of perturbative estimates of these functions. The need to carefully correct the simulation data obtained for the low density states is established in [6.101]. The accuracy of the perturbative approaches of van der Waals type is also confirmed by *Rull* et al. [6.102]. The validity of various approximations for the calculation of the rdf for systems of particules interacting by a hard core potential with Yukawa tail, is discussed by *Konior* and *Jedrzejek* [6.103]. The statistical distribution of the values of the components of the forces in LJ fluids has a quasi-exponential form. *Powles* and *Fowler* [6.104] propose an approximate interpretation of this empirical fact.

The gas-liquid equilibrium of the LJ fluid which has been determined by free energy computations in the canonical and grand canonical ensembles, is computed by the Gibbs ensemble procedure as a test of its efficiency [6.20, 23]. An excellent agreement is obtained with the previous calculations and with the results of the GC simulations, except in the vicinity of the critical point. There is no simple explanation of these significant differences other than an insufficient sampling of the configurations of the di-phasic system in the GC simulations. In any case, the works [6.20, 23] establish the ability of the Gibbs ensemble method to determine accurately and efficiently, the phase equilibria of fluid systems.

The gas-liquid transition of a "two-Yukawa fluid", modelling the LJ fluid, has been studied by *Rudisill* and *Cummings* [6.105]. The motivation of this unusual calculation is to prove the similarity of the two systems, in order to apply to the LJ fluid and other fluids, the analytical approximations developed for the two-Yukawa fluid.

Quantum effects are generally small in the rare gases Ar, Kr or Xe, but they cannot be neglected in theoretical or numerical calculations that wish to achieve a quantitative agreement with experimental results. Using the LJ potential as a model of the rare gas interactions, *Barocchi* et al. [6.106] estimate the quantum corrections up to the order h^6 for these fluids. These corrections are obtained by MC calculations of the ensemble averages of the n-body correlation functions associated with the h expansion. If this expansion converges, the method avoids the use of path-integral MC simulations. The quantum effects contribute $\sim 10\%$ to the pressure and kinetic energy in Ne, D_2 and He at all densities in the temperature domain 50–100 K. Significant displacements of the main peak of $g(r)$ are obtained, which are confirmed by comparison with path integral simulations in the domain of values of r where the h expansion is convergent.

An original MD method is proposed by *Chapela* et al. [6.107] for the systems with pair potentials. It consists of using a discretised form of the potentials to enable the computation of the particle trajectories by a generalisation of the MD algorithm for the "hard sphere plus square well" particles. The method is tested on the LJ fluid and seems a viable alternative to MC or continuous MD simulations.

From the comparison between MC simulation data and the results of integral equations and perturbation theories for a helium-like system at high pressures and high temperatures, *Talbot* et al. [6.108] conclude that these theories faithfully predict the properties of atomic fluids in the full domain of the gas and liquid phases. For the atomic fluids, *Reatto* et al. [6.109] prove the validity of a procedure devised for the computation of two-body effective potentials from the experimental static structure factor $S_{ex}(k)$ and from the experimental rdf $g_{ex}(r)$. The method is an iterative predictor-corrector process which begins with an estimate of the two-body potential by inverting $S_{ex}(k)$ using an integral equation theory. For this first potential, $g(r)$ and $S(k)$ are obtained by MC simulation and are used to calculate, from their differences with $g_{ex}(r)$ and $S_{ex}(k)$, a corrected potential. The process is continued until the $g(r)$ and $S(k)$ computed by simulation are within their statistical uncertainties equal to $g_{ex}(r)$ and $S_{ex}(k)$.

The results of the chemical potential computations for the LJ fluid are given in [6.22, 31, 33, 34, 37, 110, 111]. The technical details of the sampling procedures have already been discussed. In [6.111], Stapleton and Panagiotopoulos combine the Gibbs ensemble method extended to mixtures, and the so-called EVMS technique derived in [6.33]. In [6.36], the values of μ have been found to depend on the cut-off of the LJ potential tail used in the simulations, this dependence has not been obtained in other calculations (cf. [6.37]). In [6.22], the generalisation of the formulas (6.7) and (6.8) for the Gibbs ensemble is performed and used to calculate μ along the gas–liquid coexistence line. A careful comparison between the different ways to compute μ is presented in [6.33].

The melting and crystallisation of the LJ systems is analysed in detail by *Chokappa* and *Clancy* [6.112], and *Nosé* and *Yonezawa* [6.113]. Both studies are performed with the MD isothermal-isobaric algorithm. In [6.112], the limits of the mechanical stability of the fluid and solid phases are determined. The hysteresis loops enclosing liquid, solid, superheated solid and supercooled liquid are obtained in the enthalpy-temperature phase diagram. The arithmetic average of the temperatures of the extreme stable superheated solids and supercooled liquids is empirically shown to give the temperature of the thermodynamic liquid–solid equilibrium. Nosé and Yonezawa report very interesting qualitative modifications of the structural order during heating and cooling. The importance of the attractive tail of the LJ potential for the localisation of the melting line is investigated by *de Kuijper* et al. [6.114]. They show that the attractive tail displaces the line towards higher pressure if T is fixed, and at low temperature, increases the volume variation between the two phases. The solid phases of the Yukawa systems are determined by *Robbins* et al. [6.115] who give an approximate localisation of the melting line.

Stillinger and collaborators [6.116–118] have analysed the structural order of the liquid phase in terms of the arrangements of minimal energy obtained by a steepest descent procedure starting from a liquid configuration. This arrangement is the so-called inherent structure; its analysis reveals many interesting qualitative differences in the local atomic order of simple liquids which freeze into fcc or bcc solid phases, and also the characteristic features of the arrangements of the metastable high density fluid near the glass transition.

The results of quenching process of LJ systems are studied by *Tanaka* [6.119]. The quenched configurations of the LJ fluid and the liquid–metal model systems are similar and are a result of the stabilisation of the short range order in local arrangements of minimal energies.

In the 3D well fluids, clusters can be defined by the association of atoms, in which the well of each atom overlaps at least another well. The growth of these clusters and the localisation of the percolation threshold are studied by *Chiew* and *Wang* [6.120] and *Heyes* [6.121]. From the simulation data, it is shown that the increase of the well range makes the percolation threshold less dependent on temperature, and that for the repulsive well the density of the threshold is higher than for the attractive well. The localisation of the percolation thresholds computed for the square-well fluid from the PY approximation is in good agreement with the simulation results. *Heyes* and *Melrose* [6.122] perform a similar study for the 3D LJ fluid, where two atoms are associated if their separation is smaller than a given value d_S. In spite of the arbitrary value of d_S, clustering and percolation processes of this system manifest many characteristics identical to those of the same processes in lattice systems.

6.2.4 Real Fluids

For a significant comparison with experiment, the simulations of real fluids require a good knowledge of the pair potential interactions and also an estimate of the three-body or many-body interactions. For Ne, H_2 or D_2, this comparison will be possible only after an accurate calculation of the quantum effects.

Zoppi et al. [6.123] present the results of careful measurements of the static structure factors of D_2 in the critical region where unexpected Raman spectra have been reported. From the simulations of a LJ model of D_2 they conclude that there is no enhanced ordering in the critical region that is able to induce the fine structure of the Raman experiment. *Barocchi* et al. [6.124], using the method of the calculation of the quantum corrections described in [6.106], accurately reproduce the experimental rdf of Ne at ~ 35 K, the small discrepancies being possibly due to the shortcomings of the LJ potential which models the Ne–Ne interaction. For He and Ne, *Singer* and *Smith* [6.125] compare the results of classical MC and path integral simulations to estimate the importance of the quantum effects for temperatures larger than 20 K. The efficient sampling of the GC ensemble phase space devised by *Baranyai* and *Ruff* [6.126] is applied by *Ruff* et al. [6.127] to the computation of the thermodynamic and structural properties of liquid Ar. In the dense liquid, the atoms are arranged in a distorted hexagonal close packed structure. The simulations in the GC ensemble are shown to be possible up to a density 0.88 (LJ reduced units) by inserting the particles on the vertices of the Dirichlet–Voronoi polyhedra surrounding each atoms.

The LJ potential, parametrised to reproduce the properties of Ar, is compared to a more realistic potential including pair and three-body interactions in [6.128]. The results of simulations for $T \sim 100$ K show that the two potentials give thermodynamic properties and rdf in similar agreement with experiment. The consequences of important three-body interactions in Kr and Xe are investigated by *Levesque* et al. [6.129]. These interactions give essential contributions to the pressure and internal energy; but their influence on the rdf is very small. *Broughton* and *Li* [6.130] examine the ability of the Stillinger–Weber potential [6.131] to reproduce the properties of liquid, amorphous and solid Si. An excellent agreement between the theoretical and experimental results is obtained for the states in the vicinity of the triple point, but the amorphous phase is poorly described.

The effective interionic potentials of liquid metals result from the direct interaction between ionized atoms screened by the electrons of conduction. These interactions being difficult to calculate theoretically, empirical parametrised potentials have been proposed in the literature. *Dzugutov* et al. [6.132] adjust the parameters of a potential for liquid Pb by numerical simulations. They reproduce very accurately the experimental structure factor of Pb at 1110 K. Using the inversion method derived in [6.109], *Bellissent-Funel* et al. [6.133] have determined an effective potential for liquid Ga which has been proved by simulation to be compatible with the experimental data for $S(k)$ at the temperatures 326 K and 959 K. The efficiency of the same inversion procedure is also established by *Dzugutov* [6.134], who obtains a potential for liquid Pb in good agreement with the empirical potential determined in [6.132]. *Mentz-Stern* and *Hoheisel* [6.135] make the unfounded proposal of a six-centre LJ potential to model the interaction between atoms in liquid Pb. Obviously, there are rather large discrepancies between theoretical and experimental $S(k)$, for instance the height of the main peak differs by 10%. This work shows again that an accurate agreement between theoretical and experimental $S(k)$ must be achieved to get reliable informations on the effective two-body potential in liquid metals; at the level of an accuracy of 10–15% on $S(k)$, the possible forms of the effective potentials are so different that no valuable physical information can be extracted from the experiment-simulation comparison. *Hafner* [6.136] examines the experimental $S(k)$ for liquid As, and the results of a simulation performed with a theoretical potential. The agreement is qualitative, suggesting quantitative shortcomings of the theoretical interaction. Other studies have been done by *Arnold* et al. [6.137] and *Jank* and *Hafner* [6.138] who compute the structure factors of the Ge, Si, Sn and Pb liquids. The interatomic interactions are derived from the pseudo-potential and linear response theories. The agreement between the simulation results and experiments is too insufficient for it to be a stringent test of the two-body potentials. Using another class of potentials (Finn–Sinclair many-body potential) *Holender* [6.139, 140] obtains a qualitative description of the melting properties of Cu, Ag, Au and Ni. The positions of the main peak of the rdf are in agreement with the experiment but their heights are overestimated by ~10%.

Gonzalez Miranda [6.141] gives the results of similar calculations for Li, Na, K, Rb and Cs near their melting points. He reports large changes of the structural and time dependent properties with the variation of the parameters in the effective two-body potential.

6.2.5 Mixtures of Simple Fluids

(a) Hard Core Systems. The perturbation theories of the simple fluids which rely upon the use of the monodisperse hard sphere fluid (reference system) are extended easily to the mixtures. However, this extension will be useless in practice unless an accurate equation of state of the hard spheres mixtures is known. It is the purpose of *Jackson* et al. [6.142] to provide a precise estimate of the validity of empirical equations of state of these mixtures. They consider the mixtures with diameter ratios $d_1/d_2 = 5/3, 3, 5$ and 20. They find a good agreement with the previous simulations on the same systems and report that the equation of state of *Mansoori* et al. [6.143] is fairly accurate except for high densities and large diameter ratios. At high densities, the 5/3 mixture seems to freeze for the packing fraction $\eta \sim 0.52$, for the diameter ratio 3, the solidification of the large component is observed. The mixture with $d_1/d_2 = 5$ seems to exhibit a continuous transition from solid to amorphous fluid and at $d_1/d_2 = 20$ the large component is solid for $\eta \sim 0.53$ and concentrations larger than 0.5. More precise localisations of the solid–fluid transition are obtained by *Kranendonk* and *Frenkel* [6.144] for $d_1/d_2 = 0.95$ and 0.90; the phase diagram of the first mixture is spindle-like, for the other mixture the computer results indicate the presence of an azeotropic point for the concentration of the large component equal to ~0.3. The results of the simulations are in disagreement with the predictions of the DFT.

In the perturbation theories, the hard sphere mixtures with additive diameters are a useful reference system only for the mixtures where the repulsive interactions have ranges that approximately satisfy the addivity rule. Thus, for the extension of these theories to other mixtures, it is interesting to have data on the properties of the non-additive hard sphere mixtures. Defining the non-additive character of the mixtures by $\Delta = 2d_{12} - (d_1 + d_2)$ where d_{12} is the range of the cross hard cores, both cases $\Delta < 0$ and $\Delta > 0$ have been considered. For $\Delta > 0$, unmixing processes are observed with increasing density.

Three works [6.145–147] have recently been published on the non-additive hard spheres. In [6.145], the main purpose is to compare integral equations theories with the MC data, good agreements are reported for the pressure and the rdf; an empirical equation of state is derived for the equimolar mixtures. New forms of this equation are proposed in [6.146] and excellent agreement is obtained with MC simulation results. The mixture $d_1 = d_2$ and $\Delta = 0.2$ is investigated in [6.147], using the Gibbs ensemble. For the equimolar composition, the unmixing process occurs for $\eta > 0.2$ and at $\eta \sim 0.30$, the equilibrium is obtained between almost pure phases. The MC evaluations of the cavity distributions of hard sphere mixtures, from which the direct correlation functions and the bridge functions can be deduced, are compared with theoretical expressions by *Labik* and *Smith* [6.148]; the theoretical formulas of *Boublik* [6.149] are the most accurate.

The equations of state of soft sphere mixtures ($(\sigma/r)^n$ potentials) where the σ_{12} parameters satisfy an addivity rule, are found to be qualitatively different from the equations of state of the additive hard sphere mixtures. For $n = 6$, a phase separation occurs for $\sigma_1/\sigma_2 = 3$ and equimolar mixture. The ideal mixing rule is very accurate for these mixtures [6.150]. The structural properties of additive and non-additive soft sphere mixtures are computed by *Mountain* [6.151], where in both cases an attractive or repulsive tail has been added to the soft core interactions. The rdf exhibit a compositional ordering which can be associated with the compositional ordering in the solid phases. Very extensive MC simulations of the square well fluid mixtures offer the possibility of performing stringent tests of approximate formulas of the local ordering and bring also new information on the equation of state of these fluids [6.152].

A mixture exhibiting very large non-additive hard core effects is the Widom–Rowlinson model, formed by two perfect gases with a cross hard sphere potential. *Borgelt* et al. [6.153] study a similar model where the hard sphere interaction is replaced by a $(\sigma_{12}/r)^{30}$ potential. The simulations localise the unmixing transition at the density $\rho\sigma_{12}^3 = 0.41$ for the equimolar mixture, and at 0.50 for the 75% mixture. These values are in disagreement with the PY predictions 0.57, and 0.67 respectively, for the two mixtures. The mixtures of atomic and molecular hydrogen, modelled by non-additive pair potentials, is studied by *Saumon* et al. [6.154]. A theory of perturbation is shown able to reproduce the simulation data.

(b) LJ Mixtures. The LJ mixtures are the prototype models of rare gas mixtures, for various values of the LJ parameters σ_{ij} and ε_{ij}. *Schoen* and *Hoheisel* [6.155, 156] analyse the behaviour of the static cross correlation functions defined by the sum of the local mole fraction and by the ratio between the unlike rdf and the sum of the two like rdf. They show that these functions are suitable for the distinction between stable and unstable compositions of the mixture. A long range of the cross correlation functions indicates unmixed states without ambiguity. The stability of the mixtures is studied for various values of σ_{ij} and ε_{ij}, a gas–gas phase separation is obtained for an adequate choice of these parameters. In [6.157], Nakanishi calculates the excess internal energy of mixing u_{ex} for LJ mixtures. From the variation of u_{ex}, he concludes that these mixtures are probably not athermal. The excess Gibbs free energy, excess enthalpy, excess volumes, etc., are carefully calculated by isothermal-isobaric simulation by *Shukla* and *Haile* [6.158]. The data allow testing of the first order perturbation theories based on the reference system of additive hard sphere mixtures.

In mixtures, the cavity distribution functions are computable from the potential distributions of test particles, by a generalisation of (6.7) and (6.8). As mentioned above, these functions contain information on the bridge functions. Their exact evaluations by simulation allow testing of the hypotheses made on the bridge functions in the integral equation theory. Hence, *Lee* and *Shing* [6.159] demonstrate that the well known RHNC integral equation is a very reliable theory of mixtures of simple fluids.

By isobaric Gibbs ensemble simulations, *Panagiotopoulos* et al. [6.23, 111] determine the phase equilibria of LJ systems. In [6.111], the improved insertion technique of *Deitrick* et al. [6.33] is used for exchanging the particles between the two bulk systems. This improved technique slightly modifies (a few percent) the results of [6.23]. Two of the considered mixtures exhibit spindle-like and "retrograde" phase diagrams, respectively. The application of the semi-grand ensemble technique to multicomponent systems is described in [6.24]. Simulations of the LJ mixtures are performed and the computed phase equilibria are in good agreement with the results of [6.111].

Jonsson and *Andersen* [6.160] have adapted the steepest descent procedure of [6.116–118] to the constant pressure MD calculations, and they have determined the inherent structures for the temperatures higher and lower than the temperature of the glass transition of two component LJ systems; below the glass transition the inherent structure is predominantly icosahedral.

Analytical equations of state for exponential-six mixtures are derived by *Brown* [6.161] and checked by MC simulations. The infinite dilution chemical potential $\mu(\infty)$ or Henry constant, are fundamental data for the study of solutions. The possibility of computing accurately these quantities by perturbation theories is discussed by *Shing* et al. [6.162]. They perform MC simulations for infinitely dilute LJ mixtures where the LJ parameter ratios $\varepsilon_{AB}/\varepsilon_{BB}$ and σ_{AB}/σ_{BB} are respectively smaller than 2 and 3.5. The computation of $\mu(\infty)$ is done using (6.7) at the low and moderate densities of the solvents. At higher densities the charging process method is used. The value of $\mu(\infty)$ is found to depend mainly on $\varepsilon_{AB}/\varepsilon_{BB}$ for $\varepsilon_{AB}/\varepsilon_{BB} > 1$; a good agreement with the van der Waals perturbation theory is obtained, but for $\varepsilon_{AB}/\varepsilon_{BB} < 1$, significant differences appear due to the overestimate of the values of $\mu(\infty)$ by the theory. Similar computations have been done by *Lofti* and *Fischer* [6.163] who give the results for the LJ mixtures $\varepsilon_{AB}/\varepsilon_{BB} < 1$ and $\sigma_{AB}/\sigma_{BB} < 1$. With a convenient choice of the LJ parameters, these authors determine $\mu(\infty)$ for the mixtures He/CH_4, H_2/N_2, H_2/Ar, H_2/CH_4, Ne/Ar and Ne/Kr. The comparison with the experiments is not conclusive due to the uncertainties on the values of LJ potential parameters.

The rdf functions of the solute particles in the limit of low concentration is calculated by *Petsche* and *Debenedetti* [6.164] for the mixture Ne/Xe modelled by a LJ mixture. Both cases of a weak concentration of Ne and Xe are considered, the value of the concentration being 1/863. The environment of solute particles differs near the critical point of the solvents (Ne or Xe); the Ne particle is surrounded by a Xe cluster, and in the reverse situation, the vicinity of Xe is depleted in solvent. *Vogelsang* et al. [6.165] determine the partial enthalpies of the Ar/Kr mixtures by using the method proposed in [6.53]; these mixtures are strongly non-ideal. The rdf of Ar/Kr mixtures at 135 K and 12 mPa in the composition range 20–70% are given by *Balcells* et al. [6.166]. For the Ne/Xe mixtures at ~20 kbar, *Hoheisel* [6.167] shows that a LJ model gives a good description of the fluid phase and is able to reproduce the experimental localisation of the fluid unmixing.

(c) Polydisperse Fluids. The polydisperse systems are colloidal systems where the particles in suspension have variable sizes. There are two types of such systems: the systems where the size distribution is independent of the thermodynamic state, e.g. latex particles in a solvent, and the colloidal suspensions where the size distribution varies with the thermodynamic state, e.g. the micelles in water-surfactant mixtures. For the first type of systems, it is possible to perform NVT simulations, where the sizes of the N particles are selected from a known distribution. In the second case, the determination of the composition is one of the aims of the simulations and the good ensembles are the GC ensembles or the semi-grand ensemble. The work of *Frenkel* et al. [6.168] is an example of the first type of simulations. They calculate the static structure factors of polydisperse additive hard spheres with a normal distribution of their diameters. The computation is done for several intraparticle interference factors. The MC data are found to be in good agreement with the predictions of the PY theory, except for high densities and large polydispersities.

Stapleton et al. [6.169, 170] consider the second type of polydisperse fluids. They study in [6.169], a system of hard spheres and of LJ particles; in both cases the diameters d_{ij} and the parameters σ_{ij} are additive. The simulation is done in the canonical ensemble by using the following procedure. The particles are supposed to have an a priori size distribution $f_0(d)$. Starting with a given composition selected from $f_0(d)$, the particles are moved and simultaneously new diameters are sampled from $f_0(d)$; these new positions and diameters are accepted according to the transition probability of the Metropolis algorithm. This procedure is equivalent to the semi-grand ensemble simulations and generates the equilibrium size distribution. In [6.170], the authors show how the simulations of the polydisperse fluids can be performed in the GC and Gibbs ensembles. With the latter technique, the phase equilibria of polydisperse systems are computed. This is the main result of [6.170] and is the first successful simulation of such equilibria. The "infinitely polydisperse" fluids are characterised by a logarithmic relation between the chemical potential and the sizes of the particles.

In [6.171], Kofke and Glandt show that for the "infinitely polydisperse" hard spheres the equation of state is $\beta PV/N = 4/3$ at all the densities.

The study of the fluids interacting by potentials depending strongly on the relative orientations and distances of the molecules which are seminal models of real associating fluids with hydrogen bondings or charge transfers, has been done theoretically by *Wertheim* [6.172]. On simple models (hard spheres with bonding sites), *Jackson* et al. [6.173] demonstrate by simulations the validity of the theoretical approach. In this work the effect of associations on the phase equilibria is discussed.

6.3 Coulombic and Ionic Fluids

The interactions between charged atoms or molecules are composed of a Coulomb potential and short ranged potentials which model the dispersion forces and the effects of excluded volume. This last term is needed to avoid the collapse of the systems due to the singularity of the Coulomb potential for the unlike charges, but it can be suppressed in the most simple model of Coulomb systems, the 2D and 3D one component plasma (OCP) formed by identical point charges in an uniform neutralising background. In the realistic models of ionic systems (molten salts or liquid metals) complex forms of the short range potentials must be used to obtain a quantitative agreement with experiment.

6.3.1 One-Component Plasma, Two-Component Plasma and Primitive Models of Electrolyte Solutions

(a) OCP and TCP. For the 3D OCP very accurate MC simulations data are presented in [6.174, 175]. The accuracy of the data reported in [6.175] seems the best, and the empirical equations of state reported for the fluid and solid phases are the most precise. Usually, to compare the MC and MD data, a correction is made to take into account the fixed position of the centre of mass in the MD simulations. *Ogata* and *Ichimaru* [6.176] show that this procedure is probably wrong for OCP. The large quantitative changes induced on the OCP equation of state by the polarization of the neutralising background, supposedly a degenerated electron gas, are estimated by *Totsuji* and *Takani* [6.177]. For the 3D OCP, Totsuji estimates in [6.178] the shortcomings of the approximate expressions of the triplet correlations (superposition approximation and convolution approximation).

If point charges, interacting via a $1/r$ potential, are constrained to a circular planar domain enclosing a neutralising background, the minimal energy does not seem to correspond to a neutral configuration. This point is discussed on the basis of the GCMC simulation data by *Lotte* and *Feix* [6.179]; they conjecture that the excess of charge follows from the fact that the $1/r$ potential is not the solution of the Poisson equation for the constrained geometry of the system.

The values of the fluctuation of the electric dipole moment (or dielectric susceptibility) of finite Coulomb systems depend on their boundary conditions. This question is investigated in detail by *Choquard* et al. [6.180, 181] for the specific case of the 2D OCP at the peculiar value of Γ $(=\beta e^2)=2$ where the model is exactly soluble. They show that the values of the dielectric susceptibility originate from two equal contributions, a bulk contribution and a surface contribution which is due to the long range of the correlations between particles near the surface. For the 2D OCP at $\Gamma \neq 2$ and for the 2D two component plasma (TCP) with short range repulsive interactions precluding the collapse of the system, numerical simulations show a similar behaviour of the dielectric susceptibility.

The point charge TCP is thermodynamically stable for values of $\Gamma < 2$. *Hansen* and *Viot* [6.182] study, theoretically and numerically, the variations of the short range correlations between particles when Γ increases towards the critical value $\Gamma = 2$. The MC simulations reveal a quasi "chemical" equilibrium between the tightly bound pairs and free ions, which is well described by a Bjerrum model.

On the basis of mean field and renormalisation group theories, it has been demonstrated that the TCP with short repulsive interactions exhibits, at low density, a Kosterlitz–Thouless (KT) transition between a dielectric phase and a conducting phase. *Caillol* and *Levesque* [6.183] present an investigation of this transition by simulations. The MC simulations show that, at moderate densities, the KT transition becomes a first order transition between a dielectric gas and a conducting liquid. These simulations were realised in the NPT ensemble where the charged particles are on the surface of a 3-D sphere. The charges interact by a potential which is a solution of the Poisson equation in this geometry avoiding the use of the Ewald energy.

The version of the 2D TCP where one of the charged species is constrained to stay on the sites of a triangular lattice, shows a localisation-delocalisation process of the pairs of unlike charges typical of the KT transition. In [6.184, 185], *Clérouin* and *Hansen*, and *Clérouin* et al. study this model and determine the temperature of the transition by static and dynamic diagnostics. The dynamic diagnostic is the decrease towards zero of the diffusion coefficient of the mobile charges, and the static diagnostic is the variation of the specific heat. The temperature of the transition at the low densities seems higher than in the 2D TCP. A very detailed computation of the phase diagram of a similar model where the singularity of the interaction for the unlike charges is regularised by smearing the positive charge of the lattice sites over disks, is done by *Choquard* et al. [6.186]. The model exhibits a KT transition at low temperature; but at high densities the phase diagram is very complex. For instance, at low temperature the dielectric phase undergoes an infinite succession of first and second order transitions between ferroelectric and paraelectric phases.

(b) Primitive Models. The primitive model of the electrolyte solutions in the regime of dilute ionic concentrations is believed to be satisfactorily described by the Debye–Hückel theory and its improvements. The results of the numerical simulation of *Sloth* et al. [6.187] show the limit of the validity of this theory; systematic and significant differences occur between MC and theoretical data for ionic concentration $\sim 10^{-5}$–10^{-4}. It is worth noticing that these simulations are achieved by using the minimum image convention (cf. Sect. 6.1.4).

Linse and *Andersen* [6.188] support the use of truncated potentials in numerical simulations of Coulombic systems, combined with the correction of the simulation data by an integral equation theory to account for the long tail of the interactions. This approach, in spite of agreements obtained between data corrected with the HNC (hyper-netted chain) or RHNC integral equations and results of simulation performed with an Ewald energy, must be used with caution (cf. Sect. 6.1.4); furthermore the only way to check the results of this procedure is a comparison with simulations where the treatment of the Coulomb interaction is in agreement with the laws of electrostatics. The reference [6.189] by Linse illustrates well the shortcomings of an approximate treatment of the Coulomb interactions. In this work, the simulations performed using minimum image convention give a crystalline phase and the truncated Coulomb interaction gives a fluid phase in the same domain of density and temperature.

The models of Coulomb systems where the molecules interact by effective short range potentials are studied by *Clarke* et al. [6.190]. For this purpose, they compare the results of simulations where the long tails of model potentials of the KCl molten salt, are treated by Ewald summation or are approximated by r^{-n} algebraic terms truncated at a finite range. The rdf of the two simulations agree rather well. Here again we remark that the validity of these effective short range potentials must be established for each specific case.

In [6.191], the liquid–solid transition of a 2D system with a screened Coulomb interaction, expected to be relevant for the study of a polar monolayer on water, is determined and shown to be weakly of the first order. *Vlachy* et al. [6.192] prove, on the basis of GCMC simulation results, that the MHNC integral equation is very accurate for the computation of the rdf and thermodynamic properties of screened Coulomb fluids. The same type of fluids seems able to reproduce the structural order of colloidal suspensions [6.193].

For the primitive model of electrolyte solutions composed by hard spheres of unequal diameters, *Abramo* et al. [6.194] have done an extensive comparison between MC simulations with Ewald energy and approximated integral equation theory. The HNC theory gives very reliable estimates of the internal energies. *Vlachy* et al. [6.195] perform a similar comparison for the two models of charged hard spheres and soft spheres with a large asymmetry of the sizes and charges of the spheres. The simulations are done with the minimum image convention (cf. remarks above and Sect. 6.1.4). The HNC integral equation seems reliable for the dilute solutions, but it becomes inaccurate for the large charge asymmetry.

The stability of the mixtures of three species of charged hard spheres is investigated by *Caccamo* and *Malescio* [6.196]. The spinodal decompositions and the coexistence lines are determined by using mean spherical and HNC approximations; a comparison with MC data is done in a few cases for the excess internal energies and the local ordering.

6.3.2 Realistic Ionic Systems

This section is devoted to the numerical simulations of the molten salts and the liquid metals. Most of these works use the following forms of interaction potentials,

$$z_i z_j r^{-1} + e_{ij} r^{-4} + h_{ij} r^{-n} \tag{6.19}$$

and

$$z_i z_j r^{-1} + c_{ij} r^{-6} + d_{ij} r^{-8} + a_{ij} \exp(-b_{ij} r), \tag{6.20}$$

where r is the distance between the atoms i and j, z_i and z_j are the charges of these atoms and $a_{ij}, b_{ij}, c_{ij}, d_{ij}, e_{ij}$ and h_{ij} are adjustable parameters.

Based on a careful representation of the ab initio potentials by analytical expressions, the MD calculations of *Laaksonen* and *Clementi* [6.197] give a complete description of the solid and fluid properties of NaCl. For instance, the rdf of the liquid phase seem in good agreement with the experimental data. With a simplified expression of (6.20) where c_{ij} and d_{ij} are zero, *Baranyai* et al. [6.198] calculate the internal energies and rdf of the molten salts for the associations of Li, Na, K, Rb with Cs, F, Cl, Br and I. The agreement with the available experimental data for both quantities is at the level of a few percent. *Ross* et al. [6.199] consider the case of KI and study the liquid phase in the vicinity of the melting line at the pressure of 17 kbar; they prove that the co-ordination numbers in the liquid increase accordingly to the change of the structure of the solid phase at this pressure.

The fluid phase of $ZnCl_2$ is studied by *Kumta* et al. [6.200] and by *Abramo* and *Pizzimenti* [6.201]. The first article establishes that a good description of the local order and internal energy at 1 atm and 1200 K is obtained with potentials of type (6.19). In [6.201], the interaction potentials are parametrised with $d_{ij}=0$ and $c_{ij}=0$, the Coulomb part of the potential being multiplied by a switching function. The rdf computed by simulation are in agreement with the experiments, but there are large discrepancies for the pressure and the internal energy.

Margheritis and *Sinistri* [6.202, 203] propose to evaluate in MC simulations of molten salts, the variation of the polarization energy due to the move of an atom, by assuming that this variation is a linear function of the variation of the Coulomb energy. With this they obtain [6.203] a good prediction of the thermodynamic properties (internal energies, densities) of AgCl between 750–1500 K. For AgI, these authors report the results of simulations with potentials of the form (6.19) and (6.20), parametrised to reproduce the properties of the solid phase. The simulations [6.204] show that these potentials are inadequate in the liquid phase. Similar disagreements are obtained in CuI, CuB and AgI by *Stafford* et al. [6.205].

In the equiatomic Na Pb, K Pb, Rb Pb and Cs Pb liquid alloys, the Pb atoms form tetrahedral arrangements of four ions, so-called Zintl ions. *Reijers* et al. [6.206] model the interatomic interactions by a potential form (6.20), and within the Zintl ion the Pb–Pb interaction by an harmonic potential. The full set of the potential parameters is chosen by adjusting the structure factors calculated by simulation on the experimental results. The potential form (6.20) is then shown able to reproduce the local ordering in the liquid phases of these salts. *Cheng* et al. [6.207] determine the orientational order in molten salts, described by the potentials (6.20). This order is defined by the angular distribution of the lines connecting an atom and its nearest neighbours. In molten LiF, KF and KCl, the local order structure is tetrahedral.

Ag_2Se is a superionic conductor; it has been studied extensively by simulation in [6.208]. The article is mainly devoted to the superionic state of Ag_2Se in which the interactions are modelled by potentials (6.19) with a screening of the r^{-4} term. It focuses on the interpretation of the structure factors measured by neutron diffraction and the diffusion coefficients of the mobile Ag ions. Some results are reported for the liquid state. The arrangement of the Ag atoms around the fcc lattice sites occupied by the Se atoms are shown to be tetrahedral. In their work [6.209, 210], *Vashista* et al. make simulations of the molten and amorphous $GeSe_2$. In these articles, the interactions have a form (6.19), but three-body contributions are added in [6.210]. The main features of the experimental structure factors are obtained by using the two-body interactions and the discrepancies are corrected by the use of the three-body terms.

The interior of the white dwarfs is constituted by a mixture of C and O ionized atoms in a degenerated electron gas. These mixtures are well modelled by a classical fluid of positive point charges in a neutralising background; in [6.211] *Iyetomi* determines the phase diagrams of this fluid by integral equation theory, DFT and MC simulations. The system can form an azeotropic mixture. The molten-compound semiconductor GaAs is studied by *Hafner* and *Jank* [6.212] using the same method as [6.136]; a good agreement is achieved between the experimental and theoretical static structure factors.

MD simulations of the molten phases of $NaBaF_3$ and Na_2BeF_4 with atomic interaction of the form (6.20) have been performed by *Furuhashi* et al. [6.213a]. These authors report reasonable agreement with the experimental structure factors; a detailed analysis of the local structure is effectuated by the computations of the co-ordination numbers and the angular distribution functions of the vector distances of an atom with its nearest neighbours. *Habasaki* [6.213b] adjusts potentials (6.19) on ab initio estimates of the intermolecular interactions of the Li_2CO_3 and Na_2CO_3 compounds. The simulations are performed by treating the CO_3 anions as a rigid rotor having D_{3h} symmetry. A comparative analysis of the structural order in the molten phases shows that the qualitative differences in the positions of Li and Na atoms are due to their sizes.

The complex structural organisation of the mixtures of molten salts, for instance NaCl and $AlCl_3$ or KF and YF_3, are described by *Rahman* and collaborators [6.214]. The positions of the anions around the polyvalent cations go from octahedral to tetrahedral arrangements; there are also independent moieties which become connected by linear bridges of anions and cations with the increase of the concentrations of the species such as $AlCl_3$ or YF_3.

6.4 Simulations of Inhomogeneous Simple Fluids

6.4.1 Liquid–Vapour Interfaces

The liquid–vapour interfaces have been studied in two different geometries: planar interfaces or spherical interfaces. Generally, in the first geometry the liquid–vapour phases are enclosed in a cell (slab) limited by two parallel walls in one direction and periodic in two directions. At the initial step of the simulation, the dense phase is located near one of the walls or in the middle of the cell. During the simulation, one or two stable interfaces between liquid and vapour are generated. The spherical interface is obtained by achieving the equilibrium of liquid drops in the central part of a spherical cavity confining the particles. In the slab geometry, the periodic boundary conditions contribute to localise the interfaces. In the spherical geometry the interface is stabilised only by the surface tension.

Hawley et al. [6.215] study the interaction between two liquid–vapour interfaces of the LJ fluid. They use a slab geometry periodic along two directions, with two attractive walls. The attraction stabilises the liquid phase near the walls. In the central part of the slab there are two stable liquid–vapour interfaces. If the distance between the walls decreases the two interfaces merge suddenly for a well defined separation.

In order to avoid the cumbersome initial stages of generating a liquid–vapour interface from a bulk liquid volume with periodic boundary conditions, *Chapela* et al. [6.216] propose to start the simulations from a tenuous unstable fcc configuration of particles filling all the simulation cell; if the temperature is low, by a fast spinodal decomposition a liquid film will be formed in the central part of the slab in equilibrium with a vapour phase. This procedure is tested for the square well fluid.

The estimate of the surface tension in the slab geometry for the LJ fluid is biased by the truncation of the LJ interaction tails for numerical conveniences. The estimate of the corrections due to the missing LJ tail is difficult, because its influence on the interface is obviously unknown. Using a special purpose computer, *Nijmeijer* et al. [6.217] were able to calculate the surface tension of a LJ liquid film of 10390 particles with a thickness of 16σ and a cut-off of the LJ tail at 7.33σ. They report considerable differences with the surface tensions estimated by using a cut-off of 2.5σ near the triple point. *Thomson* et al. [6.218] calculate the density profile of the LJ liquid drops for the temperature ranging from 0.7–0.8 (reduced LJ units). A very detailed analysis of the computation of the surface tension in this geometry is presented and the size effects on the value of the surface tension are carefully studied.

The liquid–vapour interface of a Ar/Kr mixture modelled by a LJ mixture with truncated tails is simulated for the relative concentrations ~ 0.5 and ~ 0.25 by *Lee* et al. [6.219]. At the concentration 0.5, the density profiles show that the interface is strongly enriched in Ar; at ~ 0.25 the density profile of Ar has a maximum in the vicinity of the interface.

The evaporation process of hot drops of LJ liquid is described in [6.220] by *Vincenti* et al. They find that the expansion of the drops is almost adiabatic and that the fragmentation occurs only if the expansion continues for densities smaller than 1/4 of the bulk liquid density.

6.4.2 Fluid–Solid Interfaces

In this section we consider three types of fluid–solid interfaces: the vapour or liquid–solid interfaces of the same substance; a fluid system, without gas–liquid transition, in contact with a solid wall, typically the hard sphere near a hard wall; and a subcritical fluid in contact with a solid.

The equilibrium between a gas or a liquid in contact with its own solid has been extensively studied by *Broughton* and *Gilmer* [6.221–223]. They calculate for both LJ vapour and liquid–solid interfaces, the excess surface free energy [6.221, 222]. For the liquid–solid interface they give a complete characterisation of the behaviour of the atoms in the interfacial region. The excess free energy for the gas–solid interface is computed by a charging process starting at temperature zero. For the liquid–solid interface, the excess free energy is calculated from the difference between the free energy of two pieces of solid and liquid separated or in contact. The gas–solid surface tension becomes independent of the type of crystal plane in contact with the vapour near the triple point, and for the liquid–solid interface the surface tension also seems independent of the crystal orientation. Another conclusion is that the density profiles give a maximal estimate of the size of the interfacial region compared to other quantities used to study the transition from the solid to the fluid phase, for instance rdf, diffusion coefficient.

For the soft spheres, *Laird* and *Haymet* [6.224] also report an unexpected sharp variation of the diffusion constants through the liquid–solid interface compared to its extension evaluated from the density profiles. Extremely detailed analysis of the structural order, energy variation, etc. in the fluid–solid interfacial region have been made [6.225]. They confirm the conclusions of previous works, namely the rather large estimate of the size of the transition domain obtained on the basis of the density profiles. These authors conclude also that the interfaces consist of islands of solid and fluids.

In generating the coexistence of the solid and liquid phases by MD simulation, *Chokappa* and *Clancy* [6.226] show that only stable thermodynamic states are produced. This procedure allows a reliable estimate of the LJ melting line to be obtained. *Schommers* et al. [6.227] analyse the melting of the surface layer of the LJ solid; they conclude that the disorder and the delocalisation of the atoms appear mainly in the direction perpendicular to the interface. Extensive simulations of the interface of hard spheres near a hard wall are performed by *Henderson* and *van Swol* [6.228]. The density profiles, surface tension and surface adsorptions are computed. The rdf's, estimated in the direction parallel to the wall near the interface, exhibit a behaviour very different from the case of the vapour–liquid interfaces.

In the hard disk or hard sphere fluid near a hard wall, the density profiles and the cavity concentration profiles are proportional. The cavity concentration profile $n(z)$ is the probability that at a point distant by z from the wall, it is possible to insert a disk or a sphere without overlap. The value of $n(0)$ allows the determination of the activity. Approximate expressions of $n(z)$ are tested and shown to be accurate by simulation for hard disks [2.229], hard spheres [2.230] and hard sphere mixtures [2.231]. In these last simulations, the density profiles show that the fluid is enriched in spheres of large sizes near the wall. *Gozdz* et al. [2.232] generalise the previous computations of $n(z)$ to the case of LJ atoms near a hard wall.

The interface between the hard sphere fluid and a "square well" adsorbing wall is studied theoretically and by simulation in [6.233] by *Kierlik* and *Rosinberg*. An effective 2D theory gives a successful interpretation of the MC results for the behaviour of the spheres in the layer in contact with the wall.

For the hard sphere system near a hard wall where a lattice of adhesive sites is located, theoretical studies predicted a first order transition for the number of occupied sites as a function of the strength of the attraction and of the bulk hard sphere density. MC simulations reported by *Caillol* et al. [6.234] show a continuous increase of the adsorption of the spheres on the sites. However, in these simulations, the finite extent of the adhesive sites can hinder the transition due to the possible overlap of two spheres adsorbed on neighbouring sites. *Kierlik* and *Rosinberg* [6.235] have made simulations for the same system where the size of the adhesive sites is reduced in order to avoid this possibility; their work seems strongly in favour of a first order transition. *Thurtell* and *Thurtell* [6.236] describe the results of MD simulation at the temperature 0.1 K of a low density LJ fluid in contact with hard smooth walls or perfectly rough walls. They report strong influence of the characteristics of the wall on the density profiles.

Two sets of simulations have been made on the wetting and drying transitions of subcritical simple fluids in contact with a wall or a solid [6.237–244].

Van Swol and *Henderson* consider the case of a square well fluid near a hard wall [6.237, 238, 240] and near a "square well" wall [6.238, 240, 241]. The first system exhibits a complete drying of the hard wall. For the system with a "square well" wall, the simulation data exhibit strong evidences for a first order wetting transition between partial wetting and complete wetting. These evidences are supported by the computation of the adsorption and the surface tensions between the fluid phases, and between the fluid phases and the wall. Detailed analysis of the behaviour of the density profiles and the shear tensor in the interfacial region is reported in [6.240]. DFT of the wetting–drying transitions are tested by simulation in [6.241]; they give an accurate prediction of the density profiles away from the drying transition.

In [6.242–244], a complete study of the wetting and drying of the LJ solid by a LJ fluid is made by simulations of large systems of ~ 11000 particles. In the two first articles, a subcritical LJ fluid located in a slab geometry periodic along two directions, is sandwiched by two solids of LJ heavy particles with lattice spacings chosen in order to avoid the nucleation of the LJ fluid. For the considered sizes of the slab, there is always a liquid–vapour interface in the system and the liquid phase is parallel to the solids. The drying and wetting of the solid interfaces are obtained by varying the ratio ε_r of the ε parameters of the LJ fluid–solid and solid–solid interactions. The localisation of the transitions can be estimated from the density profile or from the computation of the surfaces tensions between the three phases. Qualitatively the observation of the profiles and the values of the surface tension show that, in increasing ε_r, the solid interfaces go from dry to wet condition, the drying and wetting transition being first-order. However there are disagreements in the quantitative localisation of the transition from these two classes of data. In [6.244], the authors attempt to localise the transition by a direct measurement of the contact angles between the three phases. In this computation, the liquid is arranged perpendicular to the solid with periodic boundary condition along one direction. Careful estimates of the contact angles give a wetting transition for $\varepsilon_r = 0.7$ and a drying transition for $\varepsilon_r = 0.2$.

The wetting and prewetting phenomena are observed in numerical simulations by *Finn* and *Monson* [6.245] and *Sokolowski* and *Fischer* [6.246]. These phenomena appear in the domain of the pressure–temperature phase diagrams where the thickness of the adsorbed film of fluid on the solid substrate exhibits, along an isotherm, a discontinuous finite jump with the increase of the pressure. At the saturation pressure, for low temperatures, the bulk vapour condenses on the substrate and the jump is infinite; for a higher temperature T_w (the wetting) temperature, the jump is finite. At T_w, the substrate can accomodate two types of adsorbed films. This situation persists until a temperature T_{pw}, the prewetting temperature, where the jump disappears and the thickness of the film increase continuously with the pressure at constant temperature. *Finn* and *Monson* perform isothermal–isobaric MC simulations for a LJ fluid slab with truncated potential and a LJ 9-3 interaction between the smooth wall and the fluid. They compute the density profiles and the adsorption of the fluid which are in full agreement with the scenario of the wetting and prewetting phenomena described above. The wetting and prewetting temperatures are estimated to be 0.84 and 0.94 (LJ reduced unit). The result of *Finn* and *Monson* is confirmed by *Sokolowski* and *Fisher* who perform constant temperature MD simulations for the same system without truncating the LJ tails. They compute the chemical potential of the fluid phase in order to determine the pressure in the system. The adjunction of the LJ tail displaces T_w from 0.84 to 0.9. But *Sokolowski* and *Fischer* were unable to localise T_{pw}. From DFT, *Velasco* and *Tarazona* [6.247] have found a very low value of T_w for the LJ system considered in

[6.245]. The two simulations [6.245, 246] giving values of T_w in close agreement, the effect of the cut-off of the LJ tail seems weak. The origin of the discrepancy remains unclear.

The MD simulation method at constant pressure devised for the homogeneous systems, is extended to the case of inhomogeneous systems by *Lupkowski* and *van Swol* [6.248]. The liquid–liquid interface is investigated by MD simulation in [6.249] by *Meyer* et al. The characterisation of such interfaces of LJ fluids is made for temperatures and pressures near the triple point. The interfaces are generated by using a cross LJ interaction which is unfavourable for the miscibility of the two liquids. As in the case of the liquid–solid interfaces the thickness of the interfacial region differs following the choice of the observed quantities.

6.4.3 Interfaces of Charged Systems

A complete analytical and numerical study of the surface properties of Coulomb system is presented by *Choquard* et al. [6.250]. This analysis completes the work of [6.152] (cf. Sect. 6.3.1). The size effects in 2D OCP systems of 5–220 particles are carefully computed by simulation to support the results of the theoretical analysis. The surface properties of the 3D OCP are calculated by *Totsuji* [6.251] in a slab geometry periodic along two directions. The stability of the system is ensured by the rigid neutralising background localised in the centre of the slab. The computation is done by using an Ewald energy; at high temperature the density profiles and the surface energies differ from previous estimates for different geometries.

In [6.252], Heyes presents the results of simulation of bulk and inhomogeneous molten salt systems. For the inhomogeneous case, numerical techniques are proposed to simplify the computation of the double sum on **k** vectors and pairs of particles in the Ewald energy. The liquid–solid interfaces in these systems are very narrow, there is no charge separation in the surface region and in presence of an electric field the mobility of the ion in the interface is less than in the bulk.

The computation of the interface properties of simple metals is made by *Harris* et al. [6.253–255]. These authors use rather involved interaction potentials between the atoms, which include the contribution of the electrons in a sophisticated way. The computed density profiles of the pure Na and Cs oscillate near the interface. In Na/Cs mixtures the density profiles reveal a stratified arrangement of the interfacial region [6.255].

The system of mobile charges in a dielectric continuum between two charged parallel walls, is a model of electrolyte solutions at the contact of electrodes or ions in the solvent of colloidal suspensions.

MC simulations have been performed to determine the properties of this simple model and to compare the results with approximate theories. Examples of such works are [6.256–263]. *Jönsson* and collaborators [6.256–258] evaluate the force f_e between charged planar surfaces in the presence of an electrolyte represented by a primitive model. In [6.256] the system is an array of cells formed by periodic replicas, along two directions of the simulation cell containing the positive mobile charges. The neutrality of the system is assured by the surface charge of the walls which limit the array. The force f_e is the derivative of the free energy with respect to the distance of the walls. The simulations reported in this work have the shortcoming that they evaluate the Coulomb interaction by the minimum image convention. In spite of this defect the computed osmotic pressures are in agreement with the Poisson–Boltzmann theory and reveal that the force f_e is attractive. In the system used in [6.257], immobile charged aggregates, modelled by large spherical molecules with a surface charge, are added to a neutral ionic mixture. The force between the wall and these aggregates can be also attractive.

Bratko et al. [6.258] consider the system of two negatively charged parallel walls enclosing a fluid of neutralising mono and divalent counterions. On these walls there is a dielectric discontinuity, characteristic of the interfaces between two dielectric continua. The contribution to the interaction between the ions due to the image charges is limited to the first term and minimum image convention in the calculation of the Coulomb energy is corrected by the method used in [6.259]. The image charge contributions to the interaction increase the attraction between the double layers and repell the ions from the walls.

The interface between a charged plane and the charged hard spheres has been studied by *Valleau* and collaborators. The simulation cell is periodic in two directions and is limited by a charged hard wall and a neutral hard wall. The equilibrium configurations of the ions are generated by keeping constant an excess of positive ions, in order to neutralise the charge of the wall. The structure of the ionic arrangements near the wall has been studied in detail for charged spheres of equal diameters and identical ion-wall interactions (cf. [6.1, 2]).

In [6.259], *Valleau* and *Torrie* report the density profiles and the mean electrostatic potential drop across the interfacial region for a mixture of charged spheres with an asymmetric interaction between the wall and the two species of ions. The computation is done with, and without, image forces due to an external dielectric medium. The comparison with the Gouy–Chapman theory shows that the theoretical predictions are qualitatively correct. The effects of asymmetrically sized charged spheres on the double layer is examined by *Carnie* et al. [6.260]. These effects are small. The double layers of 2:2 and 2:1 primitive models are calculated by *Torrie* et al. [6.261]. They consider the case of an uncharged wall with attractive and repulsive image forces. The effects of these forces are considerable for the 2:2 electrolyte. For the 2:1 system with a neutral wall, the charge asymmetry induces a double layer which is strongly modified by the attractive image forces. Canonical MC simulations of similar systems are done by *Svensson* et al. [6.262]. They evaluate the density profiles, the osmotic pressures for different concentrations of the ions and for different surface charge of the wall. They compare their simulation data with the Poisson–Boltzmann theory and conclude that it is accurate for monovalent ions and qualitatively wrong for divalent ions.

The system formed by two reverse micelles is modelled in [6.263] by two spheres with a superficial charge enclosing neutralising counterions. *Luzar* and *Bratko* in this article calculate the density profiles of the ions inside the sphere, which are in good agreement with the Poisson Boltzmann theory. The intermicellar forces due to the ionic correlations are also computed and shown comparable in magnitude with van der Waals interaction in micellar systems. *Vlachy* and *Haymet* [6.264] establish that the HNC approximation correctly predicts the density profiles of a screened Coulomb system in the vicinity of a hard wall.

6.4.4 Fluids in Narrow Pores

The behaviour of gas and liquid in porous media is important information for practical applications. The simulations are very well suited for the study of these systems in a complete and realistic way because the size of the volumes and the number of particles involved in these systems have values generally used in numerical computations.

The simulations of a fluid in slit-like pores have been done for the LJ fluid by *Van Megen* and *Snook* [6.265], who consider this fluid between plane surfaces. The surface fluid potential is formed of a short range repulsion and a r^{-4} attractive tail. At two supercritical and subcritical temperatures they evaluate, by GC MC simulations, the adsorption of the fluid on the walls for different values of their separation. For the subcritical temperature, the adsorption increases continuously with the pressure to an almost constant value independent of pressure, corresponding to the fact that the liquid phase has filled the pore. For the supercritical temperature, the adsorption increases to a maximum. This maximum is followed by a rather fast decrease of the adsorption with an increase of the pressure. This last feature is a consequence of the very rapid increase of the bulk density for the large pressures. A similar study using MD simulations is performed by *Magda* et al. [6.266]. They compute, for the supercritical isotherm considered in [6.265] and several values of the wall separation, the density profiles, the solvation forces, the diffusion coefficients, etc. All the equilibrium properties have oscillatory variations with the size of the slit. The diffusion coefficient calculated for the displacements of the atoms parallel to the walls is almost independent of the slit size.

Antonchenko et al. [6.267, 268] report the results of GC MC and MD simulations of hard spheres between hard walls and inside a hard cylindrical pore. The pressure is defined by the value of the hard sphere density at the contact of the surface. Its variation with the separation of the walls and the radius of the pore is oscillatory and reflects the layered arrangement of the hard spheres in the slit.

A good concordance between the predictions of the continuum mechanical theory and the results of simulations is obtained by *MacElroy* and *Suh* [6.269] for the single component fluid and the mixture of hard spheres inside cylindrical pores.

The articles [6.270, 271] by *Schoen* et al. give a detailed analysis of the structural order of a LJ fluid inside a slit. The slit is limited by smooth surfaces interacting with the fluid, by the mean field potential of the LJ crystal, or by two fcc LJ solids at zero temperature. The strong influence of the surface-fluid interaction on the local order in the fluid layers parallel to the surface is revealed by the two-body correlations inside these layers. The structured wall can induce a quasi-solid arrangement in the first layers of the fluid; this fact is not unexpected because the LJ solid surface can induce the nucleation of the LJ fluid. The diffusion coefficients are calculated in [6.272], the diffusion coefficient parallel to the interface oscillates with the surface separation.

For the LJ fluid, *Heinbuch* and *Fischer* [6.36, 273] verify that the chemical potential is constant inside the volume of a cylindrical pore and show the validity of the BGY integral equation for the prediction of the density profiles. Also for a cylindrical pore, *Bratko* et al. [6.274] compare the density profiles computed by GC MC simulations and the results of the PY integral equation theory; the agreement is good. A similar agreement is obtained by *Sloth* [6.275] for hard spheres inside spherical pores. *Adams* et al. [6.276] derive theoretical approximations, for the density profiles of fluids confined in very narrow pores, which are proved very accurate by simulation.

The coexistence of low density and high density fluid states in narrow pores has been predicted theoretically by DFT and seems needed also for the interpretation of experimental data of the adsorption of gas in porous glasses. In the three articles [6.277–279], *Gubbins* and collaborators analyse the data of computer simulations of a LJ fluid in cylindrical pores which exhibit clearly this coexistence. In [6.277] the MD simulations are performed for 1560 atoms filling a cylindrical pore of radius 5σ and length 50σ periodic along the main direction. Below $T_c^* = 1$ (LJ reduced unit), inside the cylinder the average density profiles are typical of a liquid or a gas in sub-domains separated by a spherical meniscus. The chemical potentials computed in the sub-domains are equal, indicating a thermodynamically stable coexistence. These results are confirmed in [6.278] by GC MC calculations of adsorption isotherms, which show phenomena of hysteresis. By MD simulations on a system of 2848 particles [6.279], it is possible to observe simultaneously in the pore, ten different domains of gas and liquid. At low temperature, with increasing pressure or chemical potential, the adsorption of a fluid in a cylindrical pore exhibits successive layering transitions which are determined for a LJ fluid in [6.280]. The smooth wall of the pore interacts with the fluid by the "smeared" potential of a LJ solid. On this structureless wall, the first adsorbed layer becomes crystalline and the transitions from 1 to 2 and from 2 to 3 layers seem first order and weakly first order, respectively.

The hysteresis of the adsorption isotherms has been reproduced in the very extensive GC MC calculations of *Schoen* et al. [6.281]. In this article, the dependence of the phenomena on the size of the pore and on the nature of the solid surface is discussed. Clearly, the hysteresis will be strongly dependent on the radius of the cylindrical pore or the size of the slit-like pore. For large radius or size, the behaviour of large systems (cf. [6.204, 205]) should be recovered.

Tan and *Gubbins* [6.282] investigate the adsorption of methane and ethylene in carbone micropores. GC MC simulations are performed for the two fluid–solid interactions which neglect and include the substrate periodicity. The effect of the structure of the wall on density profile is small. The non-local DFT results are in good agreement with the simulation data. For a fluid in a narrow pore, the comparison of the local and nonlocal DFT and the simulation results is made in detail by *Peterson* et al. [6.283]; the non-local theory gives an accurate prediction of the phase diagram and the fluid structure in the pore.

6.5 Molecular Liquids: Model Systems

This and the following sections deal with the fluid properties of anisotropic molecules. We distinguish between model systems, introduced primarily to test methodological aspects or analytical theories and more realistic systems designed to make closer contact with real life.

6.5.1 Two-Dimensional Systems

The interest of simulating 2D molecular systems is both practical (e.g. monolayers of physically adsorbed molecules, thin freely suspended liquid crystal films etc.) and theoretical (e.g. influence of the spatial dimensionality on the nature of the liquid crystalline transitions [6.77]).

The simplest model of a 2D molecular system is perhaps fused disks (planar dumbbells). The equation of state [6.284–286] and site–site pair distribution functions [6.284] have been determined by the MC method for a range of elongations from 0.3σ to 1.4σ (σ-disk diameter) and the results compared with various theoretical predictions based on scaled particle theory and modifications [6.285–288], y-expansion [6.289], generalised Flory–Huggins theory [6.290] and reference interaction site model (RISM) theory [6.291]. In [6.286], the pressure has been obtained in a rather unusual way, namely from the contact value of the density profile of an inhomogeneous system of fused disks ($l = 1\sigma$) in contact with a hard wall. This represents a simplification over the traditional virial approach to the pressure which requires knowledge of angular correlations between neighbouring molecules [6.292].

A further application of the MC method has been the study of orientational order in 2D liquid crystal models. Most continuous 2D systems lack long range order (LRO) at finite temperature. This has been proven rigorously, for instance, for the translational order in crystals [6.293]. However, such systems can have a phase transition from a high temperature disordered phase to a low temperature phase that exhibits quasi long range order (QLRO), characterised by an algebraic decay of the relevant correlation functions. This transition can be first order [6.294] or of the Kosterlitz–Thouless (KT) type [6.295] (unbinding of disclinations).

A rigorous result concerning long range orientational order in 2D nematics has been obtained by *Straley* [6.296] for separable intermolecular potentials (i.e. of the form $V(r,\theta)=\sum_n f_n(r)g_n(\theta)$, where r and θ are the relative positional and angular coordinates of a pair of molecules). In that case, no true long range order can exist [6.296] and the isotropic-nematic transition is believed to be of the KT type [6.297, 298]. Realistic potentials are, however, nonseparable.

A continuum elastic description of orientational fluctuations in 2D nematics predicts that the mean square angular displacement diverges logarithmically with number of particles as the system size gets infinite. As a consequence, the order parameter $q=\langle\cos 2\theta\rangle$ and the angular correlation functions $g_{2l}(r)=\langle\cos\{2l[\theta(0)-\theta(r)]\}\rangle$, where $\theta(r)$ characterises the molecular orientation with respect to a fixed axis, decay algebraically in the thermodynamic limit [6.299]. By analysing MC simulation results of these quantities for a system of infinitely thin hard rods of length L, *Frenkel* and *Eppenga* [6.300] conclude that their calculations are compatible with the existence of a nematic phase with algebraic order at densities $\rho L^2 \geqq 7.5$ and an isotropic phase for densities $\rho L^2 \leqq 7$. The results do not give evidence for a first order isotropic–nematic transition but rather point to the occurrence of a KT disclination-unbinding transition around $\rho L^2 \cong 7.25$ (although, as pointed out by these authors, the transition density may shift to higher densities with increasing system size). A qualitatively similar continuous phase transition has also been found for hard ellipses of aspect ratio $\kappa=6$ [6.301]. In contrast, a first order isotropic–nematic transition is observed for $\kappa=4$ and, at $\kappa=2$ the system exhibits no nematic phase [6.301].

A simulation study by *Gingras* et al. [6.302, 303] of a 2D system of particles interacting via separable and non separable anisotropic Van der Waals forces focuses on the coupling between the molecular orientational field (MOF) and the nearest neighbour bond orientational field (BOF). According to a theoretical investigation by *Nelson* and *Halperin* [6.299] on a thin smectic film model, if such coupling exists, any order in the molecular orientation field induces ordering in the BOF and, if positional correlations are short ranged, the MOF and BOF must display quasi long range order. The MC calculations of *Gingras* et al. support these ideas when the potential is nonseparable. In this case, the transition from the isotropic to the nematic phase is accompanied by the development of bond orientational order having rectangular symmetry. The decay of the BOF correlations is consistent with a power law behaviour, although, within statistical error, an exponential decay to a finite value (characteristic of true long range order) cannot be excluded. The situation is entirely different for the separable system, for in this case bond orientational order is found to be short ranged [6.302, 303].

6.5.2 Convex Molecules (Three-Dimensional)

The simplest models for studying the effects of molecular shape are undoubtedly hard convex molecules such as spherocylinders, ellipsoids or hard gaussian overlap (HGO) models. Although highly idealised, these model systems are expected to play for complex fluids a role analogous to that played by the hard sphere system for simple fluids in elucidating the effect of excluded volume on thermodynamic properties and phase transitions. In addition, they serve as prototype systems for testing theoretical approaches (e.g. integral equation theories) as well as reference systems for perturbation treatment of more realistic molecular interactions.

In a MC simulation, the choice of a hard core shape to model a molecular fluid is largely dictated by the ease of finding a suitable contact function for deciding whether or not two hard cores overlap. While such a function is trivial to find for fused hard spheres and relatively simple for prolate spherocylinders (distance between two line segments), the decision for overlap gets already quite complicated for oblate spherocylinders [6.304] and, for ellipsoids, a numerically convenient algorithm has been proposed only quite recently [6.305, 306] improving on the one derived by *Vieillard–Baron* [6.307]. The problem of finding an overlap criterion is completely eluded in the gaussian overlap model [6.308]. This circumstance makes the model computationally very attractive. In its hard core form, the interaction between two cylindrically symmetric molecules is simply a hard sphere interaction with collision diameter depending on the directions of the molecular principal axes, the direction of the line connecting the molecular centres and a characteristic shape parameter.

(a) Virial Coefficients and the Equation of State. Third to fifth virial coefficients (B_3–B_5) have been calculated by the MC method [6.309, 310] for prolate [6.311, 312] and oblate [6.312] hard ellipsoids of revolution, prolate [6.313] and oblate [6.304, 314, 315] spherocylinders and the hard gaussian overlap model [6.316, 317]. Several aspects were considered in these studies. An initial aspect concerns the validity of Onsager's theory [6.318] of the isotropic-nematic transition of needle shaped molecules. The existence of such a transition rests on the assumption that reduced virial coefficients higher than the second vanish for extremely non-spherical shapes. This has been verified for the reduced virial coefficients B_n/B_2^{n-1} ($n = 3, 4, 5$) of both ellipsoids [6.311] and spherocylinders [6.313]. However, it was also apparent that for shapes typical of liquid crystal forming molecules (aspect ratio $\kappa \sim 3$–5) the asymptotic behaviour of the virial coefficients was far from beeing reached so that Onsager's theory fails to be applicable to realistic situations. In passing, we can remark that evaluation by the Monte Carlo method of virial coefficients of extremely non-spherical shapes requires special care [6.313].

A second aspect considered is the dependence of virial coefficients on molecular shape. For hard convex bodies a shape factor α can be conveniently defined by [6.319] $\alpha = RS/3v_m$, where R is the mean radius of curvature, S the surface area and v_m the volume of the body. Quite remarkably, the reduced second virial coefficient of any convex body depends only on α,

$$\frac{B_2}{Nv_m} = 3\alpha + 1, \tag{6.21}$$

and a question is raised whether this remains true also for higher virial coefficients. The answer is no, as demonstrated, for instance, by comparing virial coefficients of prolate and oblate ellipsoids [6.312] (ellipsoids with aspect ratio κ and $1/\kappa$ have the same α), or those of prolate and oblate spherocylinders [6.313, 304, 315]. However, it is apparent [6.317] that, to a good approximation, prolate ellipsoids, spherocylinders or gaussian overlap molecules having similar α also have similar virial coefficients and the same is true for oblate molecules although to a lesser degree for the HGO model [6.317].

From the previous considerations one might expect that the equations of state of prolate and oblate bodies are also quite different. Surprisingly, the asymmetry is rather small even for quite anisometric molecules (see e.g. [6.320] for ellipsoids). Equation of state results have been obtained for ellipsoids (κ=3–1/3) [6.320], spherocylinders ($\kappa = 5$ [6.321], $\kappa = 1, 3$ [6.322]), oblate spherocylinders ($\kappa = 0.5$–2.5) [6.304] and HGO model ($\kappa = 1.5$–0.667) [6.317]. For ellipsoids the aspect ratio $\kappa = a/b$ where $2a$ is the length of the symmetry axis and $2b$ the length of an axis perpendicular to the symmetry axis; for spherocylinders $\kappa = L/D$, where L is the length of the cylindrical part and D the diameter of the spherical caps; for oblate spherocylinders $\kappa = l/\sigma$, where l and σ are the diameter and thickness of the disk core, respectively). Theoretical equations of state derived by *Boublik* [6.323, 324] and *Song* and *Mason* [6.325] are the most successful in reproducing this data [6.304, 325, 326]. The former equation is an extension of the Carnahan–Starling [6.327] equation of state for hard spheres; it depends only on α (and density) and thus cannot take account of the small differences between prolate and oblate molecules. The latter equation takes advantage of the knowledge of the third and fourth virial coefficients and remains accurate up to high values of α.

(b) Pair Distribution Function. The most widely used description of orientational structure in molecular systems is an expansion of the pair distribution function (pdf) $g(r, \boldsymbol{\omega})$ on a set of spherical harmonics. Expansion coefficients have been obtained for oblate spherocylinders [6.304] and prolate ellipsoids of revolution [6.328]. The purpose of the latter work was to test the hypernetted-chain and Percus–Yevick integral equation theories, which, incidentally, turn out to be very accurate. In practice, an expansion of the pdf in spherical harmonics is, however, not very useful because of its slow convergence for distances at which steric correlations are strong (in particular near or at contact) [6.329]. To overcome this shortcoming it has been proposed to expand the pdf on non-spherical surfaces [6.330, 331] (e.g. equipotential surfaces). Application of this method to gaussian overlap models [6.330, 331] and hard spherocylinders [6.332] indeed shows improved convergence properties. Nevertheless expansions of this type may be quite impractical for theoretical studies. Still, a different representation, useful for hard core bodies, is obtained by expressing the pdf as a function of the surface-to-surface distance and three angles measured with respect to the normal vector specifying the minimum surface–surface distance [6.333]. Such a representation has been adopted in a simulation study of the structural properties of hard ellipsoids [6.334]. A disadvantage of this approach is the difficulty to determine the minimum surface-to-surface distance in the computer simulation.

(c) Phase Transitions. One of the more fascinating aspects of the Monte Carlo simulations of hard core molecules is the evidence given for the existence of stable liquid-crystal-like phases like nematic, smectic or columnar phases [6.311, 335, 336]. These simulations show that a sufficiently high degree of non-sphericity is a sufficient condition for formation of orientationally ordered phases, although other factors, for instance, anisotropic attractive forces or the presence of flexible tails certainly play a decisive role in stabilising these phases [6.337].

The most complete phase diagram has been obtained for spherocylinders [6.313, 321, 322, 338–340]. *Onsager* [6.318] was the first to show that this model presents a transition from an isotropic fluid phase to an orientationally ordered (nematic) phase in the limit $L/D \rightarrow \infty$ (infinitely thin spherocylinders). Extensive Monte Carlo calculations by *Frenkel* and collaborators have revealed that a thermodynamically stable nematic phase and also a smectic phase occur already for aspect ratios L/D larger than 3 [6.322] (and quite certainly for L/D equal to 5 [6.338]). They conjecture the existence of an isotropic-smectic-solid-triple point for L/D somewhat larger than 3 and an isotropic-nematic-smectic triple point at still a slightly higher value of L/D. A more detailed investigation of the range of elongations $3 < L/D < 5$ seems clearly necessary to reach more firm conclusions. The simulations seem to indicate that the nematic-to-smectic transition is a continuous phase transition (large pretransitional smectic order parameter fluctuations) although the finite size of the system does not allow us to distinguish unambiguously between a continuous and a weak first order transition [6.313, 338]. The isotropic-nematic and smectic-solid transitions are found to be first order (hysteresis effects).

It seems worthwhile to point out that the precise location of the thermodynamic coexistence densities in a first order phase transition is a formidable task requiring accurate absolute free energy calculations of the coexisting phases. In Frenkel's work the free energy of a given phase is obtained by constructing a reversible path to a reference state for which the free energy is known (for example the Einstein crystal for the solid free energy [6.42, 338]). Once the absolute free energies are known, the phase coexistence densities are determined as those for which the pressures and chemical potentials are simultaneously equal.

Small differences between molecular shapes, as for instance between spherocylinders and ellipsoids of revolution can lead to qualitatively different phase behaviour at high density. The phase diagram of hard ellipsoids, for example, does not show a smectic phase [6.341, 320]. A nematic phase occurs for aspect ratios $\kappa = a/b > 2.75$ and $\kappa \leqq 1/2.75$ and a plastic solid phase for small anisotropies $(1.25 \geqq \kappa \geqq 0.8)$. The phase diagram is remarkably symmetric under interchange of κ and $1/\kappa$ [6.320] similar to that which has already been observed for the pressure.

The limiting case $\kappa = 0$ corresponding to infinitely thin hard platelets has been considered by *Eppenga* and *Frenkel* [6.342]. A weak first order isotropic-nematic transition ($\Delta\rho/\rho < 2\%$) is found in marked contrast with the strong first order transition ($\Delta\rho/\rho \cong 26\%$) predicted by *Onsager* [6.318] for thin hard rods ($\kappa = \infty$).

Restriction of orientational phase space can also affect the phase diagram considerably. For parallel spherocylinders, a small nonsphericity ($L/D = 0.5$) is already sufficient for a smectic phase to be thermodynamically stable [6.339, 340]. But more remarkably, for $L/D > 3$, a first order phase transition occurs from the smectic to a columnar phase [6.340]. In contrast, parallel ellipsoids of revolution have phase behaviour similar to hard spheres [6.343] and thus do not exhibit liquid crystal like phases.

Evidence for a columnar phase has also been given in a Monte Carlo simulation of freely rotating plate-like molecules (cut-sphere model). Cut spheres of thickness $L/D = 0.1$ (D-diameter) are found to exhibit a nematic phase for reduced densities (with respect to close packing) < 0.35 and columnar ordering at a density $\rho > 0.45$ [6.336].

Intermediate between 2D and 3D systems are monolayers of molecules which can rotate freely in 3D space but have their centres confined to a plane. MC simulations of monolayers of oblate ellipsoids [6.344] and 'UFO's' (hard spheroidlike model defined to be the intersection volume of two identical spheres) [6.345] indicate that these systems are always orientationally ordered (except at zero density). By increasing the pressure from zero value, the spheroids first tip up with their symmetry axes tending to lie in the plane; then the long axes align, apparently in a continuous way (the precision of the calculations is not precise enough to decide whether a thermodynamic phase transition occurs).

More realistic than the hard gaussian overlap model considered above is the (6–12) gaussian overlap model originally proposed by *Berne* and *Pechukas* [6.308] and subsequently modified by *Gay* and *Berne* [6.346] to remedy several unrealistic features. In these models the molecules interact by a (12–6) Lennard-Jones potential having orientation dependent strength $\varepsilon(\boldsymbol{\omega}_1, \boldsymbol{\omega}_2)$ and range $\sigma(\boldsymbol{\omega}_1, \boldsymbol{\omega}_2)$ parameters. Simulation results for the thermodynamic and structural properties of the (6–12) gaussian overlap model with constant strength parameter (this is a reasonable simplification since ε depends only weakly on molecular orientation) have been reported in [6.347, 330, 348] and for the Gay–Berne potential in [6.349, 350]. The results covering a range of aspect ratios (anisotropy parameter) from 0.5 (benzene-like) to 1.55 (ethane-like) to 1.83 have been compared with those for site–site potentials [6.347] and with perturbation theories [6.348, 350, 351]. In addition to the thermodynamic properties, the phase diagram of the Gay–Berne model with anisotropy parameters $\kappa = 3$ (aspect ratio) and $\kappa' = \varepsilon_\parallel/\varepsilon_\perp = 5$ ($\varepsilon_\parallel$ and $\varepsilon_\perp$ strength parameters for side-by-side and end-to-end configurations, respectively) has been investigated in some detail. Thus the liquid-vapour coexistence curve and an estimate of the critical temperature and density have been obtained [6.352] using the Gibbs ensemble MC method [6.20]. In the high density region, evidence has been given for the existence of a nematic phase [6.353] and the densities of the coexisting isotropic and nematic phases determined as a function of temperature by performing absolute free energy calculations [6.352]. The isotropic-nematic transition is found to be weakly first order [6.354].

A further potential model, which shares with the gaussian overlap model the advantage that it involves only one interaction site to describe anisotropic molecular shape, is the Kihara core model [6.355]. In this model the potential is defined in terms of the shortest distance between convex hard cores (e.g. spherocylinders or ellipsoids). More flexible than the gaussian overlap potential, it has nevertheless the disadvantage that the calculation of the shortest distance between two convex cores may be quite time consuming. Monte Carlo calculations for linear rod-like molecules interacting through the Kihara potential [6.356] have served mainly to test analytical equations of state [6.357] and perturbation theory [6.358].

6.5.3 Site–Site Potentials

The use of site–site potentials is presently the most common way to describe molecular interactions. Model systems, as for instance fused hard spheres or site–site Lennard–Jones potentials, involving a small number of sites have been extensively studied in the past [6.1, 2] and have benefitted from relatively little new developments. Recent MC studies include simulations of the equation of state, site–site correlation functions and site–site spherical harmonic expansion coefficients of homonuclear [6.359, 360] and heteronuclear hard diatomics [6.361], linear symmetric hard triatomics [6.361], tetrahedral hard penta-atomics [6.362], mixtures of heteronuclear diatomics and linear triatomics with hard spheres and with each other [6.361], and mixtures of non-linear triatomic molecules with either heteronuclear dumbbells or linear symmetric triatomics [6.363]. A critical compilation of the structural results of hard core bodies can be found in [6.364]. Further MC simulations of hard diatomics are devoted to the calculation of the background correlation function [6.365, 366], bridge function [6.366] and fluid-plastic crystal coexistence curve [6.367]. In [6.365], it is shown that the background correlation function can be obtained from the knowledge of the chemical potential and the probability of inserting a cavity of position $\boldsymbol{r}_1$ and orientation $\boldsymbol{\omega}_1$ into a system consisting of $N-1$ molecules and a cavity with coordinates $(\boldsymbol{r}_2, \boldsymbol{\omega}_2)$. In fact, since for hard core particles any of the N molecules can be regarded as cavity 2 (because it does not overlap with the remaining $N-1$ particles) one needs to calculate only the probability of inserting a single cavity. This method is much faster than the procedure used by *Torrie* and *Patey* [6.368], but it is clearly restricted to hard core systems and requires the chemical potential as input. Also its accuracy depends on density and separation. *Labik* and *Malijevsky* [6.365] present radial slices of $y(r_{12}, \boldsymbol{\omega}_1, \boldsymbol{\omega}_2)$ at selected orientations, whereas *Lomba* et al. [6.366] calculate the spherical harmonic expansion coefficients. These, when combined with the corresponding coefficients for the pair correlation function enable to estimate the bridge function [6.366]. The fluid-plastic crystal coexistence of hard dumbbells has been investigated by *Singer* and *Mumaugh* [6.367]. From absolute free energy calculations of both phases they conclude that a stable plastic crystal phase exists for reduced bond lengths $L/\sigma < 0.4$. Extensive Helmholtz free energy calculations have also been performed for two-site [6.369], triangular three-site and hexagonal six-site Lennard–Jones (12–6) potentials [6.370] with the purpose of determining the vapour-liquid coexistence curves of nitrogen- and oxygen-like systems and their critical densities and temperatures [6.369] and testing perturbation expansions [6.370].

6.5.4 Chain Molecules

Chain molecules have received considerable attention in recent years, especially the problem of developing theoretical equations of state, in part due to its importance to the chemical and petroleum industries. The simplest model for chain molecules is the freely jointed tangent hard sphere chain. Although this model neglects variation of the bond length and does not impose constraints on the bond and torsional angles, it captures essential features as the intra- and intermolecular site–site excluded volume interactions, molecular connectivity etc. The connectivity constraint makes theoretical approaches considerably more complex for chain molecules than for simple molecules so that simplifying assumptions are generally necessary to make theory tractable. In this respect, computer simulations appear as an indispensable tool to control these approximations.

The equation of state of short chains composed of 4, 8 and 16 tangent hard spheres has first been obtained in a Monte Carlo simulation of a system bounded in one direction by two hard walls. The pressure (divided by kT, T-temperature) is then simply given by the density of sites (segments) in contact with the wall [6.371] provided the system size is sufficiently large that the central part of the system behaves as a homogeneous bulk fluid. Subsequent molecular dynamics computations [6.372] where the pressure is calculated from a time average of the collisional exchange of momentum, besides extending the pressure results to chain lengths of 32 spheres, confirmed the MC values and demonstrated the reliability of the method. These simulation results served to test a number of analytic equations of state, in particular, generalisations of the Flory and Flory–Huggins lattice-based equations to continuum space [6.373], the generalised Flory-dimer (GF-D) equation of *Honnell* and *Hall* [6.374], thermodynamic theory of polymerisation by *Wertheim* [6.375] and *Chapman* et al. [6.376], integral equation approaches based on the reference interaction site model (RISM) formalism [6.377] and particle–particle descriptions [6.378] as well as extensions of fused hard sphere equations of state [6.379]. Simulations of longer chains involving up to 200 [6.380] and even 400 [6.381] segments (sites) have been performed by *Gao* and *Weiner* [6.380] and *Kremer* and *Grest* [6.381–383]. In both studies the non-covalent interaction is a shifted, purely repulsive LJ potential and the covalent bonds are represented by either stiff harmonic springs

[6.380] or the so-called FENE (finite extensible nonlinear elastic) potential [6.381–383]. Some of the conclusions reached in [6.380] are that the equation of state of chains up to 200 sites is accurately reproduced by the GF-D theory, that the compressibility factor shows little dependence on chain length when expressed in terms of monomer density (for a given packing fraction), and that covalent interactions make a large contribution to the pressure (30% at liquid like packing fractions) in accord with similar findings for 4- and 8-bead hard sphere chains [6.384]. The impressive work of *Kremer* and *Grest* [6.381–383] is mainly concerned with dynamical aspects of dense polymer melts, covering the range from the short non-entangled (Rouse) regime to the highly entangled (reptation) regime. As far as structural properties are concerned, simulations of intra- and intermolecular correlation functions of tangent hard sphere chains of 4 and 8 spheres [6.384] and longer chains of 50 and 200 monomers, interacting by continuous potentials, made it possible to test integral equation approaches based on the RISM theory with Yukawa [6.385] and Percus–Yevick [6.386] closures and on particle–particle description of the system in the context of the PY theory [6.387] and to critically assess the validity of the approximations inherent in these theories. For more details on chain molecule simulations see also Chap. 9 of the present book.

6.5.5 Dipolar Systems

The presence, in computer simulations, of dipolar interactions entails new methodological problems which have their origin in the long range nature of these interactions. A convenient summary of these problems which concern most crucially the calculation of dielectric properties, has been given in a recent review article by *de Leeuw* et al. [6.388]. The two basic difficulties encountered are

a) the choice of an adequate boundary condition to eliminate finite size effects,
b) adequate sampling of phase space.

The two boundary conditions most frequently used in computer simulations are periodic boundary conditions (p.b.c.) and reaction field (RF) boundary condition. In p.b.c. the simulation cell is surrounded by a finite array of copies of itself and the region outside the array approximated by a dielectric continuum of dielectric constant ε'; in the RF method the dipole–dipole interaction is spherically truncated and the effect of long range interactions included by surrounding the truncation sphere with a polarizable dielectric continuum. Whatever boundary conditions are used, it is essential to realise that the electrical potential ψ entering the electrostatic part of the Hamiltonian of the system,

$$H_{\mathrm{ES}} = \frac{1}{2} \sum_i \sum_j \boldsymbol{\mu}_i \cdot \nabla_{r_i} \nabla_{r_j} \psi(\boldsymbol{r}_i, \boldsymbol{r}_j) \cdot \boldsymbol{\mu}_j, \tag{6.22}$$

must be a solution of Poisson's equation with the chosen boundary condition [6.388]. Only if this condition is satisfied can a meaningful relation between the dielectric constant and the sample dipole moment fluctuations be derived.

From this point of view the use of p.b.c. has to be recommended in preference to the RF boundary conditions. The latter method is moreover plagued with technical problems such as energy drift arising from the discontinuity of the dipolar interaction at the truncation sphere, or the difficulty in evaluating the reaction field when the dipole moment has finite extent (i.e. represented by a distribution of point charges) etc. [6.388]. The use of RF boundaries has often been motivated by the assertion that it is less time consuming than periodic boundary conditions. This however, is only marginally true, as fast algorithms are available to evaluate lattice sums [6.388, 389]. The second difficulty which arises in the simulation of dipolar fluids and which is particularly relevant for calculating dielectric constants, is related to sampling of phase space. From the very long molecular dynamics runs (up to 1 million time steps) by *Neumann* [6.390, 391] on water-like models, and by *Kusalik* [6.392] on dipolar soft spheres (0.5 million time steps), it is apparent that for high dipole moments, the running average of the dielectric constant ε presents long time oscillatory behaviour which precludes a precise determination of ε. *Kusalik* [6.392] observes that the period of oscillations is largest when $\varepsilon' = \infty$ (the so-called tinfoil boundary conditions) but is much smaller (and the oscillations much less regular) when ε' is chosen close to the actual value ε of the dielectric constant. If this result is confirmed for dipolar potentials other than soft spheres, it would indeed be more appropriate to use a value $\varepsilon' \approx \varepsilon$ in conjunction with p.b.c. Most existing estimates of ε for dipolar hard spheres and Stockmayer particles (cf. e.g. [6.388]) are based on relatively short runs (a few ten thousand time steps in MD or a few million configurations generated in MC calculations) and for large dipole moments likely to be not converged or affected by much larger error than generally quoted. (Earlier calculations of ε may also be in error due to use of incorrect fluctuation formulas [6.388].) As these model systems are primarily designed to provide benchmarks for testing theories it would seem desirable to perform new calculations of ε with increased accuracy.

Another requirement for obtaining a meaningful estimate of ε is that the simulation sample is sufficiently large so that the average polarization of the sample (in the presence of an external field) can be identified with a macroscopic polarization. System size dependence of ε for highly dipolar systems ($\mu^{*2} \gtrsim 2$) has been investigated in most detail by *Kusalik* [6.392] using 4–2000 particles. Converged results are obtained for particle numbers $N \gtrsim 256$ if $\varepsilon' \sim \varepsilon$. Convergence is however found to be much slower when $\varepsilon' = \infty$ [6.392]. For small dipole moments ($\mu^* = 1$) independence of ε on system size is already noted with 108 particles [6.393]. Specific results for dipolar model systems are the following: thermodynamic, dielectric and structural properties of dipolar soft spheres have been calculated by Kusalik for two thermodynamic states characterised by a reduced density $\rho^* = 0.8$, a reduced temperature $T^* = 1.35$ and reduced dipole moments $\mu^* = 1.65$ [6.394] and $\mu^* = 2$ [6.392]. For the lowest dipole moment comparison is made with results for the full Stockmayer potential and the reference hypernetted chain theory [6.394]. The liquid-vapour equilibria for Stockmayer fluids with dipole moments $\mu^* = 1$ and $\sqrt{2}$ have been investigated by *Smit* et al. [6.395] using Monte Carlo simulations in the Gibbs ensemble (cf. Sect. 6.1.3). The critical densities and temperatures are estimated. Only the critical temperature varies noticeably (increase) with dipole strength. *Cummings* and *Blum* [6.396] calculate the dielectric constant of mixtures of dipolar hard spheres and *de Leeuw* et al. [6.397, 398] calculate the excess mixing properties and dielectric constants of mixtures of Lennard–Jones and Stockmayer molecules of various compositions and dipolar strengths. The latter results suggest that for reduced dipolar strength $\mu^* > \sqrt{3.15}$, phase separation occurs into a phase rich in polar component and an almost pure LJ fluid [6.398]. The influence of non-sphericity of the short range forces on dielectric behaviour has been investigated for two-centre Lennard–Jones fluids [6.399, 400] and the gaussian overlap model with embedded dipole moment parallel to the molecular symmetry axis [6.401]. An increase of elongation or aspect ratio of the molecules results in a reduction of the dielectric constant due to a change in short range angular correlations imposed by the steric forces. In contrast, if one increases the density of dipolar hard dumbbells at constant elongation and dipole strength a maximum is observed in the dielectric constant [6.400].

6.5.6 Quadrupolar Systems

The applicability of the Gibbs ensemble Monte Carlo method to strong orientationally dependent forces is demonstrated by *Stapleton* et al. [6.402] and *Smit* and *Williams* [6.403], who, independently, determine the vapour-liquid coexistence curves of quadrupolar LJ fluids for various values of the quadrupolar moment and give an estimate of the critical temperatures and densities. Both sets of results are in good agreement except for the somewhat higher critical density found in [6.403]. *Stapleton* et al. [6.402] extend their calculations to the phase coexistence properties of a mixture of LJ quadrupolar fluids having (reduced) quadrupole moments $Q^* = 1$ and -1. At low temperatures, a weak azeotrope is found as observed experimentally in the CO_2/C_2H_2 mixture. The effect of competition between excluded volume effects and electrostatic interactions on the thermodynamics and orientational structure of quadrupolar diatomics [6.404–407] and mixtures of quadrupolar diatomics [6.408] has been investigated by systematically varying elongation and quadrupolar moment of the molecules. A special study was devoted to the melting transition of quadrupolar diatomics at constant density [6.409]. *Shing* [6.410] uses a special implementation of Widom's potential distribution theorem [6.411] to calculate infinite-dilution activity coefficients of quadrupolar LJ mixtures.

6.5.7 Polarizable Polar Fluids

Polarization (or induction) energy is associated with the distortion of the charge distribution of a molecule in the electric field of its neighbours. It is clearly not pairwise additive and therefore more difficult to incorporate in computer simulations. The effect is generally significant (see, for example, the case of water, Sect.

6.6.6) and it is more and more apparent that explicit incorporation of non-additivity in the intermolecular interactions is required for an accurate representation of gas and condensed phase properties of polarizable systems [6.412–425].

Within the context of a polarizability model for describing many body effects, the polarization energy is obtained from the dipole moments induced by the local electrical field acting on the molecule. These dipole moments satisfy a set of coupled equations which have to be solved self-consistently, either by matrix inversion or an iterative solution method [6.426]. This step is responsible for an increase in computer time of the order of 30% and time saving procedures have been envisaged, e.g. by maintaining the individual dipole moments fixed for several configurations [6.427] or using a predictive algorithm for the induced dipole moments [6.420]. However, not fully self consistent algorithms should be applied with great care as they may lead to errors in the average total energy [6.419] or cause instabilities [6.420].

The Monte Carlo method is less efficient than the molecular dynamics method to treat polarizability effects. The reason for this is that for each particle move, the electric field needs to be calculated. The efficiency of the MC method can however be improved by moving several particles simultaneously [6.412].

There have been few simulations of polarizable model systems (generalised Stockmayer potentials) [6.428, 429]. These were concerned with testing a self-consistent mean field theory [6.430] which reduces the many-body energy of a polarizable system to a sum of pairwise additive interactions, characterised by an effective permanent dipole moment.

6.6 Molecular Liquids: Realistic Systems

In this section we give a brief review of simulations which employ realistic potential models. By realistic we merely mean that the potential function has either been obtained from *ab initio* quantum mechanical calculations or that some effort has been put in adjusting potential parameters in order to reproduce experimental data (empirical or semi-empirical potentials). Thus realistic does not mean that the potential under consideration necessarily gives an overall faithful description of the physical system. In fact this is rarely the case. More often, several potential functions are available for the same system with accuracy depending on what properties are considered (thermodynamics, crystal structure, diffusion coefficient, reorientational times etc.).

In the construction of empirical (or semi-empirical) potentials, the monomer is generally represented by a set of interaction sites. Sites on different molecules interact via short range forces (e.g. of the LJ or exp-6 type) and/or long range electrostatic forces (charges, dipoles etc.). The site–site interactions have generally been assumed isotropic although there is no need (except computational convenience) for doing so; in fact it is recognised more and more that anisotropic (repulsive and electrostatic) site–site potentials greatly improve the accuracy of potential models (see e.g. the review article by *Stone* and *Price* [6.431]). The positions of the sites within the molecule, as well as the parameters of the various interactions, are considered as adjustable parameters (e.g. the range and depth parameters of the LJ potential, the magnitude of the charges or dipole moments). Ideally, some of these parameters can be fixed to reproduce known monomer and dimer properties (dipole moment, quadrupole moment, dimerisation energies etc.) and the remaining parameters adjusted by fitting liquid or solid state properties available from experiment. The positions of the interaction sites need not to be rigidly distributed in the monomer but allowance can be made for variation of the bond lengths, bond and torsional angles (flexible molecules). Also many-body effects can be accounted for through inclusion of polarizable sites and bonds.

A second route to interaction potentials is to perform ab initio quantum mechanical computations. The strategy of this approach is well documented in a recent review article by Clementi and coworkers [6.432]. The cost of the computations depends largely on the complexity of the molecule, the level of sophistication of the calculations, and the number of dimer, trimer, etc. configurations needed to obtain an accurate analytical representation of the two-, three-, etc. body potential surfaces.

6.6.1 Nitrogen (N_2)

Several potential models of the 2-centre LJ type, effective or fitted to ab initio calculations [6.433] have been tested in the pressure range 0–3 kbar [6.434]. None is very satisfactory over the whole pressure range. The limitation of spherical site–site potential models (LJ or exp-6 plus point charges) is apparent from the difficulty in accurately fitting ab initio potential surfaces [6.433–435]. A simple anisotropic atom–atom potential has been shown to give a far more satisfactory fit [6.436]. Nonetheless an isotropic site–site model of *Etters* et al. [6.437] reproduces the experimental pressure (at room temperature) to within 0.3% up to pressures of about 10 kbar [6.438]. This model has been used in MC simulations to predict thermodynamic properties of N_2 over the pressure range 2–300 kbar and temperature range 300–3000 K [6.439, 440] as well as the melting point at 300 K [6.438]. The latter work presents an accurate new method to compute the absolute free energy of the solid phase (necessary to locate the fluid-β solid phase transition densities). This method proceeds in two steps: first each particle is coupled to its corresponding equilibrium lattice site (method of [6.42]), then the system is expanded to zero density so that it becomes equivalent to a lattice site-coupled solid without interparticle interactions. For harmonic coupling this is an Einstein crystal.

6.6.2 Halogens (Br_2, Cl_2, I_2)

Halogens are examples where the use of anisotropic repulsive interactions greatly improves the description of solid and liquid state properties [6.441–443]. Earlier calculations with isotropic 2-centre LJ potentials, even with inclusion of a central quadrupole moment, failed to reproduce the observed lattice structures of the solid phases, and were not fully successful in describing the structure of the liquid phase, in particular the asymmetry of the first peak and the relative heights of the first and second peaks [6.444–447].

6.6.3 Benzene (C_6H_6)

Several simulations have been performed for benzene, all, with one exception [6.448], treating the molecule as a rigid planar hexagon [6.449–457]. The main conclusions of these studies, as regards potential modelling, is that a realistic treatment of the electrostatic interaction is essential in determining the structural properties and that account has to be taken of the steric effects associated with the hydrogen atoms. Two types of models satisfying these requirements proved satisfactory in reproducing both liquid and crystal properties of benzene: an isotropic 12-site model with fractional charges on the carbon and hydrogen atoms [6.452, 455] and an anisotropic 6-site model with distributed-quadrupole representation of the electrostatic forces (BENZ6) [6.453]. The (12–6–1) potential of *Jorgensen* and *Severance* [6.455] has been fitted to the observed heat of vaporisation and density at 25° C and 1 atm, whereas the (12–10–6–2) potential of *Bartell* et al. [6.452] is a simplified representation of ab initio quantum calculations of *Karlström* et al. [6.458], with parameters rescaled to reproduce the binding energy and molar volume of crystalline benzene. A previous fit of *Karlström*'s ab initio calculations by *Linse* [6.450] was less satisfactory, predicting too low a lattice energy and too small a molar volume at 0 K. The BENZ6 potential of *Yashonath* et al. [6.453] has been fitted to both forms, orthorombic and monoclinic, of solid benzene. The structural properties predicted by these models are in agreement with Narten's X-ray measurements of the carbon–carbon correlation functions and seem compatible with a preferred T-shaped nearest neighbour orientation in accord with recent pulsed neutron scattering experiments [6.457]. The structural properties do not appear to be particular sensitive to the flexibility of the molecule [6.448].

6.6.4 Naphthalene ($C_{10}H_8$)

Naphthalene has a shape of doubled carbocyclic hexagons, each of which is quite similar to that of a benzene molecule. Computer simulations by *Sediawan* et al. [6.459] compare two models: one based on a 10-site LJ (12–6), the other on an anisotropic two-site pair potential with the interactions between these sites modelled using the modified gaussian overlap potential. Both potentials give a quite good description of the energy and density of saturated liquid naphthalene. Additional quadrupolar interactions influence mainly the structural properties, enhancing, as already remarked for benzene, the preference of T orientations also observed in neutron scattering experiments [6.457]. These potentials remain, however, to be tested against solid state properties.

6.6.5 *n*-Alkanes: $CH_3(CH_2)_{n-2}CH_3$

Recent simulations of normal alkanes were primarily concerned with liquid state conformational properties, though vibrational [6.460] and solid state properties (mainly crystal structures) [6.461, 462] were investigated as well. *n*-butane, being the simplest hydrocarbon capable of exhibiting trans-gauche rotational isomeration, has been considered most oftenly. Of particular interest is the question of trans-gauche equilibrium shift when going from the gas to the liquid phase (at constant temperature). Due to the relatively unfrequent barrier crossings (at low temperature), a conclusive answer to this question can be given only by performing long simulation results. The most precise result presently available [6.463], obtained with a potential model where bond lengths and bond angles are rigidly fixed at their equilibrium values and the torsional potential given by the expression of *Ryckaert* and *Bellemans* [6.464], indicates that the effect is extremely small (about 1% for the model considered). Most other simulation studies [6.465–468], based on much shorter runs, agree with this result within their generally large (sometimes underestimated) error bars. The sensitivity of liquid butane to intra- and intermolecular potential function changes has been investigated by *Toxvaerd* [6.469]. Conformational changes seem most sensitive to dihedral potential variation, whereas pressure, barrier crossing rate and diffusion constant are most affected by bond-angle vibrations [6.469]. A simulation of one isolated decane molecule [6.470], with and without bond length and bond angle constraints, further shows that for this larger alkane molecule, bond angle flexibility decreases the ratio between gauche and trans states by an amount which is bigger, or of the same order of magnitude, as the solvent effect from other decane molecules in a corresponding liquid state and dramatically changes the trans-gauche transition rate. In contrast, freezing of the bond lengths has negligible effect [6.470]. It is well established that treating the degrees of freedom of bond stretching and bond angle bending in chain molecules as rigid constraints or allowing them to vary leads to different partition functions, even in the limit as the force constants for the vibrations become infinitely large. The reason for this is that in both cases, different phase space is sampled. As a consequence conformational distributions [6.471, 472] or pressure [6.292] will differ according to whether a rigid constraint or infinitely stiff flexible model is invoked. A Monte Carlo study for butane shows, however, that the differences on trans-gauche equilibrium are small and totally negligible for the pressure [6.468].

Toxvaerd [6.473] addresses the question whether traditional potential models which treat methylene groups as one isotropic interaction unit can reproduce the experimental pressure of fluid *n*-alkanes. He concludes that this is not possible within the assumption of an effective Van der Waals radius for a CH_2 group identical for all alkanes as suggested by experiment. An improved description can be obtained by using either a full atomic model as done for studying crystal and melting properties of alkanes [6.461, 462, 474] or, more economically, an anisotropic model in which the methylene (methyl) group is still treated as one unit but the centre of force shifted from the carbon atom to the geometrical centre of the $CH_2(CH_3)$ groups [6.473, 475, 476].

Simulations by *Rigby* and *Roe* [6.477–479] of *n*-alkane-like chains having up to 50 CH_2 segments interacting through bond stretching, bending and torsional potentials and non-bonding truncated LJ potentials, investigate the influence of density on the population of transconformers [6.479]. No appreciable dependence is found; only at the highest densities considered does a gauche enhancement occur among bonds at the chain ends. The major purpose of the work was, however, to investigate the glass transition.

6.6.6 Water (H_2O)

Recent progress in the simulation of water concerns the development of refined potential models, including flexibility and polarization, and extension of the domain of investigation of the water properties to high pressure, dielectric properties, density fluctuations etc. In view of the considerable effort which has been devoted to the development of water potentials it is somewhat frustrating to note that an entirely satisfactory water model reproducing both static (pressure, structure, dielectric properties) and dynamic (diffusion constant, reorientational times, spectroscopic data etc.) properties over an appreciable density-pressure range is still lacking; however the direction in which to proceed for achieving improvement is now more clearly delineated [6.431].

The most successful and widely used effective pair potentials for water are the TIP4P potential of *Jorgensen* et al. [6.480] and the SPC potential of *Berendsen* et al. [6.481]. The TIP4P potential is a rigid 4-site model with parameters optimised to give a correct description of the thermodynamic properties of water in the vicinity of 298 K and 1 atm. Since its original application near normal water conditions both temperature ($T = 243$–373 K, $p = 1$ atm [6.482]; $T = 273$–1140 K, $\rho = 0.1$–$0.999\,g\,cm^{-3}$ [6.483]) and pressure dependence ($\rho = 0.999$–$1.149\,g\,cm^{-3}$, $p = 1$ bar–5 kbar [6.484]; $T = 268$ K, 298 *K*, $p = 1$ bar–5 kbar [6.485]; $T = 273$ K, $p = 1$–10 kbar [6.486]; $T = 340$ K, $p = 0$–18 kbar [6.487]) of the thermodynamic and structural properties have been obtained. The calculated densities are in reasonable agreement with experiment up to pressures of about 10 kbar and over the temperature range -30–100 K (at 1 atm). The TIP4P model has also been applied to study the glassy state [6.488], dynamical properties [6.489–492], free energy [6.493–495], dielectric properties [6.391].

The SPC potential of *Berendsen* et al. [6.481] is a simpler model than the TIP4P potential, having only three interaction sites. It is quite successful in predicting the energy and second nearest neighbour peak structure of liquid water, but gives virial coefficients and diffusion coefficients which are too large and a density at 1 atm and 298 K which is too small. *Berendsen* et al. [6.496] have pointed out that although effective pair potentials include polarization in an average way, a polarization self-energy term has systematically been omitted when adjusting potential parameters to reproduce experimental heat of vaporisation data. A new effective pair potential (termed SPC/E) giving much improved values for the density and diffusion constant is determined [6.496] by including this self-energy term and reparametrising the SPC model. Allowance for a polarization self-energy correction, evaluated in a self-consistent mean field approximation, [6.430] has also been made by *Watanabe* and *Klein* [6.497] in their determination of an effective two-body potential (WK) adjusted to fit experimental thermodynamic, structural and dielectric properties of liquid water at room temperature.

Concerning ab initio potentials, the shortcomings of the MCY potential have been recognised [6.432, 421] to relate to the limited sampling of the repulsive region of the energy surface and the insufficient flexibility of the analytical form selected to fit the calculated dimer energies. Several ab initio two-body potentials based on quantum mechanical computations at various levels of sophistication have since been obtained [6.498–503]. They are not expected to give an accurate description of liquid water or ice where many body effects are important. Ab initio 3- and 4-body interactions have been derived as well, and their contribution to static [6.504–506] and dynamic properties [6.507, 508] evaluated. The effect of 3- and 4-body interactions is to lower the energy (by -0.85 and $-1.22\,\mathrm{kcal\,mol^{-1}}$, respectively, for normal water), increase the amplitude of the peaks of the pair correlation functions (by 0.2 and 0.14, respectively, for the oxygen–oxygen correlation function) and shift their positions to smaller r-values. The apparent lack of convergence of the expansion of the potential in 2-, 3- and 4-body contributions and the difficulty in implementing 3- and 4-body terms explicitly in a computer simulation are the likely reasons why potential models have been sought which include polarization in the interaction potential at the two-body level. The most straightforward way to do this is to simply write the potential energy as the sum of the usual two-body terms and a non-additive induction term,

$$U_{\mathrm{ind}} = -\frac{1}{2}\sum_{i}^{N_m}\sum_{\lambda}^{N_d} \boldsymbol{E}^0_{i\lambda}\cdot\boldsymbol{\alpha}_{i\lambda}\cdot\boldsymbol{E}_{i\lambda}, \tag{6.23}$$

where N_m is the number of molecules, N_d the number of polarization centres in a molecule, $\boldsymbol{E}_{i\lambda}$ the total electric field at the polarization centre (created by the site charges and the induced dipole moments $\boldsymbol{p}_{i\lambda}=\boldsymbol{\alpha}_{i\lambda}\cdot\boldsymbol{E}_{i\lambda}$) and $\boldsymbol{E}^0_{i\lambda}$ the electric field created by the charges only. The polarization centres are most frequently located at the oxygen atoms or on the OH bonds [6.421]. The potential parameters, including those of the two-body terms, can again be adjusted by fitting experimental data [6.419, 420] or ab initio calculations [6.421–425]. Potentials of this type have not yet been obtained in optimised form and, in the liquid phase, do no better than existing effective pair potentials. However, they provide a much better overall description of all three gas, liquid and crystal phases.

A quite different approach to treat many-body polarization effects has been proposed by *Sprik* and *Klein* [6.509] who model polarization by four point charges with a fixed position (in the molecular frame) but a varying magnitude.

This procedure has technical advantages but appears to be somewhat more time consuming than the iteration scheme for calculating induced moments in the polarizable atom models described above.

Most water potentials are constructed under the assumption of a fixed geometry of the molecules (the experimental gase phase geometry). If one is interested in high frequency motions or geometrical changes in the liquid phase these models need to be modified to treat the internal vibrations of the water molecules. Several flexible water potentials have been proposed [6.510–517]. They combine traditional intermolecular potentials (as for instance the MCY, SPC or central force potentials) with harmonic or anharmonic potentials for the covalent interactions. As the original empirical potentials may contain already flexibility in an average way the potential parameters are sometimes readjusted.

The effects of intramolecular vibrations on the thermodynamic and structural properties have been demonstrated most clearly in a computer simulation by *Barrat* and *McDonald* [6.515]. The lowering of potential energy, drop in pressure, sharpening of the first peak in the oxygen–oxygen correlation function, decrease in the diffusion constant and increase of the reorientational correlation times are correlated in a rational way with the enhanced strength of the dipole–dipole interactions (there is a 0.16D increase of the dipole moment when going from the rigid [6.481] to the flexible [6.516] SPC model). The results contrast with those of [6.514], where much larger effects of flexibility are reported, some of which may be difficult to rationalise. In the preceding treatment of flexibility coupling of the inter- and intramolecular degrees of freedom is neglected. The importance of such coupling has been brought to attention in a recent paper by *Dinur* [6.518].

A commonly used criterion for judging the reliability of a potential model is the comparison of the site–site correlation functions obtained from computer simulations with those derived from X-ray and neutron scattering experiments. Several sets of experimental partial pair correlation functions are now available [6.519–522] showing, unfortunately, qualitative discrepancies among themselves. This fact expresses the difficulty in extracting pair correlation functions from experiment. X-ray scattering is dominated by oxygen–oxygen correlations and up to quite recently, oxygen–oxygen pair correlations were directly compared to the experimental results of *Narten* and *Levy* [6.519]. However, it has been pointed out [6.523] that this is not entirely satisfactory as X-ray scattering also contains significant contributions from oxygen–hydrogen and hydrogen–hydrogen correlations. Neutron scattering, on the other hand, provides a weighted sum of the three partial correlation functions g_{OO}, g_{OH} and g_{HH} so that at least three independent measurements are necessary to disentangle these correlations functions. It should be stressed, however, that before this can be done, the experimental differential cross section has to be corrected (among others) for multiple and inelastic scatterings, and the incoherent terms have to be removed. Inelastic scattering which is quite large for hydrogen scatterers affects primarily the incoherent (self) terms and the intramolecular contribution to the coherent term and its removal is a major source of error. To circumvent the necessity of applying inelasticity corrections, *Soper* and *Phillips* [6.521] apply a hydrogen/deuterium subtraction procedure which in principle removes the incoherent background and directly yields the HH partial structure factor and from it the HH pair correlation function by Fourier transform. However, in order to further obtain the remaining OH and OO correlation functions, an effective mass model is invoked for the dynamic scattering law to subtract the single atom (self) scattering from the diffraction data. The hydrogen–hydrogen and oxygen–oxygen pair correlation functions are certainly more accurate [6.524] than the earlier results of Narten and coworkers and are now widely used for testing water potentials. There is a troublesome difference in the (main) peak height of g_{OO} (3.09), as compared to the X-ray scattering result (2.31). In the diffraction experiments of Soper and Phillips the concentration of H_2O in the H_2O/D_2O mixtures was high. As explained in [6.522], the accuracy of the results obtained from the isotopic substitution method can be improved by using low concentrations of H_2O. This reduces incoherent scattering, inelasticity corrections and minimises the difference of H and D as regards quantum effects. This approach has been adopted recently [6.522] and, except for fine details, gives good agreement with the data of Soper and Phillips for g_{OH} and g_{HH}. Quite annoyingly, however, a substantially lower peak height (2.12) is found for g_{OO}, although the peak positions do agree. This value of the peak height is also lower than the X-ray scattering result. Further experimental work, combining the neutron scattering data with high quality X-ray measurements is planned [6.525] with the hope of resolving the ambiguity concerning the correlations between oxygen pairs. Both the SPC and TIP4P potentials reproduce the experimental [6.521] g_{OH} and g_{HH} functions very accurately, the SPC model being slightly superior (note, however, that there is some ambiguity with the SPC model as the charge sites do not coincide with the hydrogen positions). Other potential models such as the WK [6.497], SPC/E [6.496] and the flexible SPC/F [6.516] potentials as well as the most recent polarizable models [6.421, 422, 424, 425], give somewhat less good agreement. None of the potential models, with the exception of the polarizable CPC potential of [6.425] and the WK potential [6.497], adjusted to do so, gives the correct position of the main peak of g_{OO}. The position is generally shifted to smaller r-values. Although it is presently not worthwhile to make detailed comparison of the peak heights (because of the large differences in experimental results) it is quite interesting to remark that *all* simulation results give peak heights of g_{OO} in the range 2.5–3.2 and thus higher than the experimental results of [6.519] and [6.522].

The calculation of the dielectric constant provides a further test for the quality of a water model. An accurate prediction of the dielectric constant seems indeed a prerequisite for a water model to be applicable in a reliable way to electrolyte solution studies. As discussed in Sect. 6.5.5, an accurate determination of ε is hampered by the slow convergence of the mean square dipole moment of the sample. The long simulation results of Neumann for the MCY [6.390] and TIP4P [6.391] potentials showed quite convincingly that at least 5×10^5 time steps are necessary in a MD simulation (or an equivalent number of moves per particle in a MC simulation) to obtain dielectric constants with a precision of a few percent and that runs of about 100 000 moves per particle can give poor estimates of ε. In some calculations (e.g. [6.516, 526]) the error bars seem manifestly underestimated. The room temperature dielectric constants of the most frequently used water potentials are summarised in Table 6.1. The TIP4P and especially the MCY potentials are seen to considerably underestimate ε. The values for the SPC and SPC/E potentials are closer to the experimental result but still somewhat too small; only the SPC/F and the WK potentials give about the right value (within large error bars). The temperature intensity dependence of ε has not yet been investigated in a systematic way, but from the available results for the TIP4P potential over the temperature range 300–3000 K [6.391. 529] it would seem that the Kirkwood g_K factor decreases with temperature in accord with experiment. The apparent increase of g_K for the MCY and TIP4P potentials observed by *Neumann* [6.391] over the very limited temperature range (293–373 K) is most probably the result of large statistical error.

Table 6.1. Dielectric constant of water at 1 g cm^{-3} obtained from molecular dynamics simulations for various potential models. RF denotes reaction field boundary conditions and EW periodic boundary conditions (see text). The experimental value for the dielectric constant [6.528] is given in parantheses

Water potential	Number of molecules	Total time (ps)	Temperature (K)	Boundary condition	Dielectric constant		Ref.
		2000	292	RF $\varepsilon' = \infty$, $R_c = 9.85$ Å	35	(80)	
MCY	256	1400	292	RF $\varepsilon' = 30$, $R_c = 9.85$ Å	34 ± 1.5		[6.390]
		1000	373	RF $\varepsilon' = \infty$, $R_c = 9.85$ Å	30 ± 1.5	(56)	
		2000	293		53 ± 2		
TIP4P	256			RF $\varepsilon' = \infty$, $R_c = 9.85$ Å			[6.391]
		1000	373		43 ± 2		
TIP4P	216	1400	293	RF $\varepsilon' = \infty$, $R_c = 9.30$ Å	47.3 ± 3.9		
TIP4P	345	630	293	RF $\varepsilon' = \infty$, $R_c = 10.85$ Å	51.7 ± 10.8		[6.527]
SPC	216	1455	350	RF $\varepsilon' = 70$, $R_c = 8.0$ Å	59.7 ± 1.9	(61.8)	
SPC	256	1019	300	RF $\varepsilon' = 70$, $R_c = 9.85$ Å	67.8 ± 6.7	(77.6)	
WK	216	700	298		80 ± 8	(78.4)	[6.497]
				EW $\varepsilon' = \infty$			
SPC	216	700	298		72 ± 7		
	216	800	298		70.7 ± 0.8		
SPC/E				RF $\varepsilon' = \infty$, $R_c = 9.3$ Å			[6.526]
	216	800	373		51 ± 0.35		
		175	259		109 ± 5	(94.1)	
SPC/F	125	200	300	EW $\varepsilon' = \infty$	82.5 ± 4		[6.516]
		150	350		70.5 ± 4		

In [6.530] and [6.527] it is pointed out that if a switching function is used to smoothly truncate the water–water potential near the simulation cell boundary, account has to be taken of the effects of this smoothing function in the reaction field energy and in the relation between ε and the mean square dipole moment. Otherwise the dielectric constant could be seriously underestimated [6.527].

The excess Helmholtz free energy (with respect to the ideal gas) of water at 27 °C and 1 atm has been calculated by *Hermans* et al. [6.493] for several water models, including SPC, SPC/E and TIP4P using the thermodynamic integration method [6.531]. For SPC they find a value $\Delta F = -5.8\,\mathrm{kcal\,mol^{-1}}$ in agreement with calculations by *Mezei* [6.532] using a similar method (earlier calculations by *Mezei* [6.531] were in error) and with the experimental value $-5.74\,\mathrm{kcal\,mol^{-1}}$. A grand canonical ensemble Monte Carlo simulation by *Mezei* [6.95] gives $-5.47 \pm 0.1\,\mathrm{kcal\,mol^{-1}}$. For TIP4P, *Hermans* et al. [6.493] find $-5.3\,\mathrm{kcal\,mol^{-1}}$, *Li* and *Scheraga* [6.494], using a recursion method, $-5.60 \pm 0.09\,\mathrm{kcal\,mol^{-1}}$. A more indirect method of *Jorgensen* et al. [6.495] calculating the difference in free energy of hydration of water and methane and combining with the free energy of hydration of methane yields $-5.50\,\mathrm{kcal\,mol^{-1}}$. For the SPC/E potential model an excess free energy of $-5.5\,\mathrm{kcal\,mol^{-1}}$ is found (if allowance is made for a polarization self-energy correction) [6.493] whereas the MCY potential gives the much too low value -4.30 ± 0.08 [6.494]. *Sussman* et al. [6.533] compare the efficiency of ratio overlap and umbrella sampling methods in calculating free energy differences for water at different temperatures.

Hydrogen bond network analysis of low temperature water in terms of distribution of polygons has been performed by *Speedy* et al. [6.534, 535] and *Belch* and *Rice* [6.536]. The aim was to test proposals which identify the unusual behaviour of water just above the normal freezing point and in the supercooled region with changes in the distribution of rings of hydrogen bonded molecules.

Finally we would like to mention a recent MC simulation [6.537] of the liquid–vapour coexistence curve of SPC water using the Gibbs ensemble method. The estimated critical temperature is 314 °C and the estimated critical density $0.27\,\mathrm{g\,cm^{-3}}$.

6.6.7 Methanol (CH_3OH)

Another hydrogen-bond forming fluid which has benefitted recently from a number of investigations is methanol [6.538–546]. Both rigid 3-site potential models of the Lennard–Jones plus Coulomb form [5.538, 540, 541] and flexible 3-site [6.542] and 6-site [6.544] central force type potentials yield structural properties in fair agreement with X-ray and neutron scattering experiments. The structural features result from a competition of packing requirements of the methyl group and hydrogen bond network formation of the hydroxyl groups. An average number of about two nearest neighbours and 1.85 hydrogen bonds per molecule (at room temperature) is consistent with formation of a chain structure [6.538, 541, 542, 545, 546], in marked contrast to the tetrahedral structure of water. The mean chain size is found to be ~15 at 300 K and ~76 at 200 K [6.546], the mean life time of a hydrogen bond 1–2 [6.541] to 5–7 ps [6.546].

6.6.8 Other Polar Systems

Simulation studies of polar systems, two examples of which have been given in the preceding sections—water and methanol—are numerous and cannot be reviewed here in detail. We limit ourselves to giving a brief list, Table 6.2, which will help the interested reader to find his way to the relevant papers.

6.6.9 Mixtures

Computer simulations of mixtures of anistropic molecules are still scarce. Excess thermodynamic and structural properties of Ar/N_2, Ar/O_2 and N_2/O_2 mixtures are reported by *Gupta* and co-workers [6.578–580]. The main motivation of this work is to test perturbation theories. In [6.581], integral equation theory for the site–site correlation functions of a C_6F_6/C_6H_6 mixture is compared with computer simulations.

The liquid mixture CCl_4/CS_2 is a simple mixture in the sense that the excess free energy and enthalpy are very small at room temperature and the components miscible at all concentrations. *Mittag* et al. [6.582, 583] show that 5-centre and 3-centre LJ (12–6) potentials for CCl_4 and CS_2, respectively, in combination

Table 6.2. A compilation of simulation studies of polar systems

hydrogen chloride	HCl	[6.415, 547, 548, 412]
hydrogen fluoride	HF	[6.549]
ammonia	NH_3	[6.415, 417, 418, 550–552]
hydrogen sulfide	H_2S	[6.533, 410]
carbon dioxide	CO_2	[6.554–556]
carbon disulphide	CS_2	[6.557, 416, 556, 555]
sulfur dioxide	SO_2	[5.558–560]
methyl chloride	CH_3Cl	[6.556, 413, 561]
methylene chloride	CH_2Cl_2	[6.562, 563]
chloroform	$CHCl_3$	[6.564–566]
dichloroethane	$C_2H_4Cl_2$	[6.567]
methylfluoride	CH_3F	[6.556]
fluoroform	CHF_3	[6.556]
1,1-difluoroethane	CH_3CHF_2	[6.568]
Chlorodifluoromethane	$CHClF_2$	[6.569, 570]
dichlorodifluoromethane	CCl_2F_2	[6.570]
methyl cyanide	CH_3CN	[6.556, 571–573, 414]
ethylene oxyde	C_2H_4O	[6.574]
ethanol	C_2H_6O	[6.538]
glycerol	$C_3H_8O_3$	[6.575]
pyridine	C_6H_5N	[6.576]
formamide	CH_3NO	[6.577]

with the Lorentz–Berthelot mixing rules give an excellent description of the structural properties of the mixture as compared with neutron scattering experiments.

In the CH_4/CF_4 mixture, on the contrary, the excess thermodynamic properties are large [6.584]. Excellent agreement with experimental data is obtained by using ab initio dimer potentials represented by 5-site models [6.584]. This work, as well as that of *Schoen* et al. [6.585], suggests a tendency for phase separation.

The pressure-composition curve of the CO_2/C_2H_6 mixture has been determined by *Fincham* et al. [6.586] using simple 2-centre LJ (12–6) potentials (without quadrupole) and Lorentz–Berthelot mixing rules. Quite remarkably, a positive azeotrope is found in qualitative agreement with experiment. In contrast, the excess thermodynamic properties agree poorly with experiment [6.587].

Computer simulations are performed by *Kovaks* et al. [6.588] for liquid mixtures of chloroform and acetonitrile. As far as static properties are concerned, the authors are primarily interested in the change of structural order entailed by a variation of the concentration of acetonitrile.

Ferrario et al. [6.589] study aqueous mixtures containing methanol, ammonia or acetone using effective pair potentials parametrised to properties of the pure liquids and simple combining rules for the cross interactions. Special emphasis is given to structure and hydrogen-bonding patterns in an effort to elucidate the anomalous behaviour displayed by water–methanol and water-acetone mixtures. A mixture of water and CCl_4 molecules has been studied by *Evans* [6.590].

6.7 Solutions

The value of computer simulations as a means for providing insight into the microscopic structure and dynamics of solutions, including those of chemical and biological interest is now fully accepted and testified to by the large body of simulation studies which appeared in the field during the last few years. These comprise aqueous and non-aqueous solutions of ionic, polar and non-polar molecules at infinite dilution or finite concentration. Here we have space only to give a brief enumeration of the most representative of these studies. They are generally based on analysis of energetic, hydrogen-bonding, solute–solvent and solvent–solvent structural properties.

6.7.1 Infinite Dilution

Typical examples of aqueous hydration of ionic or polar molecules at infinite dilution include Na^+, K^+, Li^+, F^-, Cl^-, Br^- ions [6.495, 591–599]; NH_4^+ [6.600]; Ca^{2+}, Ni^{2+} [6.599]; Fe^{2+}; Fe^{3+} [6.601–603] acetate anion and methyl ammonium cation [6.604]; benzene [6.605–607]; methanol [6.605]; ethanol

[6.608]; fluoroalcohols [6.609]; and formamide [6.610]. Efficient methods for calculating free energy differences, based on thermodynamic perturbation theory and thermodynamic integration (cf. Sect. 6.1.3) have been applied to calculate free energies of hydration of bromide and chloride anions [6.611, 6.12], ethane, propane, methanol and ethanol [6.613, 45], acetone and dimethylamine [6.614, 615], methylamine and methanol [6.616], methylamine and acetonitrile [6.617], dimethyl phosphate anion in the various photodiester conformations [6.618], alkyl- and tetra-alkylammonium ions, amines and aromatic compounds [6.619] as well as conformational equilibrium of 1,2-dichloroethane in methylchloride [6.620], and relative pK_a's for organic acids [6.621]. The equilibrium constant of ionic dissociation of water has been calculated [6.32] using test particle methods [6.622].

Insight into the association or dissociation of ion pairs in solution can be gained from the free energy of an ion as a function of ion–ion separation (potential of mean force). The free energy curve for oppositely charged ions in water is generally found to have two local minima corresponding to contact and solvent separated ion pairs. For like charged ions the results are variable. For example, contact ion pairs have been found for tert-butyl cation and chloride ion [6.623] or chloride–chloride ion pairs [6.624, 625]. Stability of these anionic pairs is attributed to bridging structure in which water molecules hydrogen bond to both chloride ions simultaneously. In contrast, tetramethylammonium ions show no tendency for self-association in water [6.626]. The influence of solvent on oppositely charged contact and near contact pairs has been investigated in [6.627], and the relaxation of a realistic polar solvent near a rapidly dissociating Li^+F^- pair in [6.628].

The extent to which the simulation results are affected by the finite size of the systems, the commonly applied cutoff of the long range ion–dipole and dipole–dipole interactions between solute–solvent and solvent–solvent particles, and the lack of explicit accounting for polarization of the ions and water molecules, has hardly been touched upon [6.629]. The importance of many-body interactions on ionic hydration has been noted for Li^+ [6.593], Fe^{2+} and Fe^{3+} [6.601], Li^+ and Na^+ [6.596] and Cl^- [6.597]. It is observed, in particular, that many-body effects (three-body exchange and polarization) tend to lower the co-ordination numbers [6.601, 596, 597]. For the divalent transition metal ions Fe^{2+} and Fe^{3+} in water, inclusion of three-body energies apparently brings the co-ordination numbers into agreement with the experimental values. This suggests that ab initio two-body potentials are not the most appropriate for studying these systems. On the other hand, effective pair potentials for the solvent may not be quite appropriate either as they consider polarization effects in a mean field approximation which may not be a good approximation in inhomogeneous systems, as for instance, near solutes.

Solvation of ions in nonaqueous solvents has benefitted from relatively little computer work. Examples are Li^+ in ammonia [6.630], Li^+ and Cl^- in methanol, ammonia and methylamine [6.631] Na^+, K^+, Br^+, Cl^- in methanol and dimethyl [6.632], allyl cations in hydrogen fluoride [6.633].

Hydrophobic interactions between nonpolar molecules or nonpolar groups on polar molecules in solution play an important role in stabilisation of protein and nucleic acids structures, micelle formation, stability of lipid bilayers etc. To get some insight into these processes, several simulation studies on simple prototype systems of nonpolar solutes in water have been performed to learn about the thermodynamics and structural properties of such solutions at infinite dilution. Topics of interest include the hydration of inert gas atoms in aqueous solutions [6.634–636] and solute size dependence [6.637], hydration of N_2 in water [6.638], absolute free energy of binding of two methane-like molecules at their contact separation and potential of mean force [6.46], differences in free energy of hydration between normal alkanes, hydration numbers of various alkanes in water [6.639], the influence of solvent environment on the conformations of alkanes and hydrocarbons [6.640–642].

The application of computer simulations to thermodynamic properties of biological macromolecules, in particular to free energy calculations for various processes in solutions is just explosive and cannot be reviewed in detail here. We limit ourselves to giving a few illustrative examples:

—free energy of association of amides [6.643], nucleic acid bases [6.644, 645], guanidinium–guanidinium ion pair [6.646], 18-crown-6: K^+ complex [6.647],
—relative change in binding free energy of a protein-inhibitor complex [6.648], related ligand-receptor pairs, halide ions to a cryptand [6.649], 18-crown-6 complexes with alkali-metal cations in water [6.650] and methanol [6.651],
—free energy changes of conversion of glycine → alanine, alanine → phenylalanine [6.47], Asp G1 (99) $\beta \rightarrow$ Ala [6.652],
—differences in free energy of hydration of Ala dipeptide in different conformations [6.642],
—effect of solvation on the conformational equilibrium of N-methylacetamide in water [6.653],
—influence of solvent on nucleic acid base associations [6.654],
—hydration of cyclen [6.655], muscarine [6.656], amides [6.657, 658], urea [6.659], β-polyalanine [6.660], double helix Z-DNA [6.661], decaalanine [6.662],
—simulations of proteins in solutions: bovine pancreatic trypsin inhibitor [6.663], parvalbumin [6.664, 665], silver pheasant ovomucoid [6.666], B-DNA [6.667–669],
—relative stabilities of nucleic acid base tautomers [6.670],
—binding of adenine to a molecular tweezer [6.671].

Review articles on simulations of proteins and complex molecules are found in [6.672–675].

6.7.2 Finite Concentration

Simulation studies of solutions at finite concentration were mainly concerned with aqueous electrolyte solutions, Examples are NaCl [6.676–679], KCl

[6.680], LiCl [6.681–683], $NaClO_4$ [6.684], $CaCl_2$ [6.685–687], $BeCl_2$ [6.688], $SrCl_2$ [6.689], and $LaCl_3$ [6.690]. Most calculations were performed for 1–2 molal solutions and the structural arrangement of the water molecules in the vicinity of the ions discussed on the basis of ion–water and water–water radial distribution functions, hydration numbers and orientation of the water molecules in the hydration shells of the ions. A good summary of simulation results prior to 1986 can be found in two review articles by *Heinzinger* [6.691] and *Heinzinger* et al. [6.692]. At concentrations of the order of 1–2 M severe convergence problems are encountered for the evaluation of ion–ion properties, due in part to the small number of ions involved (about 10 of each species in a typical simulation with a few hundred molecules), and in part to the low ionic diffusion rate. This shortcoming has been demonstrated most convincingly in a MC simulation of a simple ion–dipole mixture [6.693] where it has been shown that 10^5 trial moves per particle are necessary for a 864 particle system to give accurate ion–ion energies and pair correlation functions and that simulation runs of a few million configurations can be seriously in error [6.694]. The problem is less dramatic for highly concentrated electrolyte solutions. The influence of ionic concentration on solutions structure has been shown for 13.9 M and 18.5 M solutions of LiCl [6.681–683]. A simulation of a supersaturated (7.01 M) solution of NaCl [6.678] has been performed to give evidence for precursors of nucleation.

The effect of pressure on the ionic hydration shells is considered in [6.676, 677], whereas in [6.685, 687, 688, 690] the effect of valency and ion size is examined.

A weakness of most ionic solution studies are the uncertainties on the ion–ion and ion–water potentials, especially the pairwise additivity assumption. A simulation study of a 18.5 molal LiCl aqueous solutions reveals that a change of the ion–ion interactions affects not only the ion–ion, but also the ion–water radial distribution functions in a significant way. A study of $BeCl_2$ [6.695] seems to indicate that 3-body forces are necessary to obtain the correct hydration number for Be^{2+}.

A subject of considerable interest for chemical and biological processes, and also for testing interaction potentials, is the modification of the water dielectric properties by the presence of the ions. The dielectric constant of an ionic solution, as defined from the low frequency behaviour of the complex frequency dependent dielectric constant, contains both an equilibrium and a dynamic contribution, the latter surviving even in the low frequency limit [6.696]. The relevant expressions which relate the dielectric constant to the (shape dependent) correlations of polarization and electrical current have been derived for non-polarizable [6.697] and polarizable [6.698] ions and solvent molecules and boundary conditions most commonly used in MC or MD simulations. Similar to what has already been remarked for pure water, the practical evaluation of the dielectric constant remains problematic because of the slow convergence of sample dipole moment fluctuations [6.693]. The only serious attempt at obtaining dielectric properties of a realistic ionic solution was made in a simulation of the concentration dependence of the (equilibrium) dielectric constant of an aqueous NaCl solution [6.679].

Finally we would like to mention three simulation studies of model electrolytes with solvent modelled by simple dipolar hard spheres [6.693, 699] and LJ spheres with embedded point dipole and tetrahedral quadrupole moments [6.700]. The purpose of this work was primarily to test integral equation theories such as the reference hypernetted chain equation [6.701, 702] and mean spherical approximation [6.703] theories.

Simulations of non-electrolyte solutions include aqueous solutions of tert-butyl alcohol [6.704], methanol [6.705], urea [6.706], and quinuclidine solution in benzene [6.707].

6.7.3 Polyelectrolytes and Micelles

Simulation studies of polyelectrolytes are still based on very simplified models. The most widely used, invoking local stiffness of the polyions, consists of a single infinitely long and uniformly charged impenetrable cylinder immersed in a dielectric continuum solvent, with neutralising counterions and added salt represented by charged hard spheres or repulsive soft potentials [6.708–715]. A convenient Monte Carlo simulation setup for this system consists of a cylindrical simulation cell with polyion lying along the symmetry axis; the infinite extent in the axial direction is mimicked by using periodic boundary conditions. An important point in these simulations, stressed by *Murthy* et al. [6.709] is to take proper account of long range interactions beyond the simulation cell. This can be achieved [6.709, 710] by a self-consistent mean field procedure analogous to that used previously by *Valleau* and coworkers [6.716] in their study of electrical double layers. Most calculations focused on thermodynamic properties and ion distribution around the polyion. Ionic environment in DNA solutions, for example, plays an important role in determining its conformational stability [6.717]. Thermodynamic coefficients (mean ionic activity coefficient, preferential interaction coefficient etc.) for the infinite cylinder model (with parameters chosen to mimic an aqueous DNA–NaCl solution) have been determined by *Mills* et al. [6.713], *Paulsen* et al. [6.714] and *Vlachy* and *Haymet* [6.711] using grand canonical MC simulations. Radial distribution functions for counterions and co-ions have been obtained as a function of salt concentration in [6.709–711]. In order to resolve an apparent conflict between experimental observations and MC predictions for the salt dependence of the number of counterions in the close vicinity of DNA, a simulation has been performed [6.712] with a helical charge distribution. The effect of this more realistic charge distribution (as compared to a uniform distribution) is, however, small [6.712] in accord with findings of [6.708]. The importance of end effects on thermodynamic properties and ion distributions in oligoelectrolyte solutions has been demonstrated by *Olmsted* et al. [6.715] by carrying out grand canonical MC simulations for model DNA oligomers of varying numbers of phosphate charges ($8 \leq N \leq 100$). A major objective of the cited MC work was to test thermodynamic and structural predictions of approximate analytical theories. These include the Poisson–Boltzmann mean field theory, the counterion condensation theory of *Manning* [6.718, 719] and the hypernetted chain integral equation theory.

Conformational properties of an isolated polyelectrolyte chain (mean square end-to-end distance, mean square radius of gyration, finite-chain persistence length, segment density distribution functions etc.) have been investigated by *Christos* et al. [6.720, 721] using a flexible chain model made of hard spheres connected by rigid bonds of fixed length and angle. In these calculations no explicit account is taken of counterions and added salt, and the charged units on the chain are assumed to interact via a potential of mean force given by Debye–Hückel theory. Both fully [6.720] and partially ionised [6.721] polyelectrolytes have been considered. In subsequent work by *Christos* and *Carnie* [6.722] and *Valleau* [6.723] counterions and salt ions are included explicitly and the results for this more complete model compared with those based on a screened Coulomb interaction. *Valleau* [6.723] shows, in particular, that the charge distribution around a polyion is very unlike that around a free ion, a result which sheds some doubt on the validity of the screening model. Such an approximation seems more reasonable for low polyion ionisation and high added salt concentrations [6.722]. Semi-dilute polyelectrolyte solutions have been investigated recently by *Christos* and *Carnie* [6.724].

In none of the preceding work has solvent structure been accounted for explicitly. Calculations including water molecules focused on the short ranged structure around a small segment of polyion (DNA) [6.725].

Åkesson et al. [6.726] perform MC simulations to study the double layer interaction between two charged aggregates (modelled as planar charged surfaces) in the presence of neutralising monovalent point counterions connected via bonding potentials to form flexible polyelectrolytes. There is no additional salt. For this particular model, the double layer interaction is attractive for a wide range of the parameters characterising the polyelectrolytes, in contrast to the case where the counterions are unconnected [6.726]. A polyelectrolyte Poisson–Boltzman theory developed by the authors reproduces this behaviour in a qualitatively correct manner.

The most detailed simulation study of an ionic surfactant micelle, a sodium octanoate micelle consisting of 15 monomers in aqueous solution, has been reported by *Watanabe* et al. [6.727, 728] and *Jönsson* et al. [6.729]. The micelle is found to be very nonspherical, undergoing appreciable shape fluctuations on a time scale of about 30 ps [6.728]. The presence of these fluctuations calls into question earlier model simulations in which solvent is absent and the head groups constrained to a spherical shell [6.730, 731]. An appreciable amount of water is found in the micelle core region [6.727–729], in disagreement with a recent analysis of SANS data [6.732]. Further topics investigated are the conformations of the alkyl chains, counterion condensation near the carboxylate head groups and distribution functions for the carbon atoms. The latter differ significantly in [6.729] and [6.727], indicating some sensitivity of the results to potential model and treatment of the long range Coulomb interactions (spherical cutoff in [6.729], Ewald sums in [6.727]).

In this context we also want to cite a molecular dynamics simulation [6.733] of a sodium-decanoate/decanol/water system in which all the constituents are treated in full atomic detail (except for the methyl and methylene groups). This system can be thought of as a model for a biological membrane.

6.8 Interfaces in Molecular Systems

Recent computer simulations of inhomogeneous molecular systems were mainly devoted to the interface properties of polar molecules and chain-like molecules. Typical examples are the liquid–vapour interface of water, water in contact with ice, a hydrophobic or metallic surface or long *n*-alkane chains near impenetrable surfaces. We shall first consider the case of polar systems.

6.8.1 Polar Systems

As one might expect the simulation of inhomogeneous polar systems poses difficulties similar to those encountered in the study of bulk polar systems. Ideally one would like to find boundary conditions which do not appreciably distort the dielectric properties (orientational structure, polarization, etc.) from those of a macroscopically large system (e.g. those considered in theory). For homogeneous systems contained in a cubic (rectangular, octahedral etc.) box, the most appropriate boundary conditions to take proper account of the long range of the dipolar forces appear (at least at present time) to be periodic boundary conditions, in which the simulation cell is reproduced an unlimited number of times in all directions of space. (cf. Sect. 6.5.5). The electrostatic interactions can then be evaluated using lattice sum techniques. In the presence of a (planar) interface the system has, however, to be considered finite in the direction perpendicular to the interface and the periodicity restricted to the remaining two directions of space. The lattice sum technique applies of course equally well to these quasi 2D systems (cf. Sect. 6.1.4) but its use is (computationally) more expensive and for this reason has been employed only in very few cases. Instead, approximate mean field estimates of the contribution of dipolar interactions from particles outside the simulation cell have been applied [6.734–736] ("dipole sheet" method [6.734], approximate reaction field method [6.735]). Such methods ignore the short-wavelength structure in the fluid near the simulation box.

An alternative to periodic boundary conditions are the so-called spherical boundaries in which the molecules are enclosed in a spherical container. This type of boundary conditions has been applied successfully to dipolar hard spheres near inert [6.737] or adsorbing [6.738] hard walls and to the liquid–vapour interface of drops of Stockmayer particles [6.739–742]. This geometry is however not without limitations either. First, the necessity to use relatively large system sizes in order to minimise curvature effects and the requirement not to truncate the particle interactions may lead to fairly time consuming energy calculations. Second, it is obvious that the spherical geometry does not allow simulations of charged surfaces.

The effect of boundary conditions (including truncation of the dipole–dipole interaction as usually done) on orientational structure has not yet been investigated in a systematic way. Experience with bulk systems would indicate that these questions are of importance for interpreting polarization effects, in particular surface potentials.

(a) Model Systems. The simplest models to evaluate the effect of dipole moment on interfacial properties of polar molecules, are dipolar hard spheres or Stockmayer particles. These models have been applied in computer simulations to determine various properties such as the density profile, orientational profile, polarization, surface tension, dielectric constant etc. near neutral inert [6.737, 743] and adsorbing [6.738] walls, charged walls [6.743, 744] and in the liquid–vapour interface [6.739–741, 745]. Near neutral walls a clear layering of the orientational structure is observed [6.737] (typically three layers) with preferential ordering of the dipoles parallel to the wall in the immediate vicinity of the wall, and perpendicular in the regions corresponding to maxima of the density oscillations. The density profile is mainly determined by steric effects except very close to the wall, where the dipole interactions cause a slight lowering of the density profile. This behaviour is easily understood by remarking that the dipolar interactions lower the pressure in the "bulk" region of the system [6.737] and that the density at contact (for a hard wall) is directly related to the bulk pressure ($\rho(0) = p/kT$). In [6.743], a tendency for formation of chain-like structures is observed in the layer next to the wall. The effect of turning on an electric field is to align the dipoles in the direction of the field. This alignment causes the density profile to decrease near the wall and the oscillations to get more pronounced as a result of the increased attraction between the particles [6.743]. The polarisation density is found to oscillate in phase with the local density. The modification of the density and polarisation profiles with wall separation (in the absence and presence of a strong electric field) with particular emphasis of solvation forces has been given in [6.744]. In [6.743, 744], long range dipolar interactions were handled as within 3D periodic boundary conditions.

The influence of boundaries on the liquid–vapour interface of Stockmayer molecules can be assessed by comparing simulation results for spherical [6.739] (liquid drop) and planar [6.745] geometries. Although a quantitative comparison is not possible due to different cutoffs of the dipolar forces, it is apparent that drop sizes much larger than a thousand particles need to be considered, for the liquid and vapour bulk densities and the surface tension to attain their planar limits. The density profiles are qualitatively similar for both geometries (for drop sizes of ~ 1000 particles) and a tendency of the dipole moments to orient parallel to the surface is noticed in both cases [6.739, 745]. *Powles* and *Fowler*

[6.742] characterise the dielectric properties of the liquid–vapour interface of a drop of Stockmayer particles by means of a local dielectric constant $\varepsilon(R)$, calculated from the mean square dipole moment fluctuations $\langle \boldsymbol{M}^2(R) \rangle$ inside a spherical region of radius R centred on the centre of the drop. *Powles* and coworkers [6.741] raise the question as to whether a description of the dielectric properties of a microscopic inhomogeneous region by means of a local dielectric constant can be given a meaning other than purely formal. By comparing the results of independent routes to the dielectric constant they conclude that it is a reasonable approximation for the inhomogeneities present in the liquid–vapour interface, but would probably be inadequate for the more strongly oscillating inhomogeneities encountered near hard walls.

The model systems considered in this section are sufficiently simple to be treated by integral equation or perturbation theories. Monte Carlo calculations have served to test theories for dipolar hard spheres near an uncharged wall [6.746] and the liquid–vapour interface of Stockmayer particles [6.747].

(b) Realistic Systems. Among the computer simulations of interface properties performed with "realistic" potentials those for water occupy a privileged position, obviously, because of its determinant role in electrochemical and biological processes. The topics which have been considered most frequently are water near large hydrophobic surfaces, water near charged and metallic surfaces and the liquid–vapour and crystal–melt interfaces. Due to its ability to form strong hydrogen bonds, water has unique behaviour and it is quite unlikely that its interfacial properties can be simply inferred from the results obtained with the idealised models considered above.

The orientation of water molecules in an interfacial region can generally be understood in terms of the ability of water to engage a maximum number of hydrogen bonds. Near a flat hydrophobic surface (e.g. a hydrocarbon solid) a layered orientational structure with alternative pattern occurs [6.748]. In the layer nearest to the surface the structure is dominated by configurations in which one hydrogen-bonding group is directed toward the hydrophobic surface. Such a structure results from the compromise to increase the density near the surface (packing effect) while keeping a maximal number of hydrogen bonds with neighbouring molecules [6.748]. This orientational structure reflects in the density profile as small and broad peaks which damp out quickly, in marked contrast with simple LJ or Stockmayer particles [6.748, 734]. A consistent interpretation of the structural properties is obtained by considering that the effect of the inert surface is to organise the water molecules into a layer of ice I with the c-axis normal to the surface [6.748, 734]. Simulations of water near a spherical hydrophobic surface confirm this picture [6.749]. Similar, but weaker structural features have been observed for water near a flexible hydrophobic surface (liquid benzene interface [6.750]) with the exception that longitudinal ordering is now absent.

Wallqvist and *Berne* [6.751] consider the hydrophobic interaction between a (spherical) methane molecule and a flat paraffin surface in the presence of water (limiting case of 2 solutes, one being very large). The surprising result they find is that near the surface, a solvent-separated configuration appears to be more stable than a configuration where the methane molecule is in contact with the wall.

When the hydrophobic surface is charged, the density and polarisation profiles do not differ markedly from the uncharged case for low charges, but an abrupt change (sharp increase of density and polarisation near the surface) occurs when the surface charge density exceeds some critical value [6.736]. This result contrasts with the findings of *Gardner* and *Valleau* [6.752]. However, in the latter work the electric field is that of an ionic double layer (surface charges + diffusive layer ions) in the mean field approximation and is, in fact, strongly reduced by the high value of the dielectric constant of water.

Near flat metallic walls, the simplest being the ideal classical metallic wall characterised by an infinite dielectric constant, polarisation of the metal surface by the water charges can be accounted for through image charges [6.752]. The effect of polarization is found [6.752] to attract the water particles strongly towards the wall and to orient one of the hydrogen atoms towards the surface. The dipole moments of the water layers near the surface are enhanced: In the layer closest to the wall the average dipole moment is directed toward the surface (in contrast to the inert surface) whereas in the next layer it points away from it leaving an essentially zero net polarisation at the surface. The fact that hydrogen atoms are situated next to the metallic surface (and not the oxygen atoms) contradicts experimental observation and may be a consequence of an oversimplified wall-water interaction. Computer simulations using an additional LJ (9-3) wall-water interaction [6.753] or the more realistic Pt-water interaction predict orientations of the water molecules in agreement with experiment [6.754].

Water between two ideal classical metallic walls is a special case for which the long range dipolar forces can be evaluated by means of 3D Ewald sums [6.753]. To see this, it suffices to consider a unit cell twice as large as the original cell containing both the original water molecules and one set of image charges. This larger cell can be replicated periodically in 3D space and the energy calculated using 3D lattice sums [6.753].

As already indicated briefly above, a rather sophisticated model, taking explicit account of the corrugation of the metallic surface, has been used by *Spohr* [6.753] in a computer simulation of water in contact with the (100) lattice plane of the platinum crystal. The Pt–Pt interaction is modelled by a harmonic nearest neighbour potential, the water-metal interaction is obtained from quantum chemical cluster calculations [6.755] and the water–water interaction described by the flexible *Bopp–Jancso–Heinzinger* [6.511] potential. The main conclusions of this work are that the orientational structure in the surface layer is dominated by the water–water interactions yielding a preferential ordering of the molecular plane nearly parallel to the surface (i.e. a lone pair directed toward the metal surface). With increasing distance from the wall, different preferential directions occur with a broad distribution about the average values.

Only the first layer of water molecules exhibits a significant net dipole moment and contributes to the potential drop.

Foster et al. [6.756] extend the calculations of *Spohr* to lower temperature (275 K) using the same Pt-water potential fitted to a convenient analytical form [6.755] and a rigid water potential (SPC/E [6.496]). They conclude that flexibility has little effect on the structural features and that the water layer adjacent to the metal surface has a crystal-like structure in two dimensions parallel to the surface. At the temperature of 275 K, the orientational ordering persists over a distance of ~ 10 Å from the surface.

Also worth mentioning is a self-consistent MC simulation undertaken to determine the effect of an adsorbed layer of dipolar molecules on the longitudinal density distribution of a liquid metal [6.757].

The interface of ordered ice-1 h and liquid water has been investigated by *Karim* and *Haymet* [6.758] using a solid sample of 3456 molecules coexisting on two sides with 2304 molecules in a liquid state. All the molecules interact by a smoothly truncated TIP4P potential. The interface region is found to be 10–15 Å wide, the orientational order induced by the crystal propagating somewhat deeper into the liquid phase than translational order. No system size dependence is observed (1440 molecules) [6.759] and no significant difference by using the SPC model for water (instead of TIP4P) [6.760].

The liquid–vapour interface of water was first simulated for a spherical cluster of 1000 ST2 water particles by *Townsend* et al. [6.761] and later in more detail by *Wilson* et al. [6.762–764] for TIP4P water and by *Matsumoto* and *Kataoka* [6.765] for the Caravetta–Clementi water potential [6.499]. The orientational orders found in these calculations seem to agree: on the vapour side of the interface a probable orientation of the water molecules has one OH bond pointing toward the vapour while on the liquid side there is a preference for the OH bonds to orient parallel to the interface with the two H atoms slightly directed toward the liquid phase [6.762, 765]. A net dipole moment pointing from the vapour to the liquid phase is observed only on the liquid side of the interface.

Attempts have been made to calculate the surface tension and the surface potential of the water liquid–vapour interface [6.762, 765]. The results are not in good agreement with experiment. For the TIP4P water model, the surface tension is twice as large as the experimental value and the surface potential has only the right magnitude but not the right sign [6.762]. With the Caravetta–Clementi potential a positive surface potential is obtained [6.765] but it has been pointed out [6.764] that the results of [6.765] and [6.762] are based on different expressions for the surface potential.

The preceding results for the liquid–vapour interface of water can be contrasted with those for methanol, another strongly hydrogen-bonding system. Due to the presence of the hydrophobic methyl group, the orientational ordering in methanol is much stronger than in water. The methyl group not participating in hydrogen bonding directs toward the vapour phase thus allowing the liquid phase to maximise the number of hydrogen bonds near the surface for energetic stabilisation [6.766]. The dipole moment is almost parallel to the surface with a slight tendency to point towards the vapour phase (negative surface potential).

Simulations of the solid-electrolyte solution interface are still sparse. The reason for this lies in the difficulty of obtaining converged ion–ion interfacial properties at realistic ionic concentrations (a 2 M solution contains about 35 ions of each species for a thousand particles). Hence the structure modifications in an aqueous solution of LiI ions by a LJ (12–6) wall have been entirely characterised by the ion–water and water–water structure [6.667]. The hydration shell of the iodide ion is more affected than that of the more strongly hydrated Li^+ ion.

6.8.2 Chain Molecules Confined by Hard Plates

The structure of chain molecules (modelling hydrocarbons, polymers etc.) can be understood as resulting from the competition between the tendency of packing of the molecules against the wall and the compensating tendency of depletion near the wall due to the reduction in configurational entropy of chains located there. The latter tendency prevails at low densities and large chains [6.286, 371, 768, 769]. A Monte Carlo simulation [6.768] of 2885 *n*-tridecane ($C_{13}H_{28}$) chains confined between two hard walls separated by 51 segment diameters and interacting through a rather realistic model provides the following picture for the chain arrangements near the walls: chains having their centres of mass nearest to the wall preferentially align along the wall with all units contained within a 4 Å thick layer from the surface. As the distance from the surface increases, the preferential alignment disappears rapidly. Molecules having their centres of mass at ~ 5 Å from the wall show practically unperturbed distribution of segment units along the direction perpendicular to the wall. However, the end sections of the chains prefer to be located near the wall, thus increasing the total density and fraction of terminal units in the region immediately adjacent to the wall [6.768]. Previous MC calculations with a smaller number of chains [6.769] (1250 chains of freely jointed beads) gave essentially similar results. The effects of variations in polymer chain length and plate separation on the microscopic conformation of polymer chains in thin films were further investigated in [6.770], and the effect of wall-chain interaction in [6.771]. Monte Carlo simulations of athermal chains in the presence of hard walls provide a means for determining their bulk pressure through use of the contact rule $p = kT\rho(0)$ which relates the pressure in the homogeneous part of the system to the segment density at contact [6.371]. This method has been used to determine the equation of state of chains of 4, 8 and 16 freely jointed hard spheres [6.371] and of 2D tangent hard disks [6.286].

The effect of solvent on the chain structure has been investigated by considering a mixture of chains (freely jointed hard spheres) and monomers (hard spheres) [6.772]. The chain density profile in the mixture is qualitatively similar to the pure chain fluid (at the same volume fraction); however, in the mixture the chains are more depleted near the wall and the oscillations in the chain profile more pronounced [6.772]. This work considers also the effect of wall separation (2–10 hard sphere diameters).

We finally mention a computer simulation study of a system of linear triatomic molecules, modelling carbon disulphide, between two solid surfaces [6.773, 774]. The novelty of this work is to examine the effect of the walls on the molecular geometry, both in the absence [6.773] and presence [6.774] of an external electric field (the molecules are assumed flexible and polarisable).

References

6.1 D. Levesque, J.J. Weis, J.P. Hansen: In *Monte Carlo Methods in Statistical Physics* ed. by K. Binder, Topics Current Phys., Vol. 7 (Springer, Berlin, Heidelberg 1979, 2nd ed. 1986)
6.2 D. Levesque, J.J. Weis, J.P. Hansen: In *Monte Carlo Method in Statistical Physics* ed. by K. Binder, Topics Curr. Phys., Vol. 36 (Springer, Berlin, Heidelberg 1984, 2nd ed. 1987)
6.3 M.P. Allen, D.J. Tildesley: *Computer Simulation of Liquids* (Clarendon Press, Oxford 1987)
6.4 G. Bhanot: Rep. Prog. Phys. **51**, 429 (1988)
6.5 D.D. Frantz, D.L. Freeman, J.D. Doll: J. Chem. Phys. **93**, 2769 (1990)
6.6 A.F. Voter: J. Chem. Phys. **82**, 1890 (1985)
6.7 M. Mezei: J. Comp. Phys. **68**, 237 (1987)
6.8 S.C. Harvey, M. Prabhakaran: J. Phys. Chem. **91**, 4799 (1987)
6.9 S. Goldman: J. Comput. Phys. **62**, 441 (1986)
6.10 W. Chapman, N. Quirke: Physica B **131**, 34 (1985)
6.11 J. Kolafa: Mol. Phys. **63**, 559 (1988)
6.12 J. Kolafa: Mol. Phys. **59**, 1035 (1986)
6.13 M. Bishop, S. Frinks: J. Chem. Phys. **87**, 3675 (1987)
6.14 M.S.S. Challa, J.H. Hetherington: Phys. Rev. A **38**, 6324 (1988)
6.15 A. Caliri, M.A.A. da Silva, B.J. Mokross: J. Chem. Phys. **91**, 6328 (1989)
6.16 D.P. Fraser, M.J. Zuckermann, O.G. Mouritsen: Phys. Rev. A **42**, 3186 (1990)
6.17 M.S. Shaw: J. Chem. Phys. **89**, 2312 (1988)
6.18 N.G. Parsonage: Chem. Phys. Lett. **127**, 594 (1986)
6.19 D.A. Kofke, E.D. Glandt: J. Chem. Phys. **87**, 4881 (1987)
6.20 A.Z. Panagiotopoulos: Mol. Phys. **61**, 813 (1987); **62**, 701 (1987)
6.21 B. Smit, Ph. de Smedt, D. Frenkel: Mol. Phys. **68**, 931 (1989)
6.22 B. Smit, D. Frenkel: Mol. Phys. **68**, 951 (1989)
6.23 A.Z. Panagiotopoulos, N. Quirke, M. Stapleton, D.J. Tildesley: Mol. Phys. **63**, 527 (1988)
6.24 D.A. Kofke, E.D. Glandt: Mol. Phys. **64**, 1105 (1988)
6.25 N.A. Seaton, E.D. Glandt: J. Chem. Phys. **84**, 4595 (1986)
6.26 N.A. Seaton, E.D. Glandt: J. Chem. Phys. **86**, 4668 (1987)
6.27 W.G.T. Kranendonk, D. Frenkel: Mol. Phys. **64**, 403 (1988)
6.28 S. Nosé: J. Chem. Phys. **81**, 511 (1984)
6.29 C.L. Cleveland: J. Chem. Phys. **89**, 4987 (1988)
6.30 J. Jellinek, R.S. Berry: Phys. Rev. A **38**, 3069 (1988); **40**, 2816 (1989)
6.31 B. Guillot, Y. Guissani: Mol. Phys. **54**, 455 (1985)
6.32 Y. Guissani, B. Guillot, S. Bratos: J. Chem. Phys. **88**, 5850 (1988)
6.33 G.L. Deitrick, L.E. Scriven, H.T. Davis: J. Chem. Phys. **90**, 2370 (1989)
6.34 K.K. Mon, R.B. Griffiths: Phys. Rev. A **31**, 956 (1985)
6.35 K.K. Han, J.H. Cushman, D.J. Diestler: J. Chem. Phys. **93**, 5167 (1990)
6.36 U. Heinbuch, J. Fischer: Mol. Sim. **1**, 109 (1987)
6.37 B. Smit, D. Frenkel: J. Phys. Condens. Matter: **1**, 8659 (1989)
6.38 P. Sloth, T.S. Sorensen: Chem. Phys. Lett. **173**, 51 (1990)
6.39 P. Sloth, T.S. Sorensen: Chem. Phys. Lett. **143**, 140 (1988)
6.40 B.R. Svensson, C.E. Woodward: Mol. Phys. **64**, 247 (1988)
6.41 K.S. Singh, S.T. Chung: J. Phys. Chem. **91**, 1674 (1987)
6.42 D. Frenkel, A.J.C. Ladd: J. Chem. Phys. **81**, 3188 (1984)
6.43 J.F. Lutsko, D. Wolf, S. Yip: J. Chem. Phys. **88**, 6525 (1988)

6.44 K.K. Mon: Phys. Rev. A **31**, 2725 (1985)
6.45 W.L. Jorgensen, C. Ravimohan: J. Chem. Phys. **83**, 3050 (1985)
6.46 W.J. Jorgensen, J.K. Buckner, S. Boudon, J. Tirado-Rives: J. Chem. Phys. **89**, 3742 (1988)
6.47 U.C. Singh, F.K. Brown, P.A. Bash, P.A. Kollman: J. Am. Chem. Soc. **109**, 1607 (1987)
6.48 P.A. Bash, U.C. Singh, R. Langridge, P.A. Kollman: Science **236**, 564 (1987)
6.49 J.A. McCammon: Science **238**, 486 (1987)
6.50 D.J. Tobias, C.L. Brooks, III: Chem. Phys. Lett. **142**, 472 (1987)
6.51 J. Van Eerden, W.J. Briels, S. Harkema, D. Feil: Chem. Phys. Lett. **164**, 370 (1989)
6.52 A.A. Chialvo: J. Chem. Phys. **92**, 673 (1990)
6.53 P. Sindzingre, G. Ciccotti, C. Massobrio, D. Frenkel: Chem. Phys. Lett. **136**, 35 (1987)
6.54 A. Branka, M. Parrinello: Mol. Phys. **58**, 989 (1986)
6.55 O. Edholm, H.J.C. Berendsen: Mol. Phys. **51**, 1011 (1984)
6.56 G. Aloisi, R. Guidelli, R. A. Jackson, P. Barnes: Chem. Phys. Lett. **133**, 343 (1987)
6.57 Z.Q. Wang, D. Stroud: Phys. Rev. A **41**, 4582 (1990)
6.58 P.G. Debenedetti: J. Chem. Phys. **86**, 7126 (1987)
6.59 P.G. Debenedetti: J. Chem. Phys. **88**, 2681 (1988)
6.60 S.W. de Leeuw, J.W. Perram, E.R. Smith: Proc. R. Soc. London Ser. A **373**, 27 (1980)
6.61 S.W. de Leeuw, J.W. Perram, E.R. Smith: Proc. R. Soc. London Ser. A **373**, 57 (1980)
6.62 B. Cichocki, B.U. Felderhof, K. Hinsen: Phys. Rev. A **39**, 5350 (1989)
6.63 B. Cichocki, B.U. Felderhof, Physica A **158**, 706 (1989)
6.64 Y.J. Rhee, J.W. Halley, J. Hautman, A. Rahman: Phys. Rev. B **40**, 36 (1989)
6.65 S. Kuwajima, A. Warshel: J. Chem. Phys. **89**, 3751 (1988)
6.66 E.R. Smith: Mol. Phys. **65**, 1089 (1988)
6.67 B. Cichocki, B.U. Felderhof: Mol. Phys. **67**, 1373 (1989)
6.68 M. Plischke, F.F. Abraham: J. Stat. Phys. **52**, 1353 (1988)
6.69 W. Schreiner, K.W. Kratky: Chem. Phys. **89**, 177 (1984)
6.70 J.J. Erpenbeck, M. Luban: Phys. Rev. A **32**, 2920 (1985)
6.71 M.R. Reddy, S.F. O'Shea: Can. J. Phys. **64**, 677 (1986)
6.72 C. Bruin, A.F. Bakker, M. Bishop: J. Chem. Phys. **80**, 5859 (1984)
6.73 R.R. Singh, K.S. Pitzer, J.J. de Pablo, J.M. Prausnitz: J. Chem. Phys. **92**, 5463 (1990)
6.74 M. Rovere, D.W. Hermann, K. Binder: Europhys. Lett. **6**, 585 (1988)
6.75 M. Rovere, D.W. Hermann, K. Binder: J. Phys. Condens. Matter **2**, 7009 (1990)
6.76 D.R. Nelson, B.I. Halperin: Phys. Rev. B **19**, 2457 (1979)
6.77 K.J. Strandburg: Rev. Mod. Phys. **60**, 161 (1988)
6.78 A.F. Bakker, C. Bruin, H.J. Hilhorst: Phys. Rev. Lett. **52**, 449 (1984)
6.79 K.J. Strandburg, J.A. Zollweg, G.V. Chester: Phys. Rev. B **30**, 2755 (1984)
6.80 C. Udink, J. Van der Elsken: Phys. Rev. B **35**, 279 (1987)
6.81 C. Udink, D. Frenkel: Phys. Rev. B **35**, 6933 (1987)
6.82 D.M. Heyes, J.R. Melrose: Mol. Phys. **68**, 359 (1989)
6.83 J.J. Erpenbeck, W.W. Wood: J. Stat. Phys. **35**, 321 (1984)
6.84 R.D. Groot, J.P. van der Eerden, N.M. Faber: J. Chem. Phys. **87**, 2263 (1987)
6.85 S. Labik, A. Malijevsky: Mol. Phys. **67**, 431 (1989)
6.86 A.E.F.M. Haffmans, I.M. de Schepper, J.P.J. Michels, H. Van Beijeren: Phys. Rev. A **37**, 2698 (1988)
6.87 Y. Rosenfeld, D. Levesque, J.J. Weis: J. Chem. Phys. **92**, 6818 (1990)
6.88 J. Tobochnik, P.M. Chapin: J. Chem. Phys. **88**, 5824 (1988)
6.89 J.P.J. Michels, N.J. Trappeniers: Phys. Lett. A **104A**, 425 (1984)
6.90 M.L. Frisch, N. Rivier, D. Wyler: Phys. Rev. Lett. **54**, 2061 (1985)
6.91 M. Luban, J.P.J. Michels: Phys. Rev. A **41**, 425 (1990)
6.92 P. de Smedt, J. Talbot, J.L. Lebowitz: Mol. Phys. **59**, 625 (1986)
6.93 B.J. Yoon, H.A. Scheraga: J. Chem. Phys. **88**, 3923 (1988)
6.94 N.A. Seaton, E.C. Glandt: J. Chem. Phys. **87**, 1785 (1987)
6.95 M. Mezei: Mol. Phys. **61**, 565 (1987); **67**, 1207 (1989)
6.96 J.L. Barrat, J.P. Hansen, G. Pastore: Mol. Phys. **63**, 747 (1988)
6.97 U. Balucani, R. Vallauri: Chem. Phys. Lett. **166**, 77 (1990)
6.98 D.M. Heyes, L.V. Woodcock: Mol. Phys. **89**, 1369 (1986)
6.99 R. Vogelsang, C. Hoheisel: Mol. Phys. **53**, 33 (1984)
6.100 R. Vogelsang, C. Hoheisel: Mol. Phys. **53**, 1355 (1984)
6.101 R. Vogelsang, C. Hoheisel: Mol. Phys. **55**, 1339 (1985)
6.102 L.F. Rull, F. Cuadros, J.J. Morales: Phys. Lett. A **123**, 217 (1987)

6.103 J. Konior, C. Jedrzejek: Mol. Phys. **63**, 655 (1988)
6.104 J.G. Powles, R.F. Fowler: Mol. Phys. **62**, 1079 (1987)
6.105 E.N. Rudisill, P.T. Cummings: Mol. Phys. **68**, 629 (1989)
6.106 F. Barrochi, M. Neumann, M. Zoppi: Phys. Rev. A **36**, 2440 (1987)
6.107 G.A. Chapela, L.E. Scrivenand, T. Davis: J. Chem. Phys. **91**, 4307 (1989)
6.108 J. Talbot, J.L. Lebowitz, E.M. Waisman, D. Levesque, J.J. Weis: J. Chem. Phys. **85**, 2187 (1986)
6.109 L. Reatto, D. Levesque, J.J. Weis: Phys. Rev. A **33**, 3451 (1986)
6.110 J.G. Powles: Chem. Phys. Lett. **125**, 113 (1986)
6.111 M.R. Stapleton, A.Z. Panagiotopoulos: J. Chem. Phys. **92**, 1285 (1990)
6.112 D.K. Chokappa, P. Clancy: Mol. Phys. **61**, 597 (1987); **61**, 617 (1987)
6.113 S. Nosé, F. Yonezawa: J. Chem. Phys. **84**, 1803 (1988)
6.114 A. de Kuijper, J.A. Schouten, J.P.J. Michels: J. Chem. Phys. **93**, 3515 (1990)
6.115 M.O. Robbins, K. Kremer, G.S. Grest: J. Chem. Phys. **88**, 3286 (1988)
6.116 R.A. La Violette, F.H. Stillinger: J. Chem. Phys. **83**, 4079 (1985)
6.117 T.A. Weber, F.H. Stillinger: J. Chem. Phys. **81**, 5089 (1984)
6.118 F.H. Stillinger, T.A. Weber: J. Chem. Phys. **83**, 4767 (1985)
6.119 M. Tanaka: J. Phys. Soc. Jpn. **54**, 2069 (1985)
6.120 Y.C. Chiew, Y.H. Wang: J. Chem. Phys. **89**, 6385 (1988)
6.121 D.M. Heyes: J. Phys. Condens. Matter **2**, 2241 (1990)
6.122 D.M. Heyes, J.R. Melrose: Mol. Phys. **66**, 1057 (1989)
6.123 M. Zoppi, R. Magli, W.S. Howells, A.K. Soper: Phys. Rev. A **39**, 4684 (1989)
6.124 F. Barrochi, M. Neumann, M. Zoppi: Phys. Rev. A **29**, 1331 (1984); **31**, 4015 (1985)
6.125 K. Singer, W. Smith: Mol. Phys. **64**, 1215 (1988)
6.126 A. Baranyai, I. Ruff: J. Chem. Phys. **85**, 365 (1986)
6.127 I. Ruff, A. Baranyai, G. Palinkas, K. Heinzinger: J. Chem. Phys. **85**, 2169 (1986)
6.128 a) N. Ohtomo, Y. Tanaka: J. Phys. Soc. Jpn **56**, 2801 (1987)
b) Y. Tanaka, N. Ohtomo: J. Phys. Soc. Jpn **56**, 2814 (1987)
6.129 a) D. Levesque, J.J. Weis, J. Vermesse: Phys. Rev. A **37**, 918 (1988)
b) D. Levesque, J.J. Weis: Phys. Rev. A **37**, 3967 (1988)
6.130 J.Q. Broughton, X.P. Li: Phys. Rev. B **35**, 9120 (1987)
6.131 F.H. Stillinger, T.A. Weber: Phys. Rev. B **31**, 5262 (1985)
6.132 M. Dzugutov, K.E. Larssan, I. Ebbsjö: Phys. Rev. A **38**, 3609 (1988)
6.133 M.-C. Bellissent-Funel, P. Chieux, D. Levesque, J.J. Weis: Phys. Rev. A **39**, 6310 (1989)
6.134 M. Dzugutov: Phys. Rev. A **40**, 5434 (1989)
6.135 R. Mentz-Stern, C. Hoheisel: Phys. Rev. A **40**, 4558 (1989)
6.136 J. Hafner: Phys. Rev. Lett. **62**, 784 (1989)
6.137 A. Arnold, N. Mauser, J. Hafner: J. Phys. Condens. Matter **1**, 965 (1989)
6.138 W. Jank, J. Hafner: Phys. Rev. B **41**, 1497 (1990)
6.139 J.M. Holender: J. Phys. Condens. Matter **2**, 1291 (1990)
6.140 J.M. Holender: Phys. Rev. B **41**, 8054 (1990)
6.141 J.M. Gonzalez Miranda: Physica B **144**, 105 (1987)
6.142 G. Jackson, J.S. Rowlinson, F. van Swol: J. Phys. Chem. **91**, 4907 (1987)
6.143 G. Mansoori, N.F. Carnahan, K.E. Starling, T.W. Leland: J. Chem. Phys. **54**, 1523 (1971)
6.144 W.G.T. Kranendonk, D. Frenkel: J. Phys. Condens. Matter **1**, 7735 (1989)
6.145 P. Ballone, G. Pastore, G. Galli, D. Gazzillo: Mol. Phys. **59**, 275 (1986)
6.146 D. Gazzillo, G. Pastore: Chem. Phys. Lett. **159**, 388 (1989)
6.147 J.G. Amar: Mol. Phys. **67**, 739 (1989)
6.148 S. Labik, W.R. Smith: J. Chem. Phys. **88**, 1223 (1988)
6.149 T. Boublik: Mol. Phys. **59**, 775 (1986)
6.150 G.I. Kerley: J. Chem. Phys. **91**, 1204 (1989)
6.151 R.D. Mountain: Mol. Phys. **59**, 857 (1986)
6.152 R.J. Lee, K.C. Chao: Mol. Phys. **61**, 1431 (1987)
6.153 P. Borgelt, C. Hoheisel, G. Stell: J. Chem. Phys. **92**, 6161 (1990)
6.154 D. Saumon, G. Chabrier, J.J. Weis: J. Chem. Phys. **90**, 7395 (1989)
6.155 M. Schoen, C. Hoheisel: Mol. Phys. **53**, 1367 (1984)
6.156 M. Schoen, C. Hoheisel: Mol. Phys. **57**, 65 (1986)
6.157 K. Nakanishi: Physica B **139** & **140**, 148 (1986)
6.158 K.P. Shukla, J.M. Haile: Mol. Phys. **64**, 1041 (1988)
6.159 L.L. Lee, K.S. Shing: J. Chem. Phys. **91**, 477 (1989)
6.160 H. Jonsson, H.C. Andersen: Phys. Rev. Lett. **60**, 2295 (1988)

6.161 W.B. Brown: J. Chem. Phys. **87**, 566 (1987)
6.162 K.S. Shing, K.E. Gubbins, K. Lucas: Mol. Phys. **65**, 1235 (1988)
6.163 A. Lofti, J. Fischer: Mol. Phys. **66**, 199 (1989)
6.164 I.B. Petsche, P.G. Debenedetti: J. Chem. Phys. **91**, 7075 (1989)
6.165 R. Vogelsang, C. Hoheisel, P. Sindzingre, G. Ciccotti, D. Frenkel: J. Phys. Condens. Matter **1**, 957 (1989)
6.166 M. Balcells, A. Giro, J.A. Padro: Physica A **35**, 414 (1986)
6.167 C. Hoheisel: Mol. Phys. **62**, 385 (1987)
6.168 D. Frenkel, R.J. Vos, C.G. de Kruif, A. Vrij: J. Chem. Phys. **84**, 4625 (1986)
6.169 M.R. Stapleton, D.J. Tildesley, T.J. Sluckin, N. Quirke: J. Phys. Chem. **92**, 4788 (1988)
6.170 M.R. Stapleton, D.J. Tildesley, N. Quirke: J. Chem. Phys. **92**, 4456 (1990)
6.171 D.A. Kofke, E.D. Glandt: J. Chem. Phys. **90**, 439 (1989)
6.172 M.S. Wertheim: J. Stat. Phys. **36**, 19 (1984); **35**, 35 (1984); **42**, 459 (1986); **42**, 477 (1986)
6.173 G. Jackson, W.G. Chapman, K.E. Gubbins: Mol. Phys. **65**, 1 (1988)
6.174 H.L. Helfer, R.L. McCrory, H.M. Van Horn: J. Stat. Phys. **37**, 577 (1984)
6.175 G.S. Stringfellow, H.E. DeWitt, W.L. Slattery: Phys. Rev. A **41**, 1105 (1990)
6.176 S. Ogata, S. Ichimaru: Phys. Rev. A **36**, 5451 (1987)
6.177 H. Totsuji, K. Tokami: Phys. Rev. A **30**, 3175 (1984)
6.178 H. Totsuji: Phys. Rev. A **29**, 314 (1984)
6.179 Ph. Lotte, M.R. Feix: Phys. Rev. A **29**, 2070 (1984)
6.180 P. Choquard, B. Piller, R. Rentsch: J. Stat. Phys. **43**, 197 (1986)
6.181 P. Choquard, B. Piller, R. Rentsch: J. Stat. Phys. **46**, 599 (1987)
6.182 J.P. Hansen, P. Viot: J. Stat. Phys. **38**, 823 (1985)
6.183 J.M. Caillol, D. Levesque: Phys. Rev. B **33**, 499 (1986)
6.184 J. Clérouin, J.P. Hansen: Phys. Rev. Lett. **54**, 2277 (1985)
6.185 J. Clérouin, J.P. Hansen, B. Piller: Phys. Rev. A **36**, 2793 (1987)
6.186 P. Choquard, B. Piller, R. Rentsch, J. Clérouin, J.P. Hansen: Phys. Rev. A **40**, 931 (1989)
6.187 P. Sloth, T.S. Sorensen, J.B. Jensen: J. Chem. Soc. Faraday Trans. 2, **83**, 881 (1987)
6.188 P. Linse, H.C. Andersen: J. Chem. Phys. **85**, 3027 (1986)
6.189 P. Linse: J. Chem. Phys. **93**, 1376 (1990)
6.190 J.H.R. Clarke, W. Smith, L.V. Woodcock: J. Chem. Phys. **84**, 2290 (1986)
6.191 H. Cheng, P. Dutta, D.E. Ellis, R. Kalia: J. Chem. Phys. **85**, 2232 (1986)
6.192 V. Vlachy, C. Pohar, A.D.J. Haymet: J. Chem. Phys. **88**, 2066 (1988)
6.193 R.O. Rosenberg, D. Thirumalai: Phys. Rev. A **33**, 4473 (1986)
6.194 M.C. Abramo, C. Caccamo, M. Malescio, G. Pizzimenti, S.A. Rogde: J. Chem. Phys. **80**, 4396 (1984)
6.195 V. Vlachy, C.H. Marshall, A.D.J. Haymet: J. Am. Chem. Soc. **111**, 4160 (1989)
6.196 C. Caccamo, G. Malescio: J. Chem. Phys. **90**, 1091 (1989)
6.197 A. Laaksonen, E. Clementi: Mol. Phys. **56**, 495 (1985)
6.198 A. Baranyai, I. Ruff, R.L. Mc Greevy: J. Phys. C Solid State Phys. **19**, 453 (1986)
6.199 M. Ross, F.J. Rogers, G. Zerah: J. Phys. C, Solid State Phys. **21**, 1877 (1988)
6.200 P.N. Kumta, P.A. Deymier, S.H. Risbud: Physica B **153**, 85 (1988)
6.201 M.C. Abramo, G. Pizzimenti: Physica B **154**, 203 (1989)
6.202 C. Margheritis, C. Sinistri: Z. Naturforsch. **43a**, 129 (1988)
6.203 C. Margheritis, C. Sinistri: Z. Naturforsch. **43a**, 751 (1988)
6.204 C. Margheritis: Z. Naturforsch. **44a**, 567 (1989)
6.205 A.J. Stafford, M. Silbert, J. Trullas, A. Giro: J. Phys. Condens. Matter **2**, 6631 (1990)
6.206 H.T.J. Reijers, W. Van der Lugt, M.L. Saboungi: Phys. Rev. B **42**, 3395 (1990)
6.207 Z. Cheng, X. Lou, J. Ma, J. Shao, N. Chen: J. Chem. Phys. **91**, 4278 (1989)
6.208 J.P. Rino, Y.M.M. Hornos, G.A. Antonio, I. Ebbsjö, R.K. Kalia, P. Vashishta: J. Chem. Phys. **89**, 7542 (1988)
6.209 P. Vashishta, R.K. Kalia, I. Ebbsjö: Phys. Rev. B **39**, 6034 (1989)
6.210 P. Vashishta, R.K. Kalia, G.A. Antonio, I. Ebbsjö: Phys. Rev. Lett. **62**, 1651 (1989)
6.211 H. Iyetomi, S. Ogata, S. Ichimaru: Phys. Rev. B **40**, 309 (1989)
6.212 J. Hafner, W. Jank: J. Phys. Condens. Matter **1**, 4235 (1989)
6.213 a) K. Furuhashi, J. Habasaki, I. Okada: Mol. Phys. **59**, 1329 (1986)
b) J. Habasaki: Mol. Phys. **69**, 115 (1990)
6.214 a) M.L. Saboungi, A. Rahman, M. Blander: J. Chem. Phys. **80**, 2141 (1984)
b) M. Blander, M.L. Saboungi, A. Rahman: J. Chem. Phys. **85**, 3995 (1986)
6.215 D.W. Hawley, J.M.D. MacElroy, J.C. Hajduk, X.B. Reed, Jr.: Chem. Phys. Lett. **117**, 154 (1985)

6.216 G.A. Chapela, S.E. Martinez-Casas, C. Varea: J. Chem. Phys. **86**, 5683 (1987)
6.217 M.J.P. Nijmeijer, A.F. Bakker, C. Bruin, J.H. Sikkenk: J. Chem. Phys. **89**, 3789 (1988)
6.218 S.M. Thomson, K.E. Gubbins, J.P.R.B. Walton, R.A.R. Chantry, J.S. Rowlinson: J. Chem. Phys. **81**, 530 (1984)
6.219 D.J. Lee, M.M. Telo da Gama, K.E. Gubbins: Mol. Phys. **53**, 1113 (1984)
6.220 A. Vicentini, G. Jacucci, V.R. Pandharipande: Phys. Rev. C **31**, 1783 (1985)
6.221 J.Q. Broughton, G.H. Gilmer: J. Chem. Phys. **84**, 5741 (1986)
6.222 J.Q. Broughton, G.H. Gilmer: J. Chem. Phys. **84**, 5749 (1986)
6.223 J.Q. Broughton, G.H. Gilmer: J. Chem. Phys. **84**, 5759 (1986)
6.224 B.B. Laird, A.D.J. Haymet: J. Chem. Phys. **91**, 3638 (1989)
6.225 R.J. Galejs, H.J. Raveche, G. Lie: Phys. Rev. A **39**, 2574 (1989)
6.226 D.K. Chokappa, P. Clancy: Mol. Phys. **65**, 97 (1988)
6.227 W. Schommers, P. Von Blanckenhagen, U. Romahn: Mod. Phys. Lett. B **2**, 1131 (1988)
6.228 J.R. Henderson, F. van Swol: Mol. Phys. **51**, 991 (1984)
6.229 W.R. Smith, R.J. Speedy: J. Chem. Phys. **86**, 5783 (1987)
6.230 S. Labik, W.R. Smith, R.J. Speedy: J. Chem. Phys. **88**, 1944 (1988)
6.231 S. Labik, W.R. Smith: J. Chem. Phys. **88**, 1223 (1988); **88**, 3893 (1988)
6.232 W. Gozdz, A. Patrykiejew, S. Sokolowski: Phys. Lett. A **145**, 279 (1990)
6.233 E. Kierlik, M.L. Rosinberg: Mol. Phys. **68**, 867 (1989)
6.234 J.M. Caillol, D. Levesque, J.J. Weis: J. Chem. Phys. **87**, 6150 (1987)
6.235 E. Kierlik, M.L. Rosinberg: J. Phys. Condens. Matter **2**, 3081 (1990)
6.236 J.H. Thurtell, G.W. Thurtell: J. Chem. Phys. **88**, 6641 (1988)
6.237 F. van Swol, J.R. Henderson: Phys. Rev. Lett. **53**, 1376 (1984)
6.238 J.R. Henderson, F. van Swol: Mol. Phys. **56**, 1313 (1985)
6.239 F. van Swol, J.R. Henderson: J. Chem. Soc., Faraday Soc. 2, **82**, 1685 (1986)
6.240 J.R. Henderson, F. van Swol: J. Chem. Phys. **89**, 5010 (1988)
6.241 F. van Swol, J.R. Henderson: Phys. Rev. A **40**, 2567 (1989)
6.242 J.H. Sikkenk, J.O. Indekeu, J.M.J. Van Leeuwen, E.O. Vossnack: Phys. Rev. Lett. **59**, 98 (1987)
6.243 J.H. Sikkenk, J.O. Indekeu, J.M.J. Van Leeuwen, E.O. Vossnack, A.F. Bakker: J. Stat. Phys. **52**, 23 (1988)
5.244 M.J.P. Nijmeijer, C. Bruin, A.F. Bakker, J.M.J. Van Leeuwen: Physica A **160**, 166 (1989)
6.245 J.E. Finn, P.A. Monson: Phys. Rev. A **39**, 6402 (1989); **42**, 2458 (1990)
6.246 S. Sokolowski, J. Fischer: Phys. Rev. A **41**, 6866 (1990)
6.247 E. Velasco, P. Tarazona: Phys. Rev. A **42**, 2454 (1990)
6.248 M. Lupkowski, F. van Swol: J. Chem. Phys. **93**, 737 (1990)
6.249 M. Meyer, M. Mareschal, M. Hayoun: J. Chem. Phys. **89**, 1067 (1988)
6.250 P. Choquard, B. Piller, R. Rentsch, P. Vieillefosse: J. Stat. Phys. **55**, 1185 (1989)
6.251 H. Totsuji: J. Phys. C Solid State Phys. **19**, L573 (1986)
6.252 D.M. Heyes: Phys. Rev. B **30**, 2182 (1984)
6.253 J.G. Harris, J. Gryko, S.A. Rice: J. Chem. Phys. **86**, 1067 (1987)
6.254 J.G. Harris, J. Gryko, S.A. Rice: J. Chem. Phys. **87**, 3069 (1987)
6.255 J.G. Harris, J. Gryko, S.A. Rice: J. Stat. Phys. **48**, 1109 (1987)
6.256 L. Guldbrand, B. Jönsson, H. Wennerström, P. Linse: J. Chem. Phys. **80**, 2221 (1984)
6.257 B. Svensson, B. Jönsson: Chem. Phys. Lett. **108**, 580 (1984)
6.258 D. Bratko, B. Jönsson, H. Wennerström: Chem. Phys. Lett. **128**, 449 (1986)
6.259 J.P. Valleau, G.M. Torrie: J. Chem. Phys. **81**, 6291 (1984)
6.260 S.L. Carnie, G.M. Torrie, J.P. Valleau: Mol. Phys. **53**, 253 (1984)
6.261 G.M. Torrie, J.P. Valleau, C.W. Outhwaite: J. Chem. Phys. **81**, 6296 (1984)
6.262 B. Svensson, B. Jönsson, C.E. Woodward: J. Phys. Chem. **94**, 2105 (1990)
6.263 A. Luzar, D. Bratko: J. Chem. Phys. **92**, 642 (1990)
6.264 V. Vlachy, A.D.J. Haymet: Chem. Phys. Lett. **146**, 32 (1988)
6.265 W. Van Megen, I.K. Snook: Mol. Phys. **54**, 741 (1985)
6.266 J.J. Magda, M. Tirrell, H.T. Davis: J. Chem. Phys. **83**, 1888 (1985)
6.267 V.Y. Antonchenko, V.V. Ilyin, N.N. Makovsky, A.N. Pavlov, V.P. Sokhan: Mol. Phys. **52**, 345 (1984)
6.268 V.Y. Antonchenko, V.V. Ilyin, N.N. Makovsky, V.M. Khryapa: Mol. Phys. **65**, 1171 (1988)
6.269 J.M.D. MacElroy, S.H. Suh: Mol. Phys. **60**, 475 (1987)
6.270 M. Schoen, D.J. Diestler, J.H. Cushman: J. Chem. Phys. **87**, 5464 (1987)
6.271 M. Schoen, J.H. Cushman, D.J. Diestler, C.L. Rhykerd Jr.: J. Chem. Phys. **88**, 1394 (1987)
6.272 C.L. Rhykerd Jr., M. Schoen, D.J. Diestler, J.H. Cushman: Nature **330**, 461 (1987)

6.273 U. Heinbuch, J. Fischer: Chem. Phys. Lett. **135**, 587 (1987)
6.274 D. Bratko, L. Blum, M.S. Wertheim: J. Chem. Phys. **90**, 2752 (1989)
6.275 P. Sloth: J. Chem. Phys. **93**, 1292 (1990)
6.276 P. Adams, J.R. Henderson, J.P.R.B. Walton: J. Chem. Phys. **91**, 7173 (1989)
6.277 G.S. Heffelfinger, F. van Swol, K.E. Gubbins: Mol. Phys. **61**, 1381 (1987)
6.278 B.K. Peterson, K.E. Gubbins: Mol. Phys. **62**, 215 (1987)
6.279 G.S. Heffelfinger, F. van Swol, K.E. Gubbins: J. Chem. Phys. **89**, 5202 (1988)
6.280 B.K. Peterson, G.S. Heffelfinger, K.E. Gubbins, F. van Swol: J. Chem. Phys. **93**, 679 (1990)
6.281 M. Schoen, C.L. Rhykerd Jr., J.H. Cushman, D.J. Diestler: Mol. Phys. **66**, 1171 (1989)
6.282 Z. Tan, K.E. Gubbins: J. Phys. Chem. **94**, 6061 (1990)
6.283 B.K. Peterson, K.E. Gubbins, G.S. Heffelfinger, U.M.B. Marconi, F. van Swol: J. Chem. Phys. **88**, 6487 (1988)
6.284 J. Talbot, D.J. Tildesley: J. Chem. Phys. **83**, 6419 (1985)
6.285 L. Lajtar, S. Sokolowski: Czech. J. Phys. B **38**, 1214 (1988)
6.286 K.G. Honnell, C.K. Hall: Mol. Phys. **65**, 1281 (1988)
6.287 T. Boublik: Mol. Phys. **29**, 421 (1975)
6.288 T. Boublik: Mol. Phys. **63**, 685 (1988)
6.289 B. Barboy, W. M. Gelbart: J. Chem. Phys. **71**, 3053 (1979)
6.290 R. Dickman, C.K. Hall: J. Chem. Phys. **85**, 4108 (1986)
6.291 D. Chandler, H.C. Andersen: J. Chem. Phys. **57**, 1930 (1972)
6.292 K.G. Honnell, C.K. Hall, R. Dickman: J. Chem. Phys. **87**, 664 (1987)
6.293 N.D. Mermin: Phys. Rev. **176**, 250 (1968)
6.294 E. Domany, M. Schick, R.H. Swendsen: Phys. Rev. Lett. **52**, 1535 (1984)
6.295 J.M. Kosterlitz, D.J. Thouless: J. Phys. C **6**, 1181 (1973)
6.296 J.P. Straley: Phys. Rev. A **4**, 675 (1971)
6.297 J.V. José, L.P. Kadanoff, S. Kirkpatrick, D. R. Nelson: Phys. Rev. B 16, 1217 (1977)
6.298 D. Stein: Phys. Rev. B **18**, 2397 (1978)
6.299 D.R. Nelson, B.I. Halperin: Phys. Rev. B **21**, 5312 (1980)
6.300 D. Frenkel, R. Eppenga: Phys. Rev. A **31**, 1776 (1985)
6.301 J.A. Cuesta, D. Frenkel: Phys. Rev. A **42**, 2126 (1990)
6.302 M.J.P. Gingras, P.C.W. Holdsworth, B. Bergersen: Europhys. Lett. **9**, 539 (1989)
6.303 M.J.P. Gingras, P.C.W. Holdsworth, B. Bergersen: Phys. Rev. A **41**, 6786 (1990)
6.304 M. Wojcik, K.E. Gubbins: Mol. Phys. **53**, 397 (1984)
6.305 J.W. Perram, M.S. Wertheim, J.L. Lebowitz, G.O. Williams: Chem. Phys. Lett. **105**, 277 (1984)
6.306 J.W. Perram, M.S. Wertheim: J. Comp. Phys. **58**, 409 (1985)
6.307 J. Vieillard-Baron: Mol. Phys. **28**, 809 (1974)
6.308 B.J. Berne, P. Pechukas: J. Chem. Phys. **56**, 4213 (1972)
6.309 F.H. Ree, W.G. Hoover: J. Chem. Phys. **40**, 939 (1964)
6.310 M. Rigby: J. Chem. Phys. **53**, 1021 (1970)
6.311 D. Frenkel: Mol. Phys. **60**, 1 (1987): Erratum Mol. Phys. **65**, 493 (1988)
6.312 M. Rigby: Mol. Phys. **66**, 1261 (1989)
6.313 D. Frenkel: J. Phys. Chem. **91**, 4912 (1987); Erratum J. Phys. Chem. **92**, 5314 (1988)
6.314 I. Nezbeda: Czech. J. Phys. B **35**, 752 (1985)
6.315 W.R. Cooney, S.M. Thompson, K.E. Gubbins: Mol. Phys. **66**, 1269 (1989)
6.316 V.R. Bhethanabotla, W. Steele: Mol. Phys. **60**, 249 (1987)
6.317 M. Ribgy: Mol. Phys. **68**, 687 (1989)
6.318 L. Onsager: Proc. N.Y. Acad. Sci., **51**, 627 (1949)
6.319 A. Isahara: J. Phys. Soc. Japan **6**, 46 (1951)
6.320 D. Frenkel, B.M. Mulder: Mol. Phys. **55**, 1171 (1985)
6.321 D. Frenkel: J. Phys. Chem. **92**, 3280 (1988); Erratum J. Phys. Chem. **92**, 5314 (1988)
6.322 J.A.C. Veerman, D. Frenkel: Phys. Rev. A **41**, 3237 (1990)
6.323 T. Boublik: Mol. Phys. **42**, 209 (1981)
6.324 T. Boublik: Mol. Phys. **59**, 371 (1986)
6.325 Y. Song, E.A. Mason: Phys. Rev. A **41**, 3121 (1990)
6.326 T. Boublik, M. Diaz Pena: Mol. Phys. **70**, 1115 (1990)
6.327 N.F. Carnahan, K.E. Starling: J. Chem. Phys. **51**, 635 (1969)
6.328 J. Talbot, A. Perera, G.N. Patey: Mol. Phys. **70**, 285 (1990)
6.329 W.B. Streett, D.J. Tildesley: Proc. R. Soc. A **348**, 485 (1976)
6.330 V.N. Kabadi, W.A. Steele: Ber. Bunsenges. Phys. Chem. **89**, 9 (1985)
6.331 V.N. Kabadi: Ber. Bunsenges. Phys. Chem. **90**, 332 (1986)

6.332 F. Ghazi, M. Rigby: Mol. Phys. **62**, 1103 (1987)
6.333 B. Kumar, C. James, G.T. Evans: J. Chem. Phys. **88**, 7071 (1988)
6.334 J. Talbot, D. Kivelson, M.P. Allen, G.T. Evans, D. Frenkel: J. Chem. Phys. **92**, 3048 (1990)
6.335 M.P. Allen, D. Frenkel, J. Talbot: Comp. Phys. Rep. **9**, 301 (1989)
6.336 D. Frenkel: Liquid Crystals **5**, 929 (1989)
6.337 G. Vertogen, W.H. de Jeu: *Thermotropic Liquid Crystals: Fundamentals* (Springer, Berlin, Heidelberg 1988)
6.338 D. Frenkel, H.N.W. Lekkerkerker, A. Stroobants: Nature **332**, 822 (1988)
6.339 A. Stroobants, H.N.W. Lekkerkerker, D. Frenkel: Phys. Rev. Lett. **57**, 1452 (1986)
6.340 A. Stroobants, H.N.W. Lekkerkerker, D. Frenkel: Phys. Rev. A **36**, 2929 (1987)
6.341 D. Frenkel, B.M. Mulder, J.P. McTague: Phys. Rev. Lett. **52**, 287 (1984)
6.342 R. Eppenga, D. Frenkel: Mol. Phys. **52**, 1303 (1984)
6.343 J.L. Lebowitz, J.W. Perram: Mol. Phys. **50**, 1207 (1983)
6.344 P. Siders: Mol. Phys. **68**, 1001 (1989)
6.345 M. He, P. Siders: J. Phys. Chem. **94**, 7280 (1990)
6.346 J.G. Gay, B.J. Berne: J. Chem. Phys. **74**, 3316 (1981)
6.347 V.N. Kabadi, W.A. Steele: Ber. Bunsenges. Phys. Chem. **89**, 2 (1985)
6.348 W.B. Sediawan, S. Gupta, E. McLaughlin: Mol. Phys. **62**, 141 (1987)
6.349 V.N. Kabadi: Ber. Bunsenges. Phys. Chem. **90**, 327 (1986)
6.350 W.B. Sediawan, S. Gupta, E. McLaughlin: Mol. Phys. **63**, 691 (1988)
6.351 U.P. Singh, U. Mohanty, S.K. Sinha: Mol. Phys. **68**, 1047 (1989)
6.352 E. Miguel, L.F. Rull, M.K. Chalam, K.E. Gubbins: Mol. Phys. **71**, 1223 (1990)
6.353 D.J. Adams, G.R. Luckhurst, R.W. Phippen: Mol. Phys. **61**, 1575 (1987)
6.354 E. Miguel, L.F. Rull, M.K. Chalam, K.E. Gubbins, F. van Swol: Mol. Phys. **72**, 593 (1991)
6.355 T. Kihara: Rev. Mod. Phys. **25**, 831 (1953)
6.356 C. Vega, D. Frenkel: Mol. Phys. **67**, 633 (1989)
6.357 Y. Song, E. A. Mason: Phys. Rev. A **42**, 4743 (1990)
6.358 T. Boublik: Mol. Phys. **32**, 1737 (1976)
6.359 S. Labik, W.R. Smith, R. Popsisil, A. Malijevsky: Mol. Phys. **69**, 649 (1990)
6.360 M.P. Allen, A.A. Imbierski: Mol. Phys. **60**, 453 (1987)
6.361 I. Nezbeda, M. R. Reddy, W.R. Smith: Mol. Phys. **55**, 447 (1985)
6.362 I. Nezbeda, H. L. Vörtler: Mol. Phys. **57**, 909 (1986)
6.363 H.L. Vörtler, J. Kolafa, I. Nezbeda: Mol. Phys. **68**, 547 (1989)
6.364 I. Nezbeda, S. Labik, A. Malijevsky: Collect. Czech. Chem. Commun. **54**, 1137 (1989)
6.365 S. Labik, A. Malijevsky: Mol. Phys. **53**, 381 (1984)
6.366 E. Lomba, M. Lombardero, J.L.F. Abascal: J. Chem. Phys. **90**, 7330 (1989)
6.367 S.J. Singer, R. Mumaugh: J. Chem. Phys. **93**, 1278 (1990)
6.368 G. Torrie, G.N. Patey: Mol. Phys. **34**, 1623 (1977)
6.369 S. Gupta: J. Phys. Chem. **92**, 7156 (1988)
6.370 S. Gupta: Mol. Phys. **68**, 699 (1989)
6.371 R. Dickman, C.K. Hall: J. Chem. Phys. **89**, 3168 (1988)
6.372 M.A. Denlinger, C.K. Hall: Mol. Phys. **71**, 541 (1990)
6.373 R. Dickman, C.K. Hall: J. Chem. Phys. **85**, 4108 (1986)
6.374 K.G. Honnell, C.K. Hall: J. Chem. Phys. **90**, 1841 (1989)
6.375 M.S. Wertheim: J. Chem. Phys. **87**, 7323 (1987)
6.376 W.G. Chapman, G. Jackson, K.E. Gubbins: Mol. Phys. **65**, 1057 (1988)
6.377 K.S. Schweizer, J.G. Curro: J. Chem. Phys. **89**, 3350 (1988)
6.378 Y.C. Chiew: Mol. Phys. **70**, 129 (1990)
6.379 T. Boublik, C. Vega, M. Diaz-Pena: J. Chem. Phys. **93**, 730 (1990)
6.380 J. Gao, J.H. Weiner: J. Chem. Phys. **91**, 3168 (1989)
6.381 K. Kremer, G.S. Grest: J. Chem. Phys. **92**, 5057 (1990)
6.382 G.S. Grest, K. Kremer: Phys. Rev. A **33**, 3628 (1986)
6.383 K. Kremer, G.S. Grest, I. Carmesin: Phys. Rev. Lett. **61**, 566 (1988)
6.384 A. Yethiraj, C.K. Hall., K.G. Honnell: J. Chem. Phys. **93**, 4453 (1990)
6.385 A. Yathiraj, C.K. Hall: J. Chem. Phys. **93**, 5315 (1990)
6.386 J.G. Curro, K.S. Schweizer, G.S. Grest, K. Kremer: J. Chem. Phys. **91**, 1357 (1989)
6.387 Y.C. Chiew: J. Chem. Phys. **93**, 5067 (1990)
6.388 S.W. de Leeuw, J.W. Perram, E.R. Smith: Ann. Rev. Phys. Chem. **37**, 245 (1986)
6.389 J.W. Perram, H.G. Petersen, S.W. de Leeuw: Mol. Phys. **65**, 875 (1988)
6.390 M. Neumann: J. Chem. Phys. **82**, 5663 (1985)

6.391 N. Neumann: J. Chem. Phys. **85**, 1567 (1986)
6.392 P.G. Kusalik: J. Chem. Phys. **93**, 3520 (1990)
6.393 C.G. Gray, Y.S. Sainger, C.G. Joslin, P. T. Cummings, S. Goldman: J. Chem. Phys. **85**, 1502 (1986)
6.394 P.G. Kusalik: Mol. Phys. **67**, 67 (1989)
6.395 B. Smit, C.P. Williams, E.M. Hendriks, S.W. de Leeuw: Mol. Phys. **68**, 765 (1989)
6.396 P.T. Cummings, L. Blum: J. Chem. Phys. **85**, 6658 (1986)
6.397 S.W. de Leeuw, C.P. Williams, B. Smit: Mol. Phys. **65**, 1269 (1988)
6.398 S.W. de Leeuw, B. Smit, C.P. Williams: J. Chem. Phys. **93**, 2704 (1990)
6.399 S.W. de Leeuw, N. Quirke: J. Chem. Phys. **81**, 880 (1984)
6.400 E. Lomba, M. Lombardero, J.L.F. Abascal: Mol. Phys. **68**, 1067 (1989)
6.401 D.J. Adams: Proc. R. Soc. London Ser. A **394**, 137 (1984)
6.402 M.R. Stapleton, D.J. Tildesley, A.Z. Panagiotopoulos, N. Quirke: Mol. Simul. **2**, 147 (1989)
6.403 B. Smit, C.P. Williams: J. Phys. Condens. Matter **2**, 4281 (1990)
6.404 M.C. Wojcik, K.E. Gubbins: J. Phys. Chem. **88**, 6559 (1984)
6.405 H.J. Böhm: Mol. Phys. **56**, 375 (1985)
6.406 M. Böhm, J. Fischer, J.M. Haile: Mol. Phys. **65**, 797 (1988)
6.407 A.A. Chialvo, D.L. Heath. P.G. Debenedetti: J. Chem. Phys. **91**, 7818 (1989)
6.408 M. Wojcik, K.E. Gubbins: Mol. Phys. **51**, 951 (1984)
6.409 V.N. Kabadi, W.A. Steele: J. Phys. Chem. **89**, 743 (1985)
6.410 K.S. Shing: J. Chem. Phys. **85**, 4633 (1986)
6.411 K. Shing: Chem. Phys. Lett. **119**, 149 (1985)
6.412 O. Steinhauser, S. Boresch, H. Bertagnolli: J. Chem. Phys. **93**, 2357 (1990)
6.413 B.J. Costa Cabral, J.L. Rivail, B. Bigot: J. Chem. Phys. **86**, 1467 (1987)
6.414 D.M.F. Edwards, P.A. Madden: Mol. Phys. **51**, 1163 (1984)
6.415 A.L. Kielpinski, K. Mansour, S. Murad: International Journal of Thermophysics **7**, 421 (1986)
6.416 S.B. Zhu, J. Lee, G.W. Robinson: Phys. Rev. A **38**, 5810 (1988)
6.417 K.A. Mansour, S. Murad: Fluid Phase Equilib. **37**, 305 (1987)
6.418 J.M. Caillol, D. Levesque, J.J. Weis, J.S. Perkyns, G.N. Patey: Mol. Phys. **62**, 1225 (1987)
6.419 J.A.C. Rullmann, P.Th. Van Duijnen: Mol. Phys. **63**, 451 (1988)
6.420 P. Ahlström, A. Wallqvist, S. Engström, B. Jönsson: Mol. Phys. **68**, 563 (1989)
6.421 U. Niesar, G. Corongiu, M.J. Huang, M. Dupuis, E. Clementi: Int. of Quantum Chemistry: Quantum Chemistry Symposium **23**, 421 (1989)
6.422 U. Niesar, G. Corongiu, E. Clementi, G.R. Kneller, D.K. Bhattacharya: J. Phys. Chem. **94**, 7949 (1990)
6.423 S. Kuwajima, A. Warshel: J. Phys. Chem. **94**, 460 (1990)
6.424 A. Wallqvist, P. Ahlström, G. Kalström: J. Phys. Chem. **94**, 1649 (1990)
6.425 P. Cieplak, P. Kollman, T. Lybrand: J. Chem. Phys. **92**, 6755 (1990)
6.426 F.J. Vesely: J. Comp. Phys. **24**, 361 (1977)
6.427 J.M. Goodfellow: Proc. Natl. Acad. Sci. **79**, 4977 (1982)
6.428 J.M. Caillol, D. Levesque, J.J. Weis, P.G. Kusalik, G.N. Patey: Mol. Phys. **55**, 65 (1985)
6.429 G.N. Patey, D. Levesque, J.J. Weis: Mol. Phys. **57**, 337 (1986)
6.430 S.L. Carnie, G.N. Patey: Mol. Phys. **47**, 1129 (1982)
6.431 A.J. Stone, S.L. Price: J. Phys. Chem. **92**, 3325 (1988)
6.432 E. Clementi, G. Corongiu, M. Aida, U. Niesar, G. Kneller: "Monte Carlo and Molecular Dynamics Simulations" in *MOTECC-90* et. by E. Clementi (ESCOM, Leiden, 1990). pp. 805–888
6.433 H.J. Böhm, R. Ahlrichs: Mol. Phys. **55**, 1159 (1985)
6.434 R. Vogelsang, C. Hoheisel: Phys. & Chem. Liq. **16**, 189 (1987)
6.435 M.S.H. Ling, M. Rigby: Mol. Phys. **51**, 855 (1984)
6.436 S.L. Price: Mol. Phys. **58**, 651 (1986)
6.437 R.D. Etters, V. Chandrasekharan, E. Uzan, K. Kobashi: Phys. Rev. B **33**. 8615 (1986)
6.438 E.J. Meijer, D. Frenkel, R. LeSar, A.J.C. Ladd: J. Chem. Phys. **92**, 7570 (1990)
6.439 R.D. Etters, J. Belak, R. LeSar: Phys. Rev. B **34**, 4221 (1986)
6.440 J. Belak, R.D. Etters, R. LeSar: J. Chem. Phys. **89**, 1625 (1988)
6.441 P.M. Rodger, A.J. Stone, D.J. Tildesley: J. Chem. Soc. Faraday Trans. II **83**, 1689 (1987)
6.442 P.M. Rodger, A.J. Stone, D.J. Tildesley: Mol. Phys. **63**, 173 (1988)
6.443 P.M. Rodger, A.J. Stone, D.J. Tildesley: Chem. Phys. Lett. **145**, 365 (1988)
6.444 K. Singer, A. Taylor, J.V.L. Singer: Mol. Phys. **33**, 1757 (1977)
6.445 C.S. Murthy, K. Singer, R. Vallauri: Mol. Phys. **49**, 803 (1983)

6.446 F.P. Ricci, D. Rocca, R. Vallauri: Chem. Phys. Lett. **110**, 556 (1984)
6.447 C. Andreani, F. Cilloco, L. Nencini, D. Rocca, R.N. Sinclair: Mol. Phys. **55**, 887 (1985)
6.448 J. Anderson, J.J. Ullo, S. Yip: J. Chem. Phys. **86**, 4078 (1987)
6.449 F. Serrano Adan, A. Banon, J. Santamaria: Chem. Phys. Lett. **86**, 433 (1984)
6.450 P. Linse: J. Am. Chem. Soc. **106**, 5425 (1984)
6.451 P. Linse, S. Engström, B. Jönsson: Chem. Phys. Lett. **115**, 95 (1985)
6.452 L.S. Bartell, L.R. Sharkey, X. Shi: J. Am. Chem. Soc. **110**, 7006 (1988)
6.453 S. Yashonath, S.L. Price, I.R. McDonald: Mol. Phys. **64**, 361 (1988)
6.454 S. Gupta, W.B. Sediawan, E. McLaughlin: Mol. Phys. **65**, 961 (1988)
6.455 W.L. Jorgensen, D.L. Severance: J. Am. Chem. Soc. **112**, 4768 (1990)
6.456 A. Baranyai, D.J. Evans: Mol. Phys. **70**, 53 (1990)
6.457 M. Misawa, T. Fukunaga: J. Chem. Phys. **93**, 3495 (1990)
6.458 G. Karlström, P. Linse, A. Wallqvist, B. Jönsson: J. Am. Chem. Soc. **105**, 3777 (1983)
6.459 W.B. Sediawan, S. Gupta, E. McLaughlin: J. Chem. Phys. **90**, 1888 (1989)
6.460 J.J. Ullo, S. Yip: J. Chem. Phys. **85**, 4056 (1986)
6.461 K. Refson, G.S. Pawley: Mol. Phys. **61**, 669 (1987)
6.462 K. Refson, G.S. Pawley: Mol. Phys. **61**, 693 (1987)
6.463 D. Brown, J.H.R. Clarke: J. Chem. Phys. **92**, 3062 (1990)
6.464 J.P. Ryckaert, A. Bellemans: Chem. Phys. Lett. **30**, 123 (1975)
6.465 R. Edberg, D.J. Evans, G.P. Morris: J. Chem. Phys. **84**, 6933 (1986)
6.466 R. Edberg, D.J. Evans, G.P. Morris: J. Chem. Phys. **87**, 5700 (1987)
6.467 P.A. Wielopolski, E.R. Smith: J. Chem. Phys. **84**, 6940 (1986)
6.468 N.G. Almarza, E. Enciso, J. Alonso, F.J. Bermejo, M. Alavarez: Mol. Phys. **70**, 485 (1990)
6.469 S. Toxvaerd: J. Chem. Phys. **89**, 3808 (1988)
6.470 S. Toxvaerd: J. Chem. Phys. **87**, 6140 (1987)
6.471 M. Fixman: Proc. Nat. Acad. Sci. **71**, 3050 (1974)
6.472 D. Chandler, B.J. Berne: J. Chem. Phys. **71**, 5386 (1979)
6.473 S. Toxvaerd: J. Chem. Phys. **93**, 4290 (1990)
6.474 J.P. Ryckaert, M.L. Klein: J. Chem. Phys. **85**, 1613 (1986)
6.475 R. Lustig, W.A. Steele: Mol. Phys. **65**, 475 (1988)
6.476 S. Gupta, J. Yang, N.R. Kestner: J. Chem. Phys. **89**, 3733 (1988)
6.477 D. Rigby, R.J. Roe: J. Chem. Phys. **87**, 7285 (1987)
6.478 D. Rigby, R.J. Roe: J. Chem. Phys. **89**, 5280 (1988)
6.479 D. Rigby, R.J. Roe: Macromolecules **22**, 2259 (1989)
6.480 W.L. Jorgensen, J. Chandrasekhar, J.D. Madura, R.W. Impey, M.L. Klein: J. Chem. Phys. **79**, 926 (1983)
6.481 H.J.C. Berendsen, J.P.M. Postma, W.F. Van Gunsteren, J. Hermans: In *Intermolecular Forces*, ed. by B. Pullman (Reidel, Dordrecht 1981) pp. 331–342
6.482 M. Ferrario, A. Tani: Chem. Phys. Lett. **121**, 182 (1985)
6.483 R.D. Mountain: J. Chem. Phys. **90**, 1866 (1989)
6.484 W.L. Jorgensen, J.D. Madura: Mol. Phys. **56**, 1381 (1985)
6.485 M.R. Reddy, M. Berkowitz: J. Chem. Phys. **87**, 6682 (1987)
6.486 J.D. Madura, B.M. Pettitt, D.F. Calef: Mol. Phys. **64**, 325 (1988)
6.487 J.S. Tse, M.L. Klein: J. Phys. Chem. **92**, 315 (1988)
6.488 J. Chandrasekhar, C.N.R. Rao: Chem. Phys. Lett. **131**, 267 (1986)
6.489 D. Bertolini, M. Cassettari, M. Ferrario, P. Grigolini, G. Salvetti, A. Tani: J. Chem. Phys. **91**, 1179 (1989)
6.490 D. Bertolini, P. Grigolini, A. Tani: J. Chem. Phys. **91**, 1191 (1989)
6.491 M.A. Ricci, D. Rocca, G. Ruocco, R. Vallauri: Phys. Rev. A **40**, 7226 (1989)
6.492 R. Frattini, M.A. Ricci, G. Ruocco, M. Sampoli: J. Chem. Phys. **92**, 2540 (1990)
6.493 J. Hermans, A. Pathiaseril, A. Anderson: J. Am. Chem. Soc. **110**, 5982 (1988)
6.494 Z. Li, H.A. Scheraga: Chem. Phys. Lett. **154**, 516 (1989)
6.495 W.L. Jorgensen, J.F. Blake, J.K. Buckner: Chem. Phys. **129**, 193 (1989)
6.496 H.J.C. Berendsen, J.R. Grigera, T.P. Straatsma: J. Phys. Chem. **91**, 6269 (1987)
6.497 K. Watanabe, M.L. Klein: Chem. Phys. **131**, 157 (1989)
6.498 E. Clementi, P. Habitz: J. Phys. Chem. **87**, 2815 (1983)
6.499 V. Carravetta, E. Clementi: J. Chem. Phys. **81**, 2646 (1984)
6.500 U.C. Singh, P.A. Kollman: J. Chem. Phys. **83**, 4033 (1985)
6.501 J.E. Carpenter, W.T. Yets III, I.L. Carpenter, W.J. Hehre: J. Phys. Chem. **94**, 443 (1990)
6.502 S. Kim, M.S. Jhon, H.A. Scheraga: J. Phys. Chem. **92**, 7216 (1988)

6.503 B.J. Yoon, K. Morokuma, E.R. Davidson: J. Chem. Phys. **83**, 1223 (1985)
6.504 M. Wojcik, E. Clementi: J. Chem. Phys. **84**, 5970 (1986)
6.505 J. Detrich, C. Corongiu, E. Clementi: Chem. Phys. Lett. **112**, 426 (1984)
6.506 E. Clementi, G. Corongiu: Int. J. Quantum Chem. Quantum Biol. Symp. **10**, 31 (1983)
6.507 M. Wojcik, E. Clementi: J. Chem. Phys. **85**, 3544 (1986)
6.508 M. Wojcik, E. Clementi: J. Chem. Phys. **85**, 6085 (1986)
6.509 M. Sprik, M.L. Klein: J. Chem. Phys. **89**, 7556 (1988)
6.510 P. Bopp, G. Jancso, K. Heinzinger: Chem. Phys. Lett. **98**, 129 (1983)
6.511 G. Jancso, P. Bopp, K. Heinzinger: Chem. Phys. **85**, 377 (1984)
6.512 K. Toukan, A. Rahman: Phys. Rev. B **31**, 2643 (1985)
6.513 G.C. Lie, E. Clementi: Phys. Rev. A **33**, 2679 (1986)
6.514 O. Teleman, B. Jönsson, S. Engström: Mol. Phys. **60**, 193 (1987)
6.515 J.L. Barrat, I.R. McDonald: Mol. Phys. **70**, 535 (1990)
6.516 J. Anderson, J.J. Ullo, S. Yip: J. Chem. Phys. **87**, 1726 (1987)
6.517 J.R. Reimers, R.O. Watts: Chem. Phys. **91**, 201 (1984)
6.518 U. Dinur: J. Phys. Chem. **94**, 5669 (1990)
6.519 A.H. Narten, H.A. Levy: J. Chem. Phys. **55**, 2263 (1971)
6.520 A.H. Narten: J. Chem. Phys. **56**, 5681 (1972)
6.521 A.K. Soper, M.G. Philipps: Chem. Phys. **107**, 47 (1986)
6.522 M.-C. Bellissent-Funel: in NATO ASI on Hydrogen bonded liquids Cargèse 1989, in press
6.523 R.W. Impey, P.A. Madden, I.R. McDonald: Chem. Phys. Lett. **88**, 589 (1982)
6.524 G.C. Lie: J. Chem. Phys. **85**, 7495 (1986)
6.525 M.-C. Bellissent-Funel: private communication
6.526 M.R. Reddy, M. Berkowitz: Chem. Phys. Letters **155**, 173 (1989)
6.527 H.E. Alper, R.M. Levy: J. Chem. Phys. **91**, 1242 (1989)
6.528 D.G. Archer, P. Wang: J. Phys. Chem. Ref. Data **19**, 371 (1990)
6.529 H. Gordon, S. Goldman. Molec. Simul. **2**, 177 (1989)
6.530 M. Neumann: Mol. Phys. **50**, 841 (1983)
6.531 M. Mezei: Mol. Phys. **47**, 1307 (1982)
6.532 M. Mezei: Mol. Phys. **67**, 1205 (1989)
6.533 F. Sussman, J.M. Goodfellow, P. Barnes, J.L. Finney: Chem. Phys. Lett. **113**, 372 (1985)
6.534 R.J. Speedy, M. Mezei: J. Phys. Chem. **89**, 171 (1985)
6.535 R.J. Speedy, J.D. Madura, W.L. Jorgensen: J. Phys. Chem. **91**, 909 (1987)
6.536 A.C. Belch, S.A. Rice: J. Chem. Phys. **86**, 5676 (1987)
6.537 J.J. de Pablo, J.M. Prausnitz, H.J. Strauch, P.T. Cummings: J. Chem. Phys. **93**, 7355 (1990)
6.538 W.L. Jorgensen: J. Phys. Chem. **90**, 1276 (1986)
6.539 J. Kolafa, I. Nezbeda: Mol. Phys. **61**, 161 (1987)
6.540 M. Haughney, M. Ferrario, I.R. McDonald: Mol. Phys. **58**, 849 (1986)
6.541 M. Haughney, M. Ferrario, I.R. McDonald: J. Phys. Chem. **91**, 4934 (1987)
6.542 G. Palinkas, E. Hawlicka, K. Heinzinger: J. Phys. Chem. **91**, 4334 (1987)
6.543 G. Palinkas, Y. Tamura, E. Spohr, K. Heinzinger: Z. Naturforsch. **43A**, 43 (1988)
6.544 E. Hawlicka, G. Palinkas, K. Heinzinger: Chem. Phys. Lett. **154**, 255 (1989)
6.545 M. Marchi, M.L. Klein: Z. Natforsch. A **44**, 585 (1989)
6.546 M. Matsumoto, K.E. Gubbins: J. Chem. Phys. **93**, 1981 (1990)
6.547 S. Murad: Mol. Phys. **51**, 525 (1984)
6.548 S. Murad, A. Papaionnou, J.G. Powles: Mol. Phys. **56**, 431 (1985)
6.549 M.E. Cournoyer, W.L. Jorgensen: Mol. Phys. **51**, 119 (1984)
6.550 R.W. Impey, M.L. Klein: Chem. Phys. Lett. **104**, 579 (1984)
6.551 K.P. Sagarik, R. Ahlrichs, S. Brode: Mol. Phys. **57**, 1247 (1986)
6.552 K.A. Mansour, S. Murad, J.G. Powles: Mol. Phys. **65**, 785 (1988)
6.553 W.L. Jorgensen: J. Phys. Chem. **90**, 6379 (1986)
6.554 S.B. Zhu, G.W. Robinson: Comp. Phys. Comm. **52**, 317 (1989)
6.555 H.J. Böhm, R. Ahlrichs: Mol. Phys. **55**, 445 (1985)
6.556 H.J. Böhm, C. Meissner, R. Ahlrichs: Mol. Phys. **53**, 651 (1984)
6.557 S.B. Zhu, J. Lee, G.W. Robinson: Mol. Phys. **65**, 65 (1988)
6.558 F. Sokolic, Y. Guissani, B. Guillot: J. Phys. Chem. **89**, 3023 (1985)
6.559 F. Sokolic, Y. Guissani, B. Guillot: Mol. Phys. **56**, 239 (1985)
6.560 F. Sokolic, Y. Guissani, G. Branovic: Chem. Phys. Lett. **131**, 513 (1986)
6.561 E.R. Smith, P.A. Wielopolski: Mol. Phys. **61**, 1063 (1987)
6.562 H.J. Böhm, R. Ahlrichs: Mol. Phys. **54**, 1261 (1985)

6.563 G.R. Kneller, A. Geiger: Mol. Phys. **68**, 487 (1989)
6.564 W. Dietz, K. Heinzinger: Ber. Bunsenges. Phys. Chem. **88**, 543 (1984)
6.565 W. Dietz, K. Heinzinger: Ber. Bunsenges. Phys. Chem. **89**, 968 (1985)
6.566 M.W. Evans: J. Molec. Liq. **30**, 165 (1965)
6.567 C. Millot, J.L. Rivail: J. of Mol. Liq. **43**, 1 (1989)
6.568 C. Vega, B. Saager, J. Fischer: Mol. Phys. **68**, 1079 (1989)
6.569 K.P. Sagarik, R. Ahlrichs: Chem. Phys. Lett. **131**, 74 (1986)
6.570 R.D. Mountain, G. Morrison: Mol. Phys. **64**, 91 (1988)
6.571 H.J. Böhm, R.M. Lynden-Bell, P.A. Madden, I.R. McDonald: Mol. Phys. **51**, 761 (1984)
6.572 W.L. Jorgensen, J.M. Briggs: Mol. Phys. **63**, 547 (1988)
6.573 D.M.F. Edwards, P.A. Madden, I.R. McDonald: Mol. Phys. **51**, 1141 (1984)
6.574 P.A. Wielopolski and E.R. Smith: Mol. Phys. **54**, 467 (1985)
6.575 L.J. Root, F.H. Stillinger: J. Chem. Phys. **90**, 1200 (1989)
6.576 Z. Gamba, M.L. Klein: Chem. Phys. **130**, 15 (1989)
6.577 K.P. Sagarik, R. Ahlrichs: J. Chem. Phys. **86**, 5117 (1987)
6.578 S. Gupta, J.E. Coon: Mol. Phys. **57**, 1049 (1986)
6.579 S. Gupta: Fluid Phase Equilibria **31**, 221 (1986)
6.580 J.E. Coon, S. Gupta, E. McLaughlin: Chem. Phys. **113**, 43 (1987)
6.581 O. Steinhauser, I. Hausleithner, H. Bertagnolli: Chem. Phys. **111**, 371 (1987)
6.582 U. Mittag, J. Samios, T. Dorfmuller: Mol. Phys. **66**, 51 (1989)
6.583 U. Mittag, J. Samios, Th. Dorfmuller, St. Guenster, M.D. Zeidler, P. Chieux: Mol. Phys. **67**, 1141 (1989)
6.584 S. Brode, I.R. McDonald: Mol. Phys. **65**, 1007 (1988)
6.585 M. Schoen, C. Hoheisel, O. Beyer: Mol. Phys. **58**, 699 (1986)
6.586 D. Fincham, N. Quirke, D.J. Tildesley: J. Chem. Phys. **84**, 4535 (1986)
6.587 D. Fincham, N. Quirke, D.J. Tildesley: J. Chem. Phys. **87**, 6117 (1987)
6.588 H. Kovaks, J. Kowalewski, A. Laaksonen: J. Phys. Chem. **94**, 7378 (1990)
6.589 M. Ferrario, M. Haughney, I.R. McDonald, M.L. Klein: J. Chem. Phys. **93**, 5156 (1990)
6.590 M.W. Evans: J. Chem. Phys. **86**, 4096 (1987)
6.591 T.P. Straatsma, H.J.C. Berendsen: J. Chem. Phys. **89**, 5876 (1988)
6.592 M. Migliore, G. Corongiu, E. Clementi, G.C. Lie: J. Chem. Phys. **88**, 7766 (1988)
6.593 G. Corongiu, M. Migliore, E. Clementi: J. Chem. Phys. **90**, 4629 (1989)
6.594 H.L. Nguyen, S.A. Adelman: J. Chem. Phys. **81**, 4564 (1984)
6.595 J. Chandrasekhar, D.C. Spellmeyer, W.L. Jorgensen: J. Am. Chem. Soc. **106**, 903 (1984)
6.596 P. Cieplak, P. Kollman: J. Chem. Phys. **92**, 6761 (1990)
6.597 M. Sprik, M.L. Klein, K. Watanabe: J. Phys. Chem. **94**, 6483 (1990)
6.598 E. Guardia, J.A. Padro: J. Phys. Chem. **94**, 6049 (1990)
6.599 D.G. Bounds: Mol. Phys. **54**, 1335 (1985)
6.600 H.J. Böhm, I.R. McDonald: J. Chem. Soc. Faraday Trans. 2, **80**, 887 (1984)
6.601 L.A. Curtiss, J.W. Halley, J. Hautman, A. Rahman: J. Chem. Phys. **86**, 2319 (1987)
6.602 A. Gonzalez-Lafont, J.M. Lluch, A. Oliva, J. Bertran: Chem. Phys. **111**, 241 (1987)
6.603 Ch.L. Kneifel, H.L. Friedman, M.D. Newton: Z. Naturforsch. **44a**, 385 (1989)
6.604 G. Alagona, C. Ghio, P. Kollman: J. Am. Chem. Soc. **108**, 185 (1986)
6.605 I.I. Shcykhct, B.Ya. Simkin, V.N. Levchuk: J. Mol. Liq. **40**, 211 (1989)
6.606 P. Linse, G. Karlström, B. Jönsson: J. Am. Chem. Soc. **106**, 4096 (1984)
6.607 G. Ravishanker, P.K. Mehrotra, M. Mezei, D.L. Beveridge: J. Am. Chem. Soc. **106**, 4102 (1984)
6.608 G. Alagona, A. Tani: Theochem **43**, 375 (1988)
6.609 K. Kinugawa, K. Nakanishi: J. Chem. Phys. **89**, 5834 (1988)
6.610 F.T. Marchese, D.L. Beveridge: Chem. Phys. Lett. **105**, 431 (1984)
6.611 C.L. Brooks, III: J. Phys. Chem. **90**, 6680 (1986)
6.612 T.P. Lybrand, I. Ghosh, J.A. McCammon: J. Am. Chem. Soc. **107**, 7793 (1985)
6.613 S.H. Fleischman, C.L. Brooks, III: J. Chem. Phys. **87**, 3029 (1987)
6.614 M. Mezei: J. Chem. Phys. **86**, 7084 (1987)
6.615 M. Mezei: Mol. Phys. **65**, 219 (1988)
6.616 P.I. Nagy, W.J. Dunn, III, J.B. Nicholas: J. Chem. Phys. **91**, 3707 (1989)
6.617 W.J. Dunn, III, P.I. Nagy: J. Phys. Chem. **94**, 2099 (1990)
6.618 B. Jayaram, M. Mezei, D.L. Beveridge: J. Am. Chem. Soc. **110**, 1691 (1988)
6.619 B.G. Rao, U.C. Singh: J. Am. Chem. Soc. **111**, 3125 (1989)
6.620 B. Bigot, B.J. Costa-Cabral, J.L. Rivail: J. Chem. Phys. **83**, 3083 (1985)
6.621 W.L. Jorgensen, J.M. Briggs: J. Am. Chem. Soc. **111**, 4190 (1989)

6.622 B. Widom: J. Chem. Phys. **39**, 2808 (1963)
6.623 W.L. Jorgensen, J.K. Buckner, S.E. Huston, P.J. Rossky: J. Am. Chem. Soc. **109**, 1892 (1987)
6.624 L.X. Dang, B.M. Pettitt: J. Am. Chem. Soc. **109**, 5531 (1987)
6.625 L.X. Dang, B.M.J. Pettitt: J. Chem. Phys **86**, 6560 (1987)
6.626 J.K. Buckner, W.L. Jorgensen: J. Am. Chem. Soc. **111**, 2507 (1989)
6.627 S. Goldman, P. Backx: J. Chem. Phys. **84**, 2761 (1986)
6.628 S.B. Zhu, J. Lee, J.B. Zhu, G.W. Robinson: J. Chem. Phys. **92**, 5491 (1990)
6.629 J.D. Madura, B.M. Pettitt: Chem. Phys. Lett. **150**, 105 (1988)
6.630 S.V. Hannongbua, T. Ishida, E. Spohr, K. Heinzinger: Z. Naturforsch. **43A**, 572 (1988)
6.631 R.W. Impey, M. Sprik, M.L. Klein: J. Am. Chem. Soc. **109**, 5900 (1987)
6.632 B.G. Rao, U.C. Singh: J. Am. Chem. Soc. **112**, 3803 (1990)
6.633 M.E. Cournoyer, W.L. Jorgensen: J. Am. Chem. Soc. **106**, 5104 (1984)
6.634 W.C. Swope, H.C. Andersen: J. Phys. Chem. **88**, 6548 (1984)
6.635 K. Watanabe, H.C. Andersen: J. Phys. Chem. **90**, 795 (1986)
6.636 T.P. Straatsma, H.J.C. Berendsen, J.P.M. Postma: J. Chem. Phys. **85**, 6720 (1986)
6.637 H. Tanaka: J. Chem. Phys. **86**, 1512 (1987)
6.638 E.S. Fois, A. Gamba, G. Morosi, P. Demontis, G.B. Suffritti: Mol. Phys. **58**, 65 (1986)
6.639 W.L. Jorgensen, J. Gao, C. Ravimohan: J. Phys. Chem. **89**, 3470 (1985)
6.640 W.L. Jorgensen, J.K. Buckner: J. Phys. Chem. **91**, 6083 (1987)
6.641 D.J. Tobias, C.L. Brooks, III: J. Chem. Phys. **92**, 2582 (1990)
6.642 M. Mezei, P.K. Mehrotra, D.L. Beveridge: J. Am. Chem. Soc. **107**, 2239 (1985)
6.643 W.L. Jorgensen: J. Am. Chem. Soc. **111**, 3770 (1989)
6.644 P. Cieplak, P.A. Kollman: J. Am. Chem. Soc. **110**, 3734 (1988)
6.645 L.X. Dang, P.A. Kollman: J. Am. Chem. Soc. **112**, 503 (1990)
6.646 S. Boudon, G. Wipff, B. Maigret: J. Phys. Chem. **94**, 6056 (1990)
6.647 L.X. Dang, P.A. Kollman: J. Am. Chem. Soc. **112**, 5716 (1990)
6.648 P.A. Bash, U.C. Singh, F.K. Brown, R. Langridge, P.A. Kollman: Science **235**, 574 (1987)
6.649 T.P. Lybrand, J.A. McCammon, G. Wipff: Proc. Natl. Acad. Sci. **83**, 833 (1986)
6.650 J. van Eerden, S. Harkema, D. Feil: J. Phys. Chem. **92**, 5076 (1988)
6.651 M.H. Mazor, J.A. McCammon, T.P. Lybrand: J. Am. Chem. Soc. **111**, 55 (1989)
6.652 J. Gao, K. Kuczera, B. Tidor, M. Karplus: Science **144**, 1069 (1989)
6.653 W.L. Jorgensen, J. Gao: J. Am. Chem. Soc. **110**, 4212 (1988)
6.654 A. Pohorille, S.K. Burt, R.D. MacElroy: J. Am. Chem. Soc. **106**, 402 (1984)
6.655 S.V. Hannongbua, B.M. Rode: J. Chem. Soc. Faraday Trans. II **82**, 1021 (1986)
6.656 C. Margheritis: Z. Naturforsch. **45a**, 1009 (1990)
6.657 W.L. Jorgensen, C.J. Swenson: J. Am. Chem. Soc. **107**, 1489 (1985)
6.658 W.L. Jorgensen, C.J. Swenson: J. Am. Chem. Soc. **107**, 569 (1985)
6.659 H. Tanaka, H. Touhara, K. Nakanishi, N. Watanabe: J. Chem. Phys. **80**, 5170 (1984)
6.660 C. Runsheng, P. Otto, J. Ladik: J. Chem. Phys. **85**, 5365 (1986)
6.661 A. Laaksonen, L.G. Nilsson, B. Jönsson, O. Teleman: Chem. Phys. **129**, 175 (1989)
6.662 F.M. DiCapua, S. Swaminathan, D.L. Beveridge: J. Am. Chem. Soc. **112**, 6768 (1990)
6.663 M. Levitt, R. Sharon: Proc. Natl. Acad. Sci. **85**, 7557 (1988)
6.664 P. Ahlström, O. Teleman, B. Jönsson: J. Am. Chem. Soc. **110**, 4198 (1988)
6.665 P. Ahlström, O. Teleman, B. Jönsson, S. Forsen: J. Am. Chem. Soc. **109**, 1541 (1987)
6.666 J. Tirado-Rives, W.L. Jorgensen: J. Am. Chem. Soc. **112**, 2773 (1990)
6.667 G.L. Seibel, U.C. Singh, P.A. Kollman: Proc. Natl. Acad. Sci. **82**, 6537 (1985)
6.668 W.F. Van Gunsteren, H.J.C. Berendsen, R.G. Geursten, H.R.J. Zwinderman: An. NY Acad. Sci. **482**, 287 (1986)
6.669 P.S. Subramanian, G. Ravishanker, D.L. Beveridge: Proc. Natl. Acad. Sci. **85**, 1836 (1988)
6.670 P. Cieplak, P. Bash, U.C. Singh, P.A. Kollman: J. Am. Chem. Soc. **109**, 6283 (1987)
6.671 J.F. Blake, W.L. Jorgensen: J. Am. Chem. Soc. **112**, 7269 (1990)
6.672 J.A. McCammon: Rep. Prog. Phys. **47**, 1 (1984)
6.673 M. Karplus: Phys. Today **40**(10) 68 (1987)
6.674 M. Karplus, G.A. Petsko: Nature **347**, 631 (1990)
6.675 P.A. Kollman, K.M. Merz, Jr.: Acc. Chem. Res. **23**, 246 (1990)
6.676 G. Jansco, K. Heinzinger, T. Radnai: Chem. Phys. Lett. **110**, 196 (1984)
6.677 G. Jancso, K. Heinzinger, P. Bopp: Z. Naturforsch **40a**, 1235 (1985)
6.678 M. Schwendinger, B.M. Rode: Chem. Phys. Lett. **155**, 527 (1989)
6.679 J. Anderson, J.J. Ullo, S. Yip: Chem. Phys. Lett. **152**, 447 (1988)
6.680 A. Migliore, S.L. Fornili, E. Spohr, G. Palinkas, K. Heinzinger: Z. Naturforsch. **41A**, 826 (1986)

6.681 P. Bopp, I. Okada, H. Ohtaki, K. Heinzinger: Z. Naturforsch. **40A**, 116 (1985)
6.682 K. Tanaka, N. Ogita, Y. Tamura, I. Okada, H. Oktaki, G. Palinkas, E. Spohr, K. Heinzinger: Z. Naturforsch. **42A**, 29 (1988)
6.683 Y. Tamura, E. Tanaka, E. Spohr, K. Heinzinger: Z. Naturforsch. **43A**, 1103 (1988)
6.684 G. Heinje, W.A.P. Luck, K. Heinzinger: J. Phys. Chem. **91**, 331 (1987)
6.685 M.M. Probst, T. Radnai, K. Heinzinger, P. Bopp, B.M. Rode: J. Phys. Chem. **89**, 753 (1985)
6.686 P. Bopp: Chem. Phys. **106**, 205 (1986)
6.687 G. Palinkas, K. Heinzinger: Chem. Phys. Lett. **126**, 251 (1986)
6.688 T. Yamaguchi, H. Oktaki, E. Spohr, G. Palinkas, K. Heinzinger, M.M. Probst: Z. Naturforsch. **41A**, 1175 (1986)
6.689 E. Spohr, G. Palinkas, K. Heinzinger, P. Bopp, M.M. Probst: J. Phys. Chem. **92**, 6754 (1988)
6.690 W. Meier, P. Bopp, M.M. Probst, E. Spohr, J.I. Lin: J. Phys. Chem. **94**, 4672 (1990)
6.691 K. Heinzinger: Physica B **131**, 196 (1985)
6.692 K. Heinzinger, Ph. Bopp, G. Jancso: Acta Chimica Hungarica **121**, 27 (1986)
6.693 J.M. Caillol, D. Levesque, J.J. Weis: Mol. Phys. **69**, 199 (1990)
6.694 K.Y. Chan, K.E. Gubbins, D. Henderson, L. Blum: Mol. Phys. **66**, 299 (1989)
6.695 M.M. Probst, K. Heinzinger: Chem. Phys. Lett. **161**, 405 (1989)
6.696 J.B. Hubbard, P. Colonomos, P.G. Wolynes: J. Chem. Phys. **71**, 2652 (1979)
6.697 J.M. Caillol, D. Levesque, J.J. Weis: J. Chem. Phys. **85**, 6645 (1986)
6.698 J.M. Caillol, D. Levesque, J.J. Weis: J. Chem. Phys. **91**, 5544 (1989)
6.699 J. Eggebrecht, P. Ozler: J. Chem. Phys. **93**, 2004 (1990)
6.700 J.M. Caillol, D. Levesque, J.J. Weis, P.G. Kusalik, G.N. Patey: Mol. Phys. **62**, 461 (1987)
6.701 P.H. Fries, G.N. Patey: J. Chem. Phys. **82**, 429 (1985)
6.702 P.G. Kusalik, G.N. Patey: J. Chem. Phys. **88**, 7715 (1988)
6.703 L. Blum, F. Vericat: In *The Chemical Physics of Solvation Part A* ed. by R.R. Dogonadze, E. Kalman, A.A. Kornyshev, J. Ulstrup (Elsevier, New York 1985) pp. 143–205
6.704 H. Tanaka, K. Nakanishi, H. Touhara: J. Chem. Phys. **81**, 4065 (1984)
6.705 S. Okazaki, T. Touhara, K. Nakanishi: J. Chem. Phys. **81**, 890 (1984)
6.706 H. Tanaka, K. Nakanishi, H. Touhara: J. Chem. Phys. **82**, 5184 (1985)
6.707 A. Maliniak, A. Laaksonen, J. Kowalewski, P. Stibbs: J. Chem. Phys. **89**, 6434 (1988)
6.708 D. Bratko, V. Vlachy: Chem. Phys. Lett. **115**, 294 (1985)
6.709 C.S. Murthy, R.J. Bacquet, P.J. Rossky: J. Phys. Chem. **89**, 701 (1985)
6.710 P. Mills, C.F. Anderson, M.T. Record, Jr.: J. Phys. Chem. **89**, 3984 (1985)
6.711 V. Vlachy, A.D.J. Haymet: J. Chem. Phys. **84**, 5874 (1986)
6.712 P. Mills, M.D. Paulsen, C.F. Anderson, M.T. Record, Jr.: Chem. Phys. Lett. **129**, 155 (1986)
6.713 P. Mills, C.F. Anderson, M.T. Record, Jr.: J. Phys. Chem. **90**, 6541 (1986)
6.714 M.D. Paulsen, B. Richey, C.F. Anderson, M.T. Record, Jr.: Chem. Phys. Lett. **139**, 448 (1987)
6.715 M.C. Olmsted, C.F. Anderson, M.T. Record, Jr.: Proc. Natl. Acad. Sci. **86**, 7766 (1989)
6.716 G.M. Torrie, J.P. Valleau: J. Chem. Phys. **73**, 5807 (1980)
6.717 H. Drew, T. Takano, S. Tanaka, K. Itakura, R.E. Dickerson: Nature **286**, 567 (1980)
6.718 G.S. Manning: Q. Rev. Biophys. **11**, 179 (1978)
6.719 G.S. Manning: Acc. Chem. Res. **12**, 443 (1979)
6.720 S.L. Carnie, G.A. Christos, T.P. Creamer: J. Chem. Phys. **89**, 6484 (1988)
6.721 G.A. Christos, S.L. Carnie: J. Chem. Phys. **91**, 439 (1989)
6.722 G.A. Christos, S.L. Carnie: J. Chem. Phys. **92**, 7661 (1990)
6.723 J.P. Valleau: Chem. Phys. **129**, 163 (1989)
6.724 G.A. Christos, S.L. Carnie: Chem. Phys. Lett. **172** (1990)
6.725 E. Clementi: J. Phys. Chem. **89**, 4426 (1985)
6.726 T. Åkesson, C. Woodward, B. Jönsson: J. Chem. Phys. **91**, 2461 (1989)
6.727 K. Watanabe, M. Ferrario, M.L. Klein: J. Phys. Chem. **92**, 819 (1988)
6.728 K. Watanabe, M.L. Klein: J. Phys. Chem. **93**, 6897 (1989)

6.728 K. Watanabe, M.L. Klein: J. Phys. Chem. **93**, 6897 (1989)

7.729 B. Jönsson, O. Edholm, O. Teleman: J. Chem. Phys. **85**, 2259 (1986)
6.730 J.M. Haile, J.P. O'Connell: J. Phys. Chem. **88**, 6363 (1984)
6.731 M.C. Woods, J. M. Haile, J.P. O'Connell: J. Phys. Chem. **90**, 1875 (1986)
6.732 E.Y. Sheu, S.H. Chen, J.S. Huang: J. Phys. Chem. **91**, 3306 (1987)
6.733 E. Egberts, H.J.C. Berendsen: J. Chem. Phys. **89**, 3718 (1988)
6.734 J.P. Valleau, A.A. Gardner: J. Chem. Phys. **86**, 4162 (1987)
6.735 N.G. Parsonage, D. Nicholson: J. Chem. Soc. Faraday Trans. II, **83**, 663 (1987)
6.736 G. Aloisi, M.L. Foresti, R. Guidelli, P. Barnes: J. Chem. Phys. **91**, 5592 (1989)

6.737 D. Levesque, J.J. Weis: J. Stat. Phys. **40**, 29 (1985)
6.738 V. Russier, M.L. Rosinberg, J.P. Badiali, D. Levesque, J.J. Weis: J. Chem. Phys. **87**, 5012 (1987)
6.739 A.P. Shreve, J.P.R.B. Walton, K.E. Gubbins: J. Chem. Phys. **85**, 2178 (1986)
6.740 J.G. Powles, R.F. Fowler, W.A.B. Evans: Chem. Phys. Lett. **107**, 280 (1984)
6.741 J.G. Powles, M.L. Williams, W.A.B. Evans: J. Phys. C: Solid State Phys. **21**, 1639 (1988)
6.742 J.G. Powles, R.F. Fowler: J. Phys. C: Solid State Phys. **18**, 5909 (1985)
6.743 S.H. Lee, J.C. Rasaiah, J.B. Hubbard: J. Chem. Phys. **85**, 5232 (1986)
6.744 S.H. Lee, J.C. Rasaiah, J.B. Hubbard: J. Chem. Phys. **86**, 2383 (1987)
6.745 J. Eggebrecht, S.M. Thompson, K.E. Gubbins: J. Chem. Phys. **86**, 2299 (1987)
6.746 W. Dong, M.L. Rosinberg, A. Perera, G.N. Patey: J. Chem. Phys. **89**, 4994 (1988)
6.747 J. Eggebrecht, K.E. Gubbins, S.M. Thomson: J. Chem. Phys. **86**, 2286 (1987)
6.748 C.Y. Lee, J.A. McCammon, P.J. Rossky: J. Chem. Phys. **80**, 4448 (1984)
6.749 A.C. Belch, M. Berkowitz: Chem. Phys. Lett. **113**, 278 (1985)
6.750 P. Linse: J. Chem. Phys. **86**, 4177 (1987)
6.751 A. Wallqvist, B.J. Berne: Chem. Phys. Lett. **145**, 26 (1988)
6.752 A.A. Gardner, J.P. Valleau: J. Chem. Phys. **86**, 4171 (1987)
6.753 J. Hautman, J. W. Halley, Y.J. Rhee: J. Chem. Phys. **91**, 467 (1989)
6.754 E. Spohr: J. Phys. Chem. **93**, 6171 (1989)
6.755 E. Spohr, K. Heinzinger: Ber. Bunsenges. Phys. Chem. **92**, 1358 (1988)
6.756 K. Foster, K. Raghavan, M. Berkowitz: Chem. Phys. Lett. **162**, 32 (1989)
6.757 Z.H. Cai, J. Harris, S.A. Rice: J. Chem. Phys. **89**, 2427 (1988)
6.758 O.A. Karim, A.D.J. Haymet: J. Chem. Phys. **89**, 6889 (1988)
6.759 O.A. Karim, A.D.J. Haymet: Chem. Phys. Lett. **138**, 531 (1987)
6.760 O.A. Karim, P.A. Kay, A.D.J. Haymet: J. Chem. Phys. **92**, 4634 (1990)
6.761 R.M. Townsend, J. Gryko, S.A. Rice: J. Chem. Phys. **82**, 4391 (1985)
6.762 M.A. Wilson, A. Pohorille, L.R. Pratt: J. Chem. Phys. **88**, 3281 (1988)
6.763 M.A. Wilson, A. Pohorille, L.R. Pratt: J. Phys. Chem. **91**, 4873 (1987)
6.764 M.A. Wilson, A. Pohorille, L.R. Pratt: J. Chem. Phys. **90**, 5211 (1989)
6.765 M. Matsumoto, Y. Kataoka: J. Chem. Phys. **88**, 3233 (1988)
6.766 M. Matsumoto, Y. Kataoka: J. Chem. Phys. **90**, 2398 (1989)
6.767 E. Spohr, K. Heinzinger: J. Chem. Phys. **84**, 2304 (1986)
6.768 M. Vacatello, D.Y. Yoon, B. Laskowski: J. Chem. Phys. **93**, 779 (1990)
6.769 S.K. Kumar, M. Vacatello, D.Y. Yoon: J. Chem. Phys. **89**, 5206 (1988)
6.770 S.K. Kumar, M. Vacatello, D.Y. Yoon: Macromolecules **23**, 2189 (1990)
6.771 I. Bitsanis, G. Hadziioannou: J. Chem. Phys. **92**, 3827 (1990)
6.772 A. Yethiraj, C.K. Hall: J. Chem. Phys. **91**, 4827 (1989)
6.773 S.B. Zhu, J. Lee, G.W. Robinson: Mol. Phys. **67**, 321 (1989)
6.774 S.B. Zhu, J.B. Zhu, G.W. Robinson: Mol. Phys. **68**, 1321 (1989)

7. Monte Carlo Techniques for Quantum Fluids, Solids and Droplets

Kevin E. Schmidt and David M. Ceperley

With 12 Figures

In this chapter we review the progress that has been made since the publication of the previous articles [7.1, 2] hereafter referred to as I and II. The major advances not covered in those two reviews are improved variational wavefunctions, path integral methods for calculating at temperatures $T > 0$, systems with interfaces such as surfaces and droplets, and the calculation of exchange frequencies in quantum crystals. We will concentrate on these areas and on the Monte Carlo techniques used. We will give results from some representative calculations.

The principle constituents of the systems we describe are the helium isotopes ^{3}He and ^{4}He. Other many-body quantum systems such as electron systems, polarised hydrogen and its isotopes, can be handled in a similar way. Systems with spin and/or isospin dependent forces like nuclei are much more difficult and require nontrivial generalisations of the techniques presented here, and are beyond the scope of this work. The interested reader is referred to the recent work by *Pieper* et al. [7.3] and *Carlson* [7.4, 5] for some current Monte Carlo attacks on the nuclear structure problem.

The Hamiltonian we consider is

$$H = -\frac{1}{2m}\sum_{i=1}^{N} \nabla_i^2 + \sum_{i<j} v_{ij}, \tag{7.1}$$

with obvious generalisations to mixtures, substrates, three-body potentials, etc. We set $\hbar = 1$, m is the mass, and $v_{ij} = v(|\boldsymbol{r}_i - \boldsymbol{r}_j|)$ is the pair potential. Throughout this work we denote the $3N$ coordinates of the N atoms by R.

We will describe, in Sect. 7.1 some wavefunctions used currently in variational studies. These include 3-body and backflow correlations in addition to the standard Bijl–Dingle–Jastrow form of the many-body wavefunction; the pairing forms for the wavefunction developed by *Bouchaud* and *Lhuillier* [7.6]; and the shadow wavefunctions of *Vitiello* et al. [7.7]. We will also discuss some useful optimisation methods. In Sect. 7.2, we will define the Green's functions Monte Carlo (GFMC) method and briefly discuss some fermion methods. The path integral Monte Carlo (PIMC) method, which has been developed and applied since I and II were published, will be described in Sect. 7.3. We will give some results of calculations of helium bulk properties in Sect. 7.4, and momentum distributions and related density matrices in Sect. 7.5. In Sect. 7.6, we will give an introduction to some work on droplets, and Sect. 7.7 give some possible future directions for Monte Carlo simulations of quantum fluids.

Techniques other than Monte Carlo simulations have made important contributions to the understanding of many systems and particularly in the development of good variational wavefunctions. However, we will only refer to these methods when they have a direct bearing on the Monte Carlo development.

Some systematic errors are common to all of the Monte Carlo calculations we describe. These include the interaction potential, size dependence, statistical errors, convergence problems and bias.

The 2-body HFDHE2 potential of *Aziz* et al. [7.8] has become the potential of choice for helium studies in part because of the good agreement between GFMC ground-state calculations and the experimental low-temperature equation of state for ^{4}He. The potential needs to be revised by softening the core [7.9, 10] and Aziz and coworkers have published a new He–He potential [7.11]. The attractive well may have to be deepened somewhat [7.11, 12], although the evidence here is much less compelling, and the HFDHE2 well depth is within estimated theoretical bounds. In any case, the HFDHE2 potential is accurate to better than 0.2 K per particle out of the 20 K potential energy in ^{4}He at equilibrium density. Three-body potential effects are not well characterised, but estimates [7.13], mostly based on the Axilrod–Teller interaction [7.14], show that they are also in this range. Work at much higher pressures will require a better understanding of these 3-body forces. Recent attempts to calculate, using the Monte Carlo method, the 2- and 3-body helium interaction potentials have been successful in the highly repulsive region [7.9], but have not been conclusive in the region of the attractive well because the statistical errors have been of order 0.1 K [7.15–17].

Size dependence of the ground state energy and local properties of bulk ^{4}He are small as stated in I. However, the momentum distribution as usually calculated, has the incorrect behaviour at long wavelength [7.18], and the specific heat curve is rounded around the lambda transition [7.19]. These are a direct consequence of the finite size of the simulations used. As we shall see in Sect. 7.3, PIMC calculations which need to sample paths with a change in the winding number, are currently limited to small systems (100 atoms). In the fermi system ^{3}He, shell effects due to the very nonspherical fermi surface can be quite large [7.20, 21], of order 0.1 K per particle even for N over 100. These will make difficult the extraction from Monte Carlo calculations of a number of interesting properties such as the spin susceptibility and the effective mass.

Statistical errors, lack of convergence, and bias of the Monte Carlo calculations can also be a major problem; witness the GFMC density profiles calculated for droplets [7.22, 23] and for the free surface [7.24]. Here the slow convergence makes difficult the calculation of the exact ground-state density from the variational and GFMC mixed estimates. Clearly, the extrapolation from the variational and GFMC mixed estimates described in I and II should only be trusted when variational and mixed estimates are close so that the neglected second order error term can be assumed small. An advantage of PIMC over GFMC is that this extrapolation is not needed. A possible combination of GFMC and PIMC can be made by replacing the closed path by a path which ends with a good variational trial wave function. The resulting method would combine, for zero temperature problems, some of the advantages and disadvantages of the GFMC, PIMC and shadow wavefunction methods.

Great care needs to be taken in all Monte Carlo calculations to eliminate the effects of bias, autocorrelation, and lack of convergence. Much of the advice given by *Binder*, *Stauffer* and *Heermann* [7.25, 26] can be applied directly to quantum simulations.

7.1 Variational Method

7.1.1 Variational Wavefunctions

The variational Monte Carlo method and its application to both Bose and Fermi systems was described in I. It consists of calculating the expectation value of operators using the Metropolies Monte Carlo method and an assumed trial wavefunction. Usually the expectation value of the Hamiltonian H is minimised with respect to parameters in the trial wavefunction,

$$E_{\mathrm{V}} = \frac{\int dR\, \Psi_{\mathrm{T}}(R) H \Psi_{\mathrm{T}}(R)}{\int dR\, \Psi_{\mathrm{T}}^2(R)} = \langle E_{\mathrm{L}}(R) \rangle \geqq E_0, \tag{7.2}$$

where E_0 is the ground-state energy, $E_{\mathrm{L}}(R)$ is the local energy $H\Psi_{\mathrm{T}}(R)/\Psi_{\mathrm{T}}(R)$, and $P(R) = \Psi_{\mathrm{T}}^2/\int dR\, \Psi_{\mathrm{T}}^2(R)$ is sampled using the *Metropolis* et al. [7.27] method. A short explanation of the Metropolis method is given in Sect. 7.3, or see I. All expectation values are calculated as averages over $P(R)$ in the variational method.

For bulk ^{3}He and ^{4}He, good trial ground-state variational wavefunctions have been constructed using 2-body, 3-body, and in the case of ^{3}He, backflow [7.28] wavefunctions. The ^{3}He case is less well understood, and in particular, the difficulty in obtaining convincing results for the energy as a function of spin polarisation has lead to the construction of pairing wave functions by *Bouchaud* and *Lhuillier* [7.6]. Similarly, the lack of translational invariance in standard trial solid wavefunctions has recently lead to the introduction of the shadow trial wavefunctions of *Vitiello* et al. [7.7, 29]

7.1.2 The Pair Product Wavefunction

The simplest useful trial wavefunction for the ground state of a bulk quantum fluid or solid is the pair product or Bijl–Dingle–Jastrow form,

$$\Psi_{\mathrm{T}} = \prod_{i<j} f_{ij} \Phi, \tag{7.3}$$

usually called the Jastrow form, where Φ is a model, one-body, wavefunction. For the ground state of a Bose liquid, Φ is a constant while for a Fermi liquid it is a Slater determinant of plane waves satisfying periodic boundary conditions.

Table 7.1. Various calculations on liquid ^{4}He at the experimental equilibrium density ($\rho = 0.02186\,\text{A}^{-3}$) using the Aziz HFDHE2 potential and at zero temperature. VMC indicates a variational calculation with the indicated wavefunction. McMillan, PPA, and OPT indicate a Jastrow factor of the McMillan, paired-phonon analysis, or optimized form. 3B indicates the addition of three-body correlations as in (7.5–7). GFMC is done with the McMillan form for the importance and starting function. PIMC is calculated at $T = 1.2$ K. n_0 is the fraction of atoms in the zero momentum state

Method	Trial function	Energy	n_0	Reference
VMC	McMillan	−5.72(2)	0.11(1)	[7.88]
VMC	PPA	−5.93(1)	0.107(1)	[7.89]
VMC	Shadow	−6.24(4)	0.045(1)	[7.29]
VMC	McMillan + 3B	−6.65(2)	0.056(1)	[7.89]
VMC	OPT + 3B	−6.79(1)		[7.56]
GFMC	McMillan	−7.12(3)	0.088(5)	[7.88]
PIMC	1.2 K	−7.18(3)	0.080(10)	[7.75]
Experiment	—	−7.14	0.10(3)	[7.36]

Table 7.2. Various calculations on liquid ^{3}He at the experimental equilibrium density ($\rho = 0.01635\,\text{A}^{-3}$) using the Aziz HFDHE2 or the Lennard–Jones potential (marked with *) and at zero temperature. The notation is the same as Table 7.1. Additionally, BF indicates the wavefunction including backflow (7.12), and BCS indicates the pairing wavefunction with singlet ($s = 0$) or triplet ($s = 1$) pairs (7.14, 15). GFMC-FN, GFMC-TE, and GFMC-MP are fixed-node, transient estimation, and mirror potential calculations using the indicated wavefunction

Method	Trial function	Energy	Reference
VMC	McMillan	−1.08(3)	[7.93]
VMC	2B + 3B	−1.61(3)	[7.93]
VMC	2B + BF	−1.55(4)	[7.93]
VMC	2B + 3B + BF	−2.15(3)	[7.21]
VMC	BCS ($s = 0$)*	−1.2	[7.47]
VMC	BCS ($s = 1$)*	−2.05	[7.47]
GFMC-FN	2B + 3B + BF	−2.37(1)	[7.21]
GFMC-MP	2B + 3B + BF	−2.30(4)	[7.21]
GFMC-TE	SB + 3B + BF	−2.44(4)	[7.21]
Experiment	—	−2.47	[7.35]

Since we are dealing with spin-independent forces and operators symmetric under particle interchange, antisymmetry can be enforced by assigning particles a particular spin and antisymmetrising the spatial part under interchange of like spin particles, leading to a determinant for up spins and a determinant for down spins. For quantum solids, Φ is often taken to be the localised form [7.30, 31]

$$\Phi = \exp\left(-C\sum_i |\boldsymbol{r}_i - \boldsymbol{z}_i|^2\right), \tag{7.4}$$

where z_i are assumed lattice positions, and C is a variational parameter. Exchange effects are ignored in this wave function since particles are assigned to specific lattice sites. The solid also loses invariance to simultaneous translation of all the particles, but the wavefunction will give an upper bound to the energy of a Bose solid. The effect of exchange is generally small in 3D solids [7.32]. In practice, including an antisymmetrised or symmetrised form using Monte Carlo sampling is not difficult [7.20, 33].

The results using the simple Jastrow wavefunction, while physically reasonable, are not very accurate as seen in Tables 7.1 and 7.2. For example, the energy of ^{4}He and ^{3}He at the experimental equilibrium density both differ by about 1.2 K from the experimental values of -7.14 K and -2.47 K [7.29, 34–36]. As the densities grow, the Jastrow results become worse.

7.1.3 Three-Body Correlations

The difference between the Jastrow form and the experimental results for ^{4}He is due mainly to the absence of three-body correlations in the trial wavefunctions. An early calculation of model 2D helium was done, using the Monte Carlo method, by *Woo* [7.37]. Later, good variational functions with 3-body correlations were developed using integral equation methods [7.38–40]. The work of Pandharipande was motivated by a correlation operator method, while Chang and Campbell minimised the energy using the convolution approximation. The 3-body correlated wavefunction used in Monte Carlo calculations is similar to these forms, and is [7.41, 42]

$$\Psi_{\mathrm{T}} = \prod_{i<j<k} f_{ijk}^{(3)} \prod_{i<j} f_{ij} \Phi, \tag{7.5}$$

where

$$f_{ijk}^{(3)} = f^{(3)}(\boldsymbol{r}_{ij}, \boldsymbol{r}_{ik}, \boldsymbol{r}_{jk}), \tag{7.6}$$

$$f_{ijk}^{(3)} = \exp\left(\sum_{\text{cyclic}} -\lambda \xi_{ij} \xi_{ik} \boldsymbol{r}_{ij} \cdot \boldsymbol{r}_{ik} \right), \tag{7.7}$$

where λ is a variational parameter, the sum is over cyclic permutations of i, j, and k, and $\xi(r)$ is chosen variationally.

This form can also be motivated by operating on the pair product wavefunction with the Hamiltonian and looking at the local energy, $E_{\mathrm{L}}(R)$ of (7.2). The exact wavefunction must have a constant local energy, so terms which are not constant should be added to the function space of the trial function. The local energy resulting from a pair product trial function has 2-body terms proportional to $\nabla_i^2 f_{ij}/f_{ij}$ or v_{ij}, and 3-body terms like $\nabla_i \log(f_{ij}) \cdot \nabla_i \log(f_{ik})$. Since no choice of a 2-body term $f(r)$ will eliminate the 3-body term, additional 3-body terms must be added to the log of the trial function. These are taken to be $\boldsymbol{r}\xi(r) \approx \nabla \log[f(r)]$ with $\xi(r)$ chosen to minimise the fluctuations in the local energy. A reasonable parameterised choice is simply a Gaussian since the short range behaviour of ξ is cut off by the repulsive 2-body potential, and the long range behaviour can also be cutoff since it should decay like r^{-4} in 3 dimensions because the long range part of the Jastrow correlation decays like r^{-2} due to zero point phonon excitations.

A nice feature of the 3-body form (7.7) is that it requires only order N^2 operations to calculate instead of order N^3. This can be seen by rewriting the correlation as

$$\prod_{i<j} f_{ij} \prod_{i<j<k} f_{ijk}^{(3)} = \prod_{i<j} \tilde{f}_{ij} \exp\left(-\frac{1}{4}\lambda \sum_{l} \boldsymbol{G}_l \cdot \boldsymbol{G}_l\right), \tag{7.8}$$

where

$$\boldsymbol{G}_l = \sum_{i \neq l} \xi_{li} \boldsymbol{r}_{li}, \tag{7.9}$$

and

$$\tilde{f}(r) = f(r) \exp\left[\frac{\lambda}{2} \xi^2(r) r^2\right]. \tag{7.10}$$

If correlations of shorter range than half the size of the simulation cell are used and an average of M particles are within range of the correlation, the 3-body correlations can be calculated in only MN operations.

Usmani et al. [7.43] have used integral equation methods to calculate the effect of a more general 3-body correlation. They included terms with contributions like

$$f_{ijk}^{(3)} = \exp\left[\sum_{\text{cyclic}} \sum_{l} \lambda_l \xi_{ij}^{(l)} \xi_{ik}^{(l)} P_l(\hat{\boldsymbol{r}}_{ij} \cdot \hat{\boldsymbol{r}}_{ik})\right], \tag{7.11}$$

where the P_l are Legendre polynomials with $l = 0, 1, 2$ and $\hat{\boldsymbol{r}}$ indicates the unit vector. The $l = 1$ term reduces to the form (7.7). They found that the $l = 0$ term contributes of order 0.1 K in ^{4}He, and the $l = 2$ term is small. Preliminary Monte Carlo calculations agree with these general conclusions. Using the spherical harmonic addition theorem, the Legendre polynomial can be split into a set of $2l + 1$ 2-body sums as in (7.8).

7.1.4 Backflow Correlations

The same arguments used above for the bose system ^{4}He can be applied to motivate 3-body correlations in ^{3}He. For a fermion trial function, the local energy for a pair product trial function includes additional terms proportional to $\nabla_i \log(f_{ij}) \cdot \nabla_i \log(\phi)$, where ϕ is a single-particle orbital in the Slater determinant. Correlations that cancel these terms can be obtained by adding *Feynman–Cohen* [7.28] backflow correlations to the Slater determinant. The backflow form for a liquid, derived by using a beautiful current conservation argument, is simply a replacement of the orbitals in the Slater determinant,

$$\exp(\mathrm{i}\boldsymbol{k}\cdot\boldsymbol{r}_j)\rightarrow\exp\left[\mathrm{i}\boldsymbol{k}\cdot\left(\boldsymbol{r}_j+\sum_{m\neq j}\boldsymbol{r}_{jm}\eta_{jm}\right)\right]. \tag{7.12}$$

The argument of the plane wave is simply changed from $\boldsymbol{r}_j$ to $\boldsymbol{r}_j+\sum_{m\neq j}\boldsymbol{r}_{jm}\eta_{jm}$. Feynman and Cohen showed that $\eta(r)\approx r^{-3}$ at large r from hydrodynamic arguments. A similar replacement can be done in inhomogenous systems. For example the orbitals of electrons in molecules should be modified by back flow effects [7.44].

The inclusion of backflow and 3-body correlations has produced very good variational wavefunctions for ^{3}He as judged by the variational energy. A disadvantage of the backflow form is that evaluation of the wave function requires order N^3 operations if $\eta(r)$ is long ranged, even if only one particle is moved at a time. As a consequence, one usually moves all the atoms simultaneously, and to achieve a reasonable acceptance ratio, the step size must be chosen to be small. Often, directed sampling techniques [7.45] are used to improve the efficiency of the Monte Carlo sampling. As in the standard fermion variational Monte Carlo method [see I], the calculation of the derivatives is straightforward. All terms can be calculated using the chain rule and the identity for the determinants of matrices $\boldsymbol{A}$ and $\boldsymbol{B}$,

$$\frac{\det(\boldsymbol{A})}{\det(\boldsymbol{B})}=\det(\boldsymbol{B}^{-1}\boldsymbol{A}). \tag{7.13}$$

7.1.5 Pairing Correlations

Bouchaud and *Lhuillier* [7.6, 46, 47] have developed variational wave functions which contain pairing correlations. In particular, they show that the singlet paired BCS [7.48] wavefunction for N particles can be written as a determinant,

$$\Psi_{\mathrm{BCS}}=\det[\phi(\boldsymbol{r}_{i\mathrm{u}}-\boldsymbol{r}_{j\mathrm{d}})], \tag{7.14}$$

where u and d stand for up and down spin particles, and $\phi(r)$ is the orbital part of the single pair BCS wavefunction. One can show using the properties of Pfaffians [7.49] or Grassman variables that the square of the triplet paired wavefunction is a determinant,

$$\Psi^2_{\mathrm{BCS}}=\det[\phi(r_{i\mathrm{u}}-r_{j\mathrm{u}})]\cdot\det[\phi(r_{i\mathrm{d}}-r_{j\mathrm{d}})]. \tag{7.15}$$

Bouchaud and Lhuillier use these triplet and singlet BCS functions as the model functions in (7.3) with a simple gaussian form for the $\phi(r)$,

$$\phi(r)=\begin{bmatrix}(\hat{\boldsymbol{n}}\cdot\boldsymbol{r})\\1\end{bmatrix}\exp\left(\frac{-r}{b}\right)^2, \tag{7.16}$$

where $\hat{\boldsymbol{n}}\cdot\boldsymbol{r}$ goes with the triplet state, 1 with the singlet state, and b is a variational parameter.

Bouchaud and Lhuillier find that the single particle momentum distribution with this pairing form is smooth, with no discontinuity at the fermi momentum in contrast to the distribution resulting from the Slater determinant (see Fig. 7.12, Sect. 7.5). The static structure $S(k)$ is reasonable and close to that calculated with the Slater-determinant [7.47] form owing to the small effect of statistics on the pair correlations in ^{3}He. Results are given in Sects. 7.4 and 7.5.

7.1.6 Shadow Wavefunctions

A new variational wavefunction called the shadow wavefunction has been devised recently by *Vitiello* et al. [7.7]. Their initial motivation was to produce variational wavefunctions for the crystalline phase with the symmetries of the Hamiltonian: translational invariance and particle symmetry. The localised forms assume a particular broken symmetry and thus inhibit studies of the phase transition, vacancies etc. They write the trial wavefunction as an integral over a function of both real particle co-ordinates denoted R, and a set of parameters called the shadow variables, S,

$$\Psi_{\mathrm{T}}(R) = \int dS\, \Xi(R,S), \tag{7.17}$$

where they take

$$\Xi(R,S) = \exp\left[-\sum_{i<j} u_{\mathrm{r}}(r_{ij}) - \sum_{k} \phi(r_k - s_k) - \sum_{l<m} u_{\mathrm{s}}(s_{lm}) \right], \tag{7.18}$$

and

$$\phi(r) = \alpha r^2 \tag{7.19}$$

with α a variational parameter, and u_{r} and u_{s} variational functions. The shadow variables are in one-to-one correspondence with the real variables and can be thought of as particles that are integrated out of the wavefunction, the shadow particles. If the u_{s} is chosen to be a classical potential that crystalises, this form looks like a localised form, (7.3, 4) around the classical crystal positions. However, particle exchange and translational invariance are maintained since the underlying classical crystal maintains these symmetries. Of course, if the shadow variables are frozen into a particular crystalline state during a Monte Carlo simulation, there is no particular advantage to this form over the original localised form.

Two other motivations for the form (7.18, 19) were given by Vitiello et al. The first is that path integral Monte Carlo simulations can give a wavefunction of the shadow form. If we look at a Feynman path in a crystal, we see that a particle is attracted indirectly to the centre of mass of its path, and this centre of mass looks like a classical particle when interacting with the other particles' paths. The crystalisation of these paths forms the quantum crystal. The shadow co-ordinates are interpreted as these path variables. A slightly different motivation comes from the projection technique used in diffusion Monte Carlo calculations (see Sect. 7.2) where a trial wavefunction of a pair product form, given by the shadow u_s term, is operated on by a many-body Green's function, $\exp(-\tau H)$ which is then approximated by the u_r and ϕ terms of (7.18). Since the starting function is a crystal, we expect to stay in the crystalline phase. The Green's function will build in the physical correlations between the particles, including some 3-body and higher correlations. The shadow procedure will always improve any variational trial function by building in additional correlations. A virtue of the shadow wavefunction is that the same form can be used for the liquid and solid phases and the trial function can undergo a liquid to solid transition with only a change in variational parameters just as a classical system freezes as the temperature is lowered.

The evaluation of expectation values with the shadow wavefunction is straightforward. Both the real particle and shadow particle integrations are done using the *Metropolis* et al. method. To calculate variational expectation values there are two sets of shadow variables, one for Ψ_T^* and another for Ψ_T.

A disadvantage of the shadow form is that since the shadow integration must be done to obtain $\Psi_T(R)$ from $\Xi(R,S)$, the energy variance will not be zero even for an exact trial function. This may be offset by the extra correlations built in by the projection. Of course, the shadow procedure can be systematically improved by either adding 3-body and other correlations to the shadow form, or by using a more accurate representation of the Green's function as will be discussed in Sect. 7.3.2. No work has been done along these lines. Another disadvantage of the shadow form is that the dynamics of the VMC or GFMC seem to be significantly slowed down by the entanglement of the real variables with the shadow variables so that Monte Carlo averages have more autocorrelation and converge slower. It is not clear whether one achieves lower errors per unit of computational effort by using the shadow form or by a direct GFMC calculation with a simpler trial function.

The shadow wavefunction has properties that lend it to excited state calculations in quantum fluids. The standard Feynman excitation spectrum is obtained by operating on the ground-state wavefunction with the excitation operator

$$\rho_k = \sum_j \exp(\mathrm{i}\boldsymbol{k}\cdot\boldsymbol{r}_j). \tag{7.20}$$

In the case of shadow wavefunctions, the excitation operator can operate on the shadow variables as well as the real variables. That is the excitation operator can be taken to be

$$\sigma_k = \sum_j \exp\{\mathrm{i}\boldsymbol{k}\cdot[\boldsymbol{s}_j + \gamma(\boldsymbol{s}_j - \boldsymbol{r}_j)]\}. \tag{7.21}$$

If it operates only on the shadow variables (i.e. $\gamma = 0$), the projection argument given above indicates that the shadow correlation will build in some state dependent effects. *Wu* et al. [7.50, 51] show that the shadow form gives backflow correlations. At the roton minimum they obtain energies within 0.5–0.8 K of experiment [7.52, 53] for a range of densities with error bars of about 0.6 K.

These calculations are difficult because of the large phase fluctuations that can occur using the excitation operator σ_k with $\gamma \neq 1$.

Another interesting result of *Wu* et al. is obtained, similarly, by applying the Feynman vortex excitation operator to the shadow variables. The vortex core no longer has a singular vorticity. The singularity is smeared out by the shadow correlations producing a much more physical vortex. Detailed numerical calculations have not yet been done.

7.1.7 Wavefunction Optimisation

With increasingly complex variational wavefunctions, better methods to determine the optimal values of variational parameters are needed. Traditionally, the expectation value of the energy is minimised to find the optimum values of the variational parameters. Although repeated calculations of the energy at various values of the variational parameters can produce good results, typically reweighting techniques are used to calculate the change in energy when optimising the parameters as described in I.

The expectation value of an operator O, with wave function ψ_{new}, can be written in terms of configurations sampled from a wavefunction ψ_{old}. These reweighting schemes often yield a much better estimate for the change in expectation value between ψ_{old} and ψ_{new} than independent calculations, since many of the fluctuations in the two calculations cancel. Often, reweighting calculations take less computer time since some parts of the calculations will be the same when using a ψ_{new} close to ψ_{old}. In any case, only independent samples from ψ_{old} are used which can further reduce the computer time required. We can write,

$$\langle O \rangle_{\text{new}} = \int dR \frac{\psi^*_{\text{new}}(R) O \psi_{\text{new}}(R)}{\psi^2_{\text{new}}(R)} w(R) P(R) \Big/ \int dR\, w(R) P(R), \tag{7.22}$$

with

$$P(R) = \frac{\psi^2_{\text{old}}(R)}{\int dR\, \psi^2_{\text{old}}(R)}, \tag{7.23}$$

and

$$w(R) = \frac{\psi^2_{\text{new}}(R)}{\psi^2_{\text{old}}(R)}, \tag{7.24}$$

and calculate $\langle O \rangle_{\text{new}}$ and $\langle O \rangle_{\text{old}}$ from the same set of samples of $P(R)$.

The variance of the energy is non-negative and equals zero only for the eigenstates of the Hamiltonian. Recent results on atomic and molecular electronic systems [7.15, 44, 54, 55] and on bulk ^{4}He [7.56] indicate that minimising the variance or a linear combination of the energy and variance [7.57] by reweighting methods and using standard minimisation techniques such as Levenberg–Marquardt, Simplex, or Newton's method, can optimise of the order of 10–50 parameters.

Two-body correlation factors in quantum fluids have long been optimised using integral equation methods such as the paired–phonon method of *Feenberg* and coworkers [7.58]. Recently, Monte Carlo techniques have been used to optimise correlation factors [7.56]. A simple method consists of writing the 2-body correlation as

$$f(r) = \sum_{n=1}^{N_b} a_n f_n(r), \tag{7.25}$$

where a_n are variational parameters, and $f_n(r)$ are the solutions to the 2-body Schrödinger equation

$$-\frac{1}{m}\nabla^2 f_n(r) + v(r) f_n(r) = \lambda_n f_n(r), \tag{7.26}$$

and N_b is the number of basis functions. Typical boundary conditions are those used with a single function by *Pandharipande* [7.59] such that $f_n(r)$ goes smoothly to 1 at $r = d$, with d either a variational parameter or fixed at a reasonable cutoff distance. Boundary conditions that match smoothly to the correct long-range tail at $r = d$ could be used [7.43]. This basis set has the advantage of automatically satisfying the 2-body Schrödinger equation at small r, and having reasonable behaviour at large r.

Schmidt and *Vitiello* [7.56] have used the Levenberg–Marquardt method and variance minimisation to calculate Monte Carlo optimised 2-body correlation factors in liquid and solid ^{4}He. They find energies about 0.1 K lower in both the liquid and solid than with analytic or integral equation forms. The method can be generalised to other correlations, and the basis can be refined by redefining $v(r)$ in (7.26) to be an effective two-body potential that gives a nearly optimum $f(r)$ as its f_0.

7.2 Green's Function Monte Carlo and Related Methods

7.2.1 Outline of the Method

The Green's function Monte Carlo (GFMC) method has been described in I and II, and a simple tutorial is available [7.60]. The interested reader is referred to these articles and references therein. Here we give only enough to define terms. The method simply projects a trial state to the ground state using, as a Green's function, the real space representations

$$G(R, R') = \langle R | \frac{E_T + E_C}{H + E_C} | R' \rangle \tag{7.27}$$

or

$$G(R, R', \tau) = \langle R | \exp[-(H - E_T)\tau] | R' \rangle, \tag{7.28}$$

where E_C is a constant chosen to make the spectrum of $H + E_C$ positive, and E_T is a trial energy adjusted to keep the population of Monte Carlo walkers constant and equals the ground state energy. The Schrödinger equation becomes,

$$\Psi(R, \tau + \Delta\tau) = \int dR' G(R, R', \Delta\tau) \Psi(R', \tau), \tag{7.29}$$

and

$$\Psi(R, n+1) = \int dR' G(R, R') \Psi(R', n), \tag{7.30}$$

for (7.28) and (7.27) respectively, which both converge to the ground state of the Hamiltonian. Since the version using $G(R, R', \tau)$ maps the Schrödinger equation in imaginary time onto a diffusion equation, it is often referred to as diffusion Monte Carlo (DMC).

In II, both exact and short time versions of the diffusion Monte Carlo method were discussed. Versions with higher order accuracy in the time step than the simple short time approximation, have also been developed [7.61].

7.2.2 Fermion Methods

While giving good results for Bose ground-states, both methods suffer from exponentially growing variance when naively applied to fermion problems. If the negative signs associated with the antisymmetric wave function are carried along as weights, the method is known as transient estimation. A common technique, the fixed-node approximation, is to solve (7.29) or (7.30) in the region where a fermi trial function is positive. See II for a review of some GFMC methods for fermions. Some progress has been made in formulating fermion algorithms. Straightforward improvements of the fixed-node approximation can be made by regulating the walkers that cross the nodes, or by introducing softer boundary conditions at the nodes. The method of mirror potentials has been developed by *Kalos* and *Carlson* [7.62]. This method attempts to improve on the fixed-node approximation by writing the antisymmetric ground-state wave-function as

$$\Psi^A = \Psi^+ - \Psi^-, \tag{7.31}$$

where Ψ^+ and Ψ^- are positive functions and the Schrödinger equation can be trivially rewritten as two coupled equations,

$$[H + c(R)\Psi^+]\Psi^- = E\Psi^-, \tag{7.32}$$

and

$$[H + c(R)\Psi^-]\Psi^+ = E\Psi^+. \tag{7.33}$$

The original Schrödinger equation is recovered by subtraction, and the function $c(R)$ can be chosen arbitrarily. The mirror potential is the additional $c(R)\Psi^\pm$ in the coupled equations, and is defined by the population of the oppositely signed walkers. Unfortunately, these walker populations are not normally dense enough in real simulations to make well defined mirror potentials, and approximate trial functions are used instead. The method then loses some of its appeal. By varying the magnitude of the function $c(R)$ in (7.32) and (7.33). the approximate mirror potential method can be made to interpolate between fixed node and transient estimation results.

Calculations for lattice fermion systems, have been made using the *Stratonovich-Hubbard* transformation [7.63, 64]. The pairwise interaction is replaced by auxiliary fields. The fermion part of the wavefunction can then be solved, since the fermions will then only interact via the auxiliary fields. These fields are averaged using the Monte Carlo method. Some attempts at formulating this technique for continuum systems have been made, for example, by *Sugiyama* and *Koonin* [7.65], and a constraint similar in spirit to the fixed-node approximation has been implemented by *Fahy* and *Hamann* [7.66]. However, much more work needs to be done before these methods can be applied to quantum liquids, where the number and magnitude of the auxiliary fields is large due to the continuous nature of the motion of the particles, and the range and strength of the interactions. It is difficult to use the insight provided by the pair-product and backflow trial functions to reduce the computational time once auxiliary fields have replaced the pair interaction.

7.2.3 Shadow Importance Functions

In both Green's function Monte Carlo and diffusion Monte Carlo, importance sampling is introduced to lower the variance in the Monte Carlo walk. As discussed in I, this is simply accomplished by replacing the iterative equation, (7.29) by

$$\Psi(R, n+1)\Psi_{\mathrm{T}}(R) = \int dR' \frac{\Psi_{\mathrm{T}}(R)}{\Psi_{\mathrm{T}}(R')} G(R, R')\Psi(R', n)\Psi_{\mathrm{T}}(R'), \tag{7.34}$$

and solving for $\Psi_{\mathrm{T}}(R)\Psi(R, n)$, using the importance sampled Green's function

$$\frac{\Psi_{\mathrm{T}}(R)}{\Psi_{\mathrm{T}}(R')} G(R, R'). \tag{7.35}$$

The use of a shadow wavefunction $\Psi_{\mathrm{T}}(R) = \int dS \Xi(R, S)$ as the importance function introduces some extra difficulty, since the integral over the shadow variables must be done by Monte Carlo. *Vitiello* et al. [7.67, 68] show that one method to accomplish this consists of making an arbitrary number M_{r} of GFMC steps with an importance function given by $\Xi(R, S)$ with fixed S, followed by an arbitrary number of *Metropolis* et al. steps M_{s} on the shadow variables S, at the fixed values of the real variables, satisfying the usual detailed balance condition,

$$\Xi(R, S')P(R, S' \to S) = \Xi(R, S)P(R, S \to S'), \tag{7.36}$$

where $P(R, S \to S')$ is the Metropolis transition probability for the shadow variables at fixed value of the real variables. The average results are independent of the values of M_r and M_s, but the variance and autocorrelations, of course, will depend on these choices. Vitiello et al. report good results with $M_r = M_s = 1$ although the use of the shadow trial function increases the computation time needed for convergence by about one order of magnitude. Other methods of sampling the shadow variables such as direct incorporation into the GFMC sampling, Langevin, or molecular dynamics methods could also be used.

7.3 Path Integral Monte Carlo Method

Path integral Monte Carlo (PIMC) has emerged in the last few years as an extremely powerful computational method for computing properties of quantum fluids and solids at non-zero temperature, complementing the GFMC methods used at zero temperature. There has been an enormous amount of work using finite temperature methods on lattice models for applications in lattice gauge theory and high temperature superconductivity which is covered elsewhere in this volume.

7.3.1 PIMC Methodology

All static properties and some dynamical properties of a quantum system in thermal equilibrium are obtainable from the density matrix

$$\rho(R,R';\beta) = \sum_{\alpha} e^{-\beta E_\alpha} \phi_\alpha^*(R)\phi_\alpha(R'), \tag{7.37}$$

a sum over the exact energy eigenstates of the Hamiltonian weighted by the Boltzmann factor. As seen by direct substitution, the density matrix satisfies the convolution identity

$$\begin{aligned}\rho(R,R',\beta) = \int \cdots \int dR_1 dR_2 dR_3 \cdots dR_{M-1}\, \rho(R,R_1,\tau) \\ \cdot \rho(R_1,R_2,\tau) \cdots \rho(R_{M-1},R',\tau),\end{aligned} \tag{7.38}$$

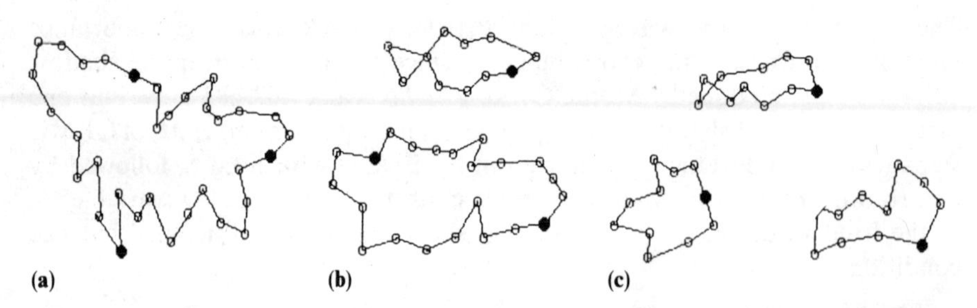

Fig. 7.1 a–c. Three types of exchange possible among three atoms. The circles represent the co-ordinates of the atoms at a given time slice with the filled circles denoting the first time slice. The lines represent the "spring" or kinetic energy part of the action and connect the same atom on neighbouring time slices

where $\tau = \beta/M$. As M is increased and τ approaches zero (the high temperature limit), one can write down an explicit approximation to the density matrix. The simplest, but by no means the recommended form, is the semiclassical or primitive form

$$\rho(R, R', \tau) = (4\pi\lambda\tau)^{-3N/2} \exp\left[-\frac{(R-R')^2}{4\lambda\tau} - \tau V(R) \right], \tag{7.39}$$

where $\lambda = \hbar^2/2m$. Substituting this approximation into the convolution identity yields the Trotter formula. A quantum observable can be obtained by averaging over the probability density: e^{-S} where the "action" is defined as

$$S = \sum_{i=1}^{M} \frac{(R_i - R_{i-1})^2}{4\lambda\tau} + \tau V(R_i), \tag{7.40}$$

for sufficiently large M. The $3N$-dimensional points R_i for $0 \leqq i \leqq M$ define the path which is periodic in the index i, that is $R_0 = R_M$. This probability distribution is similar to the classical Boltzmann distribution of a system of ring "polymers" having a harmonic spring interaction between neighbouring beads on the same polymer and with beads on different polymers interacting via the original potential function $V(r)$ as shown in Fig. 7.1a. Classical simulation techniques can be used to perform the $3NM$-dimensional path integral but one must be careful to ensure that the simulation will converge in a reasonable amount of computer time as we will discuss below.

Bose statistics introduce a very profound, yet simple, change in these paths. One can obtain the boson density matrix [7.69] from the unsymmetrised density matrix by applying a symmetrisation operator,

$$\rho(R, R', \tau) = \frac{1}{N!} \sum_{P} \rho(R, PR', \tau). \tag{7.41}$$

The sum is over the permutations P of the particle labels. This simply means that the polymers are allowed to reconnect in any manner they like, as seen in Fig. 7.1b, c for 3 atoms. For Bose systems, the Monte Carlo procedure must average both over ways of connecting the polymers and the paths for a given connection. Superfluidity is simply the formation of a "macromolecule" which stretches across the system.

There are two very important considerations in actually carrying out the simulation. First, the approximate high temperature form for the density matrix should be made as accurate as possible to reduce the value of M and, second, the dynamics of the simulation must be chosen carefully in order to have an algorithm which is effectively ergodic in permutation and path space.

7.3.2 The High Temperature Density Matrix

Although the primitive form for the action gives the right limit as the time-step goes to zero, its use in simulations is inefficient. Luckily, it is relatively easy to find a more accurate expression for the high temperature density matrix. For a soft potential, e.g. the harmonic oscillator or a lattice model, one can use an improved action derived from an $\hbar$, Wigner–Kirkwood (WK) [7.70] expansion

$$\delta S = \frac{\lambda}{12} \sum_{i,k} (\tau^2 F_{ik}^2 + 2\tau \nabla F_{ik}) + O(\lambda^2), \tag{7.42}$$

where F_{ik} is the classical force on the ith particle in "time-slice" k. Such an expansion is not appropriate for helium because the major corrections to the density matrix come from small r. For a potential like r^{-12} the WK corrections will be order r^{-26} but quantum diffraction effects tend to smooth the potential to r^{-5}. One needs an expansion appropriate to a hard sphere-like system.

One can generalise the Feynman–Kacs formula for the density matrix [7.71] to get an exact non-perturbative correction to any density matrix $\rho^{(n)}$,

$$\rho(R, R', \tau) = \rho^{(n)}(R, R', \tau) \left\langle \exp\left\{ -\int_0^\tau dt E_n[R(t), R'; \tau - t] \right\} \right\rangle_{\mathrm{DRW}}, \tag{7.43}$$

where DRW denotes an average over all drifting random walks from R to R', and E_n is the "local energy" of $\rho^{(n)}$,

$$E_n(R, R'; \tau) = \frac{1}{\rho^{(n)}(R, R', \tau)} \left(H + \frac{d}{d\tau} \right) \rho^{(n)}(R, R', \tau). \tag{7.44}$$

Starting with the free particle density matrix, and approximating the average by assuming that pairs of atoms diffuse independently of the other atoms, one arrives at the pair-product form [7.72] for the high temperature density matrix,

$$\rho^{(1)}(R, R', \tau) = \rho^{(0)}(R, R', \tau) \exp\left[-\sum_{i<j} u(r_{ij}, r'_{ij}; \tau) \right], \tag{7.45}$$

where $\rho^{(0)}$ is the free particle density matrix and u is defined to be exact for exactly two atoms. This will go over to the Jastrow wavefunction at zero temperature. To use this density matrix, one must make a numerical estimation of the exact two-atom density matrix at high temperature. It is convenient to transform the two-atom Bloch equation into spherical relative coordinates and then to solve for the radial density matrices using the matrix squaring method [7.73]. As an alternative, for hard sphere or coulomb potentials, one can use their eigenfunction expansions. The pair density matrices are functions of three co-ordinates. It is convenient [7.74] to use the variables $q = (1/2)(|r_{ij}| + |r'_{ij}|)$, $s = |r_{ij} - r'_{ij}|$ and $z = |r_{ij}| - |r'_{ij}|$. Since s and z will be order of $\sqrt{\tau}$ one can expand u as

$$u(r_{ij}, r'_{ij}; \tau) = \sum_{n,l} u_{nl}(q) s^{2n} z^{2l}. \tag{7.46}$$

Ceperley and *Pollock* [7.75] have found that if the exact two-atom density matrix is used in liquid ^{4}He, τ can be chosen as large as 0.025 K^{-1}, and one can still obtain total energies accurate to 0.1 K. Thus at least 20 time slices are needed for a simulation for superfluid ^{4}He.

In case even higher accuracy is needed, one can go to the next order by finding the local energy of $\rho^{(1)}$, and then make a trapezoidal approximation

$$\rho^{(2)}(R,R',\tau)=\rho^{(1)}(R,R',\tau)\exp\left[\frac{\lambda\tau}{3}\sum_i(\nabla_i U)^2\right], \tag{7.47}$$

where $U=\sum_{i<j}u(r_{ij})$ is the 2-body action. This density matrix, $\rho^{(2)}$, has the same functional form as the 3-body trial function discussed in Sect. 7.1.3.

7.3.3 Monte Carlo Algorithm

The numerical evaluations of the path integrals for many-body systems have been performed with either the classical Molecular Dynamics or Metropolis Monte Carlo methods. The Molecular Dynamics technique is straightforward to apply, but does not allow the possibility of making a permutational move so we will not discuss it further. Let us briefly recall the Metropolis rejection method. A Markov chain is generated based on some a priori transition probability, $T(R'|R)$, which is the probability density of sampling the trial move R' based on the old point R. Then that move is accepted with probability

$$A(R'/R)=\min\left[1,\frac{T(R|R')e^{-S(R')}}{T(R'|R)e^{-S(R)}}\right], \tag{7.48}$$

where $S(R')$ and $S(R)$ are the actions in the old and new state. In a Metropolis method it is possible to mix up moves in any blend as long as the probability of making a given type of move is independent of the system state, and each type of move individually satisfies the detailed balance relation. Eventual convergence is guaranteed by the detailed balance condition but the rate of convergence can be very slow, particularly with path integrals. A common theme in all the improved methods is that the slow convergence is a consequence of the kinetic energy term of the action in the limit of small τ. But it easy to device a way to sample this term a priori. Below we list the various types of Metropolis transitions that have been proposed. Unfortunately there is little comparison of their efficiencies in the literature.

7.3.4 Simple Metropolis Monte Carlo Method

In the simplest choice for the transition probability, a single atom at a single time slice, a "bead", is displaced uniformly inside a cube of side Δ, with Δ adjusted to achieve 50% acceptance. As M increases the random walk diffuses through configuration space very slowly because the largest displacement allowed by the free particle density matrix is order $\sqrt{(\lambda\tau)}$. Even worse, an atomic path acquires an inertia, so that in following moves, fluctuations away from the centre of mass are suppressed. As a consequence, it is difficult to achieve convergence in a reasonable number of steps.

7.3.5 Normal Mode Methods

The kinetic energy term for each atom can be diagonalised by working with the fourier coefficients,

$$Q_k = \sum_{l=1}^{M} R_l e^{-\pi i k l/M}. \tag{7.49}$$

The kinetic part of the action then has an expansion

$$S = \frac{1}{\lambda\beta}\sum_k \sin^2\left(\frac{\pi k}{m}\right)|Q_k|^2, \tag{7.50}$$

and each of the $3NM$ variables, Q_k, can be sampled independently from the resulting Gaussian distribution. In the absence of a pair potential, all moves would be accepted. When a realistic potential is present the large k modes can be sampled directly from the above distribution since they cause only a small movement of the path. Usually the long wavelength modes are moved a small amount, say $|Q'_k - Q_k| < \gamma_k$, with γ_k adjusted to get 50% acceptances. The centre of mass mode ($k=0$) is treated separately and moved as a classical particle [7.76–78].

7.3.6 Threading Algorithm

In this method, one cuts out a section of one or several atomic positions for n time slices. Then one recursively generates a new path by growing from time slice 0 to slice n with the diffusion algorithm

$$R_{i+1} = R_i - 2\tau\lambda\nabla S_{\mathrm{T}}[R_i, R_n; (n-1)\tau] + \eta_i\sqrt{(2\tau\lambda)}, \tag{7.51}$$

where η_i is a normally distributed random vector with zero mean and unit variance and S_{T} is a trial action. Note that the time argument of S_{T} will be greater than τ if more than one time-slice is being updated. The trial action is only used to guide the walk and any convenient approximation can be used for it. Inaccuracies will only affect the acceptance ratio, not the converged distribution. After the new path is generated, it is accepted or rejected in the usual Metropolis fashion based on the difference between the old and new action and on the ratio of the sampling probabilities for the old and new paths. The advantage of this method over the previously discussed ones is that a completely new path can be generated and several atoms can be simultaneously updated. The form of the diffusion can be shown to be optimal in the sense that if the trial action were exact with all atoms being moved and τ were sufficiently small, the acceptance ratio would be one. However, in practice, moves of more than a few time slices are often rejected [7.71].

7.3.7 Bisection and Staging Methods

The bisection method is closely related to the Levy method of constructing a Brownian bridge. To construct a random walk from $R(0)$ to $R(\beta)$ in time β, one begins by sampling the trajectory at the midpoint as

$$R(\beta/2) = \tfrac{1}{2}(R(0) + R(\beta)) + w_\beta \eta, \tag{7.52}$$

where η is a normally distributed random vector of zero mean and unit variance and $w_\beta = \sqrt{\lambda\beta/2}$. One then bisects the time intervals $(0, \beta/2), (\beta/2, \beta)$ to find the points $R(\beta/4)$, $R(3\beta/4)$ etc. until one has sampled all the points on the path.

In the bisection method [7.74], a move consists of several levels; the first level is the midpoint, the next level consists of the two midpoints of the midpoint etc. A decision is made at the end of each level whether to continue on to the next level or to reject the entire attempted move. Only when one reaches the final level are the co-ordinates updated. There are two ingredients in this algorithm. The first is the method of sampling the midpoints. It has been found best to use a correlated Gaussian distribution so that the midpoint is sampled according to

$$R(\beta/2) = \tfrac{1}{2}(R(0) + R(\beta)) - w_\beta^2 \nabla U + \eta, \tag{7.53}$$

where U is the interaction part of the action and η is sampled from a normally distributed random number with zero mean and covariance

$$\langle \eta\eta \rangle = w_\beta^2 (I - w_\beta^2 \nabla\nabla U). \tag{7.54}$$

This correlated normal distribution is sampled using a Choleski decomposition of the covariance matrix. The effect of atomic interactions on the a priori distribution is to push the mean position of an atom away from its free particle mean if another atom is there. This is similar to the techniques of "force-bias" Monte Carlo and "smart Monte Carlo" [7.45]. The covariance has a crucial role to play when an exchange of atoms is being attempted in order to change the permutation. Then the covariance keeps the two (or more) moving atoms out of each other's way.

The second ingredient is the rule for accepting a given level and proceeding onto the next level. For simplicity, let us consider only the acceptance of the midpoint move $R'_{n/2}$ which has been sampled from the correlated Gaussian distribution $T(R'_{n/2})$. The probability to go onto the second level is given by

$$\min\left[1, \frac{T(R_{n/2})\pi_2(R'_{n/2})}{T(R'_{n/2})\pi_2(R_{n/2})}\right], \tag{7.55}$$

where

$$\pi_2(R_{n/2}) = \frac{\rho(R_0, R_{n/2}; n\tau/2)\rho(R_{n/2}, R_n; n\tau/2)}{\rho(R_0, R_n; n\tau)}. \tag{7.56}$$

Note that one must compute the probability of sampling the old midpoint and the old action.

For multi-level sampling, until the final level, the density matrices will have time arguments greater than τ. The approximation made will only affect the rate of convergence, not the final converged values of physical quantities. But as the density matrices approach the exact ones, and the sampling $T(R_i)$ approaches its optimal value, the acceptance ratio approaches unity. In the bisection algorithm, the parameter which controls the size of the move is the number of time steps. The advantage of the bisection method over the threading method is that places of high potential energy are most likely at the midpoint. Unfavourable moves can be quickly identified and the process stopped instead of continuing on to the inevitable rejection.

The staging algorithm, which has been applied to a single electron in a classical liquid [7.79], has ideas similar to the bisection method. The first level is constructed using with the Levy algorithm and then a second level repeatedly sampled to find the action of the first level. This is not as efficient as the bisection procedure because the amount of computation per accepted move is much higher.

7.3.8 Sampling Permutations

The simplest Monte Carlo algorithm to change the permutation for a Bose system, would involve interchanging a pair of particle labels without moving the chain at all. This will not sample permutation space for small τ as the following argument demonstrates. In relative co-ordinates, a pair of ^{4}He atoms will move on the average $\sqrt{\lambda\tau} = 1$ A for a time step of 1/40 K. As can be seen in Fig. 7.2, since the relative co-ordinates are about 3 A apart because of the pair potential, there is a very small chance (about $10^{-5} = \exp[-\sigma^2/(4\lambda\tau)]$) of making an exchange. In order to fully explore the configuration space of a superfluid, it is necessary to couple permutation moves with moves of a portion of the polymer chains of the atoms being relabeled.

It can be shown that the optimal function to sample a permutation P from is $\rho(R_i, PR_{n+i}; n\tau)$. To have a reasonable acceptance probability one wants the move to be as small as possible. Let us suppose that the permutation change, P, ranges over all cyclic permutations involving 2, 3 or 4 atoms. It is important to go beyond pair interchanges in a dense liquid since it is easier for three of

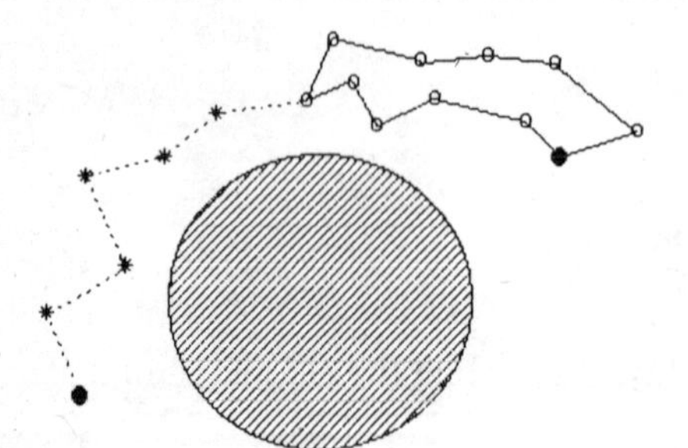

Fig. 7.2. Direct (solid) and exchange paths (dashed) for two atoms in relative co-ordinates. The shaded circle indicates the portion of phase space excluded by the interatomic potential. In relative co-ordinates an exchange path proceeds from a point Δ to a point $-\Delta$ while a direct path is a loop. The figure shows that a substantial change in the path variables are needed to move from a closed loop to an exchange path

four atoms to cyclically permute than it is for a pair. Since the exact density matrix is unknown, some approximation must be made to construct the a priori chance of a permutation. The free particle density matrix is a convenient choice. Then the probability of attempting a permutation P is proportional to

$$T(P)=\exp\left[-\frac{(R_i-PR_{n+i})^2}{4n\lambda\tau}\right]. \tag{7.57}$$

A table of these probabilities can be quickly constructed and sampled since it only involves inter-atomic distances between time slice i and slice $i+n$. If the total permutation and path are accepted, the table needs to be updated.

7.3.9 Calculation of the Energy

The most straightforward way to calculate the energy is to differentiate the partition function with respect to the temperature;

$$E=-\frac{d\ln(Z)}{d\beta}=\frac{3N}{2}\tau+\frac{1}{M}\sum_{i=1}^{M}\left\langle-\frac{(R_i-R_{i-1})}{4\lambda\tau^2}+\frac{dU_{i,i-1}}{d\tau}\right\rangle, \tag{7.58}$$

where the brackets indicate average over the walk and $U_{i,i-1}$ is the interacting part of the action as defined in (7.45) between time slices i and $i-1$. At sufficiently small τ, this reduces to the potential energy but (7.58) is exact at any τ for an exact expression for the action. In taking the β derivative, the order of an approximation to the action will be reduced by one, so if the expression for the action has an error of order τ^3, this expression for the energy is only correct to order τ^2. For that reason, it is best to include the force squared term of (7.47). If U is the solution to the two-body problem, its time derivative can be eliminated in favor of spatial derivatives. A comparison of the two will give an estimate of the systematic error of the energy [7.76].

There is a difficulty in using (7.58) to estimate the energy, namely its error grows as τ^{-1}. Thus attempts have been made to find lower variance estimators [7.71, 80, 81]. It is possible to eliminate the troublesome kinetic energy term by integrating by parts over the path variables, and using a form reminiscent of the virial expression for the pressure;

$$E=\frac{3N}{2\beta}+\frac{1}{M}\sum_{i=1}^{M}\left\langle\frac{dU_{i,i-1}}{d\tau}-\frac{F_i\Delta_i}{2}-\frac{1}{4\beta\tau\lambda}(R_{M+i}-R_i)(R_{i+1}-R_i)\right\rangle, \tag{7.59}$$

where F_i is the analog of the classical force

$$F_i=-\frac{1}{\tau}\nabla_i(U_{i,i-1}+U_{i,i+1}), \tag{7.60}$$

and Δ_i is the deviation from the particle's centroid

$$\Delta_i=\frac{1}{2M}\sum_{j=M+1}^{M-1}(R_i-R_{i+j}). \tag{7.61}$$

This virial estimator is exact (if the exact U is used) but care must be taken in interpreting terms $(R_i - R_j)$ to ensure the atoms have always continuous trajectories with periodic boundary conditions and bosonic exchange. For free particles, we can obtain the following estimator for the energy,

$$E = \frac{3N}{2\beta} - \left\langle \frac{(R_0 - R_M)^2}{4\beta^2 \lambda} \right\rangle, \tag{7.62}$$

where the difference in distances may include winding around the periodic boundaries of the box and bosonic exchange. Quantum free particles in a box can have an energy less than $\frac{3}{2}kT$; witness bose condensation. Only the primitive estimator has been applied for superfluid helium since the virial does not have lower variance for the τ and potentials used.

7.3.10 Computation of the Superfluid Density

The path integral framework translates superfluidity into very intuitive concepts. As is discussed in Feynman's papers, at approximately 2 K the polymers become long enough so that exchange becomes probable. The partition function will increase as this new phase space opens up and the second derivative of it, the specific heat, will have the familiar lambda shape. Figure 7.3 shows the probability of a given atom participating in a cyclic exchange of n atoms, as calculated with PIMC.

Superfluidity is experimentally defined by the equilibrium response of the system to a gentle motion of the walls. If the walls are rotated slowly, the normal fraction will be entrained with the walls at equilibrium while the superfluid will remain at rest. By doing a transformation from a co-ordinate frame where the walls are moving to one where they are at rest, it is possible to use PIMC to determine the dependence of the free energy on the motion of the walls. The superfluid density is proportional to the second derivative of the free energy with respect to the wall velocity at zero velocity, and in periodic boundary conditions can be shown [7.82] to be equal to the mean squared number of paths in PIMC which wind around the periodic walls,

$$\frac{\rho_{\mathrm{S}}}{\rho} = \frac{\langle W^2 \rangle}{2d\lambda\beta N}, \tag{7.63}$$

where $\mathbf{W}$ is the winding number,

$$\mathbf{W} = \sum_i \int_0^\beta dt \frac{d\mathbf{r}_i}{dt}. \tag{7.64}$$

The winding number which describes the net number of times the paths have wound around the periodic cell is an invariant of the path. For a system without boundaries such as a droplet, one can define the normal fluid mass as that part of the system which contributes to the moment of inertia.

$$\frac{\rho_{\mathrm{s}}}{\rho} = 1 - \frac{I}{I_{\mathrm{c}}} = \frac{2\langle A^2 \rangle}{I_{\mathrm{c}}\lambda\beta}, \tag{7.65}$$

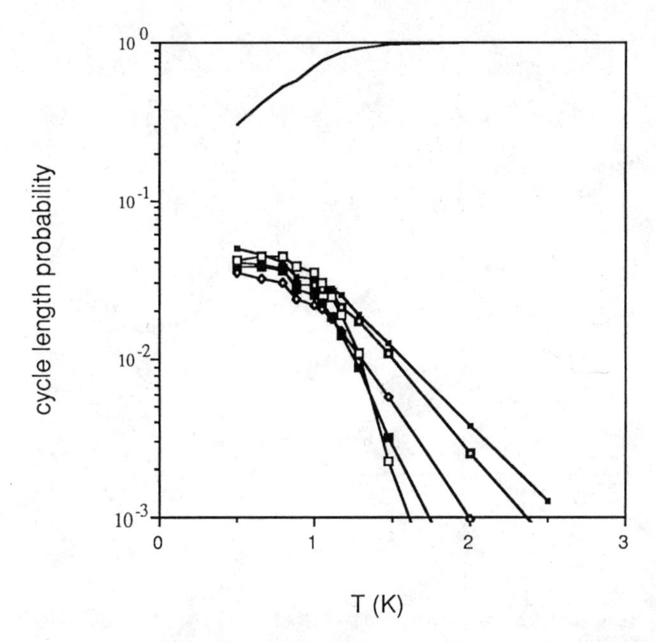

Fig. 7.3. The probability of a given atom being involved in an exchange in 2D ^{4}He as a function of temperature. The upper curve are non-exchanging atoms while the 5 lower curves are the probabilities of exchange with 2–6 other atoms. Because there are a total of 25 atoms in the simulation these probabilities approach 1/25 at low temperature. The Kosterlitz–Thouless superfluid transition occurs at a temperature of 0.7 K

where $I_c = \left\langle \sum_i r_i^2 \right\rangle$ and A is the area swept out by the paths,

$$A = \frac{1}{2} \sum_i \int_0^\beta \frac{d\boldsymbol{r}_i}{dt} \times \boldsymbol{r}_i. \tag{7.66}$$

An example of winding is shown in Fig. 7.4 and the results for the superfluid density are given in Sect. 7.4. A major difficulty in the PIMC calculations is getting the winding numbers to converge, since a change in winding number can only occur with a global change in path configuration. It has been shown recently that an estimator [7.83] based on the local diffusion of paths can give an alternative method of calculating superfluid density.

7.3.11 Exchange in Quantum Crystals

Crystal ^{3}He at millikelvin temperatures is one of the simplest and cleanest examples in nature of a lattice-spin system. Its magnetic properties result from infrequent atomic exchange, since it is only through exchange that the Pauli exclusion principle comes into play. Careful analysis of experimental data made plausible the model that the frequency of exchange of two, three, and four atoms

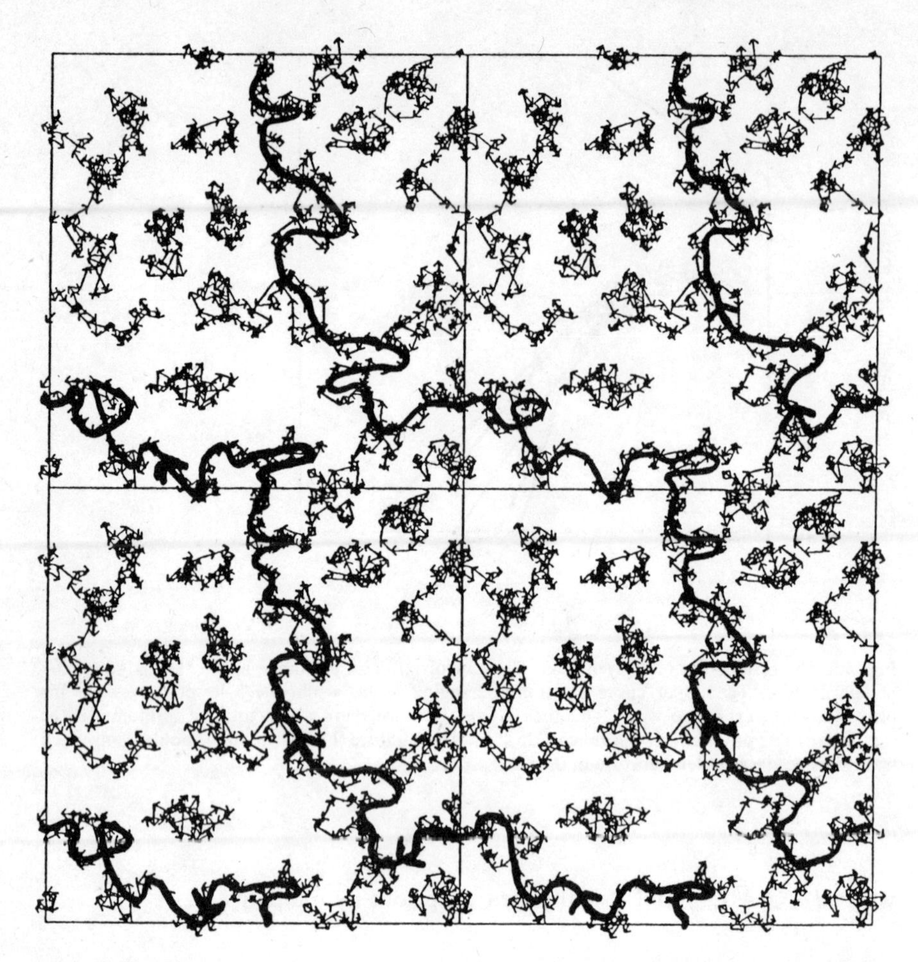

Fig. 7.4. A typical path configuration at $T = 0.8$ K in 2D liquid ^{4}He. The basic periodic cell is replicated 4 times. There are a total of 25 atoms present. The heavy line indicates a path winding across the cell in both the x and y directions

are approximately equal to each other, but calculation of exchange frequencies is necessary for the model to be verified [7.84, 85].

To define the exchange frequencies and the lattice-spin model, one assumes that most of the time the atoms are close to lattice sites. The $N!$ degeneracy of arranging N atoms onto N lattice sites is broken by the exchanges. Suppose we allow only two ways of arranging the atoms which we will denote as Z and PZ. Here P is a permutation and Z the perfect lattice vector of $3N$ positions. Then the ground state is split into an even and odd state and the frequency with which the system oscillates between the two localised states is $2J_P = E_1 - E_0$. At low temperatures, the system of spin 1/2 fermions is described by a lattice Hamiltonian acting only on the spins: $\sum_P J_P(-1)^P P_\sigma$.

The exchange frequencies are difficult to calculate since the helium atoms have large zero point motion and correlation. Variational methods fail since they require knowledge of the wavefunction in regions where it is very small. For temperatures well below the Debye temperature only the two lowest states will contribute to the density matrix, and if we consider the element of the density matrix connecting Z with PZ it can be shown that

$$\tanh[J_P(\beta - \beta_P)] = \frac{\rho(Z, PZ; \beta)}{\rho(Z, Z; \beta)}. \tag{7.67}$$

Expanding the density matrices in terms of path integrals as in (7.38) and speaking in terms of the polymer picture, the quantum exchange rate is related to the free energy necessary to "cross-link" two or more "polymers" in a "polymer crystal". It is then possible to take the computer method developed to simulate superfluid ^{4}He and thereby find these magnetic coupling constants in the crystal ^{3}He. The results are in quite good agreement with experiment and show that many exchanges are relevant in a quantum crystal and that a quantum crystal is more complex than previously thought.

This calculation is technically demanding. Bennett's method [7.86] of the two sided acceptance ratio was used to determine the free energy ratio. To calculate the above ratio we define the state of the walk to consist not only of the path, but also of the connection of the endpoints which is either P or I (the identity permutation). Just as for superfluids, the Monte Carlo moves must allow for transitions between these two states. It is actually not necessary to make moves between the two states but only necessary to compute the acceptance probabilities for proposed transitions. Then the exchange rate is not expressed as the difference between two eigenvalues but as the ratio of two rates. The error is independent of the magnitude of the exchange frequency which is on the order of 460 μK in ^{3}He and 4 μK in ^{4}He. It is quite important in this calculation to have an accurate high temperature density matrix.

7.3.12 Comparison of GFMC with PIMC

GFMC and PIMC are very closely related methods since both are based on sampling of the thermal density matrix. The density matrix is a solution of the Bloch equation

$$-\frac{d\rho(R, R', \beta)}{d\beta} = H\rho(R, R', \beta), \tag{7.68}$$

which is the "dynamics" of the GFMC algorithm. The object held in the computer's memory in GFMC is an ensemble of configuration walkers $\{R\}$ while in PIMC it is the entire imaginary time path R_i and the permutation. The dynamics in GFMC is that of a branching random walk with an assumed guiding function while in PIMC it is a generalised Metropolis method. Insight into the physics may influence the transition probability in PIMC. GFMC can calculate energies very accurately but has difficulty with other quantities. Estimates coming from PIMC have larger statistical errors, but they have less systematic error since they do not have the bias from the importance function. GFMC usually converges faster since the walks are less constrained while PIMC has a kind of critical slowing down as the time step goes to zero. But of course, one is comparing apples and oranges since GFMC is the method of choice for calculation of ground state properties and PIMC for calculations at non-zero temperatures.

7.3.13 Applications

There have been numerous applications [7.87] of PIMC to the situation where a single quantum particle (for example an electron or muon) is inserted into a classical system. Examples are an electron in gaseous classical helium, or a single electron on a protein. Without the possibility of particle exchange, getting the system to converge is not difficult, but on the other hand, to reach room temperature takes on the order of one thousand electron steps, much more than is needed for liquid helium because an electron is so much lighter than a He atom. There have also been PIMC calculations on almost classical systems like liquid argon, neon, and water. Monte Carlo methods used on those systems are straight-forward generalisations of the classical Monte Carlo or molecular dynamics methods, since quantum exchange is not taken into account. The results of the PIMC method that we will discuss in Sects. 7.4–6 are the applications to systems of liquid and solid helium.

7.4 Some Results for Bulk Helium

7.4.1 ^{4}He Results

The bulk ^{4}He liquid and solid ground-state behaviour is well characterised by the variational wavefunctions described in Sect. 7.1, and good equation of state results are obtained using the GFMC method. In Fig. 7.5, we show the equation of state of liquid ^{4}He at zero temperature from experiment [7.36], GFMC [7.88], and variational wavefunctions with 2- and 3-body correlations [7.89]. Figure 7.6 shows similar results for solid ^{4}He where there are small discrepancies between experiment [7.90] and the GFMC calculation which may be due to 3-body potential energy contributions. The variational results in the liquid and solid are qualitatively correct and in reasonable quantitative agreement. Figure 7.7 shows the GFMC and experimental [7.91] 2-body distribution function $g(r)$. The agreement between experiment and theory is quite good.

Simple shadow wave functions, (7.18), reduce the variational energy at the equilibrium liquid density of ^{4}He by about 0.31 K from a pure Jastrow trial wavefunction. This can be compared with about 0.85 K reduction for a 2- and

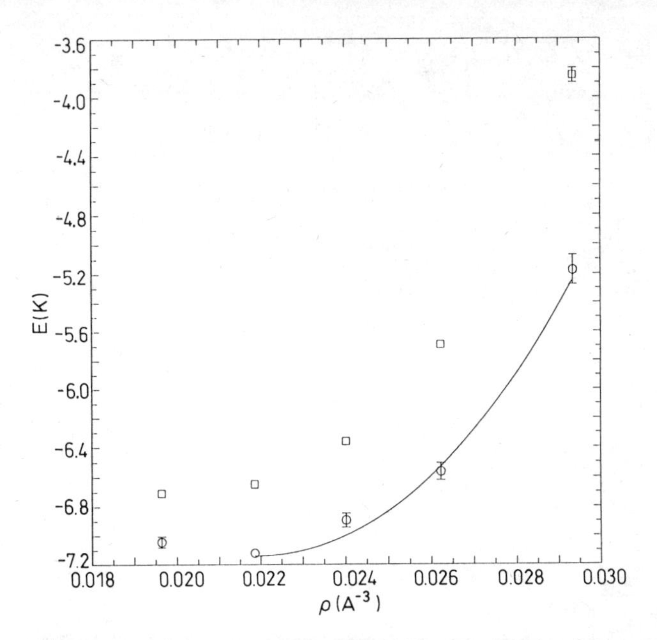

Fig. 7.5. The ground-state energy of liquid ^{4}He as a function of density. The solid line is experiment [7.36], the circles are the GFMC results [7.88], and the squares are variational results for a simple McMillan Jastrow factor and a three-body term as in (7.5–7). Points without error bars have statistical errors less than the symbol size

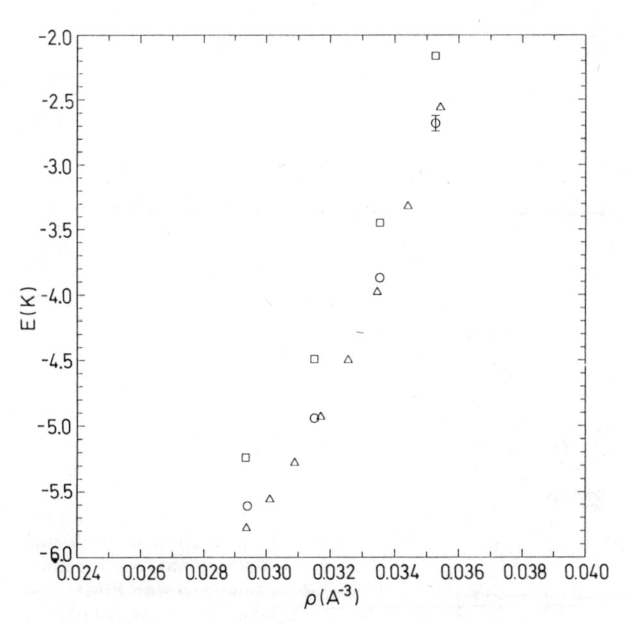

Fig. 7.6. The ground-state energy of solid ^{4}He as a function of density. The notation is the same as Fig. 7.5, except that triangles are the experimental results for the hexagonal close packed crystal [7.90]. The variational and GFMC results are for a face centred cubic crystal, and use a one-body localisation factor, (7.4), around the lattice sites

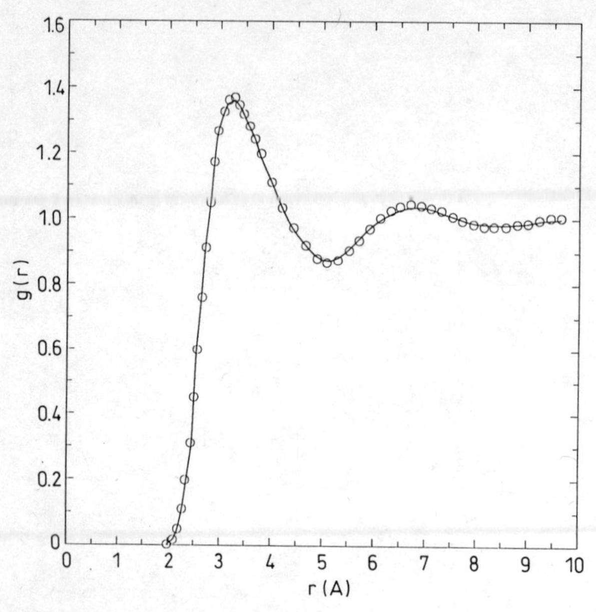

Fig. 7.7. The experimental (at 1.0 K) [7.91] (circles) and GFMC [7.88] extrapolated (solid line) two-body distribution functions $g(r)$ for liquid ^{4}He at equilibrium density

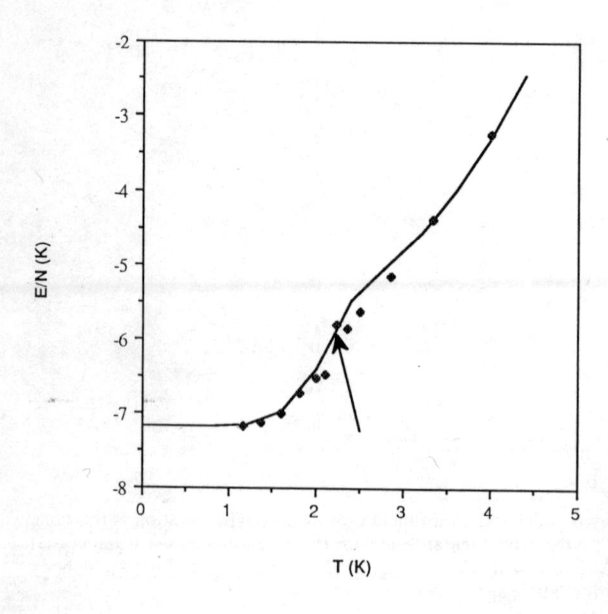

Fig. 7.8. The energy per atom of liquid ^{4}He as a function of temperature at SVP conditions as computed with PIMC (points) and measured (the curve). The arrow indicates the experimental superfluid transition temperature

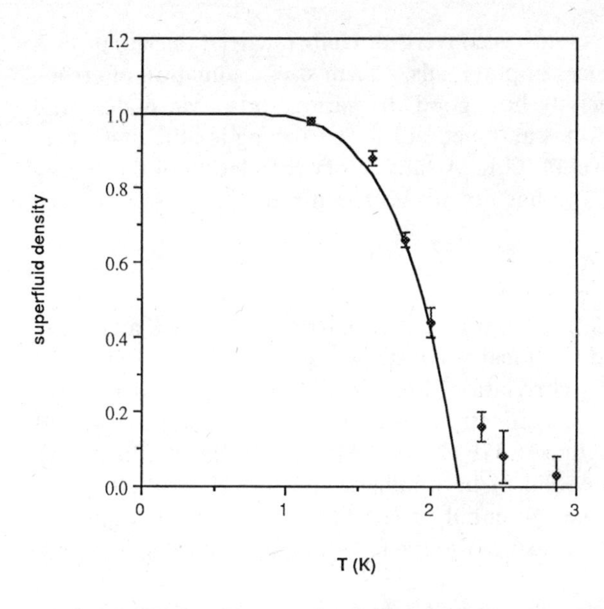

Fig. 7.9. The super-fluid density as a function of temperature at SVP conditions as computed with PIMC (points) and measured (curve)

3-body correlated wavefunction. Table 7.1 gives some other results for the liquid ^{4}He ground-state energy using various methods.

Detailed path integral Monte Carlo calculations have been performed on liquid ^{4}He [7.19, 75, 82, 92]. Figure 7.8 shows the comparison of the energy with experiment. It seems that PIMC can get energies accurate to 0.1 K except above the transition where rounding due to finite system effects occur. Path integral calculations for other properties at low temperatures agree well with GFMC results, and are in reasonable agreement with experiment. A particularly nice result is the calculated behaviour of the superfluid fraction as shown in Fig. 7.9.

7.4.2 ^{3}He Results

The ^{3}He Monte Carlo results are not nearly as accurate as those for ^{4}He. The fermi nature of ^{3}He makes even the ground-state energy difficult to calculate as described in Sect. 7.2 and in II. In Table 7.2, we show some results for the unpolarized liquid ^{3}He ground state. The 2-, 3-body and backflow correlated wavefunction produces the best variational results [7.93], although the pairing wavefunction of *Bouchaud* and *Lhuillier* [7.47] also gives a reasonable value. Unfortunately the pairing function used in that calculation has a large discontinuity at the edge of the simulation cell, thus the results are not necessarily upper bounds. The discontinuity introduces large additional size dependence that must be corrected for. Size dependence of the order of 0.1 K is a problem in all ^{3}He simulations as mentioned at the beginning of this chapter.

We also show in Table 7.2 fixed-node, approximate mirror potential, and a transient estimate result [7.21]. By comparison with experiment it appears that the transient estimate result has converged. Unfortunately, the situation is not clear-cut due to differences in potentials, system sizes, estimation of errors, etc. It is still not clear precisely how good the various proposed nodes, trial functions, potentials, algorithms and methods for correcting for the finite sizes are for liquid ^{3}He at the level of 0.1 K. A satisfactory calculation of the energy of ^{3}He versus spin polarisation has not yet been completed.

7.4.3 Solid He

In the solid phase, the shadow form at a density of 0.0329 A^{-3}, gives an energy of -3.56 ± 0.03 K compared to a localised-Jastrow result of -3.32 ± 0.02 K and -3.79 ± 0.01 K when 3-body correlations are added [7.7]. PIMC calculations in the solid phase do not have symmetry breaking terms which are introduced with the localised trial functions in GFMC or VMC. PIMC calculations have been made of the quantum solid of hard spheres and of ^{4}He with the Aziz HFDHE2 and Lennard–Jones potential [7.71, 94]. Elastic constants and the phase diagram of ^{4}He have also been extracted from these hard-sphere calculations [7.76, 95, 96].

The exchange frequencies of ^{3}He and ^{4}He atoms in the bulk solid have been calculated with PIMC [7.97]. It is found that pair interchange is most frequent, but that is followed closely by three and four atom exchange. The multiple exchange model [7.84] is strongly supported, but the calculations differ from that model in that a spectrum of exchanges is predicted to result in the experimental values of the magnetic ordering temperature and magnetic susceptibility. Even though the frequencies of some of the exchanges are small they are numerous. The importance of so many exchanges complicates the already difficult task of determining from the lattice Hamiltonian, the low temperature magnetic properties.

7.5 Momentum and Related Distributions

7.5.1 The Single-Particle Density Matrix

The single particle density matrix is

$$\rho_1(\boldsymbol{r}_1, \boldsymbol{r}_2) = \langle \psi^+(\boldsymbol{r}_1) \psi(\boldsymbol{r}_2) \rangle,$$

where the brackets indicate the thermodynamic average over the states of the system, and $\psi^+(\boldsymbol{r})$ is a creation operator for a particle at position $\boldsymbol{r}$. For a system in a normalised eigenstate Ψ,

$$\rho_1(\boldsymbol{r}_1, \boldsymbol{r}_1') = N \int d^3 r_2 \cdots d^3 r_N \, \Psi^*(\boldsymbol{r}_1, \boldsymbol{r}_2, \ldots, \boldsymbol{r}_N) \, \Psi(\boldsymbol{r}_1', \boldsymbol{r}_2, \ldots, \boldsymbol{r}_N), \tag{7.69}$$

and for a system at inverse temperature β,

$$\rho_1(\boldsymbol{r}_1, \boldsymbol{r}_1', \beta) = N \int d^3 r_2 \cdots d^3 r_N \rho(\boldsymbol{r}_1, \boldsymbol{r}_2, \ldots, \boldsymbol{r}_N; \boldsymbol{r}_1', \boldsymbol{r}_2, \ldots, \boldsymbol{r}_N; \beta), \tag{7.70}$$

where $\rho(R, R', \beta)$ is the N-particle density matrix with the symmetry or antisymmetry required by bose or fermi statistics. Equation (7.70) is just the Boltzmann average of (7.69) over the states of appropriate particle statistics. These equations have tacitly assumed spin independence of the density matrix. For partially spin polarised systems, the density matrices of the spin up and spin down particles are calculated by choosing particle 1 in (7.69) or (7.70) to be spin up or down respectively, and the factor N is replaced by the number of up or down spins.

For homogeneous systems,

$$\rho_1(\mathbf{r}_1, \mathbf{r}'_1) = \rho_1(|\mathbf{r}_1 - \mathbf{r}'_1|), \tag{7.71}$$

because of translational and rotational invariance. The fourier transform of $\rho_1(r)$ is the momentum distribution,

$$n(\mathbf{p}) = \langle a^+_{\mathbf{p}} a_{\mathbf{p}} \rangle, \tag{7.72}$$

where $a^+_{\mathbf{p}}$ is a creation operator for a particle in a state of momentum $\mathbf{p}$. Typically, $n(p)$ is calculated by Fourier transforming ρ_1, and rather than enforce the periodicity of the simulation cell, the spherically averaged $\rho_1(r)$ is extrapolated smoothly to large r, and

$$n(\mathbf{p}) = N n_0 \delta_{\mathbf{p},0} + \int e^{-i\mathbf{p}\cdot\mathbf{r}}(\rho_1(r) - \rho n_0) d^3r, \tag{7.73}$$

where n_0 is the condensate fraction, and ρ is the bulk density. The condensate fraction is directly given by the large r behaviour of ρ_1,

$$\lim_{r \to \infty} \rho_1(\mathbf{r}) = \rho n_0. \tag{7.74}$$

The kinetic energy is

$$\frac{\langle p^2 \rangle}{2m} = \frac{\int d^3p n(p) p^2}{2m \int d^3p n(p)}, \tag{7.75}$$

and should agree with direct calculations. The kinetic energy is proportional to the curvature of $\rho_1(r)$ at $r = 0$.

Calculation of ρ_1 for bosons is conceptually quite simple with PIMC. One simply cuts one of the polymer chains and measures the end-to-end distribution of the two cut ends. The condensate fraction is the value of the distribution at large end-to-end separations, divided by its value at the origin. Thus condensation is equivalent to the unbinding of the two ends of a cut polymer. Hence, momentum condensation can only occur when a macroscopic polymer is present. Several tricks can be employed to achieve a more efficient estimation of this fraction. First, it is easy to estimate ρ_1 for small r by displacing an arbitrary atom from the diagonal simulation (no cut ends), and finding the change in the action. Second, when the ends are cut, one should apply importance sampling of the end-to-end distance so the errors coming from small distance approximately equal the errors from large distances. Third, one should preferentially move and permute the cut ends of the polymer. The computation of the momentum distribution in PIMC is inconvenient since a special calculation needs to be performed.

7.5.2 y-Scaling

The main interest in calculating $n(p)$ is to understand, approximately, the dynamic structure factor of quantum many-body systems. The impulse approximation for neutron scattering at large momentum transfers follows from the assumption that a neutron scatters off a single particle, and the outgoing wavefunction is a plane wave. At large momentum transfers, with soft potentials, y-scaling holds. That is, the scattering is proportional to the impulse approximation result, and the dynamic structure factor can be written as

$$S(\boldsymbol{k},\omega) = \frac{m}{q} n_0 \delta(y) + \rho J(y), \tag{7.76}$$

where in the impulse approximation, $J(y)$ is

$$J_{\mathrm{IA}}(y) = \int n((y^2 + p_p)^{1/2}) \frac{d^2 p_p}{(2\pi)^3} = \frac{1}{\pi} \int_0^\infty \rho_1(r) \cos(yr) dr, \tag{7.77}$$

and y is the *West* scaling variable [7.98], i.e. the longitudinal momentum transfer

$$y = \frac{m}{q}\left(\omega - \frac{q^2}{2m}\right). \tag{7.78}$$

The important result is that for large momentum transfers, $S(\boldsymbol{k},\omega)$ does not depend on $\boldsymbol{k}$ and ω independently, but only on y in $J_{\mathrm{IA}}(y)$. For hard-core potentials, final state effects modify the simple impulse approximation result [7.99–102]. *Silver* [7.101, 102] has given a simple prescription for including approximately final state effects by replacing $J_{\mathrm{IA}}(y)$ by a convolution of the longitudinal momentum distribution with a function that depends on the 2-body density matrix, with one particle diagonal and the other off-diagonal,

$$\rho_2(\boldsymbol{r}_1', \boldsymbol{r}_2, \boldsymbol{r}_1, \boldsymbol{r}_2) = \langle \psi^+(\boldsymbol{r}_1')\psi^+(\boldsymbol{r}_2)\psi(\boldsymbol{r}_1)\psi(\boldsymbol{r}_2)\rangle. \tag{7.79}$$

The main correction in this theory is therefore from an additional contribution from a spectator particle carrying off a portion of the momentum. Equation (7.79) is calculated straightforwardly using either PIMC or GFMC [7.94].

Carraro and *Koonin* [7.103] have recently studied final state effects in ^{4}He directly, using a variational wavefunction and the Monte Carlo method. Instead of assuming an outgoing plane wave as in the impulsive approximation, they explicitly solve for the outgoing wavefunction using the static positions of the spectator particles, sampled from a pair product trial function. The method therefore includes multiple scattering corrections to all orders. Because only the longitudinal component of the momentum is important, their method is computationally simple since only a 1D scattering problem needs to be solved for each independent set of sampled particle positions. Their derived effective final state broadening function becomes narrower at lower momentum transfers in contrast to Silver's result. Although they used a variational wavefunction, the incorporation of their method into a path integral or Green's function Monte Carlo calculation would be feasible.

7.5.3 Momentum Distribution Results

Momentum distributions have been calculated, for both ^{3}He and ^{4}He ground-states [7.18, 47] and for ^{4}He at finite temperatures [7.19, 94]. *Kalos* et al. [7.88] and *Whitlock* and *Panoff* [7.18] have calculated the momentum distribution of liquid and solid ^{4}He using GFMC. For the equilibrium density liquid, $\rho = 0.0218\,\text{A}^{-3}$, they obtain a ground-state condensate fraction of $9.2\% \pm 0.1\%$ as compared to the experimental estimate of $10\% \pm 2.0\%$ at $T = 1.5\,\text{K}$ [7.104]. The experimental estimate may be more uncertain than this, due to the extreme difficulty of pulling out the contribution of a delta function from inelastic neutron scattering cross sections which have been broadened by final state effects, instrumental resolution and multiple scattering. The GFMC condensate fraction decreases to $3.8\% \pm 0.2\%$ at $\rho = 0.0262\,\text{A}^{-3}$ [7.88]. An interesting variational result, using the shadow wavefunction, is a condensate fraction of $4.51\% \pm 0.03\%$ at equilibrium density [7.29]. This illustrates the difficulty of drawing firm conclusions about quantities (other than the energy) with the variational method, particularly in regards to long-range order.

Ceperley and *Pollock* [7.19] calculate a condensate fraction at SVP from $T = 1$–$4\,\text{K}$. They get a value of 8% with error bars of order $\pm 1\%$ between $T = 1\,\text{K}$ and $T = 2\,\text{K}$. The condensate fraction drops rapidly above 2 K as the system goes through its lambda point, with results consistent with no condensate, as expected, for temperatures greater than 3 K. Accurate estimation of the

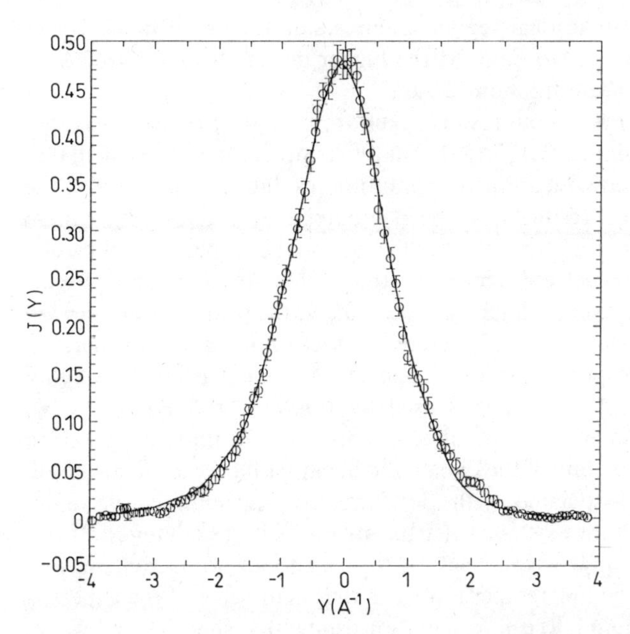

Fig. 7.10. The measured $J(y)$ of (7.76) at 0.32 K (circles) compared with the GFMC ground state result (solid line). The GFMC [7.18] result has been modified as described in the next. Data from *Sokol* et al. [7.105]

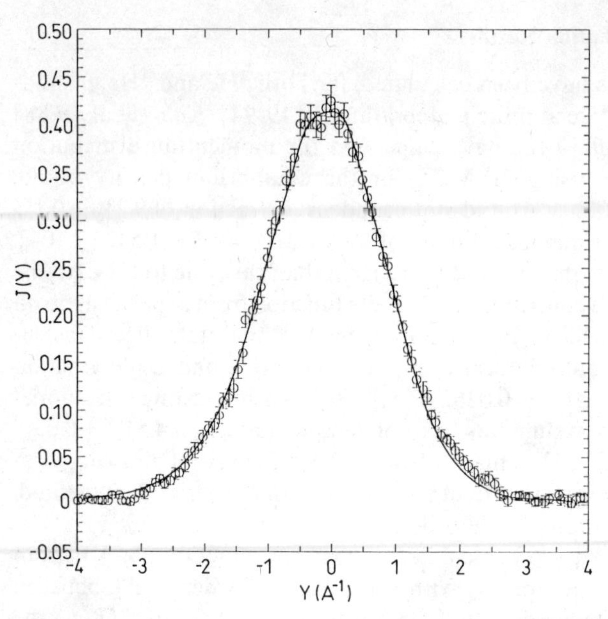

Fig. 7.11. The measured $J(y)$ of (7.76) at 3.33 K (circles) compared with the PIMC result (solid line) [75]. Data from *Sokol* et al. [7.105]

condensate fraction in the critical region requires simulations of much larger systems than have been used to date. At the higher density of $\rho = 0.0262\ \text{A}^{-3}$, PIMC predicts a condensate fraction of 2.6%.

In Figs. 7.10 and 11 we show results taken from [7.105] of the measured $J(y)$ along with the calculated $J(y)$ [7.18, 19] at two temperatures. The calculated $J(y)$ have been convoluted with an instrumentation resolution function, and the zero temperature GFMC results have been modified by convoluting with a final state effects function [7.101]. The results agree within experimental errors. Without the final state effect corrections to the GFMC results, the low temperature experiment and theory differ by about 10% at the peak of $J(y)$. *Carraro* and *Koonin* [7.103] calculate an equally good fit at these momentum transfers.

For Bose solids, the possibility of a condensate has not been thoroughly investigated. The importance functions used by *Whitlock* and *Panoff* [7.18] were of the localised form, and do not allow a condensate. At finite temperature one will have a condensate only if the two ends of a cut path become delocalised. Thus condensation is associated with the formation of bounded vacancy-interstitial pairs and with the existence of arbitrarily long ring exchanges. PIMC calculation find the exchange frequencies for three and four atom exchanges in solid ^{4}He are 1–4 μK. The low temperature and the small value of the expected condensate are discouraging. Recent calculations show that shadow trial wavefunctions do have a non-zero condensate in the solid [7.106], but numerical calculations have not extracted a non-zero value [7.29]. The $\rho_1(r)$ in the solid

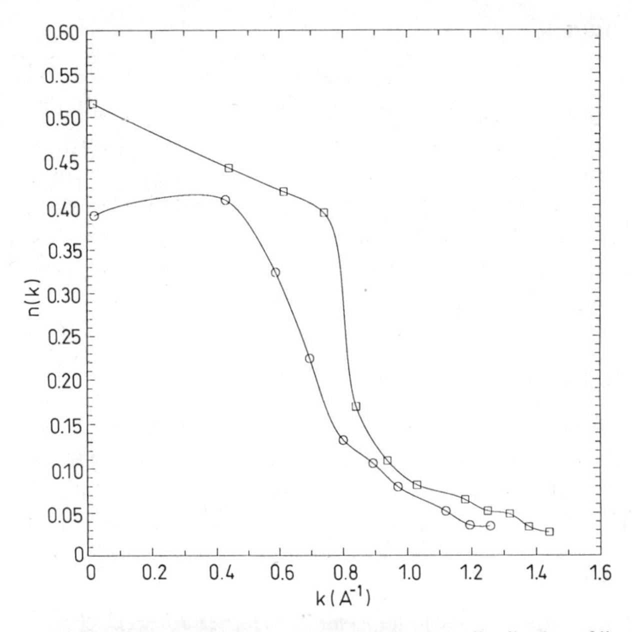

Fig. 7.12 The calculated ground-state momentum distribution of liquid ^{3}He at equilibrium density from fixed-node GFMC [7.18] (squares) and using the pairing wave function of *Bouchaud* and *Lhuillier* [7.47] (circles). The curves are drawn through the points only as a guide

phase is well represented by a gaussian; typically, the moments of ρ_1 deviate by only a few per cent from gaussian behaviour [7.29].

There are few experimental results for liquid ^{3}He and simulations are not as reliable as those for Bose systems. The momentum distributions for the ground-state of ^{3}He liquid have been calculated variationally and using fixed-node GFMC [7.18, 20, 47]. Shown in Fig. 7.12 is the momentum distribution at equilibrium density calculated variationally with the Bouchaud and Lhuillier pairing trial function and with fixed-node GFMC. The GFMC results has a fermi liquid discontinuity at the fermi momentum, but it is unlikely that this type of calculation can break away from the initial assumed symmetry of the model function, a perfect fermi gas. The Bouchaud and Lhuillier pairing form does not have a fermi surface. Unfortunately, final state effects may make ^{3}He neutron scattering experiments insensitive to this difference [7.102].

Sokol et al. report an experimental kinetic energy for ^{3}He at equilibrium density of $8.1^{+1.7}_{-1.3}$ K [7.107], while fixed-node GFMC calculations give 12.28 ± 0.04 K and transient estimates give 12.40 ± 0.10 K [7.21]. The discrepancy here is about 2.5 standard deviations. This may be due to any of the following: statistical errors, the difficulty of extracting the kinetic energy from the experimental neutron scattering, the GFMC extrapolation errors, the fixed-node approximation or lack of convergence of the transient estimation, or simulation size dependence.

7.6 Droplets and Surfaces

Although experimental results for small helium droplets are not yet available, clusters have been seen in rare gas jets [7.108, 109]. The main motivation of studying droplets is to obtain a better understanding of finite systems where physical boundary conditions can be handled exactly; to understand the transition from the finite system to the bulk limit; and to understand liquid-gas interfaces. In the future the same methods will be used to understand systems with pores and other restricted geometries [7.110].

7.6.1 Ground States of He Droplets

Calculations on ^{4}He droplets with N up to 728 were done by *Pandharipande* et al. [7.22, 111] using Green's function Monte Carlo and variational Monte Carlo. Their variational wavefunctions contained 2- and 3-body correlations as discussed in Sect. 7.1. However, for $N \leq 10$ they find that their best variational

Table 7.3. The ground-state energy (in K) and radius (in A) for N atom ^{4}He droplets calculated with various methods. VMC1, VMC2 and VMC3 are the variational results of [7.111, 112, 23] respectively. GFMC is the Green's function Monte Carlo result from [7.22], and DMC is the diffusion Monte Carlo result from [7.23]. Energy error estimates are given in parenthesis. Where more than one energy is given, we have reported the results with the lowest variational value. The PIMC result [7.113] is at a temperature of 0.5 K. The last column is an estimate of the lowest energy excitation with zero angular momentum

N	Calculation	$E_0(K)$	$r_0(A)$	$E_E - E_0(K)$
3	GFMC	−0.0391(1)	5.35	
8	VMC1	−0.5989(8)	—	
8	GFMC	−0.6165(6)	3.19	
20	VMC1	−1.5734(13)	2.27	2.85(1)
20	VMC2	−1.514(3)	2.73	3.42
20	GFMC	−1.627(3)	2.71	
40	VMC1	−2.389(2)	2.54	2.79(4)
40	VMC3	−2.196(1)	2.63	
40	GFMC	−2.487(3)	2.57	
40	DMC	−2.529(3)	2.55	3.53
64	PIMC	−2.81(5)	2.46	
70	VMC1	−3.031(3)	2.48	2.75(15)
70	VMC2	−3.005(3)	2.42	4.22
70	GFMC	−3.12(4)	2.47	
70	DMC	−3.188(2)	2.44	3.91
112	VMC1	−3.498(5)	2.43	2.63(7)
112	VMC3	−3.143(2)	2.49	
112	GFMC	−3.60(1)	2.44	
112	DMC	−3.702(3)	2.40	3.92
240	VMC1	−4.193(5)	2.37	2.64(8)
240	VMC2	−4.192(4)	2.22	2.86
768	VMC1	−4.938(5)	2.32	
∞	Experiment	−7.14	2.22	0.0

results are obtained with the 2-body correlation chosen to have the asymptotic form

$$\lim_{r \to \infty} f(r) = e^{-\kappa r} r^{-1/(N-1)}, \tag{7.80}$$

which has the effect of holding the droplet together. For $N > 10$ they add a one-body correlation between each particle and the centre of mass of the droplet to bind the droplet. Similar calculations have been done recently by *Chin* and *Krotscheck* [7.23], who did variational and DMC calculations of droplets up to $N = 192$ and by *Rama Krishna* and *Whaley* [7.112] who have calculated, using variational Monte Carlo, droplets up to $N = 240$. These latter results used the newer HFDB potential of *Aziz* et al. [7.11] while the others used the HFDHE2 potential [7.8]. This must be taken into account when comparing their results. Chin and Krotscheck used a simple 2-body McMillan, $f(r) = \exp[-\frac{1}{2}(b/r)^5]$, correlation and a gaussian 1-body factor for their importance function. Rama Krishna and Whaley used a wavefunction similar to that of *Pandharipande* et al.

In Table 7.3 we have collected ground-state energies per particle and unit radii, defined to be $\sqrt{(5/3)} r_{\rm rms} N^{-1/3}$, where $r_{\rm rms}$ is the root mean square radius from the centre of mass, for droplets up to $N = 768$. We see that there is reasonable agreement, but discrepancies between the GFMC and DMC results, particular at $N = 112$, may indicate lack of convergence. The variational results using the HFDHE2 potential (VMC1) also are remarkably close to those using the HFDB potential (VMC2). The density profiles of the droplets are much harder to calculate due to statistical errors and slow convergence of the GFMC and DMC methods. The result of Pandharipande et al. and Rama Krishna and Whaley show results consistent with a smooth profile. Chin and Krotscheck show a small oscillatory structure in their DMC density profiles for $N > 70$ which persists for large numbers of iterations. Similar oscillations were seen earlier by Pandharipande et al. In both calculations, large initial oscillations die out. It is not clear how much, if any, of this structure will remain in a completely converged calculation. Some of these effects may also be caused by the extrapolation from the mixed and variational estimates. The PIMC results at finite temperature [7.113] do not have this problem of extrapolation (but are at finite temperature though a superfluid has few excitations) and do not show the same structure as Chin's density profile.

Pandharipande et al. have fit their GFMC energies and those of the bulk [7.88] as

$$\frac{E}{N}(K) = -7.10 + 17.6x + 1.15x^2 - 30.6x^3 + 19.5x^4, \tag{7.81}$$

where $x = N^{-1/3}$. Taking the derivative and assuming a liquid drop radius, they extract a surface tension of 0.28 K A^{-2} as compared to an experimental values of 0.274 K A^{-2} [7.114] and 0.265 K A^{-2} [7.115].

7.6.2 Excitations in Droplets

All three groups have calculated an upper bound to the low energy excitations of ^{4}He droplets using the Feynman excitation operator method [7.116]. They assume the excited state wavefunction can be approximated by

$$\Psi_E = \sum_i \chi(\boldsymbol{r}_i - \boldsymbol{r}_{cm}) \Psi_0 = F \Psi_0,$$

where Ψ_0 is the exact ground-state wavefunction, and $\boldsymbol{r}_{cm}$ is the centre of mass of the droplet. χ is a function to be determined variationally. A bound on the excited state energy E_E is written as

$$E_E - E_0 = \frac{1}{2} \frac{\langle \Psi_0 | [F,[T,F]] | \Psi_0 \rangle}{\langle \Psi_0 | F^2 | \Psi_0 \rangle}, \tag{7.82}$$

where T is the kinetic energy operator. Equation (7.82) is the ratio of two ground-state expectation values and can be written in terms of ground-state distribution functions. Typically, χ is expanded is spherical harmonics and (7.82) minimised, subject to the constraint that Ψ_E is orthogonal to the lower energy states, to obtain approximate excited state wavefunctions.

The calculations of *Pandharipande* et al. [7.111] and *Rama Krishna* and *Whaley* [7.112] have used the variational wavefunction as an approximation to the ground state, while Chin and Krotscheck have used their DMC results for the distribution functions. A comparison of some excitation energies of the lowest angular momentum zero states are shown in Table 7.3. We see that there is qualitative agreement, but little quantitative agreement between the calculations. Calculations of other states have been done by each of these groups. The results of Chin and Krotscheck are probably the most reliable, because of their use of the diffusion Monte Carlo ground-state wavefunction. Calculations with backflow correlations in the excitation operator are more difficult, and have not yet been done, but would presumably give better excitation energies.

7.6.3 ^{3}He Droplets

The ground states of ^{3}He droplets have been studied by *Pandharipande* et al. [7.111] and *Lewart* et al. [7.117] using variational Monte Carlo and wavefunctions containing 2-body, 3-body, and backflow correlations of the type described in Sect. 7.1. They fit their variational ^{3}He energies to polynomials of the form of (7.81). Their results indicate that the 20 particle droplet is not bound, but is in a metastable state. The 20 particle droplet was found to be unbound in an earlier fixed-node GFMC calculations [7.118]. Fixed-node Green's function Monte Carlo calculations indicate that the 70 particle variational energy is only about 0.1 K too high per particle. The single particle density shows shell structure expected in a fermi droplet.

Lewart et al. calculate the single particle density matrix (see Sect. 7.5) of both ^{4}He and ^{3}He droplets using variational Monte Carlo. They also define and calculate the quasi-particle wavefunctions defined by

$$\phi_{qp}(r_1) \alpha \int \Psi_h^*(R) \Psi_0(R) d^3r_2 \cdots d^3r_N \tag{7.83}$$

where Ψ_h is a single hole state of $N-1$ particles, and is approximated by omitting a particle and an orbital from the model function. For bulk systems, the orbitals in the Slater determinant, the quasiparticle wavefunctions, and the natural orbitals, ψ_i, defined by diagonalising the single particle density matrix,

$$\rho_1(\boldsymbol{r}'_1, \boldsymbol{r}_1) = \sum_i n_i \psi_i^*(\boldsymbol{r}_1)\psi_i(\boldsymbol{r}'_1) \tag{7.84}$$

are all plane waves. For droplets these functions are all distinct, and related to different excitation properties. The n_i in (7.84) are the analog of the momentum distributions, and n_0 is the generalisation, in a droplet, of the condensate fraction.

7.6.4 Droplets at Finite Temperature

Helium-4 droplets of 64 and 128 atoms have been studied at temperatures below 2 K [7.113]. The specific heat shows a broadened transition from the normal liquid to the superfluid as expected. The normal (i.e. non-superfluid) density is defined as the ratio of the moment of inertia of the droplet to its classical value defined in terms of the radius of gyration as discussed in Sect. 7.3.10. The superfluid density is then calculated as proportional to the mean squared area swept out by a path. Most of the droplet becomes superfluid below 1.5 K. The density profiles are in reasonable agreement with those calculated with GFMC.

7.6.5 Surfaces and Interfaces

Helium adsorbed on substrates can be simulated directly once a helium-substrate potential is assumed. One approximation for monolayer systems is to assume that the substrate strongly binds the helium, and view the monolayer as a 2D film. A number of calculations have been done on 2D helium atoms at both zero and finite temperatures [7.0, 119–122]. *Whitlock* et al. [7.119] have found that the liquid ^{4}He freezes at a density of 0.0678 A^{-2} and melts at a density of 0.0721 A^{-2} at zero temperature. The zero pressure, zero temperature density of the liquid is 0.04356 A^{-2}. At this density the condensate fraction is approximately 0.25, much larger than in 3-dimensions due to the reduced interatomic spacing.

The PIMC method has also been applied to 2D helium [7.92]. At zero temperature, the 2D and 3D systems are similar, but at any non-zero temperature the 2D system is described by the Kosterlitz–Thouless picture of a vortex unbinding transition. According to this theory, the single particle density matrix decays algebraically with a temperature dependent power, instead of to a constant as it does in 3D. In addition, the superfluid density should jump from zero at high temperature to a finite value at the critical temperature. These effects are what are observed in the simulations, although finite-size scaling is needed to understand the rounding that occurs in a simulation of a finite system. At a density of 0.0432 A^{-2}, the superfluid transition is predicted by the PIMC calculations to occur at a temperature of 0.72 K. The vortex diameter and the core energy is estimated to be 3.7 A and 2.7 K respectively.

Approximating the effective two-body potential in a monolayer as the bare potential is equivalent to assuming that the atomic separations in the monolayer are much greater than the thickness of the layer. A somewhat better approximation can be made by assuming a substrate potential, and using its ground state to define the layer profile and the effective interaction in the 2D system. However, for a model graphite substrate, the effect on the energy difference is only about 0.05 K per particle [7.123]. A discussion of some other effects of the 2D approximation is given by *Cheng* and *Cole* [7.124].

The surface tension and surface profile has been calculated for a self-bound slab of ^{4}He by *Valles* and *Schmidt* [7.24]. They get a value for the surface tension of $0.265\,\mathrm{K\,A}^{-2}$ as compared to an experimental values of $0.274\,\mathrm{K\,A}^{-2}$ [7.114] and $0.265\,\mathrm{K\,A}^{-2}$ [7.115], and the value, given above, extracted from droplet data of $0.28\,\mathrm{K\,A}^{-2}$.

An area where much further work needs to be done is the calculation of the properties of helium on realistic substrates. An interesting calculation studying the registered phases of a monolayer of ^{3}He on graphite is given by *Abraham* and *Broughton* [7.125] using a PIMC. *Abraham* et al. [7.126] have studied with VMC and PIMC, the detailed structure of two solid players of ^{3}He on graphite in an attempt to understand the unusual phase transitions that occur at this density. They find that the proposed $\sqrt{7} \times \sqrt{7}$ registered structure is stable at low temperatures and melts at approximately 1 K.

7.7 Future Prospects

GFMC and PIMC methods have given us the ability to calculate from first principles, properties of many-body boson systems, at zero and non-zero temperature with computable errors. What are some of the challenging problems given that the basic equilibrium situation for bosons is in good shape? Some of the obvious applications for many-boson systems yet to be investigated are bosons in disordered environments or on surfaces, impurities in helium, and the study of excitations such as vortices, rotons and maxons.

One of the major unsolved problems in computational physics is to devise a method to simulate large many-fermion systems without uncontrolled approximations. The auxiliary field (Stratonovich–Hubbard transformation) method does just that for the Hubbard model at half filling on a bipartite lattice. For general models, the most successful ground state methods use a fixed-node approximation, but the bias introduced by the trial nodes restricts the applicablity of the method to systems where the long-range order is well understood. Currently the path integral fixed-node method is under investigation for the simulation of fermion continuum systems at finite temperature. One chooses a trial density matrix and throws out walks which cross the nodes of the trial density matrix. If the nodes are chosen correctly then the method is exact. This method will be useful at temperatures larger than the fermi energy where the nodes are closely related to the classical nodes, and goes over smoothly to the ground state fixed-node method. Again it is only applicable if the long-range structure of the density matrix is understood. Thus the fermion simulation problem, to calculate the properties of a many-body fermion systems in time which is a power of the number of fermions and the inverse temperature, is very largely unsolved.

A second major unsolved problem is to calculate dynamical properties of quantum systems. To calculate general dynamical properties seems very difficult. Even specifying the initial conditions would require an exponentially large amount of data. Luckily, most interesting questions come down to equilibrium dynamics in the linear response regime. In PIMC one can easily compute imaginary time correlation functions. The problem is to rotate it back into real time. A simple example is the dynamical density–density response function, $S(k, \omega)$. Its Laplace transform can be calculated in either DMC or PIMC as

$$F(k,t) = \langle \rho_k(0)\rho_{-k}(t)\rangle = \int_{-\infty}^{\infty} d\omega e^{-\omega t} S(k,\omega),$$

where $\rho_k(t)$ means the phonon operator (7.20) evaluated at imaginary time t. However the inversion to obtain S from F is numerically unstable. Recently maximum entropy methods [7.127] have been applied to this inversion procedure with some success for lattice models. These methods combine highly accurate Monte Carlo generated imaginary time response functions with any theoretical input that may be available, and effectively give the most likely dynamical response function consistent with all the available data. We anticipate that the method will have broad application in understanding equilibrium dynamics of quantum systems.

Acknowledgements. DMC is supported by the National Science Foundation through grant number NSF DMR88-08126 and the department of physics at the University of Illinois. KES is supported by the National Science Foundation through grant number NSF CHE90-15337.

References

7.1 D.M. Ceperley, M.H. Kalos: In *Monte Carlo Methods in Statistics Physics*, Topics Curr. Phys. Vol. 7, ed. by K. Binder (Springer, Berlin, Heidelberg 1979)
7.2 K.E. Schmidt, M.H. Kalos: In *Monte Carlo Methods in Statistical Physics II*, Topics Curr. Phys. Vol. 36, ed. by K. Binder (Springer, Berlin, Heidelberg 1984)
7.3 S.C. Pieper, R.B. Wiringa, V.R. Pandharipande: Phys. Rev. Lett. **64**, 365 (1990)
7.4 J. Carlson: Phys. Rev. C **36**, 2026 (1987)
7.5 J. Carlson: Phys. Rev. C **38**, 1879 (1988)
7.6 J.P. Bouchaud, C. Lhuillier: Europhys. Lett. **3**, 1273 (1987)
7.7 S. Vitiello, K.J. Runge, M.H. Kalos: Phys. Rev. Lett. **60**, 1970 (1988)
7.8 R.A. Aziz, V.P.S. Nain, J.S. Carley, W.L. Taylor, G.T. McConville: J. Chem. Phys. **70**, 457 (1974)

7.9 D.M. Ceperley, H. Partridge: J. Chem. Phys. **84**, 820 (1986)
7.10 R. Feltgen, H. Kirst, K.A. Köhler, H. Pauly, F. Torello: J. Chem. Phys. **76**, 2360 (1982)
7.11 R.A. Aziz, F.R.W. McCourt, C.C.K. Wong: Mol. Phys. **61**, 1487 (1987)
7.12 B. Liu, A.D. McLean: J. Chem. Phys. **91**, 2348 (1989)
7.13 P.A. Whitlock, D.M. Ceperley, G.V. Chester, M.H. Kalos: Phys. Rev. B **19**, 5598 (1979)
7.14 B.M. Axilrod, E. Teller: J. Chem. Phys. **11**, 293 (1943)
7.15 R.E. Lowther, R.L. Coldwell: Phys. Rev. A **22**, 14 (1980)
7.16 V. Mohan, J.B. Anderson: In *Quantum Simulations of Condensed Matter Phenomena*, ed. by J.D. Doll, J.B. Gubernatis (World Scientific, Singapore 1990)
7.17 G.J. Tawa, P.A. Whitlock, J.W. Moskowitz, K.E. Schmidt: to be published in Int. J. Sup. Appl. (1990)
7.18 P.A. Whitlock, R.M. Panoff: Can. J. Phys. **65**, 1409 (1987)
7.19 D.M. Ceperley, E.L. Pollock: Can J. Phys. **65**, 1416 (1987)
7.20 D.M. Ceperley, M.H. Kalos, G.V. Chester: Phys. Rev. B **16**, 3081 (1977)
7.21 R.M. Panoff, J. Carlson: Phys. Rev. Lett. **62**, 1130 (1989)
7.22 V.R. Pandharipande, J.G. Zabolitzky, S.C. Pieper, R.B. Wiringa, U. Helmbrecht: Phys. Rev. Lett. **50**, 1676 (1983)
7.23 S.A. Chin, E. Krotscheck: Phys. Rev. Lett. **65**, 2658 (1990)
7.24 J.L. Valles, K.E. Schmidt: Phys. Rev. B **38**, 2879 (1988)
7.25 K. Binder, D.W. Heermann: In *Monte Carlo Simulation in Statistical Physics*, Topics Curr. Phys. Vol. 7 (Springer, Berlin, Heidelberg 1988)
7.26 K. Binder, D. Stauffer: In *Monte Carlo Methods in Statistical Physics II*, Topics Curr. Phys. Vol. 36, ed. by K. Binder (Springer, Berlin, Heidelberg 1984)
7.27 N. Metropolis, A.W. Rosenbluth, H. Rosenbluth, A. Teller, E. Teller: J. Chem. Phys. **21**, 1087 (1953)
7.28 R.P. Feynman, M. Cohen: Phys. Rev. **102**, 1189 (1956)
7.29 S. Vitiello, K.J. Runge, G.V. Chester, M.H. Kalos: Phys. Rev. **42**, 228 (1990)
7.30 E. Saunders: Phys. Rev. **126**, 1724 (1962)
7.31 L.H. Nosanow: Phys. Rev. Lett. **13**, 270 (1964)
7.32 L.H. Nosanow: In *Quantum Solids and Fluids*, ed. by S.B. Trickey, E.D. Adams, J.W. Duffy (Plenum, New York 1977)
7.33 D. Ceperley, G.V. Chester, M.H. Kalos: Phys. Rev. B **17**, 1070 (1978)
7.34 R.M. Panoff: In *Condensed Matter Theories* (Plenum, New York 1987)
7.35 R.A. Aziz, P.K. Pathria: Phys. Rev. A **7**, 827 (1972)
7.36 P.R. Roach, S.B. Ketterson, C.-W. Woo: Phys. Rev. A **2**, 543 (1970)
7.37 C.-W. Woo: Phys. Rev. Lett. **29**, 1442 (1972)
7.38 C.C. Chang, C.E. Campbell: Phys. Rev. B **15**, 4238 (1977)
7.39 V.R. Pandharipande: Phys. Rev. B **18**, 218 (1978)
7.40 K.E. Schmidt, V.R. Pandharipande: Phys. Rev. B **19**, 2504 (1979)
7.41 K.E. Schmidt, M.H. Kalos, M.A. Lee, G.V. Chester: Phys. Rev. Lett. **45**, 573 (1980)
7.42 D. Levesque, C. Lhuillier: Phys. Rev. B **23**, 2203 (1981)
7.43 Q.N. Usmani, S. Fantoni, V.R. Pandharipande: Phys. Rev. B **26**, 6123 (1982)
7.44 K.E. Schmidt, J.W. Moskowitz: J. Chem. Phys. **93**, 4172 (1990)
7.45 M.P. Allen, D.J. Tildesley: *Computer Simulation of Liquids* (Oxford University Press, Oxford 1987)
7.46 J.P. Bouchaud, C. Lhuillier: J. Physique (Paris) **49**, 553 (1988)
7.47 J.P. Bouchaud, C. Lhuillier: In *Spin Polarised Quantum Systems*, ed. by S. Stringari (World Scientific 1989)
7.48 J. Bardeen, L.N. Cooper, J.R. Schrieffer: Phys. Rev. **108**, 1175 (1957)
7.49 T. Muir: *A Treatise on the Theory of Determinants* (Dover, New York 1960)
7.50 W. Wu, S.A. Vitiello, L. Reatto: To be published in the *Proceedings of the Elba International Physics Center* (Giardini, Rome 1990)
7.51 W. Wu, S.A. Vitiello, M.H. Kalos: Submitted to Phys. Rev. Lett. (1990)
7.52 W.G. Stirling: In *Proceedings of the second international conference on phonon physics*, ed. by J. Kollar (World Scientific, Philadelphia, 1985), p. 829
7.53 O.W. Dietrich, E.H. Graf, C.H. Huang, L. Passel: Phys. Rev. A **5**, 1377 (1972)
7.54 C.J. Umrigar, K.G. Wilson, J.W. Wilkins: Phys. Rev. Lett. **60**, 1719 (1988)
7.55 C.J. Umrigar, K.G. Wilson, J.W. Wilkins: In *Computer Simulation Studies in Condensed Matter Physics: Recent Developments*, ed. by D.P. Landau, K.K. Mon, H.B. Schuttler (Springer, New York 1988)

7.56 K.E. Schmidt, S. Vitiello: In *Condensed Matter Theories* (Plenum, New York 1989)
7.57 B. Bernu, D.M. Ceperley, V.W. Laster, Jr.: J. Chem. Phys. **93**, 552 (1990)
7.58 E. Feenberg: *Theory of Quantum Fluids* (Academic, New York 1969)
7.59 V.R. Pandharipande, H.A. Bethe: Phys. Rev. C **7**(4), 1312 (1973)
7.60 K.E. Schmidt: In *Models and Methods in Few-Body Physics*, ed. by A. Fonseca, Lecture Notes in Physics (Springer, Berlin, Heidelberg 1987)
7.61 S. Chin: Texas A and M preprint (1989)
7.62 M.H. Kalos, J. Carlson: Phys. Rev. C **32**, 1735 (1985)
7.63 R.L. Stratonovich: Sov. Phys. Dokaldy **2**, 416 (1957)
7.64 J. Hubbard: Phys. Rev. Lett. **3**, 77 (1959)
7.65 G. Sugiyama, S.E. Koonin: Ann. Phys. **168**, 1 (1986)
7.66 S.B. Fahy, D.R. Hamann: Phys. Rev. Lett. **65**, 3437 (1990)
7.67 W. Wu, S. Vitiello, K.J. Runge, MH. Kalos: Preprint (1990)
7.68 S. Vitiello, K.J. Runge, in *Computer Simulation Studies in Condensed Matter Physics: Recent Developments*, Eds. D.P. Landau, K.K. Mon, H.B. Schuttler (Springer, Berlin, Heidelberg 1988)
7.69 R.P. Feynman: *Statistical Mechanics* (Addison-Wesley, Reading 1972)
7.70 L.D. Landau, E.M. Lifshitz: *Statistical Physics* (Addision-Wesley, Reading 1969)
7.71 E.L. Pollock, D.M. Ceperley: Phys. Rev. B **30**, 2555 (1984)
7.72 J.A. Barker: J. Chem. Phys. **70**, 2914 (1979)
7.73 A.D. Klemm, R.G. Storer: Aust. J. Phys. **26**, 43 (1973)
7.74 D.M. Ceperley, E.L. Pollock: Path-Integral Computation Techniques for Superfluid 4He, to be published in *Proceedings of the Workshop on Monte Carlo Methods in Theoretical Physics, Isla D'Elba, Italy*, ed. by S. Fanstoni, Giordini Publishing, 1991
7.75 D.M. Ceperley, E.L. Pollock: Phys. Rev. Lett. **56**, 351 (1986)
7.76 K. Runge, G.V. Chester: Phys. Rev. B **38**, 135 (1988)
7.77 J.D. Doll, R.D. Coalson, D.L. Freeman: Phys. Rev. Lett. **55**, 1 (1985)
7.78 M. Takahashi, M. Imada: J. Phys. Soc. Jpn. **53**, 963 (1984)
7.79 M. Sprik, M.L. Klein, D. Chandler: Phys. Rev. B **31**, 4234 (1985)
7.80 E.J. Brushkin, M.F. Herman, B.J. Berne: J. Chem. Phys. **76**, 5150 (1982)
7.81 M. Parrinello, A. Rahman: J. Chem. Phys. **80**, 860 (1984)
7.82 E.L. Pollock, D.M. Ceperley: Phys. Rev. B **36**, 8343 (1987)
7.83 G.G. Batrouni, R.T. Scalettar, G.T. Zimanyi: Phys. Rev. Lett. **65**, 1765 (1990)
7.84 M. Roger, J. Hetherington, J.M. Delrieu: Rev. Mod. Phys. **55**, 1 (1983)
7.85 M.C. Cross, D.S. Fischer: Rev. Mod. Phys. **57**, 881 (1985)
7.86 C.H. Bennett: J. Comput. Phys. **22**, 245 (1976)
7.87 B.J. Berne, D. Thirumalai: Ann. Rev. Phys. Chem. **37**, 401 (1986)
7.88 M.H. Kalos, M.A. Lee, P.A. Whitlock, G.V. Chester: Phys. Rev. B **24**, 115 (1981)
7.89 P.A. Whitlock, K.E. Schmidt, M.A. Lee: unpublished
7.90 D.O. Edwards, R.C. Pandroff: Phys. Rev. (1981) A **140**, 816 (1965)
7.91 V.F. Sears, E.C. Svensson, A.D.B. Woods, P. Martel: *Atomic Energy of Canada Limited Report No. AECL-6779* (unpublished)
7.92 D.M. Ceperley, E.L. Pollack: Phys. Rev. B **39**, 2084 (1989)
7.93 K.E. Schmidt, M.A. Lee, M.H. Kalos, G.V. Chester: Phys. Rev. Lett. **47**, 807 (1981)
7.94 D.M. Ceperley: In *Momentum Distributions*, ed. by R.E. Silver, P.E. Sokol (Plenum Press, New York 1989)
7.95 K.J. Runge, G.V. Chester: Phys. Rev. B **39**, 2707 (1989)
7.96 A. Meroni, L. Reatto, K.J. Runge: In *Condensed Matter Theories* (Plenum, New York 1989)
7.97 D.M. Ceperley, G. Jacucci: Phys. Rev. Lett. **58**, 1648 (1987)
7.98 G.B. West: Phys. Rept. **18C**, 263 (1975)
7.99 V.F. Sears: Phys. Rev. B **30**, 44 (1984)
7.100 J.J. Weinstein, J.W. Negele: Phys. Rev. Lett. **49**, 1016 (1982)
7.101 R.N. Silver: Phys. Rev. B **38**, 2283 (1988)
7.102 R.N. Silver: In *Momentum Distributions*, ed. by R.E. Silver, P.E. Sokol (Plenum Press, New York 1989)
7.103 C. Carraro, S.E. Koonin: Phys. Rev. Lett. **65**, 2792 (1990)
7.104 H.A. Mook: Phys. Rev. Lett. **51**, 1454 (1983)
7.105 P.E. Sokol, T.R. Sosnick, W.M. Snow: In *Momentum Distributions*, ed. by R.E. Silver, P.E. Sokol (Plenum Press, New York 1989)
7.106 L. Reatto, G.L. Masserini: Phys. Rev. B **38**, 4516 (1988)

7.107 P.E. Sokol, K. Sköld, D.L. Price, R. Kleb: Phys. Rev. Lett. **54**, 909 (1985)
7.108 J. Fargo, M.F. de Feraudy, B. Raoult, G. Torchet: J. Physique (Paris), Colloq. **38**, C2 (1977)
7.109 O. Echt, K. Saltler, E. Rechnagel: Phys. Rev. Lett. **47**, 1121 (1981)
7.110 F.M. Gasparini, I. Rhee: to be published in *Prog. in Low Temp. Phys.* Vol. **XIII** (1990)
7.111 V.R. Pandharipande, S.C. Pieper, R.B. Wiringa: Phys. Rev. B **34**, 4571 (1986)
7.112 M.V. Rama Krishna, K.B. Whaley: J. Chem. Phys. **93**, 746 (1990)
7.113 P. Sindzingre, M.L. Klein, D.M. Ceperley: Phys. Rev. Letts. **63** 1601 (1989)
7.114 H.M. Guo, D.O. Edwards, R.E. Sarwinski, J.T. Tough: Phys. Rev. Lett. **27**, 1259 (1971)
7.115 M. Iino, M. Suzuki, A.J. Ikushima: J. Low Temp. Phys. **61**, 155 (1985)
7.116 R.P. Feynman: Phys. Rev. **94**, 262 (1954)
7.117 D.S. Lewart, V.R. Pandharipande, S.C. Pieper: Phys. Rev. B **37**, 4950 (1988)
7.118 K.E. Schmidt: In *Monte Carlo Methods in Quantum Problems*, Proceedings of the NATO-ARW, ed. by M.H. Kalos (Reidel, Dordrecht 1984)
7.119 P.A. Whitlock, G.V. Chester, M.H. Kalos: Phys. Rev. B **38**, 2418 (1988)
7.120 T.C. Padmore: Phys. Rev. Lett. **32**, 826 (1974)
7.121 K.S. Liu, M.H. Kalos, G.V. Chester: Phys. Rev. B **13**, 1971 (1976)
7.122 X.-Z. Ni, L.W. Bruch: Phys. Rev. B **33**, 4584 (1986)
7.123 P.A. Whitlock, K.E. Schmidt: to be published
7.124 E. Cheng, M.W. Cole: Submitted to Phys. Rev. B (1990)
7.125 F.F. Abraham, J.Q. Broughton: Phys. Rev. Lett. **59**, 64 (1987)
7.126 F.F. Abraham, J.Q. Broughton, P.W. Leung, V. Elser: Europhys. Lett. **12**, 107 (1990)
7.127 R.N. Silver, D.S. Sivia, J.E. Gubernatis: Phys. Rev. B **41**, 2380 (1990)

8. Quantum Lattice Problems

Hans De Raedt and Wolfgang von der Linden

With 1 Figure

In this chapter we review methods currently used to perform Monte Carlo calculations for quantum lattice models. A detailed exposition is given of the formalism underlying the construction of the simulation algorithms. We discuss the fundamental and technical difficulties that are encountered and give a concise survey of applications of quantum Monte Carlo methods to various quantum statistical lattice problems. New developments are illustrated and discussed in terms of two many-body problems, the Hubbard and the Heisenberg model.

8.1 Overview

A vast amount of Monte Carlo (MC) simulations for quantum many-body systems deals with continuum models. Research along this line has already been reviewed in the two previous volumes of this series [8.1, 2]. The discovery of high T_c materials has increased the interest in applying Monte Carlo methods to quantum many-body problems on lattices, for which it is difficult to get reliable results by other means. Two prominent quantum lattice models are the $(S=1/2)$ Heisenberg and Hubbard model. In practice, the formalism employed to convert the quantum lattice problem (described by operators) to a problem (described in terms of numbers) amenable to numerical treatment differ considerably from the methods developed for continuum systems although on an abstract level they are intimately related.

In this chapter we give a detailed account of the various MC algorithms for simulating quantum lattice models, thereby emphasising recent developments. Reviews of earlier applications of quantum Monte Carlo (QMC) methods to various lattice problems can be found elsewhere [8.3, 4]. After the writing of these reviews, several new algorithms for simulating lattice fermions and quantum spin systems have been proposed and implemented. A large part of this chapter is devoted to an exposition of these techniques.

Motivated by experiments on high-T_c compounds and made possible by the power of present day supercomputers, a number of researchers have used the quantum Monte Carlo method to gain more insight into the properties of the 2D $(S=1/2)$ Heisenberg and 2D (extended) Hubbard model. This chapter also gives a survey of the results obtained with the use of Monte Carlo simulations and it is organised as follows. The next section briefly introduces

Table 8.1. Abbreviations used in this chapter

QMC	quantum Monte Carlo
VMC	variational Monte Carlo
GFMC	Green's Function QMC
GCMC	Grand Canonical QMC
WLQMC	Worldline QMC
E	ground state energy
M	magnetic properties
sc	superconducting properties
EHM	extended Hubbard model
RG	renormalisation Group
e–ph	electron–phonon coupling
(OD) LRO	(off diagonal) long-range order
SDW	spin density wave
CDW	charge density wave
SR	short-range
LR	long-range
HS	Hubbard–Startonovich
LD	Langevin Dynamics
$g(\tau)$	imaginary time Green's function
$A(\omega)$	spectral density
$n(r)$	real-space density
$n(k)$	momentum distribution
$\sum(\omega)$	self-energy
$\psi(\boldsymbol{q})$	excitation spectrum
N	number of sites
n	number of electrons

the model Hamiltonians which will be used to illustrate the construction of QMC algorithms. Sections 8.3 and 8.4 deal with quantum spin systems and cover the topics variational Monte Carlo (VMC) and Green's function Monte Carlo (GFMC), respectively. In Sect. 8.3 trial wavefunctions are presented which are also needed in Sect. 8.4. Each section contains a concise summary of the results for the Heisenberg antiferromagnet. In Sect. 8.5 we sketch the theory on which the grand-canonical finite-temperature quantum Monte Carlo (GCMC) method for lattice fermions is based and describe the results obtained with its use. The subsequent section contains an exposition of the formalism that lies at the heart of a projector quantum Monte Carlo (PQMC) method for fermion systems, followed by a survey of results obtained by means of this technique. In Sect. 8.7 we discuss several conceptual and fundamental problems encountered in QMC simulations. This information is useful in assessing the reliability of QMC results. Abbreviations used in this chapter are given in Table 8.1. Table 8.2 provides an example of results obtained by these various methods and Table 8.3 (at the end of this chapter) gives a list of almost all publications concerning QMC calculations for the models which will be discussed in this chapter. It also gives references to useful review articles and papers containing algorithmic details for further reading.

Table 8.2. Ground state energies and staggered magnetisation for the spin $-1/2$ Heisenberg antiferromagnet. (*) refers to results obtained for $p = 2, 4, 5$ and SR (LR) stands for short-range (long-range) trial functions (see text)

Method	$E_0/(NzJ)$	m
Variational (one parameter Jastrow)		
v.d. Linden–Ziegler–Horsch (exact) [8.10]	−0.3221	0.42
Huse–Elser [8.53]	−0.332	0.42
Trivedi–Ceperley [8.56]	−0.3237	0.40
Variational (Jastrow)		
Huse–Elser [8.53]	−0.3319	0.36
Trivedi–Ceperley [8.56]	−0.3295	0.32
Horsch–v.d. Linden [8.58]	−0.327	0.33
Manousakis [8.57]	−0.3318	0.35
Variational (GWA)		
Horsch–Kaplan (paramagnetic) [8.42]	−0.275	0.00
Yokoyama–Shiba(SDW) [8.59]	−0.321	0.43
Variational (RVB)		
Liang–Doucot–Anderson (SR) [8.54]	−0.3325	0.00
(*) Liang–Doucot–Anderson (LR) [8.54]	−0.3341, −0.3344, −0.3337	0.38, 0.23, 0.15
Gros–Joint–Rice [8.43, 44]	−0.319	0.00
Worldline, Grand Canonical QMC		
Reger–Riera–Young [8.79, 80]	−0.3350	0.31
Dopf–Rahm–Muramatsu [8.94]	—	0.31
GFMC		
Barnes–Swanson [8.73]	−0.3363	—
Gros–Sanchez–Siggia [8.64]	−0.3330	0.29
Trivedi–Ceperley ($\psi^{(J,1)}$ [8.56, 63]	−0.3346	0.37
Trivedi–Ceperley ($\psi(^{(J,2)}$ [8.56, 63]	−0.3346	0.31
Carlson [8.65]	−0,3346	0.34

8.2 Models

The single-band Hubbard model is the simplest model for interacting electrons. The mere on-site interaction is certainly a crude approximation to the actual Coulomb interaction, but it already bears some of the essential features of strongly correlated electrons. The Hubbard model was originally introduced [8.5] to study the metal-insulator transition and the itinerant magnetism in narrow bands. During the last years it has gained considerable attention in connection with the high temperature superconductors. The model is defined by

$$H = t \sum_{\langle i,j \rangle} (c_{i,\sigma}^{+} c_{j,\sigma} + c_{j,\sigma}^{+} c_{i,\sigma}) + U \sum_{i} n_{i\downarrow} n_{i\uparrow}, \tag{8.1}$$

where $c_{i,\sigma}^{+}(c_{i,\sigma})$ creates (annihilates) a fermion of spin $\sigma = \uparrow, \downarrow$ at site i. t is the hopping matrix element, U represents the on-site Coulomb interaction strength, and the sum over i and j is restricted to nearest neighbours. The number of sites will be denoted by N.

In one spatial dimension many physical properties of model (8.1) are accessible via the Bethe ansatz [8.6]. In more than one dimension, only a few rigorous results are known and a variety of (uncontrolled) approximations have been employed to study the properties of the Hubbard model. It is generally accepted that in two or three dimensions the ground state of the half-filled repulsive ($\infty > U > 0$) model exhibits anti-ferromagnetic long-range order. Addition or removal of electrons (holes) destroys the long-range order. From the pioneering work of *Nagaoka* [8.7], it is known that the ground state is ferromagnetic for $U = \infty$ and one hole in the otherwise half-filled band. Whether the system is still ferromagnetic for a finite concentration of holes in the thermodynamic limit is a longstanding question [8.8–10]. In the opposite limit, at low electron concentration, the ground state is non-magnetic according to *Kanamori* [8.11]. In the attractive case ($U < 0$) the electrons form singlet bound states [8.12].

An extension of the Hubbard model, which is presently being studied most intensively in the context of the high temperature superconductivity, is the three-band Hubbard model (or Emery model) [8.13]. The corresponding Hamiltonian consists of a kinetic term, which describes the hopping between nearest neighbour copper-oxygen sites and nearest neighbour oxygen–oxygen sites, with different hopping integrals $t_{\mathrm{Cu-O}}, t_{\mathrm{O-O}}$. The interaction is still on-site with different coupling strengths for copper or oxygen sites $U_{\mathrm{Cu}}, U_{\mathrm{O}}$. An overview over present developments for the (extended) Hubbard model can be found in [8.12, 14].

The fact that the 2D (and 3D) Hubbard model has resisted accurate determination of various physical quantities by analytical approaches suggests that computer simulation might be an alternative approach to study this model. However, it turns out that the construction of such simulation algorithms poses great challenges, as a number of conceptual and technical problems appear which are difficult to overcome. Also, in this respect the Hubbard model can be considered as the simplest model offering the full spectrum of computational difficulties.

With this in mind we will give an exposition of the theoretical concepts underlying the algorithms using the single-band Hubbard model as an example. Extensions of the algorithms to other fermionic systems such as extended Hubbard models [8.15–23] or the Anderson impurity model [8.24–33] are usually straightforward although the practical realisation is not.

The Heisenberg system is a fundamental model in the theory of magnetism and is defined by the Hamiltonian

$$H = J \sum_{\langle i,j \rangle} \mathbf{S}_i \cdot \mathbf{S}_j, \tag{8.2}$$

where $S_i^\alpha (\alpha = x, y, z)$ is the α-th component of the spin-operator and J is the exchange interaction. The Bethe ansatz provides the exact solution for various physical properties of the 1D Heisenberg model. As to the spin ordering in the 1D problem, the ground state for ferromagnetic coupling ($J<0$) is the state of completely aligned spins. For $J>0$ the ground state has algebraic long-range order [8.34]. For higher dimensions, considerable insight has been obtained by recent Monte Carlo calculations. These results will be discussed later in this chapter. The $S=1/2$ Heisenberg model can be mapped onto a hard-core boson model with repulsive interaction [8.35]. This representation is very useful from the numerical and analytical point of view. The transformation $S_i^+=b_i^+, S_i^-=b_i$ and $S_i^z=1/2-b_i^+b_i$, with the hard-core constraint $b_i^{+2}=0$ leads to the hard-core boson Hamiltonian. For $J>0$ and a bi-partite lattice it is useful to perform an additional gauge transformation $b_i \rightarrow e_i \cdot b_i$, with e_i being 1 on one sublattice and -1 on the other. The hard-core boson Hamiltonian then reads

$$H=-J\sum_{\langle ij\rangle} b_i^+ b_i + J\sum_{\langle ij\rangle} n_i n_j + E_0 . \tag{8.3}$$

In the above equation, $E_0=-Jz(N-N_b)/4$, where N_b is the number of bosons and z the number of nearest neighbours. N_b is related to the z-component of the total spin via $N_b=N/2-S_0^z$. Both N_b and S_0^z are conserved quantities. It has been shown that the ground state of (8.3) is nodeless and unique [8.36, 37].

The spin$-1/2$ Heisenberg antiferromagnet (8.3) is the strong coupling limit of the half-filled Hubbard model, with $J=4t^2/U$ [8.38]. Away from half-filling, the Hubbard model can be approximated in the strong coupling limit by the $t-J$ model. It consists of a constraint hopping term in the subspace of single occupied sites and the Heisenberg model for the spin degrees of freedom. For a review we refer the reader to [8.39].

8.3 Variational Monte Carlo Method

8.3.1 Method and Trial Wavefunctions

The variational Monte Carlo (VMC) approach is the most transparent application of the importance sampling idea, which has been put to the test in many problems. A detailed review about this method and possible applications has already been presented in a previous volume of this series [8.1]. We assume that the reader is familiar with the basic ideas underlying Monte Carlo simulations and here we recall merely the key ideas to establish the notation for the discussion. Many of recent VMC calculations are performed for the Heisenberg antiferromagnet in two dimensions. The starting point for any VMC calculation is a suitable trial wavefunction $|\psi^{\mathrm{T}}\{\eta\}\rangle$, which in general depends on a set of variational parameters $\{\eta\}$. The best results are obtained by Jastrow type wavefunctions. The Gutzwiller ansatz, although very accurate for the 1D problem [8.40, 41], turned out to be not so appropriate for the 2D case [8.42–44]. One is prompted by the analogy of the Heisenberg model to the hard-core boson problem to use the Jastrow ansatz [8.45–47]. In the boson representation the wavefunction reads

$$|\psi^{\mathrm{J}}\rangle = \exp\left\{-\sum_{ij}\eta_{ij}n_i n_j\right\}|\psi_0\rangle, \tag{8.4}$$

ψ_0 is the Bose condensate in the state with zero momentum. Another way of writing ψ^{J} is

$$|\psi^{\mathrm{J}}\rangle = \sum_{\Gamma}\exp\left\{-\sum_{ij}\eta_{ij}\Gamma_i\Gamma_j\right\}|\Gamma\rangle, \tag{8.5}$$

where the summation extends over all real space configurations Γ, with $\Gamma_i = 1$ if site i is occupied and $\Gamma_i = 0$ otherwise. The expectation value for an arbitrary operator O is then

$$\langle O\rangle = \frac{\langle\psi^{\mathrm{J}}|O|\psi^{\mathrm{J}}\rangle}{\langle\psi^{\mathrm{J}}|\psi^{\mathrm{J}}\rangle} = \sum_{\Gamma} P(\Gamma)O(\Gamma) = \frac{1}{M}\sum_{l=1}^{M} O(\Gamma^{(l)}), \tag{8.6}$$

with

$$O(\Gamma) = \sum_{\Gamma'}\langle\Gamma|O|\Gamma'\rangle\cdot\exp\left\{-\sum_{ij}\eta_{ij}(\Gamma'_i\Gamma'_j - \Gamma_i\Gamma_j)\right\}. \tag{8.7}$$

The Markov chain of real space configurations is denoted by $\Gamma^{(1)}, \Gamma^{(2)} \dots \Gamma^{(M)}$ and is drawn from the probability density P,

$$P(\Gamma) = \frac{\exp\left\{-2\sum_{ij}\eta_{ij}\Gamma_i\Gamma_j\right\}}{\sum_{\Gamma'}\exp\left\{-2\sum_{ij}\eta_{ij}\Gamma'_i\Gamma'_j\right\}}. \tag{8.8}$$

Next the energy is evaluated using (8.6) for the expectation value of the Hamiltonian and minimised upon variation of the parameters η_{ij}.

In the spin representation, the Jastrow wavefunction is equivalent to a wavefunction proposed at about the same time by *Hulthén*, *Kasteleijn* and *Marschall* (HKM) [8.48–50]. Actually, the HKM wavefunction corresponds to a particular ansatz for the variational parameters in the Jastrow function:

$$\eta_{ij}^{\mathrm{SR}} = \begin{cases} \infty & \text{if} \quad i = j \text{ (hard-core condition);} \\ \eta & \text{if} \quad i,j \text{ nearest neighbours;} \\ 0 & \text{otherwise;} \end{cases} \tag{8.9}$$

where SR refers to a trial wavefunction with short-range character. Energy and spin correlation function for the corresponding trial wavefunction $\psi^{\mathrm{J}}_{\mathrm{SR}}$ have been evaluated exactly [8.51]. These results yield a good check for the quality of the VMC computation. More complex forms for η have been proposed, which lead to lower energies [8.52, 53]. Fairly low energies could be obtained by adding to η^{SR}, a long range tail to ensure the correct long wave length behavior for the magnetic structure factor:

$$\eta_{ij}^{\mathrm{LR}} = \begin{cases} \infty & \text{if } i=j \,(\text{hard-core condition}); \\ \eta & \text{if } i,j \text{ nearest neighbours}; \\ \alpha|\boldsymbol{x}_i - \boldsymbol{x}_j|^{-\beta} & \text{otherwise}. \end{cases} \tag{8.10}$$

We will call the corresponding wavefunction $\psi_{\mathrm{LR}}^{\mathrm{J}}$. A variational ansatz for the lowest excited state with momentum $\boldsymbol{q}$ has been studied in [8.51]:

$$|\psi(\boldsymbol{q})\rangle = S^{\gamma}(\boldsymbol{q})|\psi^{0}\rangle, \tag{8.11}$$

where γ stands for one of the components (x, y, z) of the magnon operator $S(\boldsymbol{q})$. Ansatz (8.11) describes a spinwave on the "exact" ground state for which the Jastrow ansatz has been used.

All these trial functions show (off-diagonal) long-range order (OD)LRO. Another class of trial wavefunctions, namely the resonant valence bond (RVB) states, has been proposed by *Liang* et al. [8.54];

$$|\psi^{\mathrm{RVB}}\rangle = \sum_{\substack{i_1,i_2\ldots \\ \in A}} \sum_{\substack{j_1,j_2\ldots \\ \in B}} \left[\prod_{l=1}^{N/2} h(i_l - j_l)\right]\left[\prod_{l=1}^{N/2} (i_l, j_l)\right]. \tag{8.12}$$

A bi-partite lattice is assumed with sublattices A and B and (i, j) represents a singlet bond. ψ^{RVB} is by construction a singlet, and fulfils the Marshall sign-rule if $h(\boldsymbol{x}) \geqq 0$ [8.50]. The evaluation of energy or spin correlation functions $\langle \boldsymbol{S}_i \cdot \boldsymbol{S}_j \rangle$ can be mapped onto the partition function of weighted closed loop on a planar graph. Standard Metropolis Monte Carlo techniques can be used to evaluate all possible loop coverages and to sample the partition function. It would be too technical to go into the details of this algorithm; the interested reader is referred to [8.54]. On changing the functional form of the function $h(\boldsymbol{x})$, two limiting cases can be covered: a) a quantum liquid with short-range antiferromagnetic order if $h(\boldsymbol{x}) = 0$ for $|\boldsymbol{x}| > r_{\max}$ and b) a state with long-range antiferromagnetic order in the case that $h(\boldsymbol{x}) \propto |\boldsymbol{x}|^{-p}$ for large distances.

The RVB state with long-range AF order (b) does actually also belong to the class of Jastrow wavefunctions with the particular choice of $\eta_{ij} = p \cdot \ln(|x_i - x_j|)$ [8.47]. It is equivalent to the singular-gauge Laughlin wavefunction [8.55] recently discussed in the framework of the fractional quantum Hall effect.

8.3.2 Results

The energies obtained with $\psi_{\mathrm{SR}}^{\mathrm{J}}$ by VMC computations are in excellent agreement [8.51, 53, 56] with the exact result [8.10]. A lower energy can be achieved by $\psi_{\mathrm{SR}}^{\mathrm{J}}$ [8.58, 59]. The results obtained by various VMC calculations are given in Table 8.2. All results, except the exact one, are extrapolated from finite cluster results. In all cases ODLRO was found with the spins tipped in the xy-plane. The staggered magnetisation

$$m = \sqrt{\frac{1}{N}\sum_{l} |\langle \boldsymbol{S}_0 \cdot \boldsymbol{S}_l \rangle|}, \tag{8.13}$$

is reduced by about 60% from its classical value of 0.5 due to quantum fluctuations. The values for the staggered magnetisation are also given in Table 8.2. The ground state is still translational invariant $\langle S_i \rangle = 0$, but has long-range spin correlations ($m \neq 0$). An infinitesimal field should therefore be sufficient to break the symmetry. This is possible since the non-degenerate singlet ground state [8.36, 37] for a finite system becomes asymptotically degenerate in the thermodynamic limit. The low-lying excitations have been identified as renormalised spinwaves. The gapless excitation spectrum is a necessary consequence of the finite staggered magnetisation [8.51], which is a generalisation of the Goldstone theorem. Also the magnetic structure factor is found to be in close agreement with spinwave theory.

Trial states with built-in (OD)LRO yield only slightly lower energies than the quantum fluid (RVB) state with $m = 0$. This again indicates that in the thermodynamic limit, the low-lying states are degenerate and among those one can find very different magnetic properties.

This reveals a fundamental problem of variational approaches: a trivial wavefunction can have a fairly low energy and nonetheless have correlation functions completely different from those of the exact ground state.

8.4 Green's Function Monte Carlo Method

8.4.1 Method

The Green's function Monte Carlo method (GFMC) has been successfully applied to ground state properties of helium, the interacting electron gas and small molecules [8.60–62]. Research along this line has already been reviewed in a previous volume of this series [8.1, 2]. The GFMC method for lattice problems is still at the very beginning. Up to now, it is only applicable to models like the Heisenberg model or the related hard-core boson model [8.56, 63–65], which have a nodeless ground state wavefunction. For these problems, however, it proved very useful and is certainly a promising development.

The heart of the GFMC scheme is the well-known power method [8.66]. From an almost arbitrary state of the system, the ground-state component is filtered out by repeated application of the Hamiltonian. The actual calculation is performed with the iteration procedure

$$|\phi^{n+1}\rangle = [1 - \tau(H - \omega)]|\phi^n\rangle = G|\phi^n\rangle. \tag{8.14}$$

The notion Green's function stems from the fact that G can be viewed as the series expansion of the imaginary time evolution operator $\exp[-\tau(H - \omega)]$ or, likewise, of the propagator $[1 + \tau(H - \omega)]^{-1}$ for small "time steps" τ. The vector iteration (8.14) converges towards the ground state if the imaginary time step obeys

$$\tau < \frac{2}{(E_{\max} - \omega)}. \tag{8.15}$$

The constant ω should be a good guess for the ground state energy. The power method is only applicable if the spectrum is bounded. The imaginary time step has to decrease in proportion to N^{-1} according to (8.15) as the spectral width of many-body problems generally increase proportional to N. This implies an increasing number of iterations with increasing system size.

A further and vital element of the GFMC scheme traces back to von Neumann and Ulam, who proposed the idea of sampling the inverse of a matrix by random walks on the index-space of the matrix [8.67–69]. To this end we expand the many-body wavefunction $|\phi\rangle$ of interest in a proper set of many-body basis states $|R\rangle$;

$$|\phi\rangle = \sum_R \phi(R)|R\rangle. \tag{8.16}$$

The expansion coefficients $\phi(R)$ are assumed to be non-negative and real, so that they can be regarded as probability densities. For concreteness, we will continue the discussion for a specific example, the hard-core boson problem, introduced in Sect. 8.1. The basis states are the real-space configurations

$$|R\rangle = \prod_{l=1}^{N_b} b_{r_l}^{+}|0\rangle, \tag{8.17}$$

where R stands for the set $\{r_l\}$ of occupied sites. It is convenient to think of R as an integer index, which numbers all possible configurations R. The iteration scheme (8.14) in these basis states reads

$$\phi^{n+1}(R) = \sum_{R'} G(R, R')\phi^n(R'), \tag{8.18}$$

where the Green's function G that propagates configuration R' to R is given by

$$G(R, R') = \langle R|[1 - \tau(H - \omega)]|R'\rangle. \tag{8.19}$$

The matrix elements are extremely simple for the Hamiltonian under consideration;

$$G(R, R') = \begin{cases} 1 - \tau[U(R) - \omega] & \text{if } R = R'; \\ \dfrac{\tau J}{2} & \text{if } R \in N(R'); \\ 0 & \text{otherwise.} \end{cases} \tag{8.20}$$

The set $N(R')$ consists of all those configurations that can be obtained from R' by moving one of the bosons to any of the available nearest-neighbour positions. $U(R) = \langle R|H_{\text{pot}}|R\rangle$ is the expectation value of the potential energy.

At this point the Monte Carlo aspect of the GFMC method pops up. To sample the iterated wavefunction $\phi^n(R)$ in (8.18), we decompose $G(R, R')$ into a stochastic matrix $P(R, R')$ and a residual weight $W(R')$,

$$G(R, R') = P(R, R') \cdot W(R'), \tag{8.21}$$

with

$$\sum_R P(R,R') = 1 \quad \text{and} \quad P(R,R') \geqq 0. \tag{8.22}$$

Starting with an initial vector ϕ^0 the probability density after n iterations is

$$\begin{aligned}\phi^n(R) &= \langle R|G^n|\phi^0\rangle \\ &= \sum_{R_0,R_1,\ldots,R_n} \delta(R,R_n)W(R_{n-1})W(R_{n-2})\cdots W(R_0) \\ &\quad \times P(R_n,R_{n-1})P(R_{n-1},R_{n-2})\cdots P(R_1,R_0) \times \phi^0(R_0).\end{aligned} \tag{8.23}$$

We define a n-step random walk (Markov chain) on the possible configurations R. With probability $\phi^0(R_0)$ the Markov chain begins with configuration R_0 and the random walk proceeds along $R_1 \to R_2 \to \cdots R_n$. The transition probability for the move $R_l \to R_{l+1}$ is given by $P(R_{l+1}, R_l)$. For each walk the cumulated weight is

$$W^n = \prod_{l=0}^{n-1} W(R_l). \tag{8.24}$$

Since the probability for one individual walk is $\prod_{l=1}^{n} P(R_i, R_{i-1}) \cdot \phi^0(R_0)$, it follows that the mean value of the cumulated weights, averaged over M independent random walks, is the desired wavefunction

$$\phi^n(R) = \lim_{M\to\infty} \frac{1}{M} \sum_{l=1}^{M} W_l \delta(R - R_l^n). \tag{8.25}$$

where R_l^n stands for the configuration of the n-th iteration in the l-th random walk. It is more economical and also more instructive to follow the whole set of walkers in parallel. The initial generation $\{R_l^0\}$, $l = 1, 2, \ldots, M$ is sampled from the trial function ϕ^0. The set $\{R^0\}$ is a representation of ϕ^0 with the property

$$\phi^0(R) = \lim_{N\to\infty} \frac{1}{M} \sum_{l=1}^{M} \delta(R - R_l^0). \tag{8.26}$$

To create the next generation, each configuration R is moved to a new configuration R' according to the transition probability $P(R', R)$ and picks up an additional weight $W(R)$. The n-th generation of walkers $\{R^n\}$ together with the cumulated weights (8.24) represents the n-th iterate of the wavefunction as given in (8.25).

The algorithm as it stands so far is unfortunately not applicable. *Hetherington* [8.70] has nicely illustrated the ideas and shortcomings of the GFMC scheme in terms (2×2) matrices. As a result of the central limit theorem, the probabilities $p_m(R)$ to find a given configuration Rm times in one random walk of length n is normally distributed $p_m(R) \propto \exp\{A[m - m_0(R)]^2/n\}$. The weight associated to this particular configuration is $W(R)^m = \exp(\alpha m)$. The contribution of configuration R to the wavefunction is given by $\sum_m W(R)^m \cdot p_m(R) \propto \sum_m \exp\{A[m - \tilde{m}_0(R)]/n\}$.

The maximum of the probability density $p_m(R)$, according to which the random walk is performed, is at a different position than that of the summand for the wavefunction. These maxima drift apart proportional to n. This implies that the variance V increases exponentially with the number of iterations, and decreases only by the usual $1/\sqrt{M}$ with a large number of samples: $V \propto \exp(\alpha n)/\sqrt{M}$.

The variance can be reduced by a) making G more like a stochastic matrix and b) by increasing the number of iterations over which the weights are cumulated. Point a) can be achieved by introducing a guiding functions ψ_G [8.2], which leads to a sort of importance sampling. To this end the wavefunction ϕ^n in (8.18) is changed to $F^n(R) = \phi^n(R) \cdot \psi_G(R)$, which modifies the iteration process (8.18) to $F^{n+1}(R) = \sum_{R'} \tilde{G}(R, R')F^n$, with the modified Green's function

$$\tilde{G}(R, R') = \psi_G(R)\langle r|[1 - \tau(H - \omega)]|R'\rangle/\psi_G(R'). \tag{8.27}$$

The initial probability density is replaced by $F^0 = \psi_G^2$, if the special choice $\phi^0 = \psi_G$ is made. The weights $W(R)$ are determined by normalising the transition probability, i.e. from $\sum_R P(R, R') = \sum_R \tilde{G}(R, R') \cdot W^{-1}(R') = 1$, hence

$$W(R) = 1 - \tau[E_L(R) - \omega]; \tag{8.28}$$

and

$$P(R, R') = \frac{1}{W(R')} \cdot \begin{cases} 1 - \tau[U(R) - \omega] & \text{if } R = R'; \\ \dfrac{\tau J}{2} \dfrac{\psi_G(R)}{\psi_G(R')} & \text{if } R \in N(R'); \\ 0 & \text{otherwise.} \end{cases} \tag{8.29}$$

E_L is the local energy $E_L(R) = \langle R|H|\psi_G\rangle/\psi_G(R)$.

In the ideal case of ψ_G being the exact ground state wavefunction and ω the corresponding energy, the variance is zero since according to (8.28) the weight is 1 and $\tilde{G}$ is a statistic matrix. A good guess for ψ_G and ω can reduce the variance significantly. And yet, unless ψ_G and ω are the exact eigensolution of the Hamiltonian under consideration, the variance increases exponentially with the number of iterations n. Remember that n grows proportional to the lattice size. *Trivedi* and *Ceperley* [8.56, 63] therefore introduced the concept of branching into GFMC algorithm for lattice problems. The variance is hereby improved by avoiding configurations with low weight. After a certain number K of iterations is performed, the wavefunction is re-expressed according to the cumulated weights in terms of real space configurations only, as in (8.26). The cumulated weight for a given configuration R is $W^K(R)$. $M(R)$ number of copies are made of this configuration according to $W^K(R)$. To obtain an integer $M(R)$ for $W^K(R)$, the identity $w = \int_0^0 dx \operatorname{int}(w + \xi)$ is employed and we obtain

$M(R) = \text{int}\,[W^K(R) + \xi]$. ξ is an additional random degree of freedom which is sampled uniformly from [0, 1]. A walker is discarded if the multiplicity M is zero. This new set of configurations $\{\tilde{R}^K\}$ is another representation of the wavefunction,

$$F^K(R) = \lim_{M \to \infty} \frac{1}{M} \sum_{l=1}^{M} W^K(R_l) \cdot \delta(R - R_l^K) = \lim_{M \to \infty} \frac{1}{M} \sum_{l=1}^{M} \delta(R - \tilde{R}_l^K) \tag{8.30}$$

On performing the branching process, the number of walkers is not generally constant and its growth or decline is controlled by the energy ω. This yields a rough estimator for the ground state energy. The optimal ω can be adjusted numerically during the first generation and in general a constant population can be achieved for subsequent generations. The final generation is distributed according to $\psi_G(R) \cdot \psi_0(R)$, product of the guiding function and the desired ground state wavefunction. The most accurate way to evaluate the energy is via the "mixed estimator";

$$E_n = \frac{\langle \phi^n | H | \psi_T \rangle}{\langle \phi^n | \psi_T \rangle} = \frac{\sum_l W(R_l^n) \cdot E_L(R_l^n) \cdot \dfrac{\psi_T(R_l^n)}{\psi_G(R_l^n)}}{\sum_l W(R_l^n)} . \tag{8.31}$$

While a simple guiding function ψ_G has been used for importance sampling, a more accurate function ψ_T is used to evaluate the energy. $E_L(R)$ is again the local energy as in (8.29). The mixed estimator yields the exact expectation value for the energy, as ϕ^n is an eigenstate of the Hamiltonian. This is not generally true for an arbitrary operator O,

$$\langle O \rangle_M \stackrel{\text{def}}{=} \frac{\langle \phi^n | O | \psi_T \rangle}{\langle \phi^n | \psi_T \rangle} \neq \langle O \rangle \stackrel{\text{def}}{=} \frac{\langle \phi^n | O | \phi^n \rangle}{\langle \phi^n | \phi^n \rangle} . \tag{8.32}$$

By assuming that the trial wavefunction ψ_T is in good agreement with the exact ground state, an "extrapolated estimate"

$$\langle O \rangle \approx 2 \cdot \langle O \rangle_M - \langle O \rangle_T , \tag{8.33}$$

can be used, where $\langle O \rangle_T$ stands for the average over the trial wavefunction and has to be determined separately.

The GFMC method does not have a finite time step error, contrary to schemes resorting to the Suzuki–Trotter decomposition. But not all that glitters is gold. The imaginary time step has to decrease proportional to $1/N$, according to (8.15), as the spectral width of many-body problems generally increases proportional to N. This implies an increasing number of iterations with increasing system size. Further shortcomings of this method are:

a) To calculate any quantity of interest, only a finite set of configurations is available [8.56].
b) Data are strongly correlated and very careful evaluation of the statistical error is required, otherwise the obtained error bars are much too small.

c) The mixed or extrapolated estimate can be very inaccurate as it depends on the trial wavefunction. There is, however, no such thing as a variational principle for observables other than the energy, and a good trial wavefunction with fairly low energy can still lead to spurious results as far as correlation functions are concerned, e.g. the RVB quantum fluid state proposed by *Liang* et al. [8.54] yields a very low energy but only short-range order.
d) The method as it stands is only applicable to systems with a real and nodeless ground state wavefunction. It is therefore not yet useful for fermionic problems. The scheme has certain similarities with the worldline technique, which suffers strongly from a severe minus-sign problem. It is likely that this minus-sign problem is detrimental for the GFMC method to be applicable to lattice fermion problems.

8.4.2 Results

The GFMC scheme for lattice problems has exclusively been applied to quantum spin problems [8.56, 63–65, 71–73]. Most of these papers are devoted to the spin 1/2 antiferromagnetic Heisenberg model in 2D. With these, in principle, exact simulations, the ground state energy could be lowered slightly as compared to the best VMC results. The differences are very small and nearly comparable to the uncertainties of the GFMC data. All ground state energies are given in Table 8.2. The lowest energy so far has been achieved by *Barnes* and *Swanson* [8.73]. In this calculation, however, the branching technique, which as discussed is important to reduce the variance, is not included. Moreover, several extrapolations have been made to get rid of systematic error inherent in the special variant of the GFMC used. It is difficult to judge the quality of these approximations.

In the main, the GFMC calculations corroborate the results obtained by the VMC computations. The ground state exhibits long-range antiferromagnetic order. In all cases the Jastrow ansatz was used as guiding function, the only exception is [8.64], where no importance sampling has been used at all. The rotational symmetry, which is broken in the Jastrow ansatz, could not be restored in the Monte Carlo simulation. This is in contradiction to the work by *Lieb* et al. [8.36, 37], who proved that the ground state for any finite system is a singlet. We have seen, on the other hand, that in the thermodynamic limit many states become degenerate with the ground state allowing the system to break the symmetry. It can be argued that for large enough systems, the guiding function serves as symmetry breaking field. The staggered magnetisation, with ths spins tipped in the xy-plane, is somewhat reduced as compared to the VMC results for the Jastrow wavefunction.

It has been found that the energy does not depend on the guiding function. Even the Néel state is sufficient to guide the random walk to obtain an accurate result. The magnetisation, via the extrapolated estimator (8.33), is strongly influenced by the choice of the trial functions. This is an unpleasant feature of

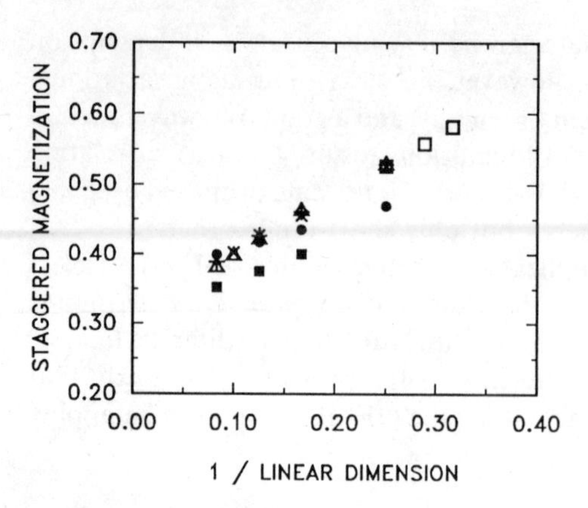

Fig 8.1. Staggered magnetization for the 2D Heisenberg antiferromagnet. The symbols stand for: □ → + = exact [8.51], □ = VMC (Jastrow) [8.52, 58], △ = WLQMC [8.79, 80], × = WLQMC [8.78], ● = GFMC (short-range trial function), ■ = GFMC (long-range trial function)

GFMC and is actually not what one expects from an "exact" Monte Carlo simulation.

The conclusions drawn from the VMC results remain valid. Within a small range of energies above the ground state are states with staggered magnetisations ranging from $0 \sim 0.3$. The results are summarised in Table 8.2. For completeness, Table 8.2 also comprises of data obtained with other MC schemes, which have not explicitly been mentioned in this section.

The existence of long range order for the spin 1/2 antiferromagnetic Heisenberg model has been confirmed by finite temperature simulations as well. Most reliable results are obtained by the Handscomb algorithm [8.74]. This method is particularly geared for the Heisenberg model in that it exploits the spin–1/2 algebra. It has no finite time-step error inherent in other path integral methods. For a review we refer the reader to [8.3, 74]. The method has been improved for the antiferromagnetic Heisenberg model [8.75–77]. The results indicate algebraic long-range order of the Kosterlitz–Thouless type for finite temperatures and an essential singularity in the correlation length for $T \to 0$.

The results given in Table 8.2 are for the thermodynamic limit $N \to \infty$. Most of them are obtained from finite size data by extrapolation, partly assuming different size dependencies. The results depend on the finite size scaling. For the ground state *Miyashita* [8.78] even concluded from a different finite size scaling that the spin correlation function exhibits only algebraic long-range order with $m = 0$. This conclusion is based on otherwise very similar finite size data as those of [8.51, 79, 80]. The results for the staggered magnetisation are plotted in Fig. 8.1 versus the inverse of the linear dimension. We leave it to the reader to judge which conclusion he thinks is more realistic for the thermodynamic limit.

8.5 Grand Canonical Quantum Monte Carlo Method

8.5.1 Method

A standard technique used to study the thermodynamic properties of systems of interacting fermions is to eliminate from the partition function, the fermionic degrees of freedom by introducing auxiliary (bosonic) fields. For fermions in continuum space, this leads to the Siegert representation [8.81] of the grand-canonical partition function. The key elements of deriving this expression are the Feynman path-integral representation for the partition function and the identity

$$\int_{-\infty}^{+\infty} \mathrm{e}^{-a\phi^2 - b\phi} d\phi = \sqrt{\frac{\pi}{a}}\, \mathrm{e}^{b^2/4a}; \quad a > 0. \tag{8.34}$$

As the auxiliary boson fields, denoted by ϕ throughout this chapter, enter Siegert's representation as a consequence of the use of mathematical identity (8.34), the ϕ's have no proper physical interpretation. Siegert's representation is, at least in principle, amenable to numerical treatment as the integration of the fields can be performed by simulation (e.g. Monte Carlo) methods.

The derivation that transformed the continuum fermion problem into a problem in terms of real numbers can be carried over to lattice fermions without much effort. Instead of starting from the Feynman path integral, one has to invoke the Trotter–Suzuki [8.4] formula to rewrite the partition function such that identity (8.34) can be applied. To illustrate this procedure, consider the partition function of the repulsive ($U > 0$) Hubbard model.

$$Z = \mathrm{Tr}\,\{\mathrm{e}^{-\beta(H-\mu N)}\} = \lim_{m \to \infty} \mathrm{Tr}\,\{(\mathrm{e}^{-\beta(K-\mu N)/m}\mathrm{e}^{-\beta V/m})^m\}, \tag{8.35}$$

where $H = K + V$ and use has been made of the Trotter–Suzuki formula to disentangle quadratic $(K - \mu N)$ and quartic (V) forms in the fermion operators.

To express $\exp(-\beta V/m)$ as an exponent of a quadratic form, we use identity (8.34) to write

$$\exp\left(-\frac{\beta U}{m}\sum_{l=1}^{N} n_{l\uparrow}n_{l\downarrow}\right) \propto \int_{-\infty}^{+\infty}\prod_{l=1}^{N} d\phi_l \mathrm{e}^{-m\phi_l^2/2\beta U - \phi_l(n_{l\uparrow}-n_{l\downarrow}) - \beta U(n_{l\uparrow}+n_{l\downarrow})/2m}, \tag{8.36}$$

where N is the number of lattice sites. For negative U both bracketed expressions on the rhs of (8.36) are replaced by $(n_{l\uparrow} + n_{l\downarrow} - 1)$. Inserting (8.36) in (8.35), the mth approximate to Z can be symbolically written as

$$Z_m = \mathrm{Tr}\,\{(e^{-\beta(K-\mu N)/m}\mathrm{e}^{-\beta V/m})^m\} = \int d\phi\, \mathrm{Tr}\,\{\mathrm{e}^{c_{i\sigma}^{+}A_{ij}^{(1)}c_{j\sigma}} \cdots \mathrm{e}^{c_{i\sigma}^{+}A_{ij}^{(2m)}c_{j\sigma}}\}, \tag{8.37}$$

where the $A_{ij}^{(k)} = A_{ij}^{(k)}(\phi, \sigma)$ are $N \times N$ matrices and the trace is over all possible configurations of 1 to N fermions. To obtain an expression for Z_m, formally identical to Siegert's representation [8.81], the trace over the fermion degrees of freedom has to be carried out explicitly. Making use of the identity

$$\mathrm{Tr}\,\{e^{c_i^+ A_{ij}^{(1)} c_j} \cdots e^{c_i^+ A_{ij}^{(p)} c_j}\} = \det(1 + e^{A^{(1)}} \cdots e^{A^{(p)}}), \tag{8.38}$$

a proof of which is given in the appendix, the partition function of the Hubbard model becomes

$$Z_m \propto \int \prod_{l,j} d\phi_{lj} e^{\phi_{l,j}^2} \det(1 + e^{-\beta\tilde{K}/m} e^{-\tilde{V}(\phi_{l,1})} \cdots e^{-\beta\tilde{K}/m} e^{-\tilde{V}(\phi_{l,m})})$$

$$\times \det(1 + e^{-\beta\tilde{K}/m} e^{-\tilde{V}(-\phi_{l,1})} \cdots e^{-\beta\tilde{K}/m} e^{-\tilde{V}(\phi_{l,m})}). \tag{8.39}$$

Here $\tilde{K} = K - (\mu - V/2)N$ is the kinetic energy matrix for a *single*-particle propagator on a lattice and $\tilde{V}(\phi_l)$ is a diagonal matrix depending on ϕ_l.

So far we have been using Gaussian fields ϕ. For some models, notably the Hubbard model, other types of fields may be used as well. In particular, *Hirsch* [8.82–84] has introduced the discrete HS transformation for $\exp\left(-\frac{\beta U}{m} \sum_l n_{l\uparrow} n_{l\downarrow}\right)$ in terms of Ising spins, i.e.

$$e^{-(\beta U/m) n_{l\uparrow} n_{l\downarrow}} = \begin{cases} \frac{1}{2} \sum_{\sigma_l = \pm 1} e^{\lambda(n_{l\uparrow} - n_{l\downarrow})\sigma_l} \cdot e^{-(U\beta/2m)(n_{l\uparrow} + n_{l\downarrow})} & \text{if } U > 0; \\ \frac{1}{2} \sum_{\sigma_l = \pm 1} e^{\lambda(n_{l\uparrow} + n_{l\downarrow} - 1)\sigma_l} \cdot e^{-(U\beta/2m)(n_{l\uparrow} + n_{l\downarrow} - 1)} & \text{if } U < 0; \end{cases} \tag{8.40}$$

where $\lambda = 2\,\mathrm{arctanh}\sqrt{\tanh(\beta|U|/2m)}$.

A comparative study of the discrete and continuous Hubbard–Stratonovich (HS) transformation has been performed in [8.82, 85] and it has been found that the discrete version is superior in quite a conclusive way.

By eliminating the fermionic degrees of freedom in favor of Gaussian or Ising fields, the partition function has been brought into a form suitable for numerical calculations, i.e., it has been expressed entirely in terms of real numbers. The resulting Ising action has much less phase space than the continuous Gaussian action and is more convenient to sample. The computation time of this algorithm increases with the cube of the system size N and linearly with the number of time-slices m. Moreover, the HS transformation replaced the many-body Hamiltonian in favour of a one particle Hamiltonian with stochastical fields. The dimension of the involved matrices is hereby drastically reduced from $\binom{N}{n} \times \binom{N}{n}$ to merely $N \times N$. The information content of the many-body problem has been cast into the summation over all field values. The number of auxiliary fields (proportional to mN) is large and statistical methods have to be invoked to obtain estimates for the relevant quantities. The importance sampling idea is well suited for this task. Unfortunately, a straightforward application of Monte Carlo or Molecular Dynamics methods leads to severe and partly fundamental problems stemming from the quantum nature of the problem. During the last years, big progress has been achieved to relieve these problems, but there is still a long way to go to make the quantum Monte Carlo schemes as reliable as the classical MC algorithms. A detailed discussion of these difficulties is relegated to Sect. (8.6).

8.5.2 Applications

The grand canonical quantum Monte Carlo (GCMC) method for lattice fermions [8.82, 86] was used by *Hirsch* and many others to investigate the properties of the Hubbard model [8.82–84, 87] in 1 to 3 dimensions. Although at low temperature ($\beta > 4$, $t = 1$, $U = 4$) the simulations techniques employed suffered from minus-sign and instability problems, more sophisticated and accurate simulations could only confirm most of Hirsch's conclusions about the properties of the 2D Hubbard model. We briefly summarise the main findings, concentrating on the repulsive case ($U > 0$) first. For a half-filled band and $T = 0$ the system is an antiferromagnet [8.88]. From the QMC point of view this is a special case because due to particle-hole symmetry, there are no minus-sign problems [8.84]. As soon as the filling deviates from half filling, the magnetic ordering disappears and the system is paramagnetic [8.88]. There seems to be no tendency to ferromagnetism nor to triplet superconductivity in the parameter regime that had been studied. Later studies revealed that there is no indication for anisotropic singlet superconductivity. Although there is some indication for enhanced d-wave pairing [8.89, 90], the GCMC simulations seems to suggest that the 2D repulsive Hubbard model does not exhibit superconductivity [8.89, 91]. GCMC data for the momentum distribution and the single-particle self-energy support this global picture [8.92]. One should bare in mind however, that these GCMC results may be misleading because the lattices one is able to study are not large enough or the lowest temperature that can be reached is still to high.

GCMC simulation results for the 2D half-filled attractive ($U < 0$) Hubbard model [8.93] suggest that the critical temperature $T_c = 0$, and that the ground state has both superconducting and charge-density-wave long-range order. Away from half-filling, $T_c > 0$ and the phase transition to the superconducting phase is of the Kosterlitz–Thouless type, charge-density-wave correlations being short ranged.

The apparent absence of superconductivity in the ground state of the one-band 2D repulsive Hubbard model has spurred the interest in performing QMC simulations of more complicated Hubbard-like models. One route to extend the Hubbard model has been to add to the on-site Coulomb interaction, terms that incorporate nearest (and other) neighbours [8.15] or further range hopping [8.15, 18, 19, 94] for which preliminary results indicate an enhancement of superconducting susceptibilities. Another extension of the Hubbard model is to include more bands, which in the context of the high-T_c materials seems necessary.

Some results for the 2D two-band Hubbard model have been reported [8.16, 95] but due to the limited range of temperature studied, the results do not seem to be conclusive. A detailed report of a GCMC study of the three-band Hubbard model (Emery model) is given in [8.23] showing that it is possible to distinguish between a charge-transfer regime with well-developed magnetic moments on particular (Cu) sites and an itinerant situation. It also addresses the question of anti-ferromagnetic long-range order, the dependence of the structure factor on doping, and the formation of local singlets. Although these GCMC calculations provide evidence for attractive pairing interaction in the extended s-wave channel, no phase transition to a superconducting state was found [8.23]. As in the one-band case, at low temperature and with doping, minus-sign problems occur and they make it difficult to obtain reliable results.

The feasibility of performing GCMC simulations for 3D fermionic systems has been examined by *Hirsch* [8.87] who studied the 3D repulsive Hubbard model on 4^3 and 6^3 lattices. The magnetic properties were studied for a large range of model parameters, and it was found that the antiferromagnetism reaches a maximum at $U \approx 10t$. The transition from itinerant to localised magnetism appears to be smooth while for intermediate U the transition to the antiferromagnetic state seem to be of first order. In view of the small lattices treated, it is difficult to extract from these simulations a reliable estimate for the transition temperature. Invoking a duality transformation, the GCMC data confirm the picture [8.96] that for the attractive 3D Hubbard model the transition from weak-coupling to strong-coupling superconductivity is a continuous function of U.

As most QMC simulations for 2D (extended) repulsive Hubbard models involve only electronic degrees of freedom and strongly suggest that such systems do not exhibit superconductivity, it is of interest to add additional degree of freedom. The GCMC technique has been employed to investigate the influence of frozen-in lattice distortions of the "breathing" type, concentration on the stability of the charge-density-wave upon doping and electron–photon coupling strength [8.21].

Another fermion lattice model which has been the subject of many GCMC studies is the (periodic) Anderson (impurity) model [8.97]. Details about the method and results for the 1D Anderson impurity model can be found in [8.24–28]. The formation of a local-moment, mixed-valence behaviour and the Kondo effect have been observed in the simulations. The impurity susceptibility agrees with the Bethe ansatz solution [8.98]. At low temperature, simulations reveal the existence of a charge-compensation sum-rule and suggest a power-law decay of correlations [8.28]. The spatial dependence of the spin correlation between an electron at the impurity site and an electron at a lattice site with which it interacts, indicates RKKY and Friedel-like oscillations with a strong antiferromagnetic correlation. A surprising finding is that the spin compensation cloud extends beyond the natural length, implying that applicability of the single impurity model would most likely be restricted to very dilute alloys. Qualitative features of the single particle spectral function have been obtained through analytic continuation of GCMC data [8.30, 31], allowing one to distinguish between the local-moment, mixed-valent, and empty-orbital regimes.

A comprehensive account of GCMC simulations of the 1D symmetric Anderson lattice (many-impurity case) is given in [8.29]. For not too large values of the on-site Coulomb repulsion, spin correlations are found to decay rapidly with distance. For large enough on-site Coulomb repulsion and above temperatures at which quantities saturate to their ground-state values, the short-distance spin correlation is well-described by a RKKY effective lattice Hamiltonian. Some implications for superconductivity in an attractive Anderson lattice are also discussed in this chapter.

8.6 Projector Quantum Monte Carlo Method

8.6.1 Method

The key idea of the projector quantum Monte Carlo method (PQMC) is similar to that of the GFMC algorithm, i.e., to filter out the ground state component from a proper trial function by applying an appropriate functional of the Hamiltonian. The numerical realisation of this idea, however, is completely different. Roughly speaking, in the PQMC method the Hamiltonian is sampled by stochastical means, while in the GFMC scheme it is the wavefunction. In this section we focus on a projector quantum Monte Carlo technique (PQMC) which uses $\exp(-\beta H)$ as a filter and limit the discussion to its application to fermion lattice problems. To fix the notation we denote the eigenvalues and corresponding eigenvectors of H by $E_0 < E_1 \leqq E_2 \leqq \cdots$ and $|\phi_0\rangle, |\phi_1\rangle, \ldots$ respectively.

Taking an arbitrary state $|\psi\rangle$ it follows from

$$\lim_{\beta \to \infty} \frac{e^{-\beta H}|\psi\rangle}{\sqrt{\langle\psi|e^{-2\beta H}|\psi\rangle}} = |\phi_0\rangle \frac{\langle\phi_0|\psi\rangle}{|\langle\phi_0|\psi\rangle|}, \tag{8.41}$$

that projection using $\exp(-\beta H)$ will yield the ground state provided that the initial state $|\psi\rangle$ is not orthogonal to $|\phi_0\rangle$, and we let $\beta \to \infty$.

In writing down (8.41), we have tacitly assumed that the ground-state is non-degenerate (i.e. $E_0 < E_1$). If the ground-state is nearly degenerate, the convergence of $\exp(-\beta H)|\psi\rangle$ to the true ground-state as a function of β will be slow [8.99], which is likewise true for the GCMC scheme. The PQMC scheme can be used to obtain the lowest energy to any given symmetry and therefore allows the determination of certain excitation energies as well. This idea has actually been exploited to compute the low-lying excitation spectrum of the 1D Heisenberg model [8.100] and led to very good agreement with the exact des-Cloiseaux and Pearson excitation spectrum. To put this projector scheme into practice, a method for computing $\exp(-\beta H)|\psi\rangle$ or more precisely, matrix elements $\langle\phi|\exp(-\beta H)|\psi\rangle$ has to be devised [8.101, 102]. As in the grand-canonical approach, the basic ingredients of such an algorithm are the Trotter–Suzuki formula and the elimination of interaction terms in favor of auxiliary fields. In the following we implicitly have in mind the application of these ideas to the Hubbard model [8.103–105].

To construct a trial state $|\psi\rangle$, it is expedient to start from a Slater determinant of single-particle states, i.e.

$$|\psi(\sigma)\rangle = \prod_{\lambda=1}^{n}\left(\sum_{l=1}^{N}\psi_{\lambda l}^{\sigma}c_{l\sigma}^{+}\right)|0\rangle, \tag{8.42}$$

where $\sigma=\uparrow,\downarrow$, the $\psi_{\lambda,l}$ are the coefficients of the λ-th single-particle state which we choose to be orthonormal $\left(\sum_{l}\psi_{\lambda,l}^{*}\psi_{\mu,l}=\delta_{\lambda,\mu}\right)$. As before we (symbolically) rewrite $\exp(-\beta H)$ as

$$\mathrm{e}^{-\beta H} = \int d\phi\,\tilde{X}(\phi,\uparrow)\tilde{X}(\phi,\downarrow), \tag{8.43}$$

with $\tilde{X}(\phi,\sigma)=\exp[c_{i\sigma}^{+}A_{ij}^{(1)}(\phi,\sigma)c_{j\sigma}]\cdots\exp[c_{i\sigma}^{+}A_{ij}^{(P)}(\phi,\sigma)c_{j\sigma}]$. The identities collected in the appendix show that the action of $\tilde{X}(\phi,\sigma)$ on a state of the form (8.42) can be written as

$$\begin{aligned}\tilde{X}(\phi,\sigma)|\psi(\sigma)\rangle &= \sum_{\{l_\lambda\}}\sum_{P}\mathrm{sign}(P)\prod_{\lambda=1}^{n}\left(\sum_{i=1}^{N}X_{l_\lambda,i}(\phi,\sigma)\psi_{P_\lambda,i}^{\sigma}\right)c_{l_1,\sigma}^{+}\cdots c_{l_N,\sigma}^{+}|0\rangle\\ &= \prod_{\lambda=1}^{n}\left(\sum_{l=1}^{N}\psi_{\lambda l}^{\sigma}(\phi)c_{l\sigma}^{+}\right)|0\rangle,\end{aligned} \tag{8.44}$$

where $X_{i,j}(\phi,\sigma)-\{\exp[A^{(1)}(\phi,\sigma)]\cdots\exp[A^{(P)}(\phi,\sigma)]\}_{i,j}$ is a $N\times N$ matrix and P denotes a permutation of the n fermions and the $\{l_\lambda\}$ label the sites. Equation (8.44) reveals that the calculation of $\tilde{X}(\phi,\sigma)|\psi(\sigma)\rangle$ amounts to multiplying each of the vectors $(\psi_{\lambda,1}\cdots\psi_{\lambda,n})$, $\lambda=1,\dots,n$ representing the single particle orbitals, by the matrix $X(\phi,\sigma)$. This leads to propagated orbitals $\psi_{\lambda l}^{\sigma}(\phi)=\sum_{i}X_{l,i}\psi_{\lambda i}^{\sigma}$. In other words, the action of an exponential of a one-particle operator $\exp(O_1)$ on a Slater-determinant retains the determinantal structure and merely modifies the orbitals. The exact many-body wavefunction has the structure

$$\psi=\mathrm{e}^{-\beta H}\det(\{\psi_{\alpha}^{\uparrow}\})\det(\{\psi_{\beta}^{\downarrow}\})=\int d\phi\,\det(\{\psi_{\alpha}^{\uparrow}(\phi)\})\det(\{\psi_{\beta}^{\downarrow}(\phi)\}). \tag{8.45}$$

The many-body character is reflected in the superposition of the respective Slater-determinants for all HS fields.

For the case at hand, $X(\phi,\sigma)$ itself is a product of sparse matrices. Hence, instead of carrying out the full matrix multiplication (for each set of fields ϕ), it is more efficient to multiply the states by $\exp[A^{P}(\phi,\sigma)]$, $\exp[A^{(p-1)}(\phi,\sigma)]\dots$, $\exp[A^{(1)}(\phi,\sigma)]$, i.e., to "propagate" the states $\{\psi_{\lambda,l}\}$. In a detailed analysis it has been shown that it is more economical for large "tight binding" matrices to use a real space *product algorithm* rather than the Fast–Fourier transform [8.99].

By construction, the initial single particle states $\psi_{\lambda,k}$ are orthonormal. The process of propagation gradually destroys the orthogonality. As the constituents of $X(\phi,\sigma)$ favour the projection of a state on their ground states, the vectors $\{X_{ll'}\psi_{\lambda,l'}\}$ have the tendency to become parallel, especially at low temperature ($\beta\to\infty$). Without special precautions, this leads to numerical instabilities in the calculation of the determinants. An important feature of the projector method (as applied to fermion problems) is that this fundamental problem can be eliminated by re-orthogonalising the single-particle states as the propagation process proceeds [8.101, 103, 104]. Orthogonalisation does not affect the value of the full wavefunction as it is equivalent to multiplication of the matrix $\{\psi_{\lambda,l}\}$ by a triangular matrix having a determinant of one [8.66].

Numerical experiments show that it is not necessary to re-orthogonalise after each propagation step. The total number of re-orthogonalisations required to achieve stability and high accuracy depends on β, the number of decompositions in the Trotter–Suzuki formula, and the model parameters, and has to be determined by monitoring the accuracy of the calculation of the wavefunction itself.

For models such as the Hubbard model, a ground-state expectation value takes the form

$$\langle A_\uparrow\rangle=\frac{\int d\phi\, d\phi'\langle\det\{\psi_\lambda^\uparrow(\phi')\}|A_\uparrow|\det\{\psi_\lambda^\uparrow(\phi)\}\rangle\langle\det\{\psi_\lambda^\downarrow(\phi')\}|\det\{\psi_\lambda^\downarrow(\phi)\}\rangle}{\int d\phi\, d\phi'\langle\det\{\psi_\lambda^\uparrow(\phi')\}|\det\{\psi_\lambda^\uparrow(\phi)\}\rangle\langle\det\{\psi_\lambda^\downarrow(\phi')\}|\det\{\psi_\lambda^\downarrow(\phi)\}\rangle}, \tag{8.46}$$

where we have assumed that $A_\uparrow$ does not contain spin-down operators, to illustrate how the matrix elements decompose according to the spin. This is a direct consequence of the use of the Hubbard–Stratonovich transformation. To compute the scalar products appearing in (8.46), four determinants have to be calculated. The orthogonalisation procedure discussed above ensures that this can be done to high accuracy. The form (8.46) has the big advantage that expectation values involving more than one pair of electron creation and annihilation operators $(c_i^+c_j)$ can be expressed via Wick's theorem in terms of the density matrix $\langle c_i^+c_j\rangle$ for which the following expression is derived in the appendix;

$$\langle\det\{\psi'_\alpha\}|c_i^+c_j|\det\{\psi_\beta\}\rangle=\langle\det\{\psi'_\alpha\}|\det\{\psi_\beta\}\rangle\cdot\sum_{\alpha\beta=1}^{n}\psi'^{*}_{\beta i}S^{-1*}_{\beta\alpha}\psi_{\alpha j}. \tag{8.47}$$

Many-body Hamiltonians, such as the Hubbard model, are represented by extremely large matrices. Diagonalisation of these matrices is impossible because of memory and CPU requirements. What has been accomplished by the manipulations that led to (8.46) is that the memory problem has been replaced by a CPU problem. Indeed, the matrices appearing in (8.46) have a dimension related to the size of the lattice, not to the size of the Hilbert space, but the price paid for this reduction is that the sums over the fields ϕ and ϕ' have to be carried out. As in the grand-canonical approach, one has to resort to the Monte Carlo simulation to compute (8.46), thereby encountering the same fundamental problem to be discussed in the next section.

8.6.2 Applications

Sorella et al. were the first to apply the projector method [8.101] to the 1D and 2D repulsive Hubbard model [8.103, 104]. The integrals over the auxiliary fields have been performed by means of a Langevin molecular dynamics technique. Results for the ground state energy and the momentum distribution were obtained. Simulations for the (finite) 1D Hubbard model showed a sharp jump in the momentum distribution away from half-filling, in disagreement with renormalisation group calculations [8.106–108]. This controversy has been resolved by several groups, using different QMC methods [8.105, 109, 110] by demonstrating that it results from subtle finite size effects. In particular it was shown [8.105, 111] that the critical length of the chain below which the continuous renormalisation group result exhibits a jump, is of the order of 10^{69} for $U = |t|$.

Up to the time of writing, PQMC simulations for the 2D single-band Hubbard model [8.112, 113] and the 2D Emery-model [8.114] have not altered the picture, obtained from GCMC calculations, with respect to the possibility for having a superconducting ground state. Upon doping the PQMC suffers from minus-sign problems akin to the ones encountered in GCMC [8.115].

To examine the interplay of strong electron correlations and a conventional pairing mechanism, a simulation study of the 2D single-band Hubbard model coupled to local anharmonic vibrations [8.116] was performed. A combination of the PQMC scheme for the electrons and the WLQMC algorithm for the phonons was used. According to *Müller* [8.117], the local anharmonic phonons are assumed to describe the salient features of the bridging oxygens in the copper–oxide superconductors. For an appropriate choice of model parameters the system was shown to be well-described by an effective attractive Hubbard model [8.118]. A transition from BCS to Bose–Einstein condensation has been observed. It was found that in all cases covered, the repulsive on-site Coulomb interaction reduces the pairing correlation function.

8.7 Fundamental Difficulties

8.7.1 The Sign Problem

A fundamental problem, usually referred to as the minus-sign problem [8.84], is encountered when the functional integrals which are to be evaluated numerically do not have a positive semi-definite measure. In general, each expectation value of interest can be symbolically written as

$$\langle A \rangle = \frac{\int A(x)\rho(x)dx}{\int \rho(x)dx}, \tag{8.48}$$

where $\rho(x)$ and $A(x)$ are known real functions of the field variables denoted by x. In general the "probability density" $\rho(x)$ is not strictly positive except in some special cases: a) 1D Hubbard model with $N_\uparrow = N_\downarrow$ [8.119]; b) the half-filled Hubbard model on a hypercube [8.83, 84] and c) the attractive Hubbard model on any lattice for $N_\uparrow = N_\downarrow$.

If $\rho(x)$ changes sign, it cannot be considered as a probability density for the importance sampling in the (Metropolis) Monte Carlo summation [8.69]. The standard trick [8.120] to avoid this difficulty is to use the probability density $\tilde{\rho}(x) = |\rho(x)|/(\int |\rho(x)|dx)$ to perform importance sampling, and to absorb the sign of $\rho(x)$ in the quantity to be measured. Symbolically,

$$\langle A \rangle = \frac{\int A(x)\cdot \mathrm{sign}(\rho(x))\cdot \tilde{\rho}(x)dx}{\int \mathrm{sign}(\rho(x))\cdot \tilde{\rho}(x)dx} = \frac{\langle A\cdot \hat{s}\rangle}{\langle \hat{s}\rangle}, \tag{8.49}$$

where $\hat{s}$ is the sign operator, as is obvious from (8.49). At first sight, this simple, mathematically correct trick might even work. However, it is important to recognise that importance sampling based on $|\rho(x)|$ is unlikely to capture the correct physics, which is described by $\rho(x)$, as it is most likely to sample predominantly unimportant regions in phase space. Except for certain 1D lattice fermion system [8.3], no general results are known about the validity of the hypothesis, that one obtains the correct results by sampling $|\rho(x)|$. In practice the breakdown of this procedure is signaled by the vanishing of the average sign $\langle \hat{s}\rangle$. If $\langle \hat{s}\rangle$ is close to unity, $\tilde{\rho}(x)$ should be as good a guiding function as $\rho(x)$ is, in cases where $\rho(x) > 0$. Otherwise, there will be large cancellations in both, $\langle A\cdot \hat{s}\rangle$ and $\langle \hat{s}\rangle$ and statistical fluctuations will have a large variance and accurate evaluation of $\langle A\rangle$ will be extremely difficult.

The importance of the minus-sign problem has been studied numerically in the frame work of the Hubbard model [8.91, 121, 122]. Numerical studies reveal that in general, apart from the above mentioned exception, the average sign falls exponentially with the inverse temperature, the system size and the coupling strength $\langle \hat{s}\rangle = \alpha \exp(-\beta N U \gamma)$. The decay constant γ depends strongly on the filling. If the numbers of electrons $N_\uparrow(N_\downarrow)$ are such that the ground state of H_{kin} is not degenerate, which is often referred to as closed shell situation, γ is fairly small, whereas in open shell cases the situation is much worse [8.91]. Qualitatively, this can be understood as the finite size gap in the closed shell situation reduces the scattering of the particles via the HS fields and corresponds effectively to a smaller U. GCMC and PQMC with Hartree–Fock trial functions have about the same decay constant γ and differ only in the prefactor α, which is bigger in the ground state algorithm (PQMC) [8.121]. On top of that, a good trial wavefunction can reduce the value of β necessary to obtain accurate ground state properties. Although we have discussed the minus-sign problem within the context of MC schemes, other simulation techniques suffer from the same difficulties, be it in a different form. It is known that simulations employing the Langevin algorithm can go out of equilibrium, when crossing nodal surfaces [8.122].

As the minus-sign problem is a major stumbling block for applying Monte Carlo methods to quantum systems, it is important to isolate the source of this problem. It is commonly believed that the minus-sign problem in the Hubbard model is related to Fermi statistics. Using ideas introduced in the frame work of the worldline technique [8.123] it is argued in [8.121] that the minus-sign problem is a consequence of the exchange of electrons. Here we want to show that it is a much more general feature directly linked to the quantum mechanical nature of the system. To this end we focus on the simplest quantum system conceivable: A Pauli spin in a magnetic field described by the Hamiltonian

$$H = a\sigma^x + b\sigma^y, \tag{8.50}$$

with the Pauli matrices

$$\sigma^x = \begin{pmatrix} 0 & 1 \\ 1 & 0 \end{pmatrix}, \quad \sigma^y = \begin{pmatrix} 0 & -i \\ i & 0 \end{pmatrix}, \quad \sigma^z = \begin{pmatrix} 1 & 0 \\ 0 & -1 \end{pmatrix}. \tag{8.51}$$

Needless to say this model is trivially solvable but let us assume that this is now known. A common step in the methods discussed above is the application of the Trotter–Suzuki formula,

$$\mathrm{e}^{-\beta H} \cong [\mathrm{e}^{-\beta a\sigma^x/m}\mathrm{e}^{-\beta b\sigma^y/m}]^m. \tag{8.52}$$

Without loss of generality we can choose the eigenvectors of σ^z as the representation to work in. Inserting complete sets of states we obtain

$$\begin{aligned} Z_m &= \mathrm{Tr}(\mathrm{e}^{-\beta a\sigma^x/m}\mathrm{e}^{-\beta b\sigma^y/m})^m \\ &= \sum_{\{\phi_l\}} \prod_{l=0}^{m-1} \langle\phi_{2l}|\mathrm{e}^{-\beta a\sigma^x/m}|\phi_{2l+1}\rangle\langle\phi_{2l+1}|\mathrm{e}^{-\beta b\sigma^y/m}|\phi_{\phi_{2l+2}}\rangle \\ &= \left[\cosh\left(\frac{\beta a}{m}\right)\cosh\left(\frac{\beta b}{m}\right)\right]^m \\ &\quad \cdot \sum_{\{\phi_l\}} \prod_{l=0}^{m-1} [\delta_{\phi_{2l},\phi_{2l+1}} + \mathrm{i}t_a(\phi_{2l} - \phi_{2l+1})] \\ &\quad \cdot [\delta_{\phi_{2l+1},\phi_{2l+2}} - t_b(1 - \delta_{\phi_{2l+1},\phi_{2l+2}})], \end{aligned} \tag{8.53}$$

whereby $|\phi_{2m}\rangle = |\phi_0\rangle$, $\phi_l = \pm 1$, $t_a = \mathrm{i}\tanh(\beta a/m)$, $t_b = \mathrm{i}\tanh(\beta b/m)$. It is easy to convince oneself that the phase of the contributions to the sum can be ± 1, $\pm\mathrm{i}$. In particular $Z_1 = \mathrm{Tr}(1 + \mathrm{i}t_a t_b\sigma^z)$ demonstrates that none of the contributions to Z_1 are real and positive. A nice feature of this almost trivial example is that it is easy to see how to fix the minus-sign problem in this particular case. Performing a rotation in spin–space such that $\sigma^y \to \sigma^z$, the Hamiltonian reads $H = a\sigma^x + b\sigma^z$ and the partition function becomes

$$Z = \sum_{\{\phi_l\}} \prod_{l=0}^{m-1} \left[\delta_{\phi_l,\phi_{l+1}} \cosh\left(\frac{\beta a}{m}\right) - (1 - \delta_{\phi_l,\phi_{l+1}}) \sinh\left(\frac{\beta a}{m}\right)\right] \mathrm{e}^{-b\beta\phi_{l+1}/m}, \tag{8.54}$$

and again periodic boundary condition in the time direction demands

$|\phi_m\rangle = |\phi_0\rangle$. All contributions are now strictly positive as the number of spinflips is even.

From this example we learn that the choice of the representation for the states, in combination with the use of the Trotter–Suzuki formula, is responsible for producing negative contributions to the partition function.

8.7.2 Numerical Instabilities

A second numerical problem encountered in fermion simulations concerns the stability of the calculation of the determinants that appear both in the grand-canonical and projection operator technique. This problem becomes most severe at low temperature. It has a deeper physical reason which becomes most transparent in the PQMC scheme. It is likewise applicable to all the other QMC methods discussed in this chapter. The electrons are treated independently by having them propagate along the imaginary time axis. Consequently all one-particle orbitals become linearly dependent as the propagation is carried further and further. They try to form a Bose condensate which would have a lower energy. The Fermi statistics enters only at the very moment when the orbitals are forced into the Slater determinant. No problems would arise if the propagated wavefunctions were given exactly. However, since the propagation is performed numerically with finite precision, the accuracy declines when the difference between the states becomes much smaller than their norm.

The reason in mathematical terms is that the spectral range of the matrices appearing in (8.39) and (8.46) increases with β and m (due to the Hubbard–Stratonovich transformation). This leads to numerical instabilities in the evaluation of matrix elements and determinants in particular. For the 2D Hubbard model, earlier grand-canonical simulations have been effectively limited to temperatures $\beta \leqq 4$ (for $t = 1$, $U = 4$) [8.84]. In the grand-canonical approach, clever rearrangements of the matrix multiplications, some of which are inspired by the singular value decomposition, have been proposed and implemented [8.90, 91, 115, 125–128]. Simulation results for temperatures down to $\beta \approx 60$ (for $t = 1$ and $U = 4$) have been reported.

In the projector operator approach, one exploits the fact that it is possible to re-orthogonalise the set of one-particle functions during the propagation (in imaginary time). In practice the set of single-particle wavefunctions is made orthogonal after a number of projection steps have been carried out. This eliminates all problems related to the numerical instability in the calculation of the determinants. The lowest temperature one can reach is limited by the number of time-slices required to keep the systematic error resulting from the Trotter–Suzuki formula below an acceptable level, not by numerical instabilities. This problem is aggravated in WLQMC as the tendency towards the bosonic ground state and the minus-sign problem are coupled. The guiding function $|\rho(x)|$ is that for a hard-core bose system and is peaked in a different region than $A(x)\cdot\rho(x)$, and it samples the tails of $A(x)\cdot\rho(x)$ [8.123]. This leads, as in the example in [8.70], to an exponential increase of the variance with the number of Trotter slices m and the inverse temperature β. To date no satisfactory solution has been found for this problem.

8.7.3 Dynamic Susceptibilities

A further fundamental problem is to determine dynamic correlations. The GCMC and PQMC calculations provide a natural framework for numerical computation of imaginary time Green's functions [8.84, 112, 119]. Equal time correlation functions and static susceptibilities are accessible. It is, however, difficult to extract the dynamic properties from the simulation data because an analytic continuation or inverse Laplace transform from imaginary time to real time is required. There have been earlier attempts to obtain dynamic susceptibilities. For a system of spinless fermions in 1D a least square fit procedure [8.129, 130] was used. The problem has also been approached by *Hirsch* [8.131] using Padé approximants. Recently, new progress has been made in this problem [8.32, 33, 132–134].

The dynamic properties of interest are given by the spectral density $A(w)$. This quantity depends in general on several quantum numbers which are irrelevant for the ensuing discussion and will therefore be suppressed. The spectral density is related to the imaginary-time Green's function by

$$g(\tau) = \int_{-\infty}^{+\infty} dw A(w) \frac{e^{-w\tau}}{1+e^{-\beta w}}, \tag{8.55}$$

The inverse transform of (8.55) from QMC data is extremely ill-posed for two reasons: a) the data for $g(\tau)$ are provided only for a limited set of imaginary times $\{\tau_l\}$ and b) these data are noisy. The spectral function is represented as a sum over δ-functions $A(\omega) = \sum_l A_l \cdot \delta(w - w_l)$ and the transform (8.55) becomes

$$g(\tau) = \sum_l A_l \frac{e^{-w_l \tau}}{1+e^{-\beta w_l}}. \tag{8.56}$$

A least square fit procedure would minimize the χ^2-measure of the data,

$$\chi^2 = \sum_l \frac{[g_d(\eta) - g_f(\tau_l)]^2}{\sigma_l^2}, \tag{8.57}$$

where $g_d(\tau_l)$ are the QMC results and $g_f(\tau_l)$ are the data which one would obtain for a given choice of A_l in (8.56). σ_l is a measure of the reliability of the QMC results for time τ_l and is usually assumed constant, independent of l. Due to the restricted number of experimental (QMC) data points for $g(\tau_l)$, large changes in A_l lead only to small changes in χ^2. Further restrictions, based on *a priori* knowledge, are needed. *White* et al. [132, 32] proposed a modified least square measure,

$$\tilde{\chi}^2 = \chi^2 + b \cdot \sum_l (A_l - A_{l+1})^2 + h \cdot \sum_{\substack{l \\ A_l < 0}} A_l^2. \tag{8.58}$$

The second term enforces smoothness and the third term allows for the positivity of the spectral density. To determine the adjustable parameters b and h, a rough prescription is given in [8.132]. The smoothness constraint b minimises the structure of $A(w)$ regardless of whether it has physical relevance or whether it is due to noisy QMC data and can lead to spurious results. This scheme demands therefore prior good knowledge about the final result. It has been applied to the Hubbard model at half filling and the Anderson impurity model.

Prompted by the similarity of the problems encountered in image reconstruction, *Silver* et al. [8.133, 33] used the maximum entropy (MaxEnt) method—well estalished in this field—as a statistical regularisation procedure. One good feature of this technique is that not only does it yield an estimate for the spectral density, but also an error for integrated densities. This gives additional information about the reliability of certain structures in the density of states, e.g. gaps. Input to the MaxEnt theory is in the first place the χ^2-measure of the data (8.57). Further input to MaxEnt is as much as possible a priori knowledge of the physical problem under consideration, like positivity of $A(w)$, sum rules etc. This information is fed into a default model $m(w)$ for $A(w)$ and is used to obtain the information theory entropy S. S is essentially the logarithm of the number of ways by which one can arrive at a given $A(w)$ in a Poisson process from the default model;

$$S = \int_{-\infty}^{+\infty} dw \left[A(w) - m(w) - A(w) \log\left(\frac{A(w)}{m(w)} \right) \right]. \tag{8.59}$$

The spectral density is the MaxEnt scheme is obtained on maximising the probability

$$P = e^{-1/2\chi^2} \cdot e^{\alpha S}. \tag{8.60}$$

Roughly speaking MaxEnt looks for the most likely $A(w)$ subject to minimising the mean square deviation (8.58) and maximising the "overlap" to our prior knowledge. The first term in (8.60) alone would lead to a mean square fit. In the opposite case of completely noisy experimental data which contain no useful information about the spectrum, the second term would lead to the default model. First applications to the Anderson impurity model have yielded encouraging results although further studies are needed to really show the reliability of the method. The MaxEnt scheme is nonetheless also biased by the choice of the default model, and the way the trade-off is made between taking the experimental structures seriously or preferring the default model.

8.7.4 Applicability

The QMC methods are specially geared to the models under consideration. For GCMC and PQMC the Hubbard Stratonovich transformation is used. This transformation is in principle applicable to any electron–electron interaction [8.68]. In practice the number of HS fields quickly becomes intractable. In the GFMC and the GCMC schemes at least arbitrary density–density type of interactions $H = \sum_{ij} V_{ij} n_i n_j$ can be treated. These schemes in turn make special use of the nearest neighbour hopping entering the Hubbard model.

For classical statistical systems, an essential condition for the Monte Carlo method to work properly is that the sequence of states generated is ergodic [9.69]. Obviously, any quantum Monte Carlo method has to be ergodic too. Although it is not difficult to devise algorithms that are ergodic in principle, in practice it is not self-evident how to sample the phase space (of the auxiliary fields) in an adequate manner [8.135]. The basic reason is that these fields lack proper physical meaning and consequently it is by no means obvious that simple local changes in the fields are sufficient to provide a good coverage of phase space.

8.8 Concluding Remarks

Quantum Monte Carlo methods have been applied to a variety of problems and it is clearly impossible to go into a discussion of all those results. Instead we give in Table 8.3, a list of all recent Quantum Monte Carlo studies along with some key words of applications and a list where to find algorithmic details. Again a selection had to be made and we list only those references which are related to the models we have discussed in this chapter. Additionally, Table 8.3 contains references to new developments which are still in their infancy.

We have given a concise overview of recent activities in the field of Quantum Monte Carlo studies for lattice problems. This field has gained in dynamics and general interest during the last couple of years and has undeniably made considerable progress. Quantum Monte Carlo calculations have shed more light on many open questions in the field of strongly correlated many-body problems. But being realistic, most of the Quantum Monte Carlo results are, however, not yet rigorous and the applicability of these methods is still limited to a small set of models. Quite a few obstacles are yet to be removed to make Quantum Monte Carlo Simulations as powerful as classical Monte Carlo methods.

8.A Appendix

Here we collect and prove some identities that have been used throughout this chapter. First consider the following n-fermion state

$$e^{c_i^+ A_{ij} c_j} c_{l_1}^+ \cdots c_{l_n}^+ |0\rangle, \tag{8.A1}$$

where A is a diagonalisable matrix and summation over i and j is implicitly assumed. By changing the representation to a basis that diagonalises A, i.e. $V^+ A V = D$ where D is a diagonal matrix, it is easily shown that

$$e^{c_i^+ A_{ij} c_j} c_{l_1}^+ \cdots c_{l_n}^+ |0\rangle = \sum_{\{l_i'\}} (e^A)_{l_1 l_1'} \cdots (e^A)_{l_n l_n'} c_{l_1'}^+ \cdots c_{l_n'}^+ |0\rangle. \tag{8.A2}$$

By induction, it then follows that for $\tilde{X} \equiv \exp[c_i^+ A_{ij}^{(1)} c_j] \cdots \exp[c_i^+ A_{ij}^{(p)} c_j^+]$,

$$\tilde{X} c_{l_1}^+ \cdots c_{l_n}^+ |0\rangle = \sum_{\{l_i'\}} X_{l_1 l_1'} \cdots X_{l_n, l_n',} c_{l_1'}^+ \cdots c_{l_n'}^+ |0\rangle, \tag{8.A3}$$

where $X_{ij} = \{\exp[A^{(1)}] \cdots \exp[A^{(p)}]\}_{ij}$.

To prove that $\mathrm{Tr}\, \tilde{X} = \det(1 + X)$, we first use (8.A3) to write

$$\mathrm{Tr}\, \tilde{X} = \sum_{n=0} \sum_{\{l_i\}} \sum_{P^{(n)}} \mathrm{sign}(P^{(n)}) X_{l_1, l_{P_1^{(n)}}} \cdots X_{l_n, l_{P_n^{(n)}}}, \tag{8.A4}$$

where $P^{(n)}$ denotes a permutation of the numbers $1, \ldots, n$, $\mathrm{sign}(P^{(n)})$ is the parity of permutation $P^{(n)}$ and the sum over $P^{(n)}$ runs over all $n!$ permutations. The r.h.s. of (8.A4) is nothing but the Sylvester expansion of $\det(\boldsymbol{1} + \boldsymbol{X})$ [8.136]. Another route to prove $\mathrm{Tr}\, \tilde{X} = \det(\boldsymbol{1} + \boldsymbol{X})$ is to use similarity transformations to write $\tilde{X}$ (which is not necessarily similar to a diagonal matrix) in its Jordan canonical form [8.66], perform the trace over all possible fermionic states and to note that

$$1 + \sum_i x_i + \sum_{i<j} x_i x_j + \cdots = \prod_i (1 + x_i). \tag{8.A5}$$

Next we derive the expression for the density matrix $\langle \det\{\psi_\alpha'\} | c_i^+ c_{i'} | \det\{\psi_\beta\} \rangle$. To this end we introduce new operators;

$$b_\alpha^+ = \sum_i \psi_{\alpha i} c_i^+, \quad b_\alpha'^+ = \sum_i \psi_{\alpha i}' c_i^+, \tag{8.A6}$$

where the $\psi_{\alpha i}$ and $\psi_{\alpha i}'$ are one-particle orbitals. To simplify the notation we have dropped the spin index. As a first step we evaluate the inner product of two Slater-determinants with different sets of one-particle orbitals. By resorting to Wick's theorem one obtains

$$\langle \det\{\psi_\alpha'\} | \det\{\psi_\beta\} \rangle = \langle 0 | \prod_{\alpha=n}^{1} b_\alpha' \prod_{\alpha=1}^{n} b_\alpha^+ | 0 \rangle = \det(\boldsymbol{O}), \tag{8.A7}$$

with the overlap matrix

$$\boldsymbol{O}_{\alpha,\beta} = \langle 0 | b_\alpha' b_\beta^+ | 0 \rangle = \sum_i \psi_\alpha'^*(i) \psi_\beta(i). \tag{8.A8}$$

The density matrix is

$$\langle \det\{\psi_\alpha'\} | c_i^+ c_{i'} | \det\{\psi_\beta\} \rangle = \delta_{i,i'} \det(\boldsymbol{O}) - \langle \det\{\psi_\alpha'\} | c_i' c_i^+ | \det\{\psi_\beta\} \rangle. \tag{8.A9}$$

Upon defining auxiliary one-particle states $b_0^{(\prime)} = c_{i^{(\prime)}}$ the last term in (8.A9) corresponds to the inner product of two Slater-determinants each with one additional orbital. According to (8.A7) and (8.A8) we obtain

$$\langle \det\{\psi_\alpha'\} | c_i' c_i^+ | \det\{\psi_\beta\} \rangle = \det(\mathscr{O}). \tag{8.A10}$$

As compared to the overlap matrix $\boldsymbol{O}$ (8.A7), $\mathcal{O}$ is enlarged by an additional 0-th column and 0-th row:

$$\mathcal{O}_{\alpha\beta} = \begin{cases} \boldsymbol{O}_{\alpha\beta} & \text{if } \alpha > 0, \beta > 0; \\ \psi_{\beta i'} & \text{if } \alpha = 0, \beta > 0; \\ \psi'^{*}_{\beta i} & \text{if } \alpha > 0, \beta = 0; \\ \delta_{ii'} & \text{if } \alpha = 0, \beta = 0. \end{cases} \tag{8.A11}$$

It is convenient to express $\det(\mathcal{O})$ in terms of the original overlap matrix $\boldsymbol{O}$ to extract the i, i' dependence. The 0-th column is modified by adding to it a linear combination of the other columns;

$$\tilde{\mathcal{O}}_{\alpha 0} = \mathcal{O}_{\alpha 0} + \sum_{\beta=1}^{n} \mathcal{O}_{\alpha\beta} \mathscr{C}_{\beta}. \tag{8.A12}$$

This leaves the determinant unchanged but gives us the opportunity to put all but the 0-th element of this column to nought ($\mathcal{O}_{\alpha 0} = 0$ for $\alpha = 1, 2, \ldots, n$). This is achieved by

$$\mathscr{C}_{\beta} = - \sum_{\gamma=1}^{n} \boldsymbol{O}^{-1}_{\beta\gamma} \mathcal{O}_{\gamma 0}. \tag{8.A13}$$

Due to the simple structure of the 0-th column the determinant is

$$\det(\mathcal{O}) = \tilde{\mathcal{O}}_0 \cdot \det(\boldsymbol{O}), \tag{8.A14}$$

where the entire i, i'-dependence is in the first factor. We reach our final goal if we now collect what we have learned from (8.A10, A11, A13, A14):

$$\langle \det\{\psi'_{\alpha}\} c_i^{+} c_{i'} | \det\{\psi_{\beta}\} \rangle = \langle \det\{\psi'_{\alpha}\} | \det\{\psi_{\beta}\} \rangle \sum_{\alpha\beta=1}^{n} \psi'^{*}_{\beta i} S^{-1*}_{\beta\alpha} \psi_{\alpha i'}. \tag{8.A15}$$

Table 8.3. List of recent publications on QMC for lattice problems. (a) Review articles and publications containing details about QMC techniques. (b) List of articles on models discussed in this chapter. Lattice dimensions are given in curly brackets. If necessary more specific characteristics as to the model are given. Abbreviations are defined in Table 8.1.

Topic	Application	References
Review Articles		
VMC, GFMC	helium, electron gas	[8.1, 2]
WLQMC, Path Integrals	general lattice problems	[8.3]
QMC general	quantum spins, fermions	[8.3, 4]
QMC general	general	[8.68]
WLQMC, QMC-RG	quantum spins	[8.148]
Technical Details		
discrete HS, algorithm	Hubbard model	[8.82]
GCMC, algorithm		[8.86]
Trotter decomposition	fermions	[8.3, 99]
PQMC, algorithm		[8.101, 102, 104, 105]
WLQMC	spinless fermions	[8.145]
WLQMC, algorithm	fermions	[8.119, 155]
WLQMC, algorithm	Hubbard model	[8.123]
GFMC, algorithm	Heisenberg model	[8.56, 63]
e–ph simulation, algorithm	Hubbard model + phonons	[8.116]
GCMC stabilization	Hubbard model	[8.115, 125–128]
studies:		
discrete vs. continuous HS	Hubbard model	[8.82, 85]
LD	Hubbard model	[8.91, 122]
GFMC	2×2 matrix	[8.70]
ergodicity (WLQMC)	Heisenberg model	[8.135]
minus-sign problem:		
GCMC, PQMC	Hubbard model	[8.103, 115, 121, 122, 167]
WLQMC	Hubbard model	[8.119, 123, 124]
dynamic susceptibility:		
$g(\tau)$		[8.84, 112, 119, 128]
MaxEnt, algorithm	Anderson impurity model	[8.133, 134]
$A(\omega)$	Anderson impurity model	[8.132]
Proposed Algorithms		
Hybrid	Hubbard model	[8.122, 151, 153, 154]
Hybrid	spinless fermions	[8.150, 163]
Hybrid	Su–Schriefer–Heeger–model	[8.152]
auxiliary bosons	Hubbard model	[8.90]
subspace diagonalization	Heisenberg model	[8.149]
decoupling cell method	Heisenberg model, XY-model	[8.157–159]

Table 8.3 (*Continued*)

Application	Method	References
{1,2} M, E	VMC(GWZ)	[8.40–44, 59]
{2} M, E, $\omega(\boldsymbol{q})$	VMC(Jastrow)	[8.52, 53, 56]
{2} M, E	VMC(RVB)	[8.54, 59]
{1,2} M, E	GFMC	[8.56, 63–65, 71–73, 156]
{2} M	Handscomb	[8.75–77]
{2} M, E	WLQMC	[8.78–80]
{1} $\omega(\boldsymbol{q})$	PQMC	[8.100]
{1} t-J-model, E, M	WLQMC	[8.112]
Hubbard model		
{1,2} M, E	VMC(GWA)	[8.40–44]
{2} M, sc, E	VMC(RVB, GWA)	[8.59, 147]
{1} EHM, $n(k)$	WLQMC	[8.146]
{1,2} EHM, M, sc	GCMC, WLQMC	[8.15, 18, 19, 144, 161, 160]
{2} e–ph, SDW, CDW, E	GCMC	[8.21]
{2} M, E	GCMC	[8.83]
{1,2} M, E, gap, $n(k)$, $U<0$, sc	GCMC	[8.84]
{3} M, neg. U, sc	GCMC	[8.87]
{2} M, sc, E, $\omega(\boldsymbol{q})$, sign	GCMC, PQMC	[8.81]
{2} $n(k)$, M, E, $\Sigma(\omega)$, gap	GCMC	[8.92, 138]
{1,2} E, $n(k)$, M	PQMC(LD)	[8.103, 104, 141]
{2} e–ph, sc, E	PQMC+WLQMC	[8.116]
{2} Emery-model, gap, M, sc	GCMC	[8.22, 23, 94, 95, 137, 139]
{2} Emery-model, M, sc, n(k)	GCMC, PQMC	[8.128]
{2} $A(\omega)$	GCMC	[8.132]
{1} EHM CDW, SDW, E	GCMC	[8.140]
{2} Emery-model sc	PQMC	[8.114]
{1,2,3} $U<0$, M, sc	GCMC	[8.15, 128, 116, 93, 87, 84]
{1,2} M, E, sc	GCQM	[8.84, 88–93, 112, 113] [8.142, 165, 166]
{1,2} $n(k)$, E	GCMC, PQMC	[8.84, 92, 103–105, 110] [8.112, 141–142]
2 EM, sc, $n(k)$, M	GCMC	[8.16, 17, 20, 143, 153, 162]
Anderson model		
{2} impurity, M, $n(r)$	GCMC	[8.24, 27, 28]
{2} impurity, M, $n(r)$	GCMC	[8.25, 131, 164]
{1} lattice, M	GCMC	[8.30, 31, 33]
{2} impurity $A(\omega)$	GCMC	[8.29]
{2} impurity $A(\omega)$, M, transport	GCMC	[8.30, 31, 33]

References

8.1 D.M. Ceperly, M.H. Kalos: In *Monte Carlo Methods in Statistical Physics*, ed. by K. Binder, Topics Curr. Phys. Vol. 7 (Springer, Berlin, Heidelberg 1979)

8.2 K.E. Schmidt, M.H. Kalos: In *Applications of the Monte Carlo Methods in Statistical Physics*, ed. by K. Binder, Topics Curr. Phys. Vol. 36 (Springer, Berlin, Heidelberg 1984)

8.3 H. De Raedt, A. Lagendijk: Phys. Rep. **127**, 233 (1985)

8.4 M. Suzuki: In *Quantum Monte Carlo Methods*, ed. by M. Suzuki, Springer Ser. Solid-State Sci. Vol. 74 (Springer, Berlin, Heidelberg 1986)

8.5 J. Hubbard: Proc. Roy. Soc. A **276**, 238 (1963)

8.6 E. Lieb, F. Wu: Phys. Rev. Lett. **20**, 1445 (1968)
8.7 Y. Nagaoka: Phys. Rev. **115**, 2 (1959)
8.8 A. Barbieri, J. A. Riera, A.P. Young: Stability of the Nagaoka State in the One Band Hubbard Model, preprint
8.9 B.S. Shastry, H.R. Krishnamurthy, P.W. Anderson: Phys. Rev. B **41**, 2375 (1990)
8.10 W. von. der Linden, D.M. Edwards: J. Phys. C **3**, 4917 (1991)
8.11. J. Kanamori: Prog. Theor. Phys. **30**, 275 (1963)
8.12 R. Micnas, J. Ranninger, S. Robaszkiewicz: Rev. Mod. Phys. **62**, 113 (1990)
8.13 V.J. Emery: Phys. Rev. Lett. **58**, 2794 (1987)
8.14 M. Rasetti (Ed.): *The Hubbard Model* (World Scientific, Singapore 1991)
8.15 Y. Zhang, J. Callaway: Phys. Rev. B **39**, 9397 (1989)
8.16 M. Imada: In *Strongly Coupled Plasma Physics*, ed. by S. Ichimaru (North Holland, Amsterdam 1990)
8.17 M.Imada: In *Mechanisms of High-T_c Superconductivity*, ed. by H. Kamimura, A. Oshiyama, Springer Ser. Mater. Sci. Vol. II (Springer, Berlin, Heidelberg 1989)
8.18 K. Saitoh, S. Takada: J. Phys. Soc. Jpn **58**, 783 (1989)
8.19 R.R. dos Santos: Phys. Rev. B **39**, 7259 (1989)
8.20 M. Imada, N. Nagaosa, Y. Hatsugai: *J. Phys. Soc. Japan* **58**, 978 (1989)
8.21 A. Muramatsu, W. Hanke: Phys. Rev. B **38**, 878 (1988)
8.22 G. Dopf, A. Muramatsu, W. Hanke: Physica C **162**, 807 (1989)
8.23 G. Dopf T. Kraft, A. Muramatsu, W. Hanke: Charge-transfer gap from a Monte Carlo simulation of three-band Hubbard Model, preprint
8.24 J.E. Gubernatis, T.C. Olson, D.J. Scalapino, R.L. Sugar: J. Stat. Phys. **43**, 831 (1986)
8.25 J.E. Hirsch, R.M. Fye: Phys. Rev. Lett. **56**, 2521 (1990)
8.26 J.E. Gubernatis: In *Quantum Monte Carlo Methods*, ed. by M. Suzuki, Springer Ser. Solid-State Sci. Vol. 74 (Springer, Berlin, Heidelberg 1986)
8.27 J.E. Gubernatis, J.E. Hirsch, D.J. Scalapino: Phys. Rev. B **35**, 8478 (1987)
8.28 J.E. Gubernatis: Phys. Rev. B **36**, 394 (1987)
8.29 R.M. Fye: Phys. Rev. B **41**, 2490 (1990)
8.30 M. Jarrell, O. Biham: Phys. Rev. Lett. **63**, 2504 (1989)
8.31 O. Biham, M. Jarrell, C. Jayaprakash: Phys. Rev. B **41**, 2639 (1990)
8.32 S.R. White, D.J. Scalapino, R.L. Sugar, N.E. Bickers: Phys. Rev. Lett. **63**, 1523 (1989)
8.33 M. Jarrell, J.E. Gubernatis, R.N. Silver, D.S. Sivia: Phys. Rev. B **43**, 1206 (1991)
8.34 A. Luther, I. Peschel: Phys. Rev. B **12**, 3908 (1975)
8.35 T. Matsubara, H. Matsuda: Prog. Theor. Phys. **16**, 569 (1956)
8.36 E. Lieb, D. Mattis: J. Math. Phys. **3**, 749 (1962)
8.37 E. Lieb, T. Schultz, D. Mattis: Ann. Phys. **16**, 407 (1961)
8.38 A.B. Harris, R.V. Lang: Phys. Rev. **157**, 295 (1967)
8.39 T.M. Rice: Helv. Phys. Acta **63**, 336 (1990)
8.40 T.A. Kaplan, P. Horsch, P. Fulde: Phys. Rev. Lett. **49**, 889 (1982)
8.41 P. Horsch, T.A. Kaplan: J. Phys. C **16**, L1203 (1983)
8.42 P. Horsch, T.A. Kaplan: Bull. Am. Phys. Soc. **30**, 513 (1985)
8.43 C. Gros, R. Joint, T.M. Rice: Phys. Rev. B **36**, 381 (1987)
8.44 C. Gros, R. Joint, T.M. Rice: Z. Phys. B **68**, 425 (1987)
8.45 A. Bijl: Physica **7**, 860 (1940)
8.46 R. Jastrow: Phys. Rev. **98**, 1479 (1955)
8.47 S. Doniach, M. Iuni, V. Kalmeyer, M. Gabay: Europhys. Lett. **6**, 663 (1988)
8.48 J.L. Hulthén: Ark. Mat. Astr. Fjs. A **26**, 1 (1938)
8.49 P.W. Kasteleijn: Physica **18**, 104 (1952)
8.50 W. Marschall: Proc. Roy. Soc. A **232**, 64 (1955)
8.51 W. von der Linden, M. Ziegler, P. Horsch: Phys. Rev. B **40**, 7435 (1989)
8.52 P. Horsch, W. von der Linden: Z. Phys. B **72**, 181 (1988)
8.53 D.A. Huse, V. Elser: Phys. Rev. Lett. **60**, 2531 (1988)
8.54 S. Liang, B. Doucot, P.W. Anderson: Phys. Lett. **61**, 365 (1988)
8.55 S.M. Girvin, A.H. MacDonald: Phys. Rev. Lett. **58**, 1252 (1987)
8.56 N. Trivedi, D.M. Ceperley: Phys. Rev. B **40**, 2737 (1989)
8.57 E. Manousakis: In *Quantum Simulations*, ed. by J.D. Doll, J.E. Gubernatis (World Scientific, Singapore 1990)
8.58 In [8.51] a slightly different Jastrow wavefunction was used in which a variational degree of freedom was introduced to suppress ferromagnetic domains according to their size

8.59 H. Yokoyama, H. Shiba: J. Phys. Soc. Jpn. **56**, 3570 (1987)
8.60 D.M. Ceperley, B.J. Alder: Phys. Rev. Lett. **45**, 566 (1980)
8.61 D.M. Ceperley, B.J. Alder: J. Chem. **81**, 5833 (1984)
8.62 R.M. Panoff, J. Carlson: Phys. Rev. Lett. **62**, 1130 (1989)
8.63 N. Trivedi, D.M. Ceperley: Phys. Rev. B **41**, 4552 (1990)
8.64 M. Gross, E. Sanchez-Velasco, E. Siggia: Phys. Rev. B **39**, 2484 (1989)
8.65 J. Carlson: In *Quantum Simulations*, ed. by J.D. Doll, J.E. Gubernatis, (World Scientific Publishing, Singapore 1990)
8.66 J.H. Wilkinson: *The Algebraic Eigenvalue Problem* (Clarendon, Oxford 1965)
8.67 G.E. Fortsythe, R.A. Leibler: MTAC **4**, 127 (1950)
8.68 J.W. Negele, H. Orland: *Quantum Many-Particle Systems*, Frontiers in Physics (Addison-Wesley, Reading MA 1987) p. 400
8.69 J.M. Hammersley, D.C. Handscomb: *Monte Carlo Methods* (Methuen, London 1964)
8.70 J.H. Hetherington: Phys. Rev. A **30**, 2713 (1984)
8.71 M.P. Nightingale, H.W.J. Böte: Phys. Rev. B **33**, 265 (1986)
8.72 T. Barnes, G.J. Daniell: Phys. Rev. B **37**, 3637 (1988)
8.73 T. Barnes, E.S. Swanson: Phys. Rev. B **37**, 9405 (1988)
8.74 D.C. Handscomb: Proc. Cambridge Phil. Soc. **58**, 594 (1962)
8.75 D.H. Lee, J.P. Joannoploulos, J.W. Negele: Phys. Rev. B **30**, 1599 (1985)
8.76 E. Manousakis, R. Salvador: Phys. Rev. Lett. **60**, 890 (1988)
8.77 E. Manousakis, R. Salvador: Phys. Rev. B **39**, 575 (1989)
8.78 J. Miyashita: J. Phys. Soc. Jpn **57**, 1934 (1988)
8.79 J.D. Reger, A.P. Young: Phys. Rev. B **37**, 5978 (1988)
8.80 J.D. Reger, J.A. Riera, A.P. Young: J. Phys. Cond. Matter **1**, 1855 (1989)
8.81 F. Wiegel: Phys. Rep. **16**, 59 (1975)
8.82 J.E. Hirsch: Phys. Rev. B **28**, 4059 (1983)
8.83 J.E. Hirsch: Phys. Rev. Lett. **51**, 1900 (1983)
8.84 J.E. Hirsch: Phys. Rev. B **31**, 4403 (1985)
8.85 G.M. Buendia: Phys. Rev. B **33**, 3519 (1986)
8.86 R. Blankenbecler, D.J. Scalapino, R.L. Sugar: Phys. Rev. D **24**, 2278 (1981)
8.87 J.E. Hirsch: Phys. Rev. B **35**, 1851 (1987)
8.88 J.E. Hirsch, S. Tang: Phys. Rev. B **62**, 591 (1989)
8.89 J.E. Hirsch, H.Q. Lin: Phys. Rev. B **37**, 5070 (1988)
8.90 S.R. White, R.L. Sugar, R.T. Scauttar: Phys. Rev. B **38**, 11665 (1988)
8.91 S.R. White, D.J. Scalapino, R. L. Sugar, E.Y. Loh, J.E. Gubernatis, R.T. Scalettar: Phys. Rev. B **40**, 506 (1989)
8.92 A. Moreo, D.J. Scalapino, R. L. Sugar, S.R. White, N.E. Bickers: Phys. Rev. B **41**, 2313 (1990)
8.93 R.T. Scalettar, E.Y. Loh, J.E. Gubernatis, A. Moreo, S.R. White, D.J. Scalapino, R.L. Sugar, E. Dagotto: Phys. Rev. Lett. **62**, 1407 (1989)
8.94 G. Dopf, J. Rahm, A. Muramatsu: In *Workshop of Interacting Electrons in Reduced Dimensions*, ed. by D. Baeriswyl, D.K. Campbell, p. 69 (Plenum Press, New York 1990)
8.95 M. Imada: J. Phys. Soc. Jpn. **57**, 3128 (1988)
8.96 P. Nozieres, S. Schmitt-Rink: J. Low Temp. Phys. **39**, 195 (1985)
8.97 P.W. Anderson: Phys. Rev. **124**, 41 (1961)
8.98 P.B. Wiegmann, A.M. Tsvelick: J. Phys. C **16**, 2281 (1983)
8.99 H. De Readt: Comp. Phys. Rep. **7**, 1 (1987)
8.100 M. Takahashi: Phys. Rev. Lett. **62**, 2313 (1989)
8.101 S.E. Koonin, G. Sugiyama, H. Friederich: *Proceedings of the International Symposium, Bad Honnef 1982*, Vol 214 (Springer, Berlin, Heidelberg 1982)
8.102 G. Sugiyama, S.E. Koonin: Annal Phys. **168**, 1 (1986)
8.103 S. Sorella, E. Tosatti, S. Baroni, R. Car, M. Parrinello: *Proceedings of the Adriatico Research Conference "Towards the Theoretical Understanding of the High-T_c Superconductors"*, ed. by S. Lundqvist, E. Tosatti, M. Tosi, L. Yu (World Scientific, Singapore 1988)
8.104 S. Sorella, S. Baroni, R. Car, M. Parrinello: Europhys. Lett. **8**, 663 (1989)
8.105 W. von der Linden, I. Morgenstern, H. De Raedt: Phys. Rev. B **41**, 4669 (1990)
8.106 J. Carmelo, D. Baeriswyl: Phys. Rev. B **37**, 7541 (1988)
8.107 J. Solyomo: Adv. in Phys. **28**, 201 (1979)
8.108 I.E. Dzyaloshinsky, A.I. Larkin: Sov. Phys. JEPT **38**, 202 (1974)
8.109 A. Parola, S. Sorella, M. Parrinello, E. Tosatti: D-waves, dimer and chiral states in the 2D Hubbard model, preprint

8.110 C. Bourbonnais, H. Nélisse, A. Reid, A.M.S. Tremblay: Phys. Rev. B **40**, 2297 (1989)
8.111 D. Baeriswyl, W. von der Linden: In *The Hubbard Model*, ed. by M. Rosetti (World Scientific, Singapore 1991)
8.112 M. Imada, Y. Hatsugai: J. Phys. Soc. Jpn **58**, 3751 (1989)
8.113 W. von der Linden, I. Morgenstern, H. de Raedt: In *Quantum Simulations*, ed. by I.D. Doll, J.E. Gubernatis (World Scientific Publishing, Singapore 1990)
8.114 M. Frick, P.C. Pattnaik, I. Morgenstern, D.M. Newns, W. von der Linden: Phys. Rev. B **32**, 2665 (1990)
8.115 E.Y. Loh, J.E. Gubernatis: *Electrons Phase Transitions*, ed. by W. Hauke, Y.V. Kopaev (Elsevier, New York 1990)
8.116 M. Frick, W. von der Linden, I. Morgenstern, H. De Raedt: Z. Phys. B **81**, 327 (1990)
8.117 K.A. Müller: Phase Transitions, **22**, 5 (1990)
8.118 H. De Raedt, T. Schneider, M.P. Sörensen: Z. Phys. B **79**, 327 (1990)
8.119 J.E. Hirsch, R.L. Sugar, D.J. Scalapino, R. Blankenbecler: Phys. Rev. B **26**, 5033 (1982)
8.120 H. De Raedt, A. Lagendijk: Phys. Rev. Lett. **46**, 77 (1981)
8.121 E.Y. Loh, J.E. Gubernatis, R.T. Scalettar, S.R. White, D.J. Scalapino, R.L. Sugar: Phys. Rev. B **41**, 9301 (1990)
8.122 S.R. White, J.W. Wilkins: Phys. Rev. B **37**, 5024 (1988)
8.123 I. Morgenstern: Z. Phys. B **70**, 291 (1988)
8.124 I. Morgenstern: Sign Problem, preprint
8.125 J.E. Hirsch: Phys. Rev. B **38**, 12023 (1988)
8.126 E.Y. Loh, J.E. Gubernatis, R.T. Scalettar, R.L. Sugar, S.R. White: *Workshop on Interacting Electrons in Reduced Dimensions*, ed. by D. Baeriswyl, D. Campbell (Plenum, New York 1989)
8.127 E.Y. Loh, J.E. Gubernatis, R.T. Scalettar, S.R. White, D.J. Scalapino, R.L. Sugar: Phys. Rev. B **40**, 506 (1989)
8.128 E.Y. Loh, J.E. Gubernatis: Stable numerical simulations of models of interacting electrons, preprint
8.129 H.B. Schüttler, D.J. Scalapino: Phys. Rev. Lett. **55**, 1204 (1985)
8.130 H.B. Schüttler, D.J. Scalapino: Phys. Rev. B **34**, 4744 (1986)
8.131 J.E. Hirsch: "Simulation of Magnetic Impurities in Metals", in *Quantum Monte Carlo Methods*, ed. by M. Suzuki, Springer Ser. Solid-State Sci. Vol. 74 (Springer, Berlin, Heidelberg 1987)
8.132 S.R. White, D.J. Scalapino, R.L. Sugar, N.E. Bickers: Phys. Rev. Lett. **63**, 1523 (1989)
8.133 R.N. Silver, D.S. Sivia, J.E. Gubernatis: *Quantum Simulations*, ed. by J.D. Doll, J.E. Gubernatis (World Scientific, Singapore 1990)
8.134 R.N. Silver, J.E. Gubernatis, D.S. Sivia, M. Jarell: Phys. Rev. B **41**, 2380 (1990)
8.135 S. Miyashita: *Quantum Simulations*, ed. by J.D. Doll, J.E. Gubernatis (World Scientific, Singapore 1990)
8.136 G. Kowalewski: *Determinantentheorie* (Chelsea, New York 1948)
8.137 J. Wagner, A. Muramatsu, W. Hanke: Charge transfer and magnetic pairing mechanisms in the extended Hubbard, preprint
8.138 N.E. Bickers, D.J. Scalapino, S.R. White: Phys. Rev. Lett. **62**, 961 (1989)
8.139 G. Dopf, A. Muramatsu: Monte Carlo simulation of a three band Hubbard model for htc superconductors, preprint
8.140 J.E. Hirsch, D.J. Scalapino: Phys. Rev. B **27**, 7169 (1983)
8.141 S. Sorella, E. Tosatti, S. Baroni, R. Car, M. Parinello: Int. J. Mod. Phys. B **1**, 993 (1988)
8.142 F.F. Assaad: Helv. Phys. Acta **63**, 580 (1990)
8.143 M. Imada, Y. Hatsugai: J. Phys. Soc. Jpn. **58**, 3752 (1989)
8.144 X. Zotos, W. Lehr, W. Webber: Z. Phys. B **74**, 289 (1989)
8.145 I. Morgenstern, D. Würz: Z. Phys. B **70**, 115 (1988)
8.146 W.R. Somsky, J.E. Gubernatis, D.K. Campbell: In *Quantum Simulations*, ed. by I.D. Doll, J.E. Gubernatis, p. 355 (World Scientific, Singapore 1990)
8.147 R.B. Jones, W. Yeung: Variational study of the 2-D Hubbard model: less than half filling, preprint
8.148 M. Suzuki: J. Sat. Phys. **43**, 883 (1986)
8.149 J.B. Bronzan, T.E. Vaughan: Phys. Rev. B **39**, 11724 (1989)
8.150 R.T. Scalettar, D.J. Scalapino, R.L. Sugar: Phys. Rev. B **34**, 7911 (1986)
8.151 M. Imada: Jpn. J. Appl. Phys. **26**, suppl. 26-3 (1987)
8.152 S. Duane, D. Olson: Nuc. Phys. B **280**, 687 (1987)
8.153 M. Imada: J. Phys. Soc. Jpn. **57**, 42 (1988)
8.154 M. Creutz: Phys. Rev. D **38**, 1228 (1988)

8.155 J.E. Hirsch, R.L. Sugar, D.J. Scalapino, R. Blankenbecler: Phys. Rev. Lett. **47**, 1628 (1981)
8.156 M.P. Nightingale, H.W.J. Blöte: Phys. Rev. B **33**, 659 (1986)
8.157 S. Homma, K. Sano, H. Matsuda, N. Ogita: In *Quantum Monte Carlo Methods*, ed. by M. Suzuki, Springer Ser. Solid-State Sci. Vol. 74 (Springer, Berlin, Heidelberg, 1987)
8.158. S. Homma, M. Matsuda, N. Ogita: Prog. Theor. Phys. **75**, 1058 (1986)
8.159 S. Homma, K. Sano, H. Matsuda, N. Ogita: Prog. Theor. Phys. **87**, 127 (1987)
8.160 H.Q. Lin, J.E. Hirsch: Phys. Rev. B **33**, 8155 (1986)
8.161 H.Q. Lin, J.E. Hirsch: Phys. Rev. B **35**, 3359 (1987)
8.162 M. Imada: J. Phys. Soc. Jpn **57**, 2689 (1987)
8.163 R.T. Scalettar, D.J. Scalapino, R.L. Sugar: Phys. Rev. B **36**, 8632 (1986)
8.164 R.M. Fye, J.E. Hirsch: Phys. Rev. B **38**, 433 (1988)
8.165 S.R. White, D.J. Scalapino, R.L. Sugar, N.E. Bickers, R.T. Scalettar: Phys. Rev. B **39**, 839 (1989)
8.166 T.A. Costi: Solid State Commun. **69**, 837 (1989)
8.167 F.F. Assaad, D. Würz: Reinvestigation of the Sign Problem in the 2D Hubbard model, IPS Research Report No. 99-03 ETH Zürich (1990)

9. Simulations of Macromolecules

Artur Baumgärtner

Reviews are based on personal experience and literature study. They are therefore limited to the extent that such experience and study are necessarily incomplete. Much has been accomplished by simulations in polymer science in recent years and the rate of publication on simulations of macromolecules is still increasing. The intention of this review is to summarise and to highlight the more recent developments, say, since the publication of the first review [9.1] in this series on "Applications of the Monte Carlo Method in Statistical Physics". Of course, several other reviews related to the simulations of polymers have appeared meanwhile, among which several are noteworthy and will be cited below.

9.1 Techniques and Models

For the sake of completeness various polymer models and Monte Carlo algorithms are summarised. Further details are presented in [9.2] (and references therein), and in particular concerning algorithms, the interested reader is referred to other chapters of the present book.

9.1.1 Polymer Models

Commonly, polymer models are classified as lattice polymer models and off-lattice polymer models. In the former models polymers are embedded on regular lattices and have to be distinguished from polymer chain models with fixed bond angles, which are not compatible with periodic lattices. Lattice models can be considered, of course, as coarse-grained versions of continuous polymer models. Intermediates do exist as well.

(a) Lattice Models. Most of the properties of lattice models have been discussed in detail by *Kremer* and *Binder* [9.3]. An interesting and useful variation of the common lattice model is the *bond fluctuation* model [9.4]. It is a *zwitter* model, sharing both properties of continuum and lattice models, but with variable bond length. In this model a polymer bead occupies a whole unit cell of a lattice (say 8 sites on a simple cubic lattice). At each step in the evolution of the polymer chain a randomly chosen bead moves to a randomly chosen nearest neighbour lattice site. The acceptance of the move is subjected to excluded volume and bond restrictions. The model has been used for 2D lattices [9.4] and in three dimensions [9.5] for diffusion in polymer mixtures.

A discretised version of the classical Edwards Hamiltonian for polymers suitable for simulations is the *Domb–Joyce model* [9.6, 7]. The Domb–Joyce Hamiltonian is given by

$$H = w \sum_{k>j} \delta(\boldsymbol{r}_k, \boldsymbol{r}_j), \tag{9.1}$$

where $\boldsymbol{r}_k, \boldsymbol{r}_j$ are the coordinates of the polymer, $\delta(x, y)$ is the Kronecker symbol with $\delta = 1$ for $\boldsymbol{r}_k = \boldsymbol{r}_j$ and zero otherwise, and $0 \leqq w$ is the excluded volume parameter. Each chain conformation has a statistical weight $P = \exp\left[-w \sum_j n_j(n_j - 1)/2\right]$, where n_j is the number of times the chain visits the lattice site j and the sum runs over the full space. For $w \approx 1$ the excluded volume exponent $\nu = 0.59$ is recovered on the cubic lattice even for comparatively short chains $N > 10$.

(b) Off-Lattice Models. The *freely jointed* polymer chain is one of the first polymer models introduced in polymer science [9.8]. It represents a random walk in continuum space with fixed step length and variable angles between two successive links. Excluded volume interactions are usually taken into account by two body potentials with short range repulsion and long range attraction. Examples of such potentials are pure hard core repulsion ("pearl-necklace chain") and Lennard–Jones types interaction. The freely jointed polymer model has found applications in single chain problems (see e.g. [9.1]) and in concentrated polymer systems including simulations at constant volume [9.9] and at constant pressure [9.10]. The freely jointed chain model has proved especially suitable for exploring topological effects on single chains [9.11, 12]. The *bead-spring* chain is an extension of the freely jointed chain where the beads are connected by harmonic springs with a finite extensibility [9.13, 1].

9.1.2 Monte Carlo Techniques

(a) Kink-Jump and Crankshaft Algorithm. The kink-jump algorithm applied to lattice polymers [9.14] consists of "flipping" a randomly chosen polymer bead from its old position $\boldsymbol{r}_n$ to a new one $\boldsymbol{r}'_n = \boldsymbol{r}_{n+1} + \boldsymbol{r}_{n-1} + \boldsymbol{r}_n$. This stochastic process simulates the Brownian motion of the polymer in the free draining limit and is in agreement with the *Rouse* model [9.15] in the absence of excluded volume interactions. However, in the presence of excluded volume it has been shown [9.16] that the model dynamics simulates a kind of "defect motion" [9.16] similar to the "reptative motion" proposed previously by *de Gennes* [9.17]. Subsequently, it has been shown that local "crankshaft" motion [9.18], in addition to the kink-jump procedure provides an adequate description of the Rouse model in the presence of excluded volume interactions. Later embellishments of the kink-jump algorithm have been many, among which the generalisation to off-lattice chains [9.19, 20] is an important one.

(b) Reptation Algorithm. In the algorithm [9.21–23], one chooses one end of the chain at random and transfers a monomer from this end to the other repeatedly. If the transferred monomer violates the excluded volume condition, the move is rejected. This scheme applied to an N-bead chain leads to a statistically independent configuration in a time of the order of N^2. The algorithm is essentially restricted to open linear polymer chains where both ends are free. It is suitable for both lattice models and off-lattice models. It does not provide a realistic model of local dynamics. There are some serious shortcomings of this algorithm for dense configurations in two dimensions ("locked-in configurations").

(c) General Reptation Algorithm. This algorithm [9.24] involves removing an arbitrary monomer and placing it at random somewhere else along the chain. The two neighbours of the removed monomer are connected by a new bond. If this bond is longer than the allowed maximum bond length, the move is rejected and the old configuration is retained. If the removal is allowed, one chooses at random a neighbouring site along the chain and connects the removed bead to the chain, thereby obeying thc constraint of the linear connectivity, i.e. replacing one bond between two successive beads i and $i + 1$ by two bonds connecting the removed bead j and i and $i + 1$. The relaxation is of the order N^2.

(d) Grand Canonical Reptation Algorithm. In the grand canonical method for single chains [9.25, 26], a fugacity β^N is assigned to an N-step walk, where β is the inverse temperature. It is possible to estimate the grand partition function and the end-to-end distance, and also the corresponding critical exponents γ and ν.

(e) Collective Reptation Method. Most of the Monte Carlo algorithms for lattice polymer models are suitable at volume fractions smaller than unity, since then the algorithms become increasingly inefficient. Two methods have been developed to overcome these difficulties, the bond breaking method [9.27, 28] and the collective motion method [9.29–31]. With the former, only polydisperse ensembles of polymers can be simulated whereas the latter one also allows the simulations of monodisperse polymers. The basic idea of the collective reptation method is to first apply conventional reptation of the kink-jump algorithm to one chain and then to specific rearrangements of other chains, consequently eliminating excluded volume violations caused by the first moved chain. Therefore the total motion loop consists of several rearranging procedures and may extend over several chains. So the collective reptation algorithm consist essentially of a list of geometrical transition rules providing a correct transition between different local chain arrangements. These rules are described in detail in

[9.29–32]. The resulting dynamics can be interpreted by collective motion or by a vacancy mechanism, but it is not clear whether the algorithm provides a realistic model of the motion of polymers at high densities [9.33].

(f) Pivot Algorithm. Starting from a given self-avoiding chain on a lattice or in the continuum, a new chain is obtained by rotating one part of a chain around a randomly selected bond [9.34–37]. This rotation can be either continuous by a randomly selected angle for off-lattice chains [9.39] or discrete in the case of lattice polymers. If segments of the transformed subchain overlap with segments of the remaining unmoved part, the new chain is rejected and the old one is retained. Originally proposed by *Lal* [9.34], the method has been carefully analysed later with respect to ergodicity and efficiency as compared to other algorithms by *Madras* and *Sokal* [9.36].

(g) Growth and Scanning Algorithms. Recently *Meirovitch* [9.40, 41] developed a new technique based on a scanning method, which is an extension and an improvement of the classical Rosenbluth [9.42–44] technique for generating polymer chains. With the scanning method, a self-avoiding walk is generated step-by-step with the help of transition probabilities; at a step k of the process the transition probability $p_k(v,b)$ for selecting a direction v is determined by scanning all different chain continuations in b future steps. Thus v_k is selected by a lottery according to the p_ks and the process continues. Once a walk i of N steps has been constructed, one knows its probability $P_i(b)$, which is the product of the N sequential transition probabilities with the directions $v_1, v_2, \ldots, v_N$ that have been chosen. For the practical value of b ($b \leq 6$ for the simple cubic lattice) the future can be scanned only partially; therefore a construction of a walk may fail. In this case, the walk is discarded and a new one is started. The efficiency of the process can be expressed by the attrition ratio $A = W_N/W_0$, where W_N is the number of walks generated and W_0 is the number of walks attempted. A normalised probability, $P'_i(b) = P_i(b)/A$, can be defined. Using $P'_i(b)$, one can define the statistical average of any property X (such as the radius of gyration) by $\langle X \rangle_b = \sum_i P'_i(b) X_i$. In particular, the approximate free energy reads

$$\langle F \rangle_b = \sum_i P'_b(n)[E_i + k_B T \ln P'_i(b)], \tag{9.2}$$

where E_i is in general an inter- or intramolecular interaction; $\langle F \rangle_b$ is expected to overestimate the correct free energy $\langle F \rangle$. More details of the scanning method are presented in [9.41, 45]. The scanning method has been applied to phase transitions of single lattice polymers [9.46, 47] and to concentrated polymer systems [9.48].

9.2 Amorphous Systems

9.2.1 Dynamics of Polymers

(a) Polymer Melts. The dynamics of highly entangled polymers in the melt has been a longstanding subject of intensive research, and the relevant theories have been reviewed in the books of *Doi* and *Edwards* [9.49] and of *Bird* et al. [9.50]. The basic concept, which is called the "tube" or "reptation" model, relies on *Edwards'* idea [9.51] of including the topological effects of mutal entanglements between different polymers by effective tubes in which each polymer performs its restrictive motion essentially along its own contour. This concept has been substantially quantified later by *de Gennes* [9.17, 52] in his "reptation model". These ideas have led to the construction of a very elegant theory for the viscoelastic behaviour of polymer melts by *Doi* and *Edwards* [9.53], which is essentially in accord with experimental findings. Later embellishments and improvements of the theory have been numerous [9.54–61].

Simulations [9.9, 62–75] have also contributed to clarify the extent of validity and the limitations of the reptation model. First simulations of dense systems of freely jointed Lennard–Jones chains [9.9, 62] did not reveal reptation characteristics for high temperature and low densities, but rather conventional Rouse dynamics. Only for low temperatures and frozen environments, like under glassy conditions, has the predicted reptative [9.17] anomalous time-dependent mean square displacement of a monomer $r^2(t) \propto t^{1/4}$ been observed. More extensive simulations of systems of pearl-necklace chains [9.20] have indicated reptative displacements $r^2(t)$ and have for the first time corrobated the predictions [9.17] for the diffusion coefficient of the centre of mass $D \propto N^{-2}$ and the reptation correlation time $\tau \propto N^3$.

Extensive simulations of lattice polymers have been carried out by *Skolnick* and coworkers [9.64, 65] continuing earlier attempts [9.66–69]. No unambiguous indications favouring or disproving reptation could be drawn from their results [9.65]; one subtle point is related to the question of whether the kink-jump algorithm applied to lattice polymers at high densities leads to artificial local dynamics which would eventually indicate an inadequacy of lattice models to describe dynamical properties of polymer melts. In particular, lattice polymers confined to straight tubes [9.69] have been simulated in order to elucidate the crossover between Rouse and reptation behaviour. It has been pointed out [9.70, 71] that excluded volume interactions in the melt could modify the reptation time to $\tau \propto N^3 \exp(\text{const.} \times N^{2/3})$, which has been corroborated by simulations of lattice polymers [9.71]. The effect of the lattice coordination number on the dynamics of models of dense polymers systems has been examined [9.72], and it has been found that a significant difference exists between simple cubic and face-centred cubic lattice models concerning the concentration dependence of the scaling exponent that describes the N dependence of the relaxation times.

Most recently, extensive molecular dynamics simulations of polymer melts have been carried out by *Kremer* and *Grest* [9.73]. For long chains ($N \leqq 400$), the chains diffuse in accord with the reptation model, while for short chains the Rouse behaviour is recovered. In particular, the time-dependent mean square displacement of a monomer exhibits the predicted $r^2(t) \propto t^{1/4}$, and the visual inspection of projections of the chain configurations [9.73] provides direct evidence that at high densities the lateral chain fluctuations are strongly suppressed and consequently the chains are trapped in effective tubes, which supports the original concepts of reptation [9.51, 17].

A different approach to simulated reptation theories has been proposed recently by *Öttinger* [9.48]. The simulation algorithm is based on the equivalence between a stochastic differential equation and the diffusion equation, which is usually employed to describe the polymer dynamics in reptation theories, either in the *Doi–Edwards* model [9.53, 49] or the *Curtiss–Bird* model [9.50]. The stochastic reformulation of the models required for developing a simulations algorithm, yields more transparent equations for the polymer dynamics, as well as a convenient tool for evaluating material functions even under non-equilibrium conditions such as under shear flow.

(b) Polymers in Flow. The behaviour of polymers in homogeneous flow is less well understood, although this situation is encountered in practice more often than the homogeneous case. Under flow, the polymer concentration profile is expected to have a gradient due either to the position dependence of the velocity gradients in bulk flow or to the presence of boundaries. Many macroscopically observable phenomena [9.76–79] are induced, which often are subtle effects due to concentration inhomogeneities on length scales comparable to polymer sizes.

Using non-equilibrium molecular dynamics simulations, the behaviour of a single Rouse chain under shear flow [9.80, 81] has been examined and compared with analytical results for the distribution of the end-to-end distance and individual chain segment extensions. Planar Couette flow of dense polymer liquids has been investigated by *Clarke* and *Brown* [9.82] using non-equilibrium molecular dynamics simulations, and they have found a significant reduction in the equilibrium viscosity due to torsional flexibility effects. *Duering* and *Rabin* [9.83] have used a general Brownian dynamics algorithm in order to study single polymers in simple shear flow between impenetrable walls. They found that the depletion layer near the wall decreases with increasing shear rate.

(c) Gel Electrophoresis. Gel electrophoresis is a valuable separation technique for biopolymers and synthetic polyelectrolytes as well. The separation is achieved due to the differing rates with which charged polymers with different molecular weights migrate through a gel under an applied electric field across the gel [9.84, 85].

The reptation theories [9.86–91] of gel electrophoresis assume that a polymer of uniform charge Q is confined to a tortuous tube of length L, formed by the topological constraints of the gel. The motion of the chain is considered to be "biased" in the field direction leading to a mobility

$$\mu = \frac{Q\langle R_x^2\rangle}{fL^2} \propto N^{2\nu-2}, \qquad (9.3)$$

where $L \propto N$, f is the friction coefficient ($f \propto N$) of the chain, and $\langle R_x^2\rangle$ is the mean square end-to-end distance of the chain along the applied field. Numerous computer simulations have been reported in the literature to unravel the mechanism of gel electrophoresis [9.91–99]. The first Monte Carlo simulation [9.92] presumes that reptation is valid in the absence of the electric field, and focuses on the characteristic time taken by a kink in a chain to propagate through the chain. This work was not extensive enough to yield any conclusions regarding the steady state limit. A subsequent simulation by *Deutsch* and *Madden* [9.95] used a Langevin equation instead Monte Carlo techniques for a chain in two dimensions with a regular array of obstacles. The simulations by *Zimm* [9.91] and *Duke* [9.97] address the consequences of field-inversion in electrophoresis. These simulations are still based on extending the original reptation model. In recent work, *Melenkevitz* and *Muthukumar* [9.99] have considered the effects of quènched randomness due to the heterogeneous structure of the gel on the polymer dynamics. They found that the mobility is significantly affected by entropic barriers caused by the random environment.

9.2.2 The Glassy State

A few Monte Carlo [9.9, 100–102] and molecular dynamics simulations [9.103, 104] have been devoted to exploring the glassy state of polymers and to judging the reliability of various theories [9.105, 106].

Initial Monte Carlo studies [9.9] of dense polymer systems with Lennard–Jones intermolecular interactions have revealed at low temperatures a glassy state where the mean square displacements of the monomers are less than the size of the monomers themselves. This was interpreted as vibrational motions in an amorphous frozen structure. Subsequent studies [9.100] on systems of pearl-necklace chains at densities close to the randomly closed packed limit have also been performed. With increasing density, a transition from liquid-like to glass-like behaviour is found, which is characterised by a "critical" minimum sphere displacement which is necessary to maintain fluidity at a particular density. It was observed that the clusters of mobile spheres grow in size and coalesce through the transition with some geometrical features of dynamic percolation.

Lattice models [9.101, 102] have been shown to be rather inadequate for tackling the glass transition, although their simplicity in interpreting "free volume", percolation effects, etc. and in comparing to theories are rather appealing. On the tetrahedral lattice for chain length $N = 32$a glass transition was found at densities of about 0.92 [9.101]; the transition point was located by means of the appearance of the immobility of the monomers. The preparation of the initial configuration is crucial, and the equilibration was achieved by applying the reptation algorithm. Subsequent diffusion was studied by a combination of 3-bond and 4-bond motions [9.107, 108]. The effect of adding free beads was also studied.

More extensive simulations of polyethylene-like systems by molecular dynamics methods have been carried out by *Rigby* and *Roe* [9.103, 104]. They have estimated the eqilibrium density (by measuring the pressure from the virial theorem) and the self-diffusion coefficient as a function of temperature. At a particular temperature the density exhibits a kink and the diffusion coefficient is lowered considerably, which was interpreted as due to a glass transition. There are considerable uncertainties in the diffusion coefficients, probably because of a pronounced slowing down of the dynamics at the transition point. It has been discovered that the glass transition of the polyethylene-like model is accompanied by short range anisotropic order due to orientational corrections among different segments of the chains. This local anisotropy is enhanced at lower temperature. The competing mechanisms of mesomorphic pattern formation and the glass transition have still to be elucidated.

9.2.3 Equation of State

Equations of state of hard-body fluids have been studied for many years. Recently, the effort has been focused on the equilibrium state of systems of associating molecules including long chain molecules. Using graph theory *Wertheim* [9.109, 110] considered the behaviour of systems of associating hard spheres forming the pearl-necklace chain, taking into account the condition of bond saturation. *Gubbins* and co-workers [9.111, 112] applied the theory of associating fluids to characterise the equilibrium behaviour of several model systems. Another approach has been followed by *Hall* and coworkers [9.113, 114], who interpreted their Monte Carlo simulations for pearl-necklace chains in terms of a generalised *Flory* and *Flory–Huggins* theory [9.115, 116]. In these generalised versions they made use of the *Carnahan–Stirling* [9.117] and *Boublik* equations [9.118] of state of hard spheres and hard dumbells, respectively. With the recent version of the equation of state *Honnell* and *Hall* [9.114] obtained very good agreement with their Monte Carlo data. However, problems are encountered if the theory is applied to systems of chains with spheres of different radii or mixtures of different chains. The generalised Flory–Huggins theory of the equation of state is also in very good agreement with molecular dynamic simulations of the freely jointed effective hard sphere model of a polymer melt [9.232] and with Monte Carlo simulations of lattice models performed by *Dickman* [9.120, 121]. In the latter approach especially, questions of how well the continuum limit is approximated by lattice models are addressed.

9.3 Disorder Effects

The problem of a polymer chain in equilibrium trapped in a quenched random environment has now been investigated for a decade. One may consider this class of problems to be divided in three parts. The first is concerned with the original problem of how the structure of a porous solid affects the configurational properties of a flexible polymer chain. This is addressed in the first subsection. The second part of the problem is related to the question of how thermodynamics are changed under the constraint of disorder, e.g. adsorption on random surfaces, decomposition in porous media, etc. The third part of the problem is related to the transport properties of polymers in a porous environment. The latter is discussed in the third subsection.

9.3.1 Polymer Chains in Random Media

The standard model for random media is the percolation model [9.122]; other models, e.g. the Gaussian model, have not been studied numerically. There are currently two basic questions: (1) What is the effect of a self-similar complex structure (e.g. percolation model at the percolation threshold $p=p_c$) on the configuration of a polymer chain? (2) Is there an effect of even a random structure, e.g. $p>p_c$ where $1-p$ is the density of the obstacles?

The first question is still a matter of controversy and various reviews and introductionary articles on this subject exist [9.123–125]. Even the most recent Monte Carlo simulations [9.126, 127] did not solve this problem, but indicated that the exponent ν of the radius of gyration is probably changed only slightly, if at all. In general, the essential difficulty in the investigations of polymers in random media is the "quenched average" over many realisations of the random media, in conjunction with the ensemble average of long chains exhibiting extensions comparable or larger than typical length scales of the media.

Progress has been made in our understanding of chains in random (i.e. in general non-self-similar) media. Discovered first by Monte Carlo simulations [9.128] and discussed intensively later [9.129–139], it has been shown that for sufficiently weak excluded volume interaction [9.132, 138, 137] the chain is strongly shrunk and resembles a "localised random walk". In the case of a Gaussian chain, the radius of gyration obeys, in three space dimensions, the crossover scaling relation

$$R^2 \propto Nlf[(1-p)N^{1/2}], \tag{9.4}$$

with the limiting cases $f(s)=\text{const.}$ for $s \ll 1$, and $f(s) \propto s^{-2}$ and hence $R^2 \propto (1-p)^{-2}$ for $s \gg 1$.

In the presence of excluded volume, the chain is still "localised" for sufficiently weak excluded volume parameter w. This has been demonstrated by simulations [9.137] of the *Domb–Joyce* model [9.6, 7] on the cubic lattice in the presence of impurities of concentration $v=1-p$ distributed according to the percolation model for $p > p_c$. The "collapse" at $v = w$ can also be understood based on a Flory–Imry–Ma type agreement [9.130, 132, 135, 140]. The free energy of the chain is given approximately by

$$F \propto \frac{N}{R^2} - v\frac{N}{R} + \frac{R^2}{N} + w\frac{N^2}{R^3}, \tag{9.5}$$

where the first and third term describe the elastic energy for compressed and swollen polymer coil sizes, respectively, whereas the second term takes into account the entropic localisation. Minimising with respect to R, one obtains three cases: for $v > w$ the chain is localised $R \cong$ const. (9.3), for $v < w$ the excluded volume interaction is prevailing and hence $R \propto N^{3/5}$; for $v = w$ there is a balance between effects due to the random potential and the repulsive energy of the coil and hence $R \propto N^{1/2}$.

9.3.2 Effect of Disorder on Phase Transitions

Since there is some evidence that for sufficiently weak self-avoidance a disorder-induced polymer collapse ("localisation") would take place [9.132, 137, 138], it is conceivable that the classical Θ-collapse of a single polymer chain is altered by the influence of a quenched disordered environment. It is expected that the effect of disorder should shift the Θ-point towards higher temperatures. It would be of interest to study the effect of structural disorder, for example of the percolation model at the percolation threshold, on the Θ-transition.

The adsorption of a Gaussian polymer chain on a planar adsorbing surface containing repulsive impurities ("chemical rough surface") has been studied recently [9.141]. It has been shown by simulation that the critical unbinding transition temperature is shifted by the concentration v of the impurities to lower values, $T_c(v) = T_c(v=0) - v$. The transition from the collapsed configuration at $T=0$, i.e. localised configuration with radius of gyration $R_\parallel \propto \sqrt{v}$, to the free Gaussian chain $R_\parallel \propto \sqrt{N}$ seems to be continuous. The simulation results are supported by a simple Flory–Imry–Ma type argument. The scaling form of the free energy of a chain confined in a 2D plane with impurities in the limit of $vN \to \infty$ is

$$F \propto \frac{N}{R_\parallel^2} + Nv \ln R_\parallel . \tag{9.6}$$

Combining this free energy with the adsorption free energy for a layer of thickness δ we get for the total free energy

$$F \propto \frac{N}{\delta R_\parallel^2} + \frac{N}{\delta} v \ln R_\parallel + \frac{N}{\delta^2} - \frac{\varepsilon N}{\delta}, \tag{9.7}$$

where ε is the adsorption energy. Minimising F with respect to both δ and $R_\parallel$ the optimum values of the layer thickness and the chain extension parallel to the surface can be obtained.

The effects arising from the competition between adsorption and localisation of a polymer chain in a disordered environment have been studied recently [9.140]. It is observed that a self-avoiding polymer trapped in a disordered array of parallel, adsorbing rods, undergoes an adsorption transition at finite temperature. In contrast, this adsorption transition is *not* observed, i.e. it takes place at zero temperature, if the rods are distributed periodically. At high temperatures, the repulsion between the hard rods and the chain leads to highly anisotropic configurations of the polymer, such that the parallel component of the radius of gyration is highly stretched along the rods, whereas the perpendicular component, $S_{\perp}^2$, is strongly shrunk and almost independent of the chain length, i.e. a localised 2D random walk. The latter effect is due to the 2D disordered arrangement of the rods which leads to the "entropic localisation" phenomena for polymers, as discussed in general above. Below the transition point, the chain is strongly adsorbed at the rods, and hence itself of rod-like shape. At the critical temperature, the two competing mechanisms, localisation and adsorption, are suppressing each other, and hence the chain exhibits conventional self-avoiding characteristics. It is interesting to note that the simulated situation of a polymer among long parallel rods can be experimentally verified, and hence the competition between adsorption and localisation could be experimentally monitored.

9.3.3 Diffusion in Disordered Media

Diffusion of atoms or small molecules in percolated media has been intensively studied numerically for more than ten years [9.142–144]. One of the main observations is the appearance of a regime of anomalous diffusion, which is a consequence of the self-similar structure of the media. The crossover between the intermediate anomalous and the long time classical diffusion is now well understood [9.145]. The crossover scaling for the mean square displacement on an infinite cluster is $R^2(t) \propto t^{2k} g(t/\xi^{1/k})$, where $\xi \propto |p - p_c|^{\nu}$ is the percolation connectedness (or correlation) length and k is the anomalous diffusion exponent [9.142]. The scaling function is $g(s) = \text{const.}$ for $t \ll \xi^{1/k}$ and $g(s) \propto s^{1-2k}$, yielding $R^2(t) \propto t\xi^{2-1/k}$ for $t \gg \xi^{1/k}$. The anomalous exponent is believed to be related to the fractal and the spectral dimensions, d_f and d_s respectively, by $2k = d_s/d_f$ [9.145–147].

Recently these investigations have been extended to the diffusion of single polymer chains in percolated media [9.128, 148–151]. In contrast to small molecules, the diffusion of a macromolecule is more complex due to various intriguing effects. Currently one distinguishes between three competing effects: the anomalous diffusion due to the structure of the material; the activated diffusion due to the presence of entropic barriers and eventually entanglement effects for very long chains.

The appearance of anomalous diffusion for polymer chains has been reported [9.148, 149], and an explanation attempted following similar arguments as for small molecules. The mean square displacement is given by

$$R^2(t) \propto D^* t^{2k} f(t/\tau), \tag{9.8}$$

where $D^* \propto N^{-x}$ is the anomalous diffusion coefficient of the centre of mass of the chain and $\tau \propto N^z \xi^{1/k}$ is the termination time for the anomalous diffusion. Assuming $f(s) \propto s^{1-2k}$ for $s \gg 1$, and $f(s) = \text{const.}$ for $s \ll 1$, one has the scaling relation $w = x + z(1-2k)$ and the diffusion coefficient $D \propto N^{-w} \xi^{2-1/k}$ for the classical diffusion $R^2(t) \propto Dt$ at $t \gg \tau$.

Entropic barriers (or "bottlenecks") in disordered media probably have a very important influence on the polymer diffusion, which is assumed to be more pronounced the longer the chain or the lower the porosity. There the chain is forced to squeeze through narrow channels in order to move to environments less crowded by obstacles and hence entropically more favourable. The motion between the different "entropic traps" is slowed down significantly by the bottlenecks connecting these regions. The diffusion coefficient D is modified with respect to the free draining value D_o by the partition coefficient $\kappa \propto D/D_o \propto \exp(-\Delta S)$ where ΔS is the entropy loss during the passage between two entropic traps via the connecting bottleneck [9.151, 152].

Recently, the importance of entropic barriers in gel electrophoresis has been stressed [9.99].

The competition of entanglement effects, on one hand, and the effects of anomalous diffusion and entropic barriers, on the other, have not yet been investigated systematically by simulations.

9.4 Mesomorphic Systems

9.4.1 Hard Rods

Computer simulations have been very illuminating in weighting the importance of excluded volume effects versus anisotropic van der Waals attraction. Qualitatively, Onsager's and Flory's prediction on the appearance of nematic-isotropic phase transitions due solely to hard core interaction is confirmed.

Frenkel and coworkers [9.153–155] indeed report results which strongly suggest that the assumptions underlying Onsager's excluded volume theory for the isotropic-nematic transition, namely that all virial coefficients higher than the second are negligible, are indeed satisfied for hard ellipsoids in the limit of $x \to \infty$, where $x = a/b$ is the ratio of the length of the major axis, $2a$, to the minor axis, $2b$. They have investigated, using Monte Carlo methods, the third, fourth and fifth virial coefficients of prolate hard ellipsoids for a range of a/b ratios between 1 and 10^5. They determined the phase diagram for various molecular shapes, from extremely oblate ellipsoids ("platelets") through spheres to extremely prolate ellipsoids ("needles"), i.e., density versus axial ratio x. Four distinct phases can be identified, the isotropic fluid, the nematic fluid, the orientationally ordered solid and orientationally disordered (plastic) solid. A conspicuous feature of the phase diagram is the high (although not perfect) degree of symmetry under the interchange of oblate and prolate ellipsoids [9.153]. The physical reason for this near symmetry at rather high densities is not understood at present. Details of the static and dynamic critical behaviour at nematic-isotropic transition have still to be investigated for various molecular shapes.

One reason, besides the difficulty of hard and time-consuming simulations, might be that hard rods do not exist in a mesomorphic state at equilibrium, but exhibit a transition from an isotropic liquid to a glassy state with local anisotropic domains only. Whether this hypothesis can be elucidated by appropriate lattice models (e.g. hard rods distributed on the cubic lattice), has still to be demonstrated; but certainly lattice models could provide some indications, and moreover, they are more easily accessible to analytical theories or exact enumeration procedures. Indeed, it has been shown that close-packed systems of hard rods of length N distributed on the square or the cubic lattice of edge length L ($N < L$) never exhibit a long-range orientational order [9.156]. Merely neighbouring rods have the tendency to accumulate in parallel order.

Usually simulations of polymeric liquid crystals are performed at constant pressure in order to distinguish between the liquid-crystal transition and the liquid–solid transition, which should exhibit a strong discontinuity of the density, in contrast to the mesogenic case. During the Monte Carlo simulation of the NpT ensemble, all intermolecular distances as well as the linear size of the basic cell (using periodic boundary conditions) are changed by a randomly chosen factor. Attempts leading to overlaps of rods are discarded. It is obvious that such simulations are very time consuming, because on one hand the size of the basic cell has to be much larger than the length of the rod, which at high densities leads to an enormous number of rods to be simulated, and on the other hand entanglements are becoming increasingly important with increasing length of the rod, which could cause very long relaxation times.

9.4.2 Semirigid Chains

Semiflexible polymers chains may be highly extended as compared to flexible randomly coiled polymers, but nevertheless possess a significant degree of flexibility and many exhibit liquid crystallinity in melt or solutions. The relation between departures from full rigidity and the characteristics of the isotropic-nematic transition is an important question. Since semiflexibility can be realised in diverse ways, one is unlikely to find a single theory encompassing all polymers of this kind. A numer of theories have been proposed [9.157, 158]. The semiflexible lattice chain introduced by *Flory* [9.157] has been widely applied to mesogenic polymers. Many of these applications go beyond the limitations of this model. It was originally introduced for illustration of the effects expected from low degrees of flexibility in long chain molecules, and not as a model for representing real chains generally. According to this model, the chain is embedded on, e.g., the cubic lattice where two successive segments of the chain gain an energy

$\varepsilon_B < 0$ if they are colinear. This difference in energy favours the rectilinear form at low temperatures. Estimates of the entropy and the existence of a phase transition given by Flory's theory has been critisised [9.159–161]. Simulations of this model at close-packed density [9.162–165] yield the result that the transformation between the coiled and the rod-like configuration has no critical point but is continuous, very similar to the transition of an isolated chain. The ground state has no long ranged orientational order, but rather consists of ordered domains.

Collective orientational fluctuations and related pretransitional phenomena are of particular importance. Pretransitional effects on liquid crystal polymers have been discussed by *de Gennes* [9.171].

9.4.3 Anisotropic Interactions

The effects of intermolecular forces fall into two categories: isotropic forces of the kind commonly treated in theories of solutions (i.e., solvent–solute and solvent–solvent interactions), and the anisotropic interactions that may contribute significantly to the stability of a nematic phase. Mean-field theories with the Maier–Saupe expression for this type of interactions have been formulated in several ways [9.166–169, 319].

An appropriate model interaction for lattice simulations of anisotropic van der Waals forces between chain segments of lattice polymers has been proposed recently [9.162, 164]. An energy $\varepsilon_S < 0$ is attributed to a local configuration of two adjacent parallel segments. Inclusion of this interaction provides an anisotropic state of semiflexible polymers at low temperatures, which is separated from the isotropic state by a phase transition which is of first order in three dimensions (cubic lattice) and of second order in two dimensions (square lattice) [9.164].

Monte Carlo simulations at constant pressure applied to a corresponding system of semiflexible continuous polymer chains (pearl-necklace polymer) decorated with appropriate generalised intra- and intermolecular interactions likewise exhibit a nematic–isotropic transition.

The characteristics of phase transitions and the orientational order in lipid monolayers have also been investigated by molecular dynamics and Monte Carlo methods and are reviewed in [9.172] and [9.173], respectively.

9.5 Networks

9.5.1 Tethered Membranes

Properties of flexible sheet polymer networks ("tethered" surfaces) are of increasing interest both from an experimental and a theoretical point of view [9.174, 175]. Such "tethered membranes" are realised in biological materials, polymer networks and polymerised Langmuir–Blodgett films and in cross-linked gels stabilised in a water-rich region between lipid bilayers. Polymerised membranes have a nonzero shear modulus in the plane and are said to be solid-like, whereas fluid membranes are not rigid in the plane and have zero in-plane shear modulus.

Fluid isotropic membranes [9.176, 177] and polymerised membranes [9.178–181] have been predicted to exhibit a high-temperature crumpled phase similar to that of coiled polymers. This has been asserted for polymerised membranes by renormalisation group arguments [9.182–184] and first pioneering Monte Carlo simulations of phantom membranes (without excluded volume) [9.185] and self-avoiding membranes [9.186] were performed by *Kantor* et al.

The simplest molecular model of polymerised membranes consists of N hard spheres each tied to six nearest neighbours to form a planar triangulated network; the network is then equilibrated by allowing it to bend and possibly crumple in three dimensions. Self-avoidance of the membrane is implemented by choosing the ratio between the diameter and the maximum bond length such that self-penetration of the surface is excluded (e.g., diameter = 1 and maximum bond length = $\sqrt{2}$).

However, subsequent extensive molecular dynamics [9.187] and Monte Carlo simulations [9.188–191] did not reveal any evidence for a crumpled state of polymerised membranes embedded in space dimensions $d = 3$; instead the membranes were found to be essentially flat and rough. The related anisotropy is characterised by the two larger eigenvalues of the inertia tensor, $\lambda_3 \sim \lambda_2 \sim N$ for the in-plane size of the membrane, and by the smallest eigenvalue $\lambda_1 \sim N^{\zeta}$ for the perpendicular size. The eigenvalues are related to the radius of gyration of the membrane by $R^2 = \lambda_3 + \lambda_2 + \lambda_1$. The roughness exponent ζ has been estimated by simulations to be between 0.65 and 0.7 [9.188, 190, 192–194]. Very recently, it has been shown [9.195] by simulating a discretised version of the generalised Edwards Hamiltonian for surfaces, that the previous estimates of ζ are probably masked by crossover effects, and the actual value should be $\zeta = 1/2$ in agreement with earlier predictions by *Nelson* and *Peliti* [9.178].

The reasons of the absence of a crumpled phase of tethered membranes are not fully understood. It has been shown by molecular dynamics techniques [9.196] that even membranes with locally vanishing self-avoidance (i.e., by assigning zero hard sphere diameter to randomly chosen spheres of the membrane) exhibit flat phases. Only below the percolation threshold is a crumpling of membranes observed.

It is expected that in the "flat" phase of polymerised membranes the classical theory of elasticity breaks down due to thermal out-of-plane fluctuations [9.180, 197]. It has been shown by Monte Carlo simulations [9.198] that the flat phase of a polymerised membrane is not described by the classical theory of elasticity. Thermal fluctuations induce important modifications to mechanical laws such as Hooke's law. Shape fluctuations of membranes under constrained boundary conditions exhibit certain "buckled" states, which have been analysed by finite size scaling relations and corresponding critical exponents. Excitations of the membrane, which are decomposed into transverse undulations $h(x)$ and internal "phonon-like" modes $u(x)$ exhibit non-classical behaviour $\langle (h(x)-h(0))^2 \rangle \propto |x|^{2-\eta}$. The exponent η has been estimated from the fluctuation of the area A projected onto the reference plane, defined by the boundary constraint [9.198],

$$\langle A^2 \rangle - \langle A \rangle^2 \propto N^{2-\eta} \tag{9.9}$$

and leads to $\eta \approx 0.75$. Experimentally one expects to observe such thermally excited states in special protein networks, such as spectrin networks of red blood cells.

Many membranes, especially those occurring in biological systems, are not polymerised, but *fluid*. Molecular models of fluid membranes can be considered as "transient" networks, where the constraint of "local" connectivity among the monomers is lifted and replaced by a "global" connectivity preserving the topology of the network. During the Monte Carlo simulations, ensembles of fluid membrane configurations are generated using a "dynamic triangulation" procedure [9.191–193, 199, 200]. It has been shown that molecular models of *fluid* membranes [9.191–193] indeed exhibit a crumpled isotropic phase where the radius of gyration is $R^2 \propto N^{0.8}$, which is in agreement with predictions based on Flory-type mean field arguments [9.201, 186]. The approximate free energy of the membrane is

$$F = \kappa R^2 + w \frac{N^2}{R^d}. \tag{9.10}$$

The first term represents the elastic (entropic) energy of a phantom membrane (up to a term ln N, which has been replaced by a constant), arising from a generalisation of the Edwards Hamiltonian for polymers,

$$H \propto \kappa \int d^D x [\nabla r(x)]^2, \tag{9.11}$$

where D is the dimension of the object, $D=1$ for polymers and $D=2$ for membranes. The second term of the free energy F is the conventional mean-field estimate of the (binary) repulsive interaction energy [9.202]. Minimising F with respect to R we obtain

$$R \propto (w/\kappa)^{1/(d+2)} N^{\nu/2}; \quad \nu = 4/(d+2). \tag{9.12}$$

This exponent has also been found in Monte Carlo simulations of fluid spherically closed membranes ("vesicles") [9.191, 193].

The thermodynamics of vesicles is of special interest in various biophysical situations. First attempts to understand the interplay between the pressure increment leading to inflated or deflated vesicles and the bending energy have been made recently, based upon Monte Carlo simulations of vesicles ("polymer rings") [9.203–206].

9.5.2 Branched Polymers and Random Networks

Over the last years, there has been considerable interest in excluded volume effects in branched polymer molecules. This was initiated by Monte Carlo methods [9.207] and by field-theoretical approaches [9.208, 209], and subsequently a good deal of work has been carried out by Monte Carlo simulations [9.210–216] and Brownian dynamics simulations [9.217, 218] and exact enumerations [9.219, 220]. They have been very useful in testing the predictions of scaling [9.221] and renormalisation group [9.222] and have raised questions about the accuracy of the renormalisation results for high values of functionality. The simulation results are in close agreement with experimental measurements [9.223]. The quantities of interest of a star or branched polymer are the mean square radius of gyration $\langle S_N^2(f)\rangle$, the mean square end-to-end length of a single branch or arm $\langle R_n^2(f)\rangle$, and the mean square distance $\langle R_{ij}^2\rangle$ between two sites i and j within the star or the related structure factor. Also of interest is the number of configurations of uniform stars s_N. Here n is the number of monomer units per branch, N is the total number of monomers in the star, and f is the functionality or number of branches with $N = nf$. If N is large, then these quantities are expected to satisfy the conventional relations $\langle S_N^2(f)\rangle \propto A(f)N^{2\nu}$ and $\langle R_n^2(f)\rangle \propto B(f)n^{2\nu}$ and $s_N \propto \lambda(f)^N N^{\gamma(f)-1}$, where ν is the usual value for linear polymers and close to 3/5. It has been shown [9.212], that $\lambda(f) \propto \mu^f$, where μ is the growth constant for a linear self-avoiding walk, and γ has been estimated by renormalisation group methods [9.222] and by exact enumeration and Monte Carlo methods [9.212, 214]. The amplitudes $A(f)$ and $B(f)$ are model dependent, but the ratios are expected to be universal for large N,

$$g(f) = \frac{\langle S_N^2(f)\rangle}{\langle S_N^2(1)\rangle} \propto f^{-\delta}, \tag{9.13}$$

where $\delta = 4/5$ has been predicted by Daoud and Cotton based on scaling considerations [9.221].

The relaxation of a self-entangled many arm star polymer has been studied using molecular dynamics simulations [9.224]. The results are consistent with the prediction that the relaxation time should scale with $N^{2\nu+1}f^{2-\nu}$, where ν is the correlation-length exponent.

The adsorption of branched polymers on hard walls has been simulated [9.225] and the specific heat has been characterised using scaling analysis. The crossover exponent has been estimated to $\phi \approx 0.714$.

Computer simulations prove to be powerful for network structure problems for end-linking systems. Not only can network structures be well resolved [9.226], but questions related to sol–gel transitions [9.227], mechanical properties [9.228, 229] and to radiation-cured networks [9.230] can also be nicely addressed.

First attempts to simulate *mechanical properties* of random networks have been made recently by Monte Carlo simulations [9.231] and by molecular dynamics [9.232]. Using Monte Carlo the effect of entanglements on the configurational properties of a micronetwork under uniaxial extension has been studied. As a model, a trifunctional network was used with 41 beads on each arm embedded on a cubic lattice without excluded volume. The three ends are randomly fixed in space and the freely moving trifunctional junction point provides information about its mean square fluctuation as a function of affine deformations. The results compared favourably with the predictions from *Ronca–Allegra* [9.233] and *Flory–Erman* [9.234] constrained junction theories.

9.6 Segregation

9.6.1 Collapse Transition

A polymer in dilute solution and in a good solvent is swollen and its size increases with length according to N^{ν}, where $\nu = 0.59$ and 3/4 in three and two dimensions respectively. As solvent conditions become poorer (e.g. at lower temperatures), a phase separation takes place [9.202] since the monomer–monomer interactions become stronger than the polymer-solvent interactions. Below a particular temperature, the Flory θ point, the effective monomer–monomer attraction prevails and the polymer collapses to a globule with size $N^{1/d}$, where d is the space dimension. The θ point has been identified by *de Gennes* [9.202] as a tricritical point with an upper critical dimension three.

This critical phenomena has been studied intensively by analytical methods [9.235–240] as well as the numerical techniques such as exact enumeration [9.241–244], and computer simulations [9.245–247] (or earlier references see for example [9.1]). While the behaviour in $d = 3$ is understood to a large extent [9.247, 248], the situation in $d = 2$ has been controversial for many years. For example, significantly different results for the critical temperature θ and of ν_t at this point of self-avoiding walks on the square lattice have been obtained by different techniques. Important progress was made by *Duplantier* and *Saleur* [9.249], who calculated, by Coulomb gas techniques and conformal invariance, the exact values of the tricritical exponents in $d = 2$, $\nu_t = 4/7$, $\gamma_t = 8/7$, $\phi = 3/7$, where γ_t is the exponent related to the free energy and ϕ_t is the crossover exponent. The first two values were significantly larger and the ϕ_t was significantly lower than previous estimates. They have also calculated the tricritical exponents of chains terminally attached to an impenetrable line. These results have been derived from a special model of self-avoiding walks on a hexagonal lattice with randomly forbidden hard hexagons (see also [9.250, 251]). However, it has been pointed out [9.246] that this model includes, in addition to nearest neighbour attraction, also a special subset of next-nearest neighbour attractions and therefore might describe a multicritical point θ'. Based on an alternative superconformal theory, different tricritical exponents have also been obtained [9.252] and supported by exact enumeration and computer simulations [9.244]. Extensive simulations [9.246] for the hexagonal lattice have indeed indicated that θ and θ' belong to different universality classes. On the other hand, transfer matrix calculations of *Duplantier* and *Saleur* [9.253] have supported their previous results. Also, exact results on the Manhattan lattice [9.254] yielded

$\nu_t = 4/7$, but $\phi_t = 6/7$. The most recent Monte Carlo simulations [9.46, 47] are in agreement with the results of Duplantier and Saleur for ν_t and γ_t, but in disagreement with respect to ϕ_t and have supported the claim that θ and θ' points belongs to different universality classes.

The competition between polymer collapse and polymer adsorption in $d = 3$ of a polymer chain terminally attached to an attractive impenetrable wall has been investigated very recently using computer simulations [9.255, 256]. It has been shown that at the adsorption temperature of a θ-polymer, the crossover exponent assumes its mean-field value $\phi = 1 - \nu_t = 1/2$ [9.255], and that the collapsed chain forms a multilayered structure on the surface [9.256].

The collapse transition of regular star polymers on the fcc lattice [9.257] have been simulated for up to $f = 12$ arms. For $f = \text{const}$ and large N, star polymers exhibit the same transition temperature as linear chains and follow the same crossover scaling. At the θ-point, evidence for residual three-body interaction in found.

9.6.2 Polymer Mixtures

One of the main questions concerning the classical Flory–Huggins theory [9.115, 116], which describes the phase separation of binary (A–B) polymer mixtures, is its reliability and range of applications [9.258, 259]. Since the derivation of the excess free energy,

$$\frac{\Delta G_M}{k_B T} = \frac{(\phi_A \ln \phi_A)}{N_A} + \frac{(\phi_B \ln \phi_B)}{N_B} + \phi_V \ln \phi_V + \chi(T)\phi_A\phi_B, \tag{9.14}$$

where ϕ is the volume fraction and V denotes either a low-molecular weight solvent or a "free volume", is based on many approximations (lattice instead continuum model, neglect of excluded volume, independence of the demixing parameter χ on ϕ, etc.), the theory has been examined extensively, among others, by Monte Carlo simulations [9.260–263]. The advantage of simulations as compared to experiments is that the statistical mechanics of the Flory–Huggins lattice model can be verified exactly, whereas comparisons with experiments are limited by the many assumptions due to the approximation of reality by lattice structures; finite size effects [9.264] due to the lattice size can be systematically included in the analysis of the Monte Carlo results.

During the simulations, chain configurations are stochastically relaxed by kink-jump and crankshaft motions (Sect. 9.1.2). In addition, a grand-canonical simulation technique [9.260–263] has been used, where the chemical potential difference $\Delta\mu$ between the two species is held fixed, rather than ϕ_A/ϕ_B. Since A-chains transform from time to time into B-chains, only the total number $n = n_A + n_B$ of chains is fixed, while $\Delta n = n_A - n_B$ fluctuates. For symmetric mixtures, the "order parameter" $\langle m \rangle = \langle |\Delta n| \rangle / n$ of the unmixing transition has been obtained from sampling Δn for $\Delta\mu$, where close to the transition temperature T_c one has $\langle m \rangle \propto (1 - T/T_c)^\beta$ for $T \leq T_c$. From the collective structure factor $S(q = 0) \propto |1 - T/T_c|^{-\gamma}$, the exponent γ has been obtained. The correlation length exponent ν of the order parameter fluctuations in the system, $\xi \propto |1-T/T_c|^{-\nu}$, has been determined from the peak positions of the collective structure factor or the specific heat. Since the peaks correspond to the case $L/\xi = \text{const}$, i.e. $L^{-1/\nu}|1 - T/T_c| = \text{const}$, the positions should be extrapolated with increasing lattice size L linearly versus $L^{-1/\nu}$. The finite size scaling analysis of the Monte Carlo data strongly indicates non-classical critical behaviour of Ising type

$$\beta \approx 0.32, \quad \nu \approx 0.63, \quad \gamma \approx 1.24, \tag{9.15}$$

which has to be compared to the classical mean field values of the Flory–Huggins approximation

$$\beta_F \approx 1/2, \quad \nu_F \approx 1/2, \quad \gamma_F \approx 1. \tag{9.16}$$

The non-classical behaviour is, according to the Ginzburg criterion [9.202, 258, 265, 266], only observable close to T_c, while for larger distances classical behaviour is expected. The non-classical region becomes smaller as the chains become longer. Therefore the simulations of relatively short chains are consistent with the expectations from the Ginzburg criterion. The discrepancy between simulations and Flory–Huggins approximations for short chains is mainly based on the overcounting of the number of contacts, which is a direct consequence of the neglect of the excluded volume effect.

In contrast to demixing processes of a melt consisting of two different polymer species, the Monte Carlo investigation of segregation processes of block copolymers has begun only very recently [9.268]. Block copolymers are composed of two covalently bounded subchains of different constituents. If the two species are mutually incompatible, a phase separation occurs at low enough temperatures. However, due to the covalent bond between the two subchains, phase separation cannot proceed to a macroscopic scale; instead, microdomains rich in each of the two components are formed (see e.g. [9.267] and references therein). This so called mesophase formation produces periodic spatial patterns. In experiments, one finds that the system can form lamellar, spherical or cylindrical structures depending on the relative chain length of the two subchains and on the amount of solvent present in the mixtures. The phase separation of block copolymers in two dimensions has been simulated recently using a Monte Carlo type dynamics for lattice polymers [9.268]. An attractice interaction is found when two monomers of the same type are nearest neighbours. The monomer concentration was equal to 0.6. Chains with $N = 10, 20, 30, 40$ have been studied. The dynamics of the chains was model by reptation and kink-jump moves (Sect. 9.1.2). It has been found that the equilibrium structure function satisfies a scaling relation $S(q) = D^2 f(qD)$, where D is a measure of the thickness of the microdomains of the resulting structure and obeys $D \propto N^{1/2}$ in agreement with previous numerical calculations of a coarse-grained model introduced by *Oono* and coworkers [9.269, 270], and with experimental findings [9.267]. Very recently, Leibler's theory of microphase separation in block copolymers has been tested by Monte Carlo simulations [9.271] and the results are in qualitative accord with the theory.

9.6.3 Dynamics of Decomposition

In recent years, the dynamics of spinodal decomposition in polymer blends have attracted much experimental [9.272, 273] and theoretical [9.274–276] interest, since in these materials it is easy to probe different regions of the phase diagram over widely varying time scales. Analytical studies of spinodal decomposition in polymer mixtures have been carried out [9.274–276] using the Cahn–Hilliard–Cook theory, formulated with a Flory–Huggins–deGennes free energy functional. Due to the same problems as encountered in the study of small molecule systems, the calculations in polymer systems have similarly dealt mainly with the early stages of the decomposition process. Also, Monte Carlo simulations have been necessarily restricted to studying the early stages of segregation in a lattice model of polymer mixtures in two [9.277] and three dimensions [9.278].

In three dimensions it has been shown [9.278] that the linear Cahn theory does not describe the initial stages of phase separation in the model, and that appreciable coil contraction occurs. In this simulation chains of length 32 on a cubic lattice with linear dimension 40 and at concentration 0.2 have been studied. A quench was performed from infinite temperature to $\varepsilon/k_B T = 0.6$, where ε is the energy which is gained if two neighbouring sites are occupied by monomers of the same type. The critical temperature in this model is $\varepsilon/k_B T_c = 0.101$ [9.261]. The chains are moved by kink-jump and crankshaft motion (Sect. 9.1.2) and their dynamics corresponds to the Rouse model [9.18].

Numerical integration [9.279] of the time evolution equation of the Flory–Huggins–deGennes free energy indicates that the growth law for the characteristic domain size is given by a modified Lifshitz–Slyozov law and the growth law exponent is independent of the quench temperature. It is found that both the pair correlation function and the structure factors show dynamical scaling at late times.

9.7 Surface and Interfaces

9.7.1 Adsorption on Rough Surfaces

The phenomenon of adsorption of polymer chains on surfaces is of wide occurrence and has attracted extensive experimental and theoretical investigations (see for example [9.280]). Adsorption of polymers takes place on various types of surfaces, planar, spherical, cylindrical, and rough. Most of the investigations have been devoted to the adsorption on planar surfaces. When a surface is *physically rough* there are three principal effects. (i) The space available for a chain to assume various configurations is larger near the top of a hill than at the bottom of a valley so that entropy considerations alone will lead to preferential adsorption on the hills. (ii) On the other hand the adsorption energy is enhanced in the valleys compared to the hills, and consequently, energy considerations alone will lead to preferential adsorption in the valleys. (iii) The potential interaction between the monomers and the surface is dependent on the local curvature which in turn affects the boundary conditions used in determining the adsorption characteristics [9.281]. The competition between the various effects could lead to very subtle phenomena [9.282], and simulations [9.141] have been informative to some extent. The case of physical roughness is modelled as an impenetrable surface with a "checker-board" corrugation. It is shown that a Gaussian chain is localised in one of the wells at low temperatures. Between the localised regime and the unbinding transition point, there exists a diffusive regime, where the chain diffuses by being shared by several potential wells. This regime is equivalent to the conduction band in the analogous model of a 1D electron in a periodic potential. In contrast to the case of a flat surface, the unbinding transition of a chain from a periodically rough surface is very sharp due to an effective anchoring of the chain in the wells. The transition resembles the discontinuous "uncorking transition" of a chain from a single well [9.141].

9.7.2 Entropic Repulsion

Rigidity, or bending energy, has been shown to play an important role in the statistical mechanics of membranes and films [9.174] and semiflexible polymers [9.283–285]. In particular, if the thermal fluctuations are governed by rigidity, the entropic repulsion between two such semirigid objects or with a hard wall, is effectively long ranged [9.286]. This steric repulsion can in fact overcome the molecular attraction acting between the membrane or polymer and the wall, and leads to an unbinding transition with special features, due to the interplay of semirigidity and the range and type of the attractive potential [9.287]. Several regimes have been identified by means of Monte Carlo methods and scaling arguments [8.283, 284]. (i) A weak-fluctuation regime for sufficiently long-ranged potentials which is well discribed by an ensemble of independent blobs; (ii) a complex intermediate fluctuation regime with infinite-order, second-order, and anomalous first-order transitions; (iii) a strong-fluctuation regime for sufficiently short-ranged interactions which is characterised by second-order transitions with universal critical behaviour.

The interaction of a polymer with a penetrable plane (neutral, attractive or repulsive) has been simulated [9.288]. It is found that the crossover exponent is in accord with the de Gennes–Bray–Moore conjecture $\phi = 1 - \nu$, where ν is the correlation length exponent of the chain.

9.7.3 Confined Polymer Melts

The structure of polymeric fluids is strongly affected by the approximity of impenetrable surfaces. The understanding of this system is of practical significance since it is relevant in the modeling of composites, lubricant films, and polymer coatings in microelectronics. Therefore, this system has been the focus of recent experimental and theoretical work. Computer simulations have been especially useful.

Initial Monte Carlo simulations on lattice [9.289, 290] for polymer chains confined between two repulsive plates have indicated the flattening of the chains in the vicinity of the surfaces, which has been predicted by mean field arguments [9.291]. They also provided indirect evidence that no long-range repulsive forces exist, in accordance with *de Gennes*' analysis [9.292]. Furthermore, the simulations indicated a weak depletion of the melt density and an enhancement of the chain ends in proximity to the hard walls. However, the chain conformations very close to the surface cannot be trusted completely because of the inherent shortcomings of lattice models for distances comparable to the lattice constant. Therefore, off-lattice Monte Carlo simulations [9.293–297] have been used to overcome these difficulties. These simulations have demonstrated that the short-range melt density variations are oscillatory, and qualitatively similar to those in simple fluids, but no quantitative comparisons have been made. These off-lattice simulations again support the absence of long-range forces by providing reliable information on the screening of the wall effect after two to three molecular diameters. Recently, molecular dynamics simulations [9.298, 299] have shown that chains with centres-of-mass close to the solid surface had a lower mobility normal to the wall and a longer relaxation time than bulk chains. The solid-melt interface was found [9.299] to be very narrow, approximately two segment diameters, and independent of chain length. The oscillations of the segment density profile were weaker and were damped faster than those of a simple fluid density profile next to the same solid surface. Finally, it should be mentioned that simulations of semidilute solutions near a wall have also been performed [9.300].

Analogous investigations of the density profiles of end-grafted polymer brushes have been performed by Monte Carlo [9.301, 302] and by molecular dynamics simulations [9.303]. For monodisperse chains, it is found [9.303, 301] that the monomer density profile can be represented by a parabolic form, except for a depeletion zone very close to the grafting surface and for not too large surface coverages. The width of this depletion zone is consistent with that predicted by scaling arguments [9.280]. The density of the free chain ends as a function of distances from the grafting surface have been calculated and did not provide any evidence for a "dead zone", i.e., the free ends are not excluded from regions near the solid surface, in agreement with numerical [9.304, 305] and analytical self-consistent field calculations [9.306–308], rather than with earlier scaling arguments [9.280, 309] that predict a plateau region for the density profile.

Systems of two end-grafted polymer brushes separated by a distance D have been studied recently by molecular dynamics simulations [9.310]. It is found that for a given chain lenth N and density ρ, the amount of interpenetration increases with decreasing separation between the brushes. In addition, the force versus separation curves have the same form if the separation D is scaled by the maximum extent of each brush, suggesting that the brushes interact as soon as their density profiles overlap. With increasing surface coverage, it is found that the forces exhibit a steeper increase with decreasing separation than is observed experimentally or is predicted by the scaling and the self-consistent field theories, since the monomer density between the surfaces was above the semidilute regime.

9.8 Special Polymers

9.8.1 Polyelectrolytes

Several Monte Carlo investigations of linear polyelectrolytes have been performed, with either Debye screening or explicit ions [9.311–323]. All located the polymer charges on the backbone, all assumed the dielectric constant to be same everywhere, all used the Metropolis Monte Carlo Method, and all except that of *Brender* et al. [9.311, 312] and *Valleau* [9.313] generated new polymer configurations from old ones by the reptation algorithm. *Brender* et al. used a cubic lattice with explicit ions, periodic boundary conditions, and a minimum image approximation in which only the closest distance between the images of two ions is used. *Valleau* also used explicit counterions, with a bead-string model for the polymer. The investigations of *Carnie* and *Christos* [9.314–316] used an off-lattice model with a Debye–Hückel potential between charges and no ions [9.315] and partially ionized polyelectrolyte chain with a distribution of salt ions around the chain [9.316]. The results from the two models were compared [9.316] and it was concluded that the two models give similar conformational properties for lower degrees of ionization of the polyion and/or higher salt concentrations. For higher degrees of ionization and/or lower salt concentrations, the counterions seem to position themselves so as to enhance their effect over the screening model counterpart, and the polyelectrolyte chains are generally much more coil-like than in the screening model, agreeing with findings of *Valleau* [9.313]. The results are not in quantitative agreement with the limiting forms in *Manning*'s counterion condensation theory [9.317]. *Victor* and *Hansen* [9.318] used a discrete 2D lattice model of a dilute polyelectrolyte solution involving screened Coulomb interactions between monomers and counterions. This model exhibits counterion condensation and a conformational behaviour, described by a pair distribution function and form factor, similar to that of neutral polymers. The predictions for the electrostatic *persistence length* of short polyelectrolytes [9.319, 320] has been examined by *C. Reed* and *W. Reed* [9.322], and they concluded that excluded volume corrections are significant and not in good agreement with theory.

The effects of polyelectrolytes serving as counterions immersed between two parallel oppositely charged plates ("electric double layer") have been studied by Monte Carlo methods [9.323]. For this particular model system it turns out that the traditional double layer repulsion becomes attractive due to an entropically driven bridging mechanism. The magnitude of attraction is significant compared to ordinary double layer with van der Waals forces.

DNA. Polyelectrolyte effects play an important role in many aspects of the structure and function of nucleic acids. The counterion atmosphere of DNA neutralises the charges of the anionic phosphates and imparts electrostatic stability to the system. The nature of the effect, "counterion condensation", is unique in polyelectrolytes compared to simple electrolyte systems [9.317, 324]. Release of condensed counterions is considered to be an important thermodynamic component of ligand and protein binding to DNA [9.325]. While the significance of counterion condensation is well documented, relatively little is known at the molecular level about the structural details of the organisation of counterions around the various forms of DNA and the detailed nature of the electrostatic stabilisation. Recently some advances have been made in this direction using Monte Carlo simulations [9.326–335]. A typical problem encountered in these studies is a high acceptance ratio for single-particle Metropolis moves and its commensurate impact on the convergence of the thermodynamic and structural quantities. In addition, the role of dielectric saturation and the influence of the dielectric boundary between the DNA and the solvent on the potential of mean force between the counterions and the anionic phosphates of duplex DNA are issues not fully resolved. Molecular dynamics simulations have also been performed [9.336–338], but have led to contradictory results for the ion distribution.

9.8.2 Proteins

(a) Protein Folding. Theoretical studies by computer simulations of the folding-unfolding process have been carried out in some simplified models of protein [9.339, 340]. In particular *Go* et al. proposed the lattice model of protein, which is especially suitable for simulating the folding–unfolding transition repeatedly over time [9.341]. They extensively studied the 2D lattice model of protein, and clarified the role of short-range and long-range interactions in the folding–unfolding process. A Monte Carlo minimisation method was proposed by *Li* and *Scheraga* [9.342] in order to overcome the multiple-minima problem in the folding process. There the Metropolis sampling, assisted by energy minimisation, surmounts intervening barriers in moving through successive discrete local minima in the multidimensional energy surface. The method has located the lowest energy minimum in a particular brain peptide, which is probably the global minimum of the energy structure. The finding seems to support the concept that protein folding may be a Markov process. Protein folding of lattice models of a β-barrel globular protein including bond angles, hydrophobic and hydrophilic effects, has been performed recently [9.343]. Starting from a high temperature state, these models undergo an all-or-none transition to a unique structure. These simulations suggest that the general rules of globular folding are rather robust in that the tertiary structure is determined by the general pattern of hydrophobic, hydrophilic, and turn-type residues, with site-specific interactions mainly involved in structural fine tuning of a given topology. These studies suggest that loops may play an important role in producing a unique native protein state. Depending on the stability of the native conformation of the long loops, the conformational transition can be described by a two-state, three-state, or even larger number of multiple equilibrium states model.

(b) Protein Dynamics. The molecular simulations of the dynamics of proteins in equilibrium states have been extensively reviewed by *McCammon* and *Harvey* [9.345] and *Karplus* [9.344]. However, this method of molecular simulations is restricted to very small molecules or very short time scales. In the molecular dynamics calculations of proteins, one step of calculation, which corresponds to the real time of 10^{-15} s, takes at least 1 s of computer time. Proteins fold within a time of the order of 1 s. Thus, in order to simulate this, 10^{15} steps of calculations are necessary, requiring 10^{15} s of computer time. Therefore, Molecular simulation can be successful only by focusing on certain selected aspects of protein dynamics.

References

9.1 A. Baumgärtner: In *Applications of the Monte Carlo Method in Statistical Physics*, ed. by K. Binder, Topics Curr. Phys. Vol.36 (Springer, Berlin, Heidelberg 1984)
9.2 K. Binder: In *Molecular Level Calculations of the Structures and Properties of Non-Crystalline Polymers*, ed. by J. Bicerano (Dekker, New York 1989)
9.3 K. Kremer, K. Binder: Comp. Phys. Rep. **7**, 259 (1988)
9.4 I. Carmesin, K. Kremer: Macromolecules **21**, 2819 (1988)
9.5 H.P. Deutsch, K. Binder: J. Chem. Phys. **94**, 2294 (1991)
9.6 C. Domb, G.S. Joyce: J. Phys. C **5**, 956 (1972)
9.7 C. Domb: J. Stat. Phys. **30**, 425 (1983)
9.8 P.J. Flory: *Statistical Mechanics of Chain Molecules* (Wiley, New York 1969)
9.9 A. Baumgärtner, K. Binder: J. Chem. Phys. **75**, 2994 (1981)
9.10 R.H. Boyd: Macromolecules **22**, 2477 (1989)
9.11 A. Baumgärtner, M.M. Muthukumar: J. Chem. Phys. **84**, 440 (1986)
9.12 V. Klenin, A.V. Vologodskii, V.V. Anshelevich, M. Frank-Kamenetskii: J. Biomol. Struct. Dyn. **6**, 707 (1989)
9.13 M. Bishop, M.H. Kalos, H.L. Frisch: J. Chem. Phys. **70**, 1299 (1979)
9.14 P.H. Verdier, W.H. Stockmayer: J. Chem. Phys. **36**, 227 (1962)
9.15 P.E. Rouse: J. Chem. Phys. **21**, 1273 (1953)
9.16 H.J. Hilhorst, J.M. Deutch: J. Chem. Phys. **63**, 5153 (1975)
9.17 P.G. de Gennes: J. Chem. Phys. **55**, 572 (1971)
9.18 M. Lax, C. Brender: J. Chem. Phys. **67**, 1785 (1977)
9.19 A. Baumgärtner, K. Binder: J. Chem. Phys. **71**, 2541 (1979)
9.20 A. Baumgärtner: Ann. Rev. Phys. Chem. **35**, 419 (1984)
9.21 F.T. Wall, F. Mandel: J. Chem. Phys. **63**, 4592 (1975)
9.22 F. Mandel: J. Chem. Phys. **70**, 3984 (1979)
9.23 A.K. Kron, O.B. Ptitsyn: Polymer Sci. USSR **9**, 847 (1967)
9.24 M. Murat, T.A. Witten: Macromolecules **23**, 520 (1990)
9.25 C. Aragao de Carvalho, S. Caracciolo: J. Physique **44**, 323 (1983)
9.26 S. Caracciolo, D.A. Sokal: J. Phys. A **20**, 2569 (1987)
9.27 M.L. Mansfield: J. Chem. Phys. **77**, 1554 (1982)
9.28 O.F. Olaj, W. Lantschbauer: Makromol. Chem. Rapid Comm. **3**, 847 (1982)
9.29 T. Pakula: Macromolecules **20**, 679; 2909 (1987)
9.30 S. Geyler, T. Pakula, J. Reiter: J. Chem. Phys. **92**, 2676 (1990)
9.31 J. Reiter, T. Eding, T. Pakula: J. Chem. Phys. **93**, 837 (1990)

9.32 J. Reiter: Macromolecules **23**, 3811 (1990)
9.33 T. Pakula, S. Geyler: In *Polymer Motion in Dense Systems*, ed. by D. Richter, T. Springer, Springer Proc. Phys. Vol. 29 (Springer, Berlin, Heidelberg 1988)
9.34 M. Lal: Mol. Phys. **17**, 57 (1969)
9.35 B. MacDonald, N. Jan, D.L. Hunter, M.O. Steinitz: J. Phys. A **18**, 2627 (1985)
9.36 N. Madras, A.D. Sokal: J. Sat. Phys. **50**, 109 (1988)
9.37 G. Zifferer: Macromolecules **23**, 3166 (1990)
9.38 M. Bishop, C.J. Saltiel: J. Chem. Phys. **88**, 6594 (1988)
9.39 S.D. Stellman, P.J. Gans: Macromolecules **5**, 516 (1972)
9.40 H. Meirovitch: Macromolecules **16**, 249; 1628 (1983)
9.41 H. Meirovitch: J. Chem. Phys. **89**, 2514 (1988)
9.42 M.N. Rosenbluth, A. W. Rosenbluth: J. Chem. Phys. **23**, 356 (1955)
9.43 J. Mazur, F.L. McCrackin: J. Chem. Phys. **49**, 648 (1976)
9.44 J. Batoulis, K. Kremer: J. Phys. A **21**, 127 (1987)
9.45 H.A. Lim, H. Meirovitch: Phys. Rev. A **39**, 4176 (1989)
9.46 H. Meirovitch, H.A. Lim: Phys. Rev. A 39, 4186 (1989)
9.47 H. Meirovitch, H.A. Lim: J. Chem. Phys. **91**, 2544 (1989)
9.48 H.C. Öttinger: Macromolecules **18**, 92 (1985)
9.49 M. Doi, S.F. Edwards: *The Theory of Polymer Dynamics*, Clarendon Press, Oxford (1986)
9.50 R.B. Bird, C.F. Curtiss, R.C. Armstrong, O. Hassager: *Dynamics of Polymeric Liquids*, Vols. 1 and 2 (Wiley, New York 1987)
9.51 S.F. Edwards: Proc. Phys. Soc. **91**, 513 (1967)
9.52 P.G. de Gennes: Macromolecules **9**, 587 (1976)
9.53 M. Doi, S.F. Edwards: J. Chem. Soc. Farady Trans. 2 **74**, 1789; 1802; 1818 (1978)
9.54 G. Ronca: J. Chem. Phys. **79**, 1031 (1983)
9.55 W. Hess: Macromolecules **19**, 1395 (1986)
9.56 W. Hess: Macromolecules **20**, 2587 (1987); **21**, 2620 (1988)
9.57 M. Rubinstein: Phys. Rev. Lett. **59**, 1946 (1987)
9.58 T.A. Kavassalis, J. Noolandi: Macromolecules **21**, 2869 (1988)
9.59 K.S. Schweizer: J. Chem. Phys. **91**, 5802; 5822 (1989)
9.60 J. des Cloizeaux: Macromolecules **23**, 3992 (1990)
9.61 M. Rubinstein, S. Zurek, T. McLeish, R. Ball: J. Physique **51**, 757 (1990)
9.62 D. Richter, A. Baumgärtner, K. Binder, B. Ewen: Phys. Rev. Lett. **47**, 109 (1981); **48**, 1695 (1982)
9.63 M. Bishop, D. Ceperley, H.L. Frisch, M.H. Kalos: J. Chem. Phys. **76**, 1557 (1982)
9.64 A. Kolinski, J. Skolnick, R. Yaris: J. Chem. Phys. **86**, 1567; 7164; 7174 (1987)
9.65 J. Skolnick: In *Reactive and Flexible Molecules in Liquids*, ed. by T. Dorfmüller (Kluwer, Dordrecht 1989)
9.66 K.E. Evans, S.F. Edwards: J. Chem. Soc. Faraday Trans. 2 **77**, 1891 (1981)
9.67 J. Deutsch: Phys. Rev. Lett. **49**, 926 (1982)
9.68 K. Kremer: Macromolecules **16**, 1632 (1983)
9.69 K. Kremer, K. Binder: J. Chem. Phys. **81**, 6381 (1984)
9.70 J. Deutsch: Phys. Rev. Lett. **54**, 56 (1985)
9.71 J. Deutsch: J. de Phys. **48**, 141 (1987)
9.72 C.C. Crabb, D.F. Hoffman, M. Dial, J. Kovac: Macromolecules **21**, 2230 (1988)
9.73 K. Kremer, G.S. Grest: J. Chem. Phys. **92**, 5057 (1990)
9.74 H.C. Öttinger: J. Chem. Phys. **91**, 6455 (1989)
9.75 H.C. Öttinger: J. Chem. Phys. **92**, 4540 (1990)
9.76 P.O. Brunn: Physica D **20**, 403 (1986)
9.77 F. Rondelez, D. Ausserre, H. Hervet: Ann. Rev. Phys. Chem. **38**, 317 (1987)
9.78 P.G. de Gennes: J. Chem. Phys. **60**, 5030 (1974)
9.79 Y. Rabin: J. Non-Newtonian Fluid Mech. **30**, 119 (1988)
9.80 P.J. Dotson: J. Chem. Phys. **79**, 5730 (1983)
9.81 H.H. Saab, P.J. Dotson: J. Chem. Phys. **86**, 3039 (1987)
9.82 J.H.R. Clarke, D. Brown: J. Chem. Phys. **86**, 1542 (1987)
9.83 E. Duering, Y. Rabin: Macromolecules **23**, 2232 (1990)
9.84 J.L. Chen, H. Morawetz: Macromolecules **15**, 1185 (1982)
9.65 D.L. Smisek, D.A. Hoagland: Macromolecules **22**, 2270 (1989)
9.86 L.S. Lerman, H.L. Frisch: Biopolymers **21**, 995 (1982)
9.87 O.J. Lumpkin, B.H. Zimm: Biopolymers **21**, 2315 (1982)
9.88 G.W. Slater, J. Noolandi: Phys. Rev. Lett. **55**, 1579 (1985)

9.89 G.W. Slater, J. Rousseau, J. Noolandi: Biopolymers **26**, 863 (1987)
9.90 M. Doi, T. Kobayashi, Y. Makino, M. Ogawa, G. W. Slater, J. Noolandi: Phys. Rev. Lett. **61**, 1893 (1988)
9.91 B. Zimm: Phys. Rev. Lett. **61**, 2965 (1988)
9.92 M. Olvera de la Cruz, J. Deutsch, S.F. Edwards: Phys. Rev. A **33**, 2047 (1986)
9.93 K.E. Evans, S.F. Edwards: J. Chem. Soc. Faraday Trans. **77**, 1891 (1981)
9.94 E.O. Shaffer, M. Olvera de la Cruz: Macromolecules **22**, 2476 (1989)
9.95 J. Deutsch, T.L. Madden: J. Chem. Phys. **90**, 2476 (1989)
9.96 J. Deutsch, J. Chem. Phys. **90**, 7436 (1989)
9.97 T.A.J. Duke: Phys. Rev. Lett. **62**, 2877 (1989)
9.98 J. Batoulis, N. Pistoor, K. Kremer, H.L. Frisch: Electrophoresis **10**, 442 (1989)
9.99 J. Melenkevitz, M. Muthukumar: Chemtracts **1**, 171 (1990)
9.100 L. Monnerie, R.D. de la Batie, A. Baumgärtner: In *Basic Features of the Glassy State*, ed. by J. Colmenero, A. Alegria (World Scientific, Singapore 1990)
9.101 R.D. de la Batie, J.L. Viovy, L. Monnerie: J. Chem. Phys. **81**, 657 (1984)
9.102 H.P. Witman, K. Kremer, K. Binder: In *Proceedings of the Second International Workshop on Non-Crystalline Solids* ed. by J. Colmenero, A. Alegria, eds, p. 225 (World Scientific, Singapore 1990)
9.103 D. Rigby, R.J. Roe: J. Chem. Phys. **87**, 7285 (1987)
9.104 D. Rigby, R.J. Roe: J. Chem. Phys. **88**, 5280 (1988)
9.105 J.H. Gibbs, E. di Marzio: J. Chem. Phys. **28**, 373 (1958)
9.106 M.H. Cohen, G.S. Grest: Phys. Rev. B **20**, 1077 (1979)
9.107 F. Geny, L. Monnerie: J. Polym. Sci. Polym. Phys. Ed. **17**, 131 (1979)
9.108 K. Kremer, A. Baumgärtner, K. Binder: J. Phys. A **15**, 2879 (1982)
9.109 M.S. Wertheim: J. Stat. Phys. **42**, 459; 477 (1986)
9.110 M.S. Wertheim: J. Chem. Phys. **87**, 7323 (1987)
9.111 G. Jackson, W.G. Chapman, K.E. Gubbins: Mol. Phys. **65**, 1 (1988)
9.112 W.G. Chapman, G. Jackson, K.E. Gubbins: Mol. Phys. **65**, 1057 (1988)
9.113 R. Dickman, C.K. Hall: J. Chem. Phys. **85**, 4108 (1986); **89**, 3168 (1988)
9.114 K.G. Honnell, C.K. Hall. J. Chem. Phys. **90**, 1841 (1989)
9.115 P.J. Flory: J. Chem. Phys. **10**, 51 (1942)
9.116 H. Huggins: Ann. N.Y. Acad. Sci. **43**, 1 (1942)
9.117 N.F. Carnahan, K.E. Starling: J. Chem. Phys. **64**, 2686 (1976)
9.118 T. Boublik: Mol. Phys. **66**, 91 (1989)
9.119 J. Gao, J.H. Wiener: J. Chem. Phys. **91**, 3168 (1989)
9.120 R. Dickman: J. Chem. Phys. **87**, 2246 (1987)
9.121 A. Hertanto, R. Dickman: J. Chem. Phys. **89**, 7577 (1988)
9.122 D. Stauffer: *Introduction to Percolation Theory* (Taylor and Francis, London 1985)
9.123 B. Derrida: Phys. Rep. **103**, 29 (1984)
9.124 J.W. Lyklema, K. Kremer: Z. Phys. B **55**, 41 (1984)
9.125 A.K. Roy, B.K. Chakrabarti: J. Phys. A **20**, 215 (1987)
9.126 S.B. Lee, H. Nakanishi: Phys. Rev. Lett. **61**, 2022 (1988)
9.127 Y. Meir, A.B. Harris: Phys. Rev. Lett. **63**, 2819 (1989)
9.128 A. Baumgärtner, M. Muthukumar: J. Chem. Phys. **87**, 3082 (1987)
9.129 S.F. Edwards, M. Muthukumar: J. Chem. Phys. **89**, 2435 (1988)
9.130 M.E. Cates, R.C. Ball: J. Physique **49**, 2009 (1988)
9.131 J.F. Douglas: Macromolecules **21**, 3515 (1988)
9.132 S.F. Edwards, Y. Chen: J. Phys. A **89**, 2963 (1988)
9.133 J.D. Honeycutt, D. Thirumalai: J. Chem. Phys. **90**, 4542 (1989)
9.134 J. Machta, R.A. Guyer: J. Phys. A **22**, 2539 (1989)
9.135 T. Nattermann, W. Renz: Phys. Rev. **40**, 4675 (1989)
9.136 S.F. Edwards: In *Molecular Basis of Polymer Networks*, ed. by A. Baumgärtner, C.E. Picot, Springer Proc. Phys. Vol. 42 (1989) p.11
9.137 A. Baumgärtner: In *Space-Time Organization in Macromolecular Fluids*, ed. by F. Tanaka, M. Doi, T. Ohta, Springer Ser. Chem. Phys. Vol 51, 141 (1989)
9.138 M. Muthukumar: J. Chem. Phys. **90**, 4594 (1989)
9.139 Th. Vilgis: J. de Phys. **50**, 3243 (1989)
9.140 A. Baumgärtner, W. Renz: J. de Phys. **51**, 2641 (1990)
9.141 A. Baumgärtner, M. Muthukumar: J. Chem. Phys. **94**, 4062 (1991)
9.142 R.B. Pandey, D. Stauffer, A. Margolina, J.G. Zabolitzky: J. Stat. Phys. **34**, 427 (1984)

9.143 S. Havlin, D. Ben-Avraham: Adv. Phys. **36**, 965 (1987)
9.144 J.W. Haus, K. W. Kehr: Phys. Rep. **150**, 263 (1987)
9.145 Y. Gefen, A. Aharony, S. Alexander: Phys. Rev. Lett. **50**, 77 (1983)
9.146 S. Alexander, R. Orbach: J. de Phys. Lett. **43**, L625 (1982)
9.147 R. Rammal, G. Toulouse: J. de Phys. Lett. **44**, 13 (1983)
9.148 A. Baumgärtner: Europhys. Lett. **4**, 1221 (1988)
9.149 A. Baumgärtner, M. Moon: Europhys. Lett. **9**, 203 (1989)
9.150 A. Baumgärtner: In *Reactive and Flexible Molecules in Liquids*, ed. by T. Dorfmüller (Kluwer, Dordrecht, 1989)
9.151 M. Muthukimar, A. Baumgärtner: Macromolecules **22**, 1941 (1989)
9.152 M. Muthukumar, A. Baumgärtner: Macromolecules **22**, 1937 (1989)
9.153 D. Frenkel: Mol. Phys. **60**, 1 (1987)
9.154 R. Eppenga, D. Frenkel: Mol. Phys. **50**, 1303 (1984)
9.155 B.M. Mulder, D. Frenkel: Mol. Phys. **55**, 1193 (1985)
9.156 A. Baumgärtner: J. de Phys. Lett. **46**, 659 (1985)
9.157 P.J. Flory: Proc. Roy. Soc. Lond. A **234**, 60 (1956)
9.158 G. Ronca, D. Y. Yoon: J. Chem. Phys. **76**, 3295 (1982)
9.159 P.D. Gujrati, M. Goldstein: J. Chem. Phys. **74**, 2596 (1981)
9.160 P.D. Gujarati: J. Stat. Phys. **28**, 441 (1982)
9.161 J.F. Nagle, P.D. Gujrati, M. Goldstein: J. Phys. Chem. **88**, 4599 (1984)
9.162 A. Baumgärtner: J. Chem. Phys. **81**, 484 (1984)
9.163 D. Y. Yoon, A. Baumgärtner: Macromolecules **17**, 2864 (1985)
9.164 A. Baumgärtner: J. Chem. Phys. **84**, 1905 (1986)
9.165 A. Kolinski, J. Skolnick, R. Yaris: Macromolecules **19**, 2550; **19**, 2560 (1986)
9.166 A. ten Bosch, P. Maissa, P. Sixiou: J. Chem. Phys. **70**, 3462 (1983)
9.167 A.R. Khokhlov, A.N. Semenov: J. Stat. Phys. **38**, 161 (1985)
9.168 G. Ronca, D.Y. Yoon: J. Chem. Phys. **80**, 925 (1984)
9.169 M. Warner, J.M.F. Gunn, A. Baumgärtner: J. Phys. A **18**, 3007 (1985)
9.170 T. Odijk: Macromolecules **19**, 2313 (1986)
9.171 P.G. de Gennes: Mol. Cryst. Liq. Cryst. (Lett.) **102**, 95 (1984)
9.172 H.J.C. Berendsen: In *Molecular Dynamics Simulations of Statistical Mechanic Systems*, ed. by C. Cicotti, W.G. Hoover (North-Holland, Amsterdam (1986) p. 496
9.173 O.G. Mouritsen: In *Molecular Description of Biological Membrane Components by Computer Aided Conformational Analysis*, ed. by R. Brasseur (CRC Press, Boca Raton, 1989)
9.174 See the reviews quoted in *Statistical Mechanics of Membranes and Surfaces*, ed. by D.R. Nelson, T. Piran, S. Weinberg (World Scientific, Singapore 1989)
9.175 F.F. Abraham, D.R. Nelson: Science **249**, 393 (1990)
9.176 W. Helfrich: J. de Phys. **46**, 1263 (1985)
9.177 L. Peliti, S. Leibler: Phys. Rev. Lett. **54**, 1690 (1985)
9.178 D.R. Nelson, L. Peliti: J. de Phys. **48**, 1085 (1987)
9.179 F. David, E. Guitter: Europhys. Lett. **5**, 709 (1988)
9.180 J.A. Aronowitz, T.C. Lubensky: Phys. Rev. Lett. **60**, 2634 (1988)
9.181 M. Paczuski, M. Kardar, D.R. Nelson: Phys. Rev. Lett. **60**, 2639 (1988)
9.182 M. Kardar, D.R. Nelson: Phys. Rev. Lett. **58**, 1289 (1987)
9.183 J.A. Aronowitz, T.C. Lubensky: Europhys. Lett. **4**, 395 (1987)
9.184 B. Duplantier: Phys. Rev. Lett. **58**, 2733 (1987)
9.185 Y. Kantor, D.R. Nelson: Phys. Rev. A **36**, 4020 (1987)
9.186 Y. Kantor, M. Kardar, D.R. Nelson: Phys. Rev. A **35**, 3056 (1987)
9.187 F.F. Abraham, W.E. Rudge, M. Plischke: Phys. Rev. Lett. **62**, 1757 (1989)
9.188 M. Plischke, D. Boal: Phys. Rev. A **38**, 4943 (1988)
9.189 J.-S. Ho, A. Baumgärtner: Phys. Rev. Lett. **63**, 1324 (1989)
9.190 D. Boal, E. Levinson, D. Liu, M. Plischke: Phys. Rev. A **40**, 3292 (1989)
9.191 A. Baumgärtner, J.-S. Ho: Phys. Rev. A **41**, 5747 (1990)
9.192 J.-S. Ho, A. Baumgärtner: Europhys. Lett. **12**, 295 (1990)
9.193 J.-S. Ho, A. Baumgärtner: Mol. Simul. **6**, 163 (1990)
9.194 S. Leibler, A.C. Maggs: Phys. Rev. Lett. **63**, 406 (1989)
9.195 R. Lipowsky, M. Girardet: Phys. Rev. Lett. **65**, 2893 (1990)
9.196 G. Grest, M. Murat: J. de Phys. **41**, 1415 (1990)
9.197 E. Guitter, F. David, S. Leibler, L. Peliti: Phys. Rev. Lett. **61**, 2949 (1988)
9.198 E. Guitter, S. Leibler, A.C. Maggs, F. David: J. de Phys. **51**, 1055 (1990)

9.199 M. Tanemura, T. Ogawa, N. Ogita: J. Comp. Phys. **51**, 191 (1983)
9.200 F. David, J.M. Drouffe: Nucl. Phys. B (Proc. Suppl.) **4**, 83 (1988)
9.201 M.E. Cates: Phys. Rev. Lett. **53**, 926 (1984); J. Physique **46**, 1059 (1985)
9.202 P.G de Gennes: *Scaling Concepts in Polymer Physics* (Cornell, Ithaca, 1979)
9.203 S. Leibler, R.P. Singh, M.E. Fisher: Phys. Rev. Lett. **59**, 1989 (1987)
9.204 M.E. Fisher: Physica D **38**, 112 (1989)
9.205 C.J. Camacho, M.E. Fisher: Phys. Rev. Lett. **65**, 9 (1990)
9.206 A.C. Maggs, S. Leibler, M.E. Fisher, C.J. Camacho: Phys. Rev. A **42**, 691 (1990)
9.207 J. Mazur, F.Mc Crackin: Macromolecules **10**, 328 (1977)
9.208 T.C. Lubensky, J. Isaacson: Phys. Rev. A **20**, 2130 (1979)
9.209 J. Isaacson, T. C. Lubensky: J. de Phys. Lett. **41**, 469 (1990)
9.210 B.H. Zimm: Macromolecules **17**, 795 (1984)
9.211 J.E.G. Lipson, S.G. Whittington, M.K. Wilkinson, J.L. Martin, D.S. Gaunt: J. Phys. A **18**, L469 (1985)
9.212 M.K. Wilkinson, D.S. Gaunt, J.E.G. Lipson, S.G. Whittington: J. Phys. A **19**, 789 (1986)
9.213 S.G. Whittington, J.E.G. Lipson, M.K. Wilkinson, D.S. Gaunt: Macromolecules **19**, 1241 (1986)
9.214 A.J. Barrett, D.L. Tremain: Macromolecules **20**, 1687 (1987)
9.215 J. Batoulis, K. Kremer: Macromolecules **22**, 427 (1989)
9.216 R.L. Lescanec, M. Muthukumar: Macromedoles **23**, 2280 (1990)
9.217 J.J. Freire, A. Rey, R. Prats: Macromolecules **19**, 452 (1986)
9.218 G. Grest, K. Kremer, T.A. Witten: Macromolecules **20**, 1376 (1987)
9.219 D.S. Gaunt, J.E.G. Lipson, S.G. Whittington, M.K. Whilkinson: J. Phys. A **19**, L811 (1986)
9.220 M.K. Wilkinson: J. Phys. A **19**, 3431 (1986)
9.221 M. Daoud, J.P. Cotton: J. de Phys. **43**, 531 (1982)
9.222 A. Miyake, K.F. Freed: Macromolecules **16**, 1228 (1983); **17**, 678 (1984)
9.223 W. Burchard: Adv. Poly. Sci. **48**, 1 (1983)
9.224 G.S. Grest, K. Kremer, S.T. Milner, T.A. Witten: Macromolecules **22**, 1904 (1989)
9.225 P.M. Lam, K. Binder: J. Phys. A **21**, L405 (1988)
9.226 Y. K. Leung, B.E. Eichinger: J. Chem. Phys. **80**, 3877; 3885 (1984)
9.227 L.Y. Shy, Y.K. Leung, B.E. Eichinger: Macromolecules **18**, 983 (1985)
9.228 B.E. Eichinger: J. Chem. Phys. **75**, 1964 (1981)
9.229 N.A. Neuburger, B.E. Eichinger: J. Chem. Phys. **83**, 884 (1985)
9.230 L.Y. Shy, B.E. Eichinger: Macromolecules **19**, 2787 (1986)
9.231 D.B. Adolf, J.G. Curro: Macromolecules **20**, 1646 (1987)
9.232 J. Gao, J.H. Weiner: Macromolecules **20**, 2525 (1987); **21**, 773 (1988)
9.233 G. Ronca, G. Allegra: J. Chem. Phys. **63**, 4990 (1975)
9.234 P.J. Flory, B. Erman: Macromolecules **15**, 900 (1982)
9.235 L. Kholodenko, K.F. Freed: J. Chem. Phys. **80**, 900 (1984)
9.236 J.F. Douglas, K.F. Freed: Macromolecules **18**, 2445; 2455 (1985)
9.237 D. Thirumalai: Phys. Rev. A **37**, 296 (1988)
9.238 H. Saleur: J. Stat. Phys. **45**, 419 (1986)
9.239 B. Duplantier: J. Chem. Phys. **86**, 4233 (1986)
9.240 B. Duplantier: Phys. Rev. A **38**, 3647 (1988)
9.241 T. Ishinabe: J. Phys. A **18**, 3181 (1985)
9.242 V. Privman: J. Phys. A **19**, 3287 (1986)
9.243 F. Seno, A.L. Stella: J. de Phys. **49**, 739 (1988)
9.244 F. Seno, A.L. Stella: Europhys. Lett. **7**, 605 (1988)
9.245 A. Kolinski, J. Skolnick, R. Yaris: J. Chem. Phys. **85**, 3585 (1986)
9.246 P.H. Poole, A. Coniglio, N. Jan, H.E. Stanley: Phys. Rev. B **39**, 495 (1989)
9.247 H. Meirovitch, H.A. Lim: J. Chem. Phys. **92**, 5144; 5155 (1990)
9.248 B.J. Cherayil, J.F. Douglas, K.F. Freed: J. Chem. Phys. **87**, 3089 (1987)
9.249 B. Duplantier, H. Saleur: Phys. Rev. Lett. **59**, 539 (1987)
9.250 A. Weinrib, S.A. Trugman: Phys. Rev. B **31**, 2993 (1985)
9.251 A. Coniglio, N. Jan, I. Majid, H.E. Stanley: Phys. Rev. B **35**, 3617 (1987)
9.252 C. Vanderzande: Phys. Rev. B **38**, 2865 (1988)
9.253 B. Duplantier, H. Saleur: Phys. Rev. Lett. **62**, 1368 (1989)
9.254 R.M. Bradley: Phys. Rev. A **39**, 3738 (1989)
9.255 F.v. Dieren, K. Kremer: Europhys. Lett **4**, 569 (1987)
9.256 M. Bishop, J.H.R. Clarke: J. Chem. Phys. **93**, 1455 (1990)

9.257 J. Batoulis, K. Kremer: Europhys. Lett. **7**, 683 (1988)
9.258 K. Binder: Coll. Poly. Sci. **265**, 273 (1987)
9.259 K. Binder: Coll. Poly. Sci. **266**, 871 (1988)
9.260 A. Sariban, K. Binder: J. Chem. Phys. **86**, 5859 (1987)
9.261 A. Sariban, K. Binder: Macromolecules **21**, 711 (1988)
9.262 A. Sariban, K. Binder: Coll. Poly. Sci. **267**, 469 (1989)
9.263 A. Sariban, K. Binder: Makromol. Chem. **189**, 2357 (1988)
9.264 K. Binder: Ferroelectrics **73**, 43 (1987)
9.265 P.G. de Gennes: J. de Phys. Lett. **38**, L44 (1977)
9.266 J.F. Joanny: J. Phys. A **11**, L117 (1978)
9.267 T. Hashimoto, K. Kowsaka, M. Shibayama, H. Kawai: Macromolecules **19**, 754 (1986)
9.268 A. Chakrabarti, R. Toral, J. Gunton: Phys. Rev. Lett. **63**, 2661 (1989)
9.269 Y. Oono, Y. Shiwa: Mod. Phys. Lett. **1**, 49 (1987)
9.270 Y. Oono, M. Bahiana: Phys. Rev. Lett. **61**, 1109 (1988)
9.271 B. Minchau, B. Dünweg, K. Binder: Polymer Commun. **31**, 348 (1990)
9.272 T. Hashimoto: Phase Transitions **12**, 47 (1988)
9.273 T. Nose: Phase Transitions **8**, 245 (1987)
9.274 P.G. de Gennes: J. Chem. Phys. **72**, 4756 (1980)
9.275 P. Pincus: J. Chem. Phys. **75**, 1996 (1981)
9.276 K. Binder: J. Chem. Phys. **79**, 6387 (1983)
9.277 A. Baumgärtner, D.W. Heermann: Polymer **27**, 1777 (1986)
9.278 A. Sariban, K. Binder: Polymer Commun. **30**, 205 (1989)
9.279 A. Chakrabarti, R. Toral, J.D. Gunton, M. Muthukumar: J. Chem. Phys. **92**, 6899 (1990)
9.280 P.G. de Gennes: Adv. Colloid Interface Sci. **27**, 189 (1987)
9.281 R.C. Ball, M. Blunt, W. Barford: J. Phys. A **22**, 2587 (1989)
9.282 D. Hone, H. Ji, P. Pincus: Macromolecules **20**, 2543 (1987)
9.283 R. Lipowsky, A. Baumgärtner: Phys. Rev. A **40**, 2078 (1989)
9.284 A.C. Maggs, D.A. Huse, S. Leibler: Europhys. Lett. **8**, 615 (1989)
9.285 A.C. Maggs, S. Leibler: Europhys. Lett. **12**, 19 (1990)
9.286 W. Helfrich: Z. Naturforschung A **33**, 305 (1978)
9.287 R. Lipowsky, S. Leibler: Phys. Rev. Lett. **56**, 2541 (1986)
9.288 K. Kremer: J. Chem. Phys. **83**, 5882 (1985)
9.289 W.G. Madden: J. Chem. Phys. **87**, 1405 (1987); **88**, 3934 (1988)
9.290 G. ten Brinke, D. Ausserre, G. Hadziioannou: J. Chem. Phys. **89**, 4374 (1988)
9.291 D.N. Theodorou: Macromolecules **21**, 1391; 1400 (1988)
9.292 P.G. de Gennes: C. R. Acad. Sci. Paris **305**, 1187 (1987)
9.293 R. Dickman, C. Hall: J. Chem. Phys. **89**, 3168 (1988)
9.294 S.K. Kumar, M. Vacatello, D.Y. Yoon: J. Chem. Phys. **89**, 5206 (1988)
9.295 S.K. Kumar, M. Vacatello, D.Y. Yoon: Macromolecules **23**, 2189 (1990)
9.296 M. Vacatello, D.Y. Yoon: J. Chem. Phys. **93**, 779 (1990)
9.297 A. Yethiraj, C.K. Hall: Macromolecules **23**, 1865 (1990)
9.298 K.F. Mansfield, D.N. Theodorou: Macromolecules **22**, 3143 (1989)
9.299 I. Bitsanis, G. Hadziioannou: J. Chem. Phys. **92**, 3827 (1990)
9.300 W.Y. Shih, W.H. Shih, I.A. Aksay: Macromolecules **23**, 3291 (1990)
9.301 A. Chakrabarti, R. Toral: Macromolecules **23**, 2016 (1990)
9.302 A.T. Clark, M. Lal: J. Chem. Soc. Faraday Trans. 2 **74**, 1857 (1975)
9.303 M. Murat, G.S. Grest: Macromolecules **22**, 4054 (1989)
9.304 T. Cosgrove, T. Heath, B. van Lent, F. Leermakers, J. Scheutjens: Macromolecules **20**, 1692 (1987)
9.305 M. Muthukumar, J.S. Ho: Macromolecules **22**, 965 (1989)
9.306 S.T. Milner, T.A. Witten, M.E. Cates: Macromolecules **21**, 2610 (1988)
9.307 J. Scheutjens, G.J. Fleer: Macromolecules **18**, 1882 (1985)
9.308 A.N. Semenov: Sov. Phys. JETP **61**, 773 (1985)
9.309 S. Alexander: J. de Phys. **38**, 983 (1977)
9.310 M. Murat, G.S. Grest: Phys. Rev. Lett. **63**, 1074 (1989)
9.311 C. Brender, M. Lax, S. Windwer: J. Chem. Phys. **74**, 2576 (1981); **80**, 886 (1984)
9.312 C. Brender: J. Chem. Phys. **92**, 4468; **93**, 2736 (1990)
9.313 J.P. Valleau: J. Chem. Phys. **129**, 163 (1989)
9.314 S.L. Carnie, G.A. Christos, T.P. Creamer: J. Chem. Phys. **89**, 6484 (1988)
9.315 G.A. Christos, S.L. Carnie: J. Chem. Phys. **91**, 439 (1989)

9.316 G.A. Christos, S.L. Carnie: J. Chem. Phys. **92**, 7661 (1990)
9.317 G.S. Manning: Quart. Rev. Biophys. **11**, 179 (1978)
9.318 J.M. Victor, J.P. Hansen: Europhys. Lett. **3**, 1161 (1987)
9.319 T. Odijk, A.C. Houwaart: J. Poly. Sci. Poly. Phys. Ed. **16**, 627 (1978)
9.320 J. Skolnick, M. Fixmann: Macromolecules **10**, 944 (1977)
9.321 S.K. Gupta, W.C. Forsman: Macromolecules **5**, 779 (1972)
9.322 C. Reed, W. Reed: J. Chem. Phys. **92**, 6916 (1990)
9.323 T. Akesson, C. Woodward, B. Jönsson: J. Chem. Phys. **91**, 2461 (1989)
9.324 M.T. Record, C.F. Anderson, T.M. Lohman: Quart. Rev. Biophys. **11**, 103 (1978)
9.325 M.T. Record, C.F. Anderson, P. Mills, M. Mossing, J.H. Roe: Adv. Biophys. **20**, 109 (1985)
9.326 M. Le Bret, B.H. Zimm: Biopolymers **23**, 287 (1984)
9.327 C.S. Murthy, R.J. Bacquet, P.J. Rossky: J. Phys. Chem. **89**, 701 (1985)
9.328 P. Mills, C.F. Anderson, M.T. Record: J. Phys. Chem. **89**, 3984 (1985)
9.329 P. Mills, M.D. Paulsen, C.F. Anderson: Chem. Phys. Lett. **129**, 155 (1986)
9.330 P. Mills, C.F. Anderson, M.T. Record: J. Phys. Chem. **90**, 6541 (1986)
9.331 M.D. Paulsen, B. Richy, C.F. Anderson, M.T. Record: Chem. Phys. Lett. **139**, 448 (1987); **143**, 115 (1988)
9.332 J. Conrad, M. Troll. B.H. Zimm: Biopolymers **27**, 1711 (1988)
9.333 P.S. Subramanian, G. Ravishanker, D. L. Beveridge: Proc. Natl. Acad. Sci. USA **85**, 1836 (1988)
9.334 P.S. Subramanian, D.L. Beveridge: Biomol. Struct. Dyn. **6**, 1093 (1989)
9.335 B. Jayaram, S. Swaminathan, D.L. Beveridge, K. Sharp, B. Honig: Macromolecules **23**, 3156 (1990)
9.336 W.K. Lee, Y. Gao, E.W. Prohofsky: Biopolymers **23**, 257 (1984)
9.337 W.F. van Gunstern, H.J.C. Berendsen, R.C. Guersten, H.R. Zwinderman: Ann. N.Y. Acad. Sci. **482**, 287 (1986)
9.338 K. Swamy, E. Clement: Biopolymers **26**, 1901 (1987)
9.339 N. Go: Ann. Rev. Biophys. Bioeng. **12**, 183 (1983)
9.340 S.I. Segawa, T. Kawai: Biopolymers **25**, 1815 (1986)
9.341 N. Go, H. Abe, H. Mizuno, H. Taketomi: In *Protein Folding*, ed. by R. Jaenicke (Elsevier, Amsterdam 1980)
9.342 Z. Li, H.A. Scheraga: Proc. Natl. Acad. Sci. USA **84**, 6611 (1987)
9.343 J. Skolnick, A. Kolinski, R. Yaris: Proc. Nat. Acad. Sci. USA **86**, 1229 (1989)
9.344 M. Karplus: Physics Today **10**, 68 (1987)
9.345 J.A. McCammon, S.C. Harvey: *Dynamics of Proteins and Nucleic Acids*(Cambridge University Press, Cambridge 1987)

10. Percolation, Critical Phenomena in Dilute Magnets, Cellular Automata and Related Problems

Dietrich Stauffer

With 2 Figures

This chapter summarises some of the progress made in the simulation of disordered systems since the completion of the second volume of this series [10.1]. As the title indicates, the selection of material is incomplete and highly subjective.

10.1 Percolation

During the 1970s it was hoped that conductivity and elasticity of random networks have the same critical behaviour. That means, we look at a large lattice where each bond between two nearst-neighbour sites is randomly either present or absent, with a concentration p of present bonds ("bond percolation"). A bond can be interpreted as an electrical wire, in which case we talk of a random resistor network. Or it may be interpreted as a chemical bond between molecules, in which case we have a model for polymer gelation. Geometrically, for p above some percolation threshold p_c, an infinite cluster of connected bonds is formed. Only if this infinite cluster is present, can an electrical current flow through the system; and only the presence of this infinite cluster ensures mechanical stability against shear or tear. Currents as well as mechanical forces are transmitted only through relatively few strands of electrically or elastically active network chains of connected bonds; isolated bond clusters or loose ends of bond chains contribute neither to the electrical conductivity nor to the elastic resistance. Thus it was hoped [10.2] that elasticity and conductivity are proportional to each other, at least near the percolation threshold (gel point), where both should vanish proportional to $(p-p_c)^\mu$ asymptotically.

Unfortunately, this picture turned out to be too simple. If the chemical bonds consist of central forces only, then even for $p=1$ a square lattice is unstable against shear forces, and on a triangular lattice simulations show [10.3] that the rigidity threshold is shifted; the critical exponents in this oversimplified model are controversial. A much better model is one that also includes bond bending forces. Thus, in the harmonic approximation, the energy comprises two terms: A spring energy proportional to the square of the change in length of the distance between nearest neighbours; and a bending energy proportional to the square of the change in the angle between two bonds emanating from the same site. (Changes are measured with respect to the equilibrium values in the undistorted lattice.)

For this more realistic model the elastic constants become critical at the geometric percolation threshold, just as the electrical conductivity vanishes there. However, the critical exponents very near the threshold are still not the same: Numerically the exponent on the disordered honeycomb lattice was found, by transfer matrix techniques, to be about 4 for the elasticity, whereas it is about 1.3 for the conductivity. The Sahimi–Roux scaling law [10.4] argues, by comparison of forces and torques, that the ratio between the electrical conductivity and the elastic modulus should diverge as the square of the connectivity length ξ; this latter length is the typical radius of finite clusters in the percolation geometry and diverges as $(p-p_c)^{-\nu}$ near the threshold. Thus the predicted elasticity exponent is 3.97 whereas the Monte Carlo estimate [10.5] is 3.96 ± 0.04, in excellent agreement, and also consistent with less precise series extrapolations [10.11]. Perhaps logarithmic corrections to scaling are important for electrical and elastic percolation properties [10.6], like factors proportional to const. $+ 1/(\log \xi) + \cdots$

Thus model is purely mechanical and ignores entropy effeccts, which at elevated temperatures may dominate the elastic behaviour of gels; such entropy effects are difficult to simulate. For laboratory experiments related to gelation elasticity we refer to recent reviews [10.7], confirming percolation theory.

The conductivity exponent μ, often also called t, has been measured very accurately by Monte Carlo simulation of narrow strips or bars ("transfer matrix technique") with the help of a special purpose computer which has *for years* performed only these simulations [10.8]. One studies the system right at the percolation threshold, where the conductivity vanishes as $L^{-\mu/\nu}$ according to finite-size scaling theory. In two dimensions the Monte Carlo result $\mu = 1.299 \pm 0.002$ is inconsistent with the *Alexander–Orbach* [10.9] hypothesis $91/72 = 1.264$ (although series expansions [10.10] agree better with that value) and close to the speculation [10.6] $\mu = \nu - \beta/4 = 187/144 = 1.299$.

Three-dimensional simulations at the "well known" [10.10] bond percolation threshold 0.2488 of the simple cubic lattice are presently under way on this special purpose computer; for a network of superconductors and insulators they have already given a divergence with an exponent of $s = 0.73 \pm 0.01$. The forthcoming results for μ should then be compared with $\mu = 2.0$–2.1 obtained on conventional computers and through simulations of diffusion on percolation lattices [10.12, 6]. (According to Einstein, diffusivity and conductivity should have the same exponent.)

Of course, not only special purpose computers can provide years of computer time. Several workstations were employed [10.13] for about two years to give very accurate Monte Carlo values for the geometrical percolation exponents like ν, β, etc., and also the thresholds, like $p_c = 0.5927461$ on the square lattice. This latter value is a rare and nice example of how, over the years, high-quality simulations have yielded ever better results, without in general contradicting the earlier estimates.

Many other percolation properties have been investigated, for example, dilute Potts antiferromagnets at zero temperature [10.14]. We now concentrate on the question of multifractality, which in some ways contradicts traditional scaling arguments for percolation.

Let s be the mass of a finite cluster in percolation, and $\sum_{c} s^q$ the qth moment of the cluster size distribution, with the sum going over all finite clusters. Then for $q = 2, 3, \ldots$ these moments diverge in a predictable way;

$$M_q \propto (p - p_c)^{-\gamma - (q-2)\Delta} \tag{10.1}$$

for p close to the threshold. This critical exponent is thus a straight line if plotted versus q, and its slope is the gap exponent $\Delta = \gamma + \beta$. This constant-gap scaling of the Hunter–Domb type means that we know all critical exponents if we have determined only two of them, because a straight line is fixed by two of its points.

If a unit voltage is applied to a random resistor network (where again each bond is either conducting or insulating, with probabilities p and $1 - p$, respectively) an analogous moment is

$$M_q = \sum_{b} V_b^q \propto (p - p_c)^{t(q)}. \tag{10.2}$$

Here the sum goes over all bonds of the infinite cluster. Now, however, the exponent $t(q)$ is not a linear or other simple function of q, and according to our present understanding there is no relation at all [10.15] between different $t(q)$. Right at the critical point the moments diverge as $L^{-t(q)/\nu}$ when the linear lattice dimension L goes to infinity. These fractal dimensions $t(q)/\nu$ thus seem all to be independent of each other, from where the name "multifractality" arises. Mandelbrot saw similar effects in turbulence. For random resistor networks, similar though not identical effects are seen if, instead of the formal moments, we take a physical system with nonlinear resistors: The voltage across a bond varies as the qth power of current [10.16].

Some of the above moments M_q have a simple meaning i.e. the limit $q \to 0$ corresponds to counting the backbone, defined as the set of current-carrying bonds, $q = 2$ is related to the total resistance of the network, $q = 4$ to its noise, and $q \to \infty$ emphasizes the "red" bonds carrying the whole current (if we omit one red bond, the whole network would no longer carry any current). Thus at present we have to admit that no known formula relates the critical exponents of the backbone, conductivity, current fluctuations, and red bonds. (Note that negative q can be pathological and should not be relied upon to argue for multifractality. See [10.17] for multifractality in cluster numbers.)

The reason for multifractal behaviour seems to be extremely rare events in the distribution of voltage or other properties. The distributions have roughly a Gaussian shape, but closer inspection shows strongly asymmetric tails, and these tails for rare cases appear to be the cause of these deviations from simple scaling behaviour. The simple scaling theory assumes that "everything" is dominated by one single length, one single cluster size, or one single typical voltage; thus simple scaling does not deal with the tails of the distributions.

10.2 Dilute Ferromagnets

Percolation is the zero-temperature limit of randomly diluted ferromagnets (see also Chap. 11). What happens at finite temperatures? To investigate this we take a nearest neighbour spin-1/2 Ising model and leave a randomly selected fraction p of all lattice sites magnetic; the remaining lattice sites are not magnetic and do not contribute to the interaction energy. Only for p above the percolation threshold does a spontaneous magnetisation appear. What are the critical exponents at the shifted Curie temperature, when we vary the temperature T at fixed concentration p above p_c but below 1.

According to the Harris criterion, at this shifted Curie temperature we should have the same exponents as for the pure ferromagnet, if the specific heat remains finite (e.g. as in the 3D Heisenberg magnet). On the other hand, a different set of exponents (also differing from percolation exponents) is expected if in the pure case the specific heat diverges. Thus in the 2D Ising model, where, according to Onsager, the specific heat diverges only very weakly (logarithmically), we expect a border-line case which is numerically very difficult to treat.

Nevertheless, extensive simulations [10.18] could clarify the behaviour of the 2D dilute Ising model. The critical exponents are not changed, but logarithmic corrections may appear. For example, the specific heat in the dilute case diverges as $\log[\log(T-T_c)]$ instead of the $\log(T-T_c)$ for the pure case. These numerical results are in agreement with some but not all theories.

In three dimensions the simulation results are less clear. Early work gave no change of critical exponents, or critical exponents varying continuously with concentration p. More recent data [10.19] showed curvature in the log–log plots leading to these exponents and thus allows for the hope that traditional universality concepts are valid; however, the exponents might also be non-universal. The (effective) exponent at $p=0.8$ in the simple cubic lattice is consistent with theoretical expectations but the error bars are large [10.19].

The methods employed in these dilute Ising simulations are also of interest: Traditional Glauber dynamics with traditional programming gave poor results. *Swendsen–Wang* cluster flip procedures [10.20] improved the quality dramatically. However, *Heuer*'s one-bit-per-spin vectorised Glauber program [10.19] made the traditional dynamics competetive again since the cluster method has not been vectorised and is a thousand times slower.

The dynamics of the phase transition in the dilute Ising model were studied mainly at or near the percolation threshold [10.21], where the statics are believed to follow standard percolation exponents. The dynamics, however, are non-standard: The logarithm of the Glauber relaxation time is no longer equal to the logarithm of the correlation length ξ, multiplied with a critical exponent z, but varies with some power of $\log \xi$.

When both ferromagnetic and antiferromagnetic bonds are present, "frustration" is possible where a spin can never obey all the bonds to its neighbours simultaneously. Spin glasses are the most famous of such systems, and in a special case have recently been treated successfully with a variant of the *Swendsen–Wang* cluster method [10.22], but are outside the scope of this review. The *Widom* model (Chap. 11). of microemulsions has been studied extensively by Monte Carlo methods [10.23]. It consists of a spin-1/2 Ising model with ferromagnetic bonds to nearest-neighbours, antiferromagnetic bonds to next-nearest neighbours, and half as strong antiferromagnetic bonds to neighbours two lattice constants away. Molecules are thought to sit on the nearest-neighbour bonds between two spins, with a pair of up spins representing water, a pair of down spins an oil molecule, and a pair of antiparallel spins meaning an amphiphilic soap molecule. For weak antiferromagnetic interactions we find ferromagnetism at low temperatures, bounded by a second-order Curie point transition at higher temperatures. Stronger antiferromagnetism leads to layered or otherwise periodic phases (liquid crystals?) at low temperatures, bounded by first-order transitions to a bicontinuous phase without long-range order at higher temperatures.

If we dilute this Widom model, i.e. if again a random fraction $1-p$ of all lattice sites are empty in the sense of not interacting with the occupied sites, we may have a primitive model for microemulsions ("brine") in a porous medium like sandstone. The Monte Carlo [10.23] results indicate that, the Curie point curve $T_c(p)$ for the ferromagnetic phase behaves very similar to that of the normal nearest neighbour Ising model. I am not aware of any patents that the oil industry holds on this subject.

10.3 Cellular Automata

In deterministic cellular automata, each lattice site carries a spin which is either up or down. The orientation of each spin at time $t+1$ is fixed by the orientation of its neighbouring spins at time t. (We may also regard the spin itself as one of the neighbours influencing it.) Thus randomness is only possible in the initial condition when the spins point up and down randomly, with probability p and $1-p$, respectively. Once this initial configuration is set up, the following simulations can be done quite fast since no further random numbers are needed. Speeds of one or more updates per nanosecond are common on fast vector computers, using one-bit-per-site techniques [10.24].

A particular case, closely related to percolation, is bootstrap [10.25] percolation. At each time step, all occupied sites (spin up) that do not have at least m neighbours occupied after the previous time step are removed; empty sites remain empty forever. Initially, a fraction p of the sites are occupied randomly. The case $m=0$ corresponds to random percolation, for $m=1$ the isolated sites are removed and for $m=2$, all dangling loopless ends also. The threshold for these cases of bootstrap percolation is not changed compared to random percolation, since the infinite cluster still remains infinite after this culling process. On the other hand, if m equals the number of neighbours, a single empty site will finally empty the whole lattice, like an infection: $p_c = 1$ similar to 1D percolation. For intermediate m, we may have a second-order phase transition for low m, similar to percolation but with different critical exponents and thresholds. For higher m, first-order phase transitions have been observed, where the final density jumps to zero if $p < p_c$.

These first-order phase transitions are a numerical challenge, since often p_c is known to be unity for infinite lattices. The approach to this limit, however, is very slow, proportional to $1/\log(\text{size})$ ($m = 3$ square lattice) or even $1/\log[\log(\text{size})]$ ($m = 4$ simple cubic lattice). Thus a computer simulation gives thresholds different from unity even for hundreds of millions of sites, and it is not clear that the exact result for infinite lattices is a better approximation to realistic systems of linear dimension 10^8 than a computer simulation with linear dimension 10^3 or 10^4. For physics, the logarithm of infinity is clearly below a hundred. More details and literature can be found in Ref. [10.25].

10.4 Multispin Programming of Cellular Automata

In the previous discussion of bootstrap percolation we have already mentioned the high-speed simulations of large systems made possible [10.24] through multispin-coding with only one bit per site. We explain this technique here for deterministic cellular automata, first with a logical OR as the rule, then with the *da Silva–Herrmann* algorithm [10.26] for all cellular automata together (Kauffman model). Here we ignore probabilistic automata like the Ising model, which are more complicated in this technique due to the continual demand for new random numbers [10.28].

Imagine a simulation of infection on a square lattice: At each time step, a site becomes infected forever if at least one of its four neighbouring sites was infected at the previous time step. If we identify infection with spin up or TRUE, then this rule is the logical OR with four inputs: The centre is infected if north OR west OR south OR east are infected. To store each site in one bit, imagine we work on a 32-bit computer with a 96*96 lattice; then three computer words are sufficient to store one line $(1, 2, \ldots, 96)$. The first word contains in its 32 bits the sites, $1, 4, 7, \ldots, 94$, the second word the spins $2, 5, 8, \ldots, 95$, the third word stores sites $3, 6, 9, \ldots, 96$. Analogously, we need LL words to store a line of length L = 32*LL.

Obviously, the neighbours to the left of the second word are stored at the corresponding bit positions of the first word, and the right-hand neighbours are in the third word. Only for the two boundary words with numbers 1 and LL = 3 do we have to take care of periodic boundary conditions, the left-hand neighbour of the first word is the last (third) word shifted circularly by one bit to the right; for after this shift the third word contains sites $96, 3, 6, \ldots, 93$, which are the left-hand neighbours of $1, 4, 7, \ldots, 94$. Similarly, the right-hand neighbour of the last word (with index LL = 3) is contained in the first word provided we shift it circularly by one bit to the left. For simplicity we store the left-hand neighbour of word 1 in an additional word 0, and the right-hand neighbour of word LL = 3 in an additional word LLP1 = 4.

These tricks are made for each of the L lines of the square lattice; to have periodic boundary conditions in the other direction too we again store information in additional buffer lines with index 0 and L. Line 0 contains the same information as line L, and line L + 1 the same as line 1; no bit shifts are needed now.

Figure 10.1 gives the full program; let us concentrate first on the essential aspects as just explained. The two buffer lines 0 and L + 1 are updated in loop 7, the two buffer words 0 and LL1(= LL + 1) in loop 8 where ISHFTC is a circular shift function with only positive arguments of the number of steps to be shifted to the left. (We work with NBIT = 32 bits per word.) We define this circular shift through the logical OR of two non-circular shifts to the left and right. The double loop 3 is the main updating part: The new variable M equals the logical OR of the four old neighbours N. Because of the buffer technique, no IF conditions are needed in this innermost FORTRAN loop, and thus it runs faster on vector computers. (Therefore we also put the longer loop over J inside the shorter loop over I to allow efficient vectorisation also for L below 10^4.) Finally, loop 5 updates the old values N with the newly calculated M.

```
      PARAMETER (LL=30,LL1=LL+1,NBIT=32,L=LL*NBIT,L1=L+1)
      DIMENSION N(0:L1,0:LP1),M(L,LL)
      ISHFTC(I,K) = IOR(ISHFT(I,K),ISHFT(I,K-NBIT))
      DATA P,MAX,ISEED /0.001,900,123456789/
      PRINT *, L,P,MAX,ISEED
      IP=(2.0*P-1.0)*2147483648.0
      IBM=2*ISEED-1
      DO 6 I=1,LL
      DO 6 J=1,L
6     N(J,I)=0
      DO 1 NB=1,NBIT
      DO 1 I=1,LL
      DO 1 J=1,L
      N(J,I)=ISHFT(N(J,I),1)
      IBM=IBM*16807
1     IF(IBM.LT.IP) N(J,I)=IOR(N(J,I),1)
      DO 2 ITIME=1,MAX
       MAG=0
       DO 4 I=1,LL
       DO 4 J=1,L
4      IF(NOT(N(J,I)).NE.0) MAG=MAG+1
       IF(MAG.EQ.0) STOP
       PRINT *, ITIME,MAG
       DO 7 I=1,LL
       N( 0,I)=N(L,I)
7      N(L1,I)=N(1,I)
       DO 8 J=1,L
       N(J, 0 )=ISHFTC(N(J,LL),NBIT-1)
8      N(J,LL1)=ISHFTC(N(J,1),   1)
       DO 3 I=1,LL
       DO 3 J=1,L
3      M(J,I)=IOR( IOR(N(J,I-1),N(J,I+1)), IOR(N(J-1,I),N(J+1,I)))
       DO 5 I=1,LL
       DO 5 J=1,L
5      N(J,I)=M(J,I)
2     CONTINUE
      END
```

Fig. 10.1. Simple example of a one-bit-per-site program: 4-input logical OR

The first part of the program in Fig. 10.1, initialises the lattice randomly with a fraction p of infected sites, using multiplication of the integer IBM (with loss of the leading bit) to produce random numbers between -2^{31} and 2^{31}. (More complicated methods exist to produce better random numbers.) Each bit of the array N is set equal to one if and only if the random number is smaller than the value corresponding in this normalisation to the concentration p.

Unfortunately, the function IOR (logical OR) and ISHFT (logical shift, with zeros filled in) are not standard FORTRAN 77 and have different names on different compilers. Also the multiplication for the random numbers causes problems on a few computers. The reader then has to find out how to do this on the particular computer at hand. I have never found out how to manipulate bits on my pocket calculator.

More complicated rules than this simple OR can be formulated by more complicated logical expressions. The rule that at least two out of the four neighbours have to be infected in order to infect the centre site can be formulated as: $M = (N1 + N2) * (N3 + N4) + N1 * N2 + N3 * N4$, where multiplication stands for AND, addition for OR, and the N for the four neighbours. Similar expressions have been applied efficiently to other cases, with typically one site treated per nanosecond on a Cray-YMP processor.

10.5 Kauffman Model and da Silva–Herrmann Algorithm

The programming and speed situation is different if we want to use *one* program to simulate all 65536 cellular automata rules possible on a square lattice with four neighbours, or a random mixture of all these rules. Then we need an automatic but unfortunately also slower method (five nanoseconds per spin), the *da Silva–Herrmann* algorithm [10.26] of Fig. 10.2. We can no longer rewrite the innermost loop for every different rule as in Fig. 10.1.

Let us assume that each site initially selects one of the many rules possible for cellular automata, and sticks to it for the rest of the simulation. Again we use the square lattice with four neighbours influencing the centre site. Each neighbouring spin can be up or down, resulting in $2^4 = 16$ possible configurations of neighbours. Thus there are $2^{16} = 65536$ possible rules since for each neighbour configuration the resulting centre spin can be either up or down. Each lattice site selects randomly one of these 65536 rules. More precisely, each lattice site initially selects, for each of the 16 neighbour configurations separately, with probability p the option to point up, and with probability $1 - p$ the option to point down. This random selection, with probability p, is made only once at the beginning; from then on the simulation is deterministic without the use of any random numbers.

In this way we have defined the *Kauffman* model [10.29] on the square lattice [10.30]; its original purpose was in genetics, as a means of studying how genes which are on or off lead to limit cycles in the dynamic behaviour. It may

```
      PROGRAM KAUFF(OUTPUT)
      PARAMETER (LL=20,L=64*LL,LP1=L+1,LLP1=LL+1)
      DIMENSION NR(0:LP1,LL,16),N(0:LP1,0:LLP1),M(L,LL)
      DATA  Q,P,MAXT,ISEED/0.5,0.27,100,1/
      PRINT *,Q,P,MAXT,L,ISEED
      CALL RANSET(1+ISEED*2)
      PP1=1.0+P
      QP1=1.0+Q
      DO 1 J=1,LL
       DO 2 I=0,LP1
2      N(I,J)=0
       DO 3 II=1,64
       DO 3 I=0,LP1
3      N( I, J) = SHIFT(OR(N(I, J) , IFIX(QP1-RANF())),1)
       DO 4 K=1,16
        DO 5 I=1,L
5       NR(I,J,K)=0
        DO 6 II=1,64
        DO 6 I=1,L
6       NR(I,J,K)=SHIFT(OR(NR(I,J,K),IFIX(PP1-RANF())),1)
4      CONTINUE
1     CONTINUE
C     END OF INITIALIZATION, START OF TIME DEVELOPMENT
      DO 7 IT=1,MAXT
       DO 8 I=1,L
8      N(I, 0 ) =SHIFT(N(I,LL),63)
       DO 9 I=1,L
9      N(I,LLP1)=SHIFT(N(I, 1), 1)
       DO 10 J=1,LL
       DO 11 I=1,L
        N1H=AND(        N(I,J+1) ,         N(I,J-1))
        N2H=AND(        N(I,J+1) ,COMPL(N(I,J-1)))
        N3H=AND(COMPL(N(I,J+1)),          N(I,J-1))
        N4H=AND(COMPL(N(I,J+1)),COMPL(N(I,J-1)))
        N1V=AND(        N(I+1,J) ,         N(I-1,J))
        N2V=AND(        N(I+1,J) ,COMPL(N(I-1,J)))
        N3V=AND(COMPL(N(I+1,J)),          N(I-1,J))
        N4V=AND(COMPL(N(I+1,J)),COMPL(N(I-1,J)))
        K1 =AND(N1H,N1V)
        K2 =AND(N1H,N2V)
        K3 =AND(N1H,N3V)
        K4 =AND(N1H,N4V)
        K5 =AND(N2H,N1V)
        K6 =AND(N2H,N2V)
        K7 =AND(N2H,N3V)
        K8 =AND(N2H,N4V)
        K9 =AND(N3H,N1V)
        K10=AND(N3H,N2V)
        K11=AND(N3H,N3V)
        K12=AND(N3H,N4V)
        K13=AND(N4H,N1V)
        K14=AND(N4H,N2V)
        K15=AND(N4H,N3V)
        K16=AND(N4H,N4V)
11      M(I,J)=AND(NR(I,J, 1), K1)+AND(NR(I,J, 2), K2)
     1        +AND(NR(I,J, 3), K3)+AND(NR(I,J, 4), K4)
     2        +AND(NR(I,J, 5), K5)+AND(NR(I,J, 6), K6)
     3        +AND(NR(I,J, 7), K7)+AND(NR(I,J, 8), K8)
     4        +AND(NR(I,J, 9), K9)+AND(NR(I,J,10),K10)
     5        +AND(NR(I,J,11),K11)+AND(NR(I,J,12),K12)
     6        +AND(NR(I,J,13),K13)+AND(NR(I,J,14),K14)
     7        +AND(NR(I,J,15),K15)+AND(NR(I,J,16),K16)
10     CONTINUE
       DO 14 J=1,LL
       DO 14 I=1,L
14     N(I,J)=M(I,J)
       DO 15 J=1,LL
       N( 0, J)=N(L,J)
15     N(LP1,J)=N(1,J)
7     CONTINUE
      MAG=0
      DO 16 J=1,LL
      DO 16 I=1,L
16    MAG=MAG+POPCNT(N(I,J))
      PRINT *,MAG
      STOP
      END
```

Fig. 10.2. Da Silva–Herrmann program for Kauffman cellular automata

also be regarded as a description of democracy: What happens if everybody sticks to his own opinion ("order out of chaos").

The innermost loop can no longer be rewritten for every different rule as in Fig. 10.1, since different lattice sites may obey different rules. Instead we program the updating of the spins in a more automatic way, using, however, the one-bit-per-site technique of Fig. 10.1. At the beginning of the innermost loop 11 in Fig. 10.2, four status variables N1H, N2H, N3H, and N4H indicate the orientation of the two horizontal neighbours N(I, J + 1) and N(I, J − 1); for example, N2H = N(I, J + 1).AND..NOT.N(I, J − 1) is true (up) if and only if the right-hand neighbour $J + 1$ is up and the left-hand neighbour $J - 1$ is down. In other FORTRAN dialects we write this expression as AND(N(I, J + 1), COMPL(N(I, J − 1))), where the complement or not-function reverses each bit of the word. Similarly, we have four status variables N1V, N2V, N3V, and N4V for the two vertical neighbours. The 4*4 = 16 combinations K1, K2, ..., K16 of these horizontal and vertical status variables characterise the whole neighbour configuration; for example, K5 = N2H.AND.N1V is true if and only if the left-hand neighbour is down and all three other neighbours are up. Of the 16 variables K1 to K16, at any moment for each site one and only one variable is true, the other 15 are false. In this way the computer knows which neighbour configuration is realised. Thus at the end of loop 11 we make the (logical or arithmetic) sum over the logical AND of the K variable with the corresponding rule variable NR; of the 16 terms in this sum only one, the described rule result, it true (one), and all others are false (zero). In this way, loop 11 automatically gives (without "thinking") the desired result for the new spin M.

The initialisation with a fraction q of spins N up and with parameter p to select the rules NR is done in loops 3 and 6 respectively, similar to Fig. 10.1. Since most of the high-speed Kauffman simulations were made on 64-bit Cray vector computers, Fig. 10.2 uses the fast and good random number generator RANF available on CDC, ETA and Cray machines; the downward rounded integer $q + 1 - \text{RANF}$ equals zero with probability $1 - q$, and one with probability q. (The compact and vectorised bit-by-bit initialisation in loop 3 may be of interest in many other applications too.) At the end, in loop 16, we evaluate the "magnetisation", i.e. the total number of up spins, with the Cray bit-counting function POPCNT. All loops are vectorised. The whole program of Fig. 10.2 updates 200 sites per microsecond on a Cray-YMP processor; the initialisation takes about a microsecond per site. This da Silva–Herrmann algorithm is slightly simplified and faster if all sites obey one single rule and if we study all 65536 possible rules in one program [10.27]; for that case the innermost loop is listed in [10.31].

Most of the research on Kauffman models assumed $q = 1/2$, i.e. half the spins point up and the other half down. As a function of p, the average stationary magnetisation varies smoothly without anomalies. More interesting are "damage spreading" studies [10.29], where we check how a small localised perturbation affects the later spin configuration; see Chap. 5. If the rule selection parameter p is close to zero or to unity, the damage produced by one "mutation" does not spread. If p is close to 1/2 (totally unbiased rule selection), then on the square and cubic lattices, but not on the honeycomb or one-dimensional lattices, the damage spreads with positive probability to infinity, in an infinite lattice after an infinite time. At some threshold p_c and $1 - p_c$ we have a second-order phase transition between spreading and nonspreading damage (chaotic and nonchaotic behaviour).

For the square lattice, finite-size effects seem to be particularly strong, and the author's published estimates started with $p_c = 0.26$. Later simulations using this algorithm made much larger sizes possible, and gave an effective threshold near 0.305 for 6976*6976-point lattices [10.32]. Thus the present value is near 0.31. At the threshold the time for the damage to spread to the boundaries varies as $L^{D'}$ in a lattice of linear dimension L; the number of sites which are damaged (at the moment the damage touches the boundary) varies as L^D, for large enough L. The fractal dimension D on the square lattice is compatible [10.32] with that for the incipient infinite cluster in 2D percolation, $D = 1.9$, as was also found for random mixtures of only two rules [10.33]. On the other hand, the time dimension D' near 1.3 or 1.4 seems to be slightly larger than the fractal dimension 1.13 for the minimal path within the incipient infinite percolation cluster. Less expensive simulation results for other lattices are cited in [10.32].

In damage spreading experiments, the initial perturbation is highly localised and can be a single spin. Instead, we may initially change a random fraction H of all sites. Now damage no longer spreads and instead we call the number of damaged sites the Hamming distance H. If the initial Hamming distance $H(t = 0)$ is small, the later Hamming distance can remain small ($p < p_c$), or grow to infinity. For $p = 0.35$, slightly above the threshold of 0.31, the final Hamming distance seems to increase logarithmically [10.31] with the initial Hamming distance $H(t = 0)$, for $H(t = 0)$ per lattice site varying between 10^{-7} and 10^{-1}. The present author knows of no explanation for this weak dependence.

Thus the Kauffman model turns out to be a numerically quite demanding task, but the results support at least partially, the universality class of random percolation. The computational techniques developed recently for these simulations also turn out to be useful for various other applications [10.27] beyond the scope of this review.

References

10.1 K. Binder (ed.): *Applications of the Monte Carlo Method in Statistical Physics*, 2nd ed., Topics Appl. Phys., Vol. 36 (Springer, Berlin, Heidelberg 1987)
10.2 P.G. de Gennes: J. Physique **37**, L1 (1976)
M. Gordon, S. Ross-Murphy: J. Phys. A **12**, L155 (1979)
D. Stauffer, A. Coniglio, M. Adam: Adv. Polym. Sci. **44**, 103 (1982)
10.3 S. Feng, P.N. Sen: Phys. Rev. Lett. **52**, 216 (1984)
10.4 M. Sahimi: J. Phys. C **19**, L79 (1986)
S. Roux: J. Phys. A **19**, L351 (1986)

10.5 J.G. Zabolitzky, D.J. Bergman, D. Stauffer: J. Stat. Phys. **44**, 211 (1986)
10.6 M. Sahimi, S. Arbabi: J. Stat. Phys. **62**, 453 (1991)
10.7 M. Adam, M. Delsanti, J.P. Munch, D. Durand: Physica A **163**, 85 (1990)
M. Kolb: In: *Correlations and Connectivity in Condensed Matter Physics*, ed. by H.E. Stanley, N. Ostrowsky (Kluwer, Dordrecht 1990) p 241
10.8 J.M. Normand, H.J. Hermann, M. Hajjar: J. Stat. Phys. **52**, 441 (1988)
J.M. Normand, H.J. Herrmann: Int. J. Modern Phys. C **1**, 207 (1990)
10.9 S. Alexander, R. Orbach: J. de Phys. **43**, L625 (1982)
10.10 J. Adler, Y. Meir, A. Aharony, A.B. Harris: Phys. Rev. B **41**, 9183 (1990)
10.11 J.S. Wang, A.B. Harris, J. Adler: preprint
10.12 H.E. Roman: J. Stat. Phys. **58**, 375 (1990), and **64**, 851 (1991)
D.B. Gingold, C.J. Lobb: Phys. Rev. B **42**, 8220 (1990)
10.13 R.M. Ziff, G. Stell: Private communication, Dec. 1988
10.14 J. Adler, R.G. Palmer, H. Meyer: Phys. Rev. Lett. **58**, 882 (1987)
H. Fried, M. Schick: Phys. Rev. B **41**, 4389 (1990)
10.15 L. de Arcangelis, S. Redner, A. Coniglio: Phys. Rev. B **31**, 4725 (1985)
10.16 R. Blumenfeld, A. Aharony: J. Phys. A **18**, L433 (1985)
10.17 J. Adler, A. Aharony: J. Stat. Phys. **52**, 509 (1988)
10.18 J.S. Wang, W. Salke, VI.S. Dotsenko, V.R. Andreichenko: Physica A **164**, 221 (1990)
10.19 J.S. Wang, M. Wöhlert, M. Mühlenbein, D. Chowdhury: Physica A **166**, 173 (1990)
H.O. Heuer: Europhys. Lett. **12**, 551 (1990), and **16**, 503 (1991)
T. Holey, M. Fähnle: Phys. Rev. B **41**, 11709 (1990)
10.20 J.S. Wang, R.H. Swendsen: Physica A **167**, 580 (1990)
10.21 D. Chowdhury, D. Stauffer: J. Phys. A **19**, L19 (1986)
S. Jain: J. Phys. A **19**, L667 (1986)
B. Biswal, D. Chowdhury: Phys. Rev. A **43**, 4179 (1991)
10.22 D. Kandel, R. Ben-Av, E. Domany: Phys. Rev. Lett. **65**, 941 (1990)
10.23 B. Widom: J. Chem. Phys. **84**, 6943 (1984)
N. Jan, D. Stauffer: J. Physique **49**, 623 (1988)
M.J. Velgakis: Physica A **159**, 167 (1989)
K.A. Dawson, B.L. Walker, A. Berera: Physica A **165**, 320 (1990)
D. Stauffer, J.S. Ho, M. Sahimi: J. Chem. Phys. **94**, 1385 (1991); D. Chowdhury, D. Stauffer: Phys. Rev. A **44**, 2247 (1991)
10.24 H.J. Herrmann: J. Stat. Phys. **45**, 145 (1986); R.W. Gerling, D. Stauffer: Int. J. Mod. Phys. C **2**, 799 (1991)
10.25 J. Adler: Physica A, **171**, 453 (1991)
10.26 L.R. da Silva, H.J. Herrmann: J. Stat. Phys. **53**, 463 (1988); T **35**, 66 (1991)
10.27 D. Stauffer: Physica A **159**, 645 (1989); Physica Scripta T **35**, 66 (1991); R.W. Gerling: Physica A **162**, 196 (1990)
10.28 P.M.C. de Oliveira: *Computing Boolean Statistical Models* (World Scientific, Singapore 1991); D. Stauffer: J. Phys. A **24**, 909 (1991)
10.29 S.A. Kauffman: J. Theor. Biol. **22**, 437 (1969)
10.30 B. Derrida, D. Stauffer: Europhys. Lett. **2**, 739 (1986)
10.31 D. Stauffer: In: *Computer Simulation Studies in Condensed Matter Physics II*, ed. by D.P. Landau, K.K. Mon, H.B. Schüttler: Springer Proc. Phys. Vol. 45 (Springer Verlag, Berlin, Heidelberg 1989)
10.32 D. Stauffer: Physica D **38**, 341 (1989)
10.33 L.R. da Silva: J. Stat. Phys. **53**, 985 (1988)

Note added in proof: Recent books on this subject are: A Bunde, S. Havlin (Eds.): *Fractals and Disordered Systems* (Springer-Verlag, Berlin, Heidelberg 1991)
D. Stauffer, A. Aharony: *Introduction to Percolation Theory*, 2nd ed. (Taylor and Francis, London 1992)

11. Interfaces, Wetting Phenomena, Incommensurate Phases

Walter Selke

With 12 Figures

Interfaces may be formed between ordered phases of different types or between ordered and disordered phases. Various interesting interfacial structures may emerge: for example, the interface can be smooth or rough; its thickness may grow continuously or in discrete steps, as one varies external parameters such as the temperature; the interface between two phases may decompose in the presence of a third phase (wetting). Correspondingly, various related critical effects may be encountered at interfaces.

Structural and critical aspects of such interfacial phenomena have been studied extensively in the last few years by using Monte Carlo techniques. Some of these recent advances will be reviewed, emphasising simulations on discrete lattice models, like Ising and Potts models, including lattice gas descriptions of adsorbates, alloys and microemulsions. A number of excellent reviews on interfaces and wetting phenomena exist, which cover the basics, analytic theories and experimental findings, e.g., [11.1–3].

Incommensurate and high-order commensurate superstructures can be thought of as consisting of arrays of spontaneously formed interacting interfaces. It therefore seems appropriate to mention new simulational results on that topic as well. An introduction to the field is given, for instance, in [11.4].

11.1 Interfaces in Ising Models

Let us first consider spin $-1/2$ Ising models in which a spin S_i, placed on the lattice site i, can only take two values, $S_i = \pm 1$. Usually, an interface then separates either a region of the lattice where the spins point predominantly "up", $+1$, from another one where the spins point predominantly "down", -1, or an ordered region from a disordered one. More complex situations may arise in the case of competing interactions between the spins which may lead to the coexistence of domains of different types separated by interfaces.

11.1.1 The Three-Dimensional Nearest-Neighbour Ising Model

The Ising spins, $S_i = \pm 1$, are arrayed on a simple cubic lattice. Neighbouring spins are assumed to interact through the ferromagnetic coupling constant,

$J > 0$. Then the ground state is two-fold degenerate, $-$ or $+$. The symmetry will be broken spontaneously below the bulk transition temperature, T_c.

In introducing interfaces, boundary conditions play the crucial role. There are several possibilities which can easily be realised in Monte Carlo studies. For the 3-dimensional (3D) model, a rich variety of interfacial phenomena has been observed by simulating systems of D layers of $L \times L$ spins, where the couplings in the surface layer, J_s, are possibly different from the bulk couplings, J, and where a surface field, H_1, is present in addition to a bulk field H. The Hamiltonian may be written in the form

$$\mathscr{H} = -J \sum_{\text{bulk}} S_i S_j - J_s \sum_{\text{surface}} S_i S_k - H \sum_{\text{bulk}} S_i - H_1 \sum_{\text{surface}} S_k. \tag{11.1}$$

By assuming H_1 to be negative and H to be positive, the magnetisation per layer may change from negative values near the surface to positive ones in the bulk, forming an interface. The distance of the interface from the surface will depend on the specific values of $J_s/J, H/J, H_1/J$ and T/J. Viewing the "$-$" region as a fluid or another adsorbate in between the surface and a gas corresponding to the "$+$" region, displacements of the interface may be regarded as wetting phenomena.

Based mainly on mean-field approximations [11.5] and phenomenological arguments [11.6], various kinds of wetting transitions had been suggested which have been checked in MC simulations [11.7–10], which will be summarised in the following.

In the case of vanishing bulk field, $H \to 0$, a conjectured phase diagram is depicted in Fig. 11.1, which gives an idea of the complexity of the expected interfacial transitions. For example, at low temperatures, the sharp interface moves away from the surface layer by layer, undergoing "layering transitions", as the magnitude of the surface field is increased. The first few transitions have been monitored in the simulations [11.9]. As depicted in Fig. 11.2, one observes pronounced first-order transitions at those values of the surface field, H_1, where the magnetisation in the first, second,... layer changes sign. The transition in the first layer is accompanied by small jumps in the magnetisation of the neighbouring layers. A similar secondary effect is seen at the points where the second layer undergoes its transition from the positive (bulk) to the negative (surface) magnetisation. To detect further layering transitions at the parameters used in Fig. 11.2, large layers are needed, otherwise the system shows a sluggish behaviour with the magnetisation in the third, fourth,... layers fluctuating back and forth between positive and negative values during a run. Based on mean-field theory and low-temperature analyses, one indeed expects a sequence of infinitely many distinct layering transitions.

At higher temperatures, the interface becomes more diffuse. At and above the roughening transition temperature, T_R, the interface would be, in the thermodynamic limit, completely delocalised; accordingly, there can be no distinct layering transitions anymore (MC estimates on the roughening transition temperature have successively been refined, giving $T_R/T_c = 0.56 \pm 0.03$ [11.11],

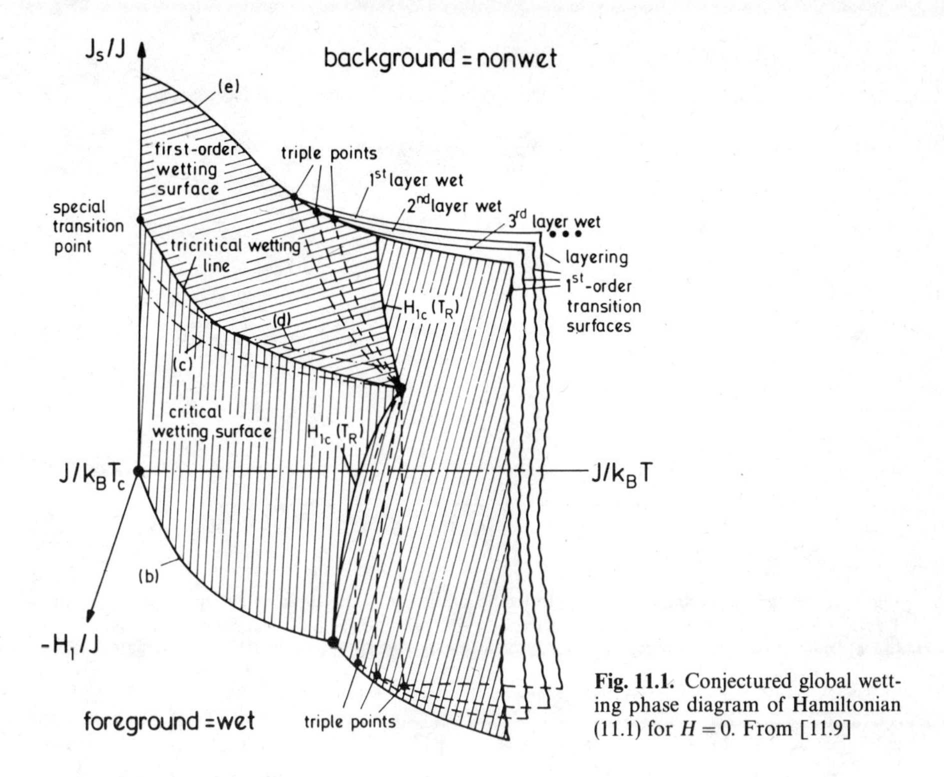

Fig. 11.1. Conjectured global wetting phase diagram of Hamiltonian (11.1) for $H = 0$. From [11.9]

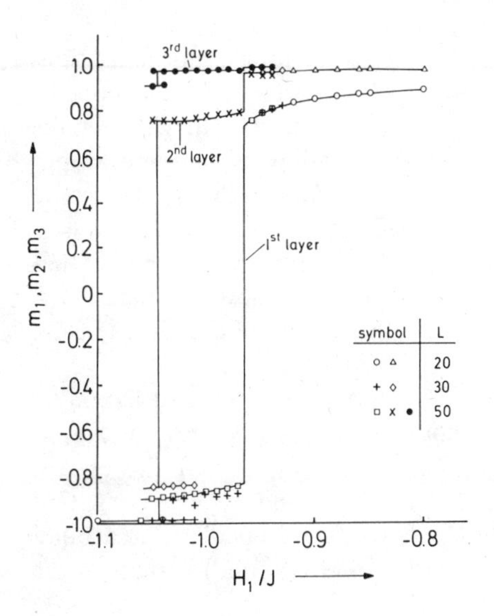

Fig. 11.2. Layer magnetisation m_1, m_2, and m_3 of the first three layers adjacent to the surface plotted vs surface field H_1, at temperature $J/k_BT = 0.45$, field $H = 0$ and surface exchange $J_s = J$. From [11.9]

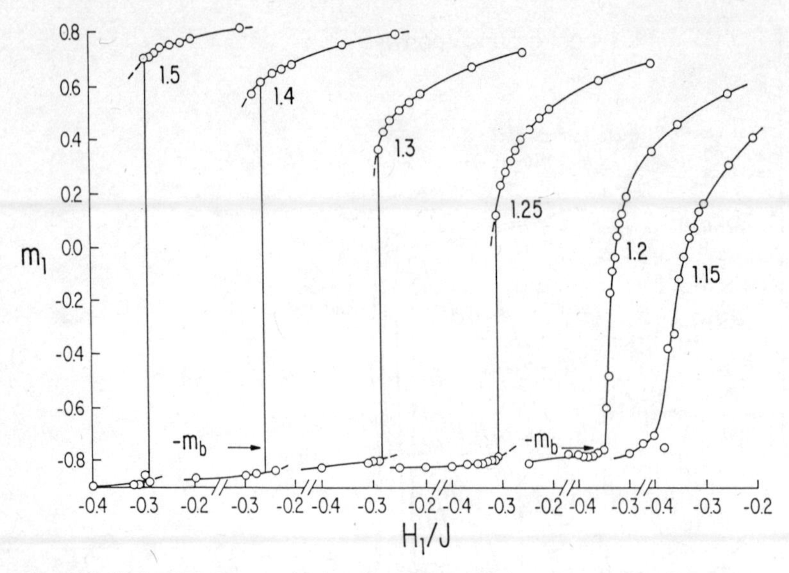

Fig. 11.3. Surface layer magnetisation plotted as a function of the surface field at $J/k_B T = 0.25$ and several values of the parameter J_s/J indicated. Arrows point to the bulk magnetisation, $-m_b$. From [11.9]

0.54 ± 0.02 [11.12, 13] and 0.542 ± 0.005 [11.14]). Only that phase transition survives, at which the magnetisation in the first layer changes sign, as one varies H_1. The transition is of first order for sufficiently strong surface coupling, J_s ("first order wetting surface" in Fig. 11.1), but it will turn into a continuous transition, if J_s/J drops below a critical value (critical wetting surface). Both types of transitions have been detected in the simulations as well as the tricitical wetting transition separating them, see Fig. 11.3.

In the case of critical (and tricritical) wetting behaviour, renormalisation-group theory indicates that mean-field theory gives the correct critical exponents above the marginal dimension $d^* = 3$. However, it is not clear, whether the critical exponents retain their mean-field values in three dimension. For example, while mean-field theory predicts that the correlation length $\xi_{\parallel}$, describing correlations in the interface, diverges at the transition with an exponent $\nu_{\parallel} = 1$, renormalisation-group work, based on isotropic continuum models, predict that $\nu_{\parallel}$ is nonuniversal, depending on the value of the dimensionless parameter $\omega = k_B/4\pi\sigma\xi_b^2$, where ξ_b is the bulk correlation length, and σ the interface tension of the free interface.

To study this issue in simulations, the magnetisation m_1 in the surface layer of the 3D Ising model (11.1) and the susceptibilities $\chi_{11} = \partial m_1/\partial H_1$ and $\chi_1 = \partial m_1/\partial H$ have been computed [11.7, 8]. It has been found that the critical exponents describing the singular behaviour of m_1, χ_1 and χ_{11} are consistent with mean-field theory. The parameter ω tends to a "universal" value around 1. However, more recent simulations on a SOS (solid on solid) variant of the

3D Ising model seem to suggest that *Binder* et al. [11.7, 8] might have not yet reached the asymptotic critical behaviour, and that ω might be somewhat smaller [11.15]. On the other hand, numerical results on the critical wetting in an interface-displacement model with short-range forces seem to indicate that deviations from mean-field theory, if present, should be visible for systems of moderate sizes [11.16]. Further work is needed to clarify the issue.

Another interesting detail of the phase diagram, Fig. 11.1, has been investigated numerically, namely the location of the tricitical wetting line which merges into a surface-bulk multicritical point at $H \to 0$, $H_1 \to 0$, the special transition point, where the bulk fluctuations and the surface fluctuations simultaneously become critical [11.10]. The results are in quantitative agreement with a phenomenological scaling theory, linking wetting behaviour in the bulk and surface critical behaviour [11.6].

So far we have assumed a vanishing bulk field H. As a function of non-zero H and temperature, one distinguishes three typical scenarios, depicted in Fig. 11.4. (i) At sufficiently strong surface field, the "$-$" (or wet) region may be

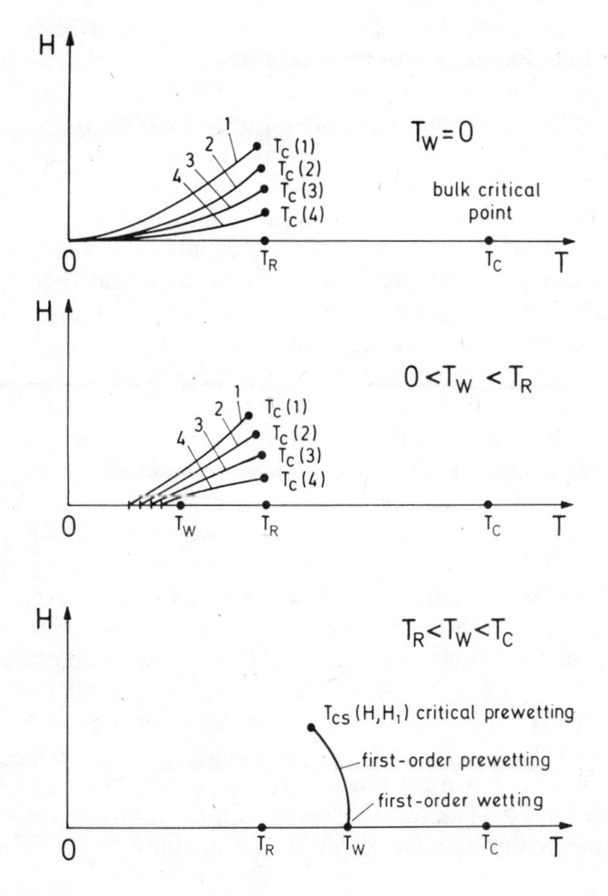

Fig. 11.4. Schematic phase diagrams of the Ising magnet (11.1) as a function of bulk field H and temperature T. Three possible scenarios are shown, as discussed in the text. From [11.9]

infinitely thick for $H \to 0$, and it will be bound to the surface only by a non-zero bulk field, giving rise to a sequence of first order layering transitions, each ending, in the H–T plane, in a critical point. The critical points accumulate, for vanishing bulk field, at T_R. (ii) At intermediate surface fields and $H \to 0$, the wet region grows with temperature (entropic effect) through a sequence of layering transitions. It becomes macroscopically thick at the wetting transition, T_w, which may be smaller than T_R. (iii) Finally, in the regime of values for T/J and J_s/J, where for vanishing bulk field a first-order surface wetting transition occurs upon variation of H_1/J, a "prewetting" transition is expected to take place for non-zero H, at which the position of the interface jumps from a place close to the surface to a more distant value. Indeed, first numerical evidence for such a prewetting transition has been provided [11.10].

Sequences of layering transitions have also been obtained in early MC studies by *Ebner* [11.17] and *Kim* and *Landau* [11.18] on lattice gas models of multilayer adsorption with long-range interactions to the substrate, expressed by layer-dependent external fields. The results suggested that the series of transitions accumulates at the roughening transition temperature, T_R, thereby refining predictions of the pioneering mean field theory [11.19], which cannot discriminate between T_R and T_c. Extensions of these studies, which have been done later, [11.20, 21], are in accordance with these results.

A first-order prewetting transition has also been identified in a recent simulation of a lattice gas confined between two parallel adsorbing walls [11.22].

To study properties of freely fluctuating interfaces away from the surface in the 3D Ising model, boundary conditions different from those in the Hamiltonian (11.1) have been considered. For instance, one may fix the spins in the top and bottom layers in opposite states, or one may apply antiperiodic boundary conditions along the, say, z direction. The latter approach has been used in simulations of tilted interfaces of systems of $L \times L \times L$ spins [11.12, 13]. By imposing, in addition, periodic boundary conditions in the y direction and a screw boundary condition in the x direction with a shift of N_θ lattice constants, a tilted interface with an angle $\theta = \tan^{-1}(N_\theta/L)$ with respect to the xy plane is introduced. By comparing the energy of this system to that of an identical system, but replacing the antiperiodic boundary condition by a periodic one, one obtains the excess interfacial energy, and from that, the interfacial free energy and the step free energy. Especially from the finite-size scaling behaviour of these quantities (which has been analysed by phenomenological scaling theories [11.23]), the characteristics of the roughening transition have been determined. Among others, it has been confirmed that the transition belongs to the Kosterlitz–Thouless universality class [11.12, 13]. At temperatures above T_R, the thickness of the interface diverges logarithmically, as the size of the system, L, goes to infinity [11.14]. The structure of the interface seems to be self-similar over a wide range of length scales, and its fractal dimension has been estimated for Ising and SOS models in two and three dimensions to be 1.58 ± 0.06 and 3.25 ± 0.05, respectively [11.24]. The value for $d = 3$ is significantly higher than those measured experimentally. This fact deserves further study.

11.1.2 Alloys and Microemulsions

Many of the wetting phenomena occurring in nature are believed to be strongly influenced by long-range interactions, such as van der Waals forces in fluids, as has been discussed, e.g., in [11.3]. Nevertheless, some materials may be approximately described by lattice gas or Ising models with short-range forces.

For instance, to study wetting in binary A_3B alloys, the following Ising Hamiltonian on a fcc lattice has been considered [11.25, 26],

$$\mathcal{H} = J\sum_{\mathrm{NN}} S_i S_j - \alpha J \sum_{\mathrm{NNN}} S_i S_j - H \sum_{\mathrm{bulk}} S_i - H_1 \sum_{\substack{\text{first}\\ \text{layer}}} S_i - H_2 \sum_{\substack{\text{second}\\ \text{layer}}} S_i, \tag{11.2}$$

where the first sum runs over nearest neighbour (NN) and the second one over next-nearest-neighbour (NNN) pairs of sites; α and J and assumed to be positive; the last three terms describe bulk and surface fields, in analogy to (11.1). The Ising model (11.2) may be transcribed into a lattice–gas model for alloys by introducing the occupation variable $c_i = (1 + S_i)/2$, where $c_i = 1(0)$ denotes an $A(B)$ atom at lattice site i. In particular, by choosing an appropriate value for $\alpha(=0.2)$, the phase diagram of (11.2) resembles fairly closely the one of Cu_3Au, as had been shown in a MC study [11.27].

Kroll and *Gompper* [11.25, 26] analysed possible wetting effects at (100) and (111) surfaces of model (11.2). Wetting occurs when the system approaches the order–disorder transition from the low-temperature, ordered A_3B phase. Near the surface, the system may start to disorder, while the bulk is still ordered ("surface induced disordering"). A measure of the width of the disordered region, i.e., of the position of the interface separating disordered and ordered regions, is given by the location of the inflection point in the profile of the order parameter, see the MC data depicted in Fig. 11.5. In agreement with mean-field and renormalisation group theory as well as experiments on Cu_3Au, the width, l, as estimated in the simulations, grows logarithmically on approach to the bulk phase transition of first order, i.e., $l \sim -\ln\tau$, where $\tau = (T^* - T)/T$, with T^* being the transition temperature. Several critical exponents describing the critical wetting behaviour, e.g., the vanishing of the order parameter at the surface, have been estimated numerically. In general, they have been shown to depend on the orientation of the surface, in accordance will predictions of mean-field theory.

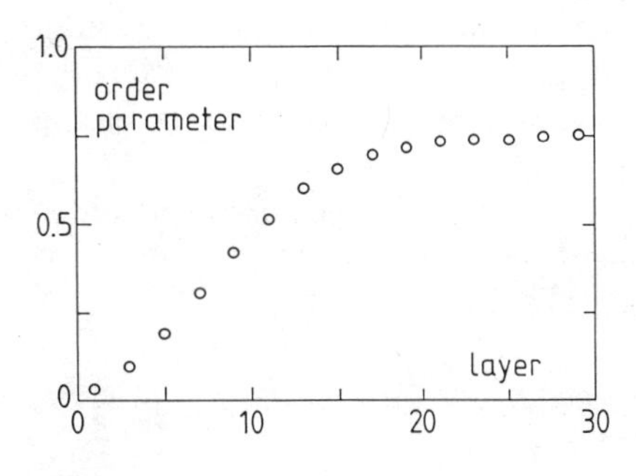

Fig. 11.5. Order parameter profile in the (100) direction for Hamiltonian (11.2) at $\alpha = 0.2$, $k_B T/J = 2.85$, $H_1/J = -2.5$ and $H_2/J = -0.12$ [11.26]

Similar features have been observed in a MC study on the Ising analog of a bcc alloy, such as the Fe–Al system [11.28]. An interface is formed between the ordered DO_3 phase and the disordered phase near the surface. The bulk phase transition is again of first order. Predictions of mean-field theory are confirmed: The thickness of the disordered layer is found to increase logarithmically and the order parameter at the (100) surface is found to vanish with a square-root power law on approach to the bulk transition.

Microemulsions or, more generally, oil–water-surfactant mixtures have come under increasing scrutiny by Statistical Physics in recent years. Several lattice models of Ising type have been proposed and analysed using, among other methods, Monte Carlo techniques. For example, in the model of *Widom* [11.29] the bending energies of the surfactants are formally identical to the configurational energies of an Ising model with short-range competing interactions. The phase diagram is extremely complicated, reminiscent of the one of the axial next-nearest-neighbour Ising (or ANNNI) model [11.4], displaying, e.g., a large number of lamellar, modulated phases [11.30]. Some parts of its phase diagram have been explored in simulations [11.31–33], which will briefly be discussed in Sect. 11.4.

In other lattice models of oil–water-surfactant mixtures, the molecules of oil and water are represented directly by Ising spin variables with $S_i = +1$ corresponding to, say, oil and $S_i = -1$ corresponding to water. A surfactant is represented by a bond variable for nearest-neighbour sites, which is unity or zero in the presence or absence of a surfactant molecule. To study the adsorption of such a model on an inert substrate, modified interactions in the first layer and an external potential coupling to the bond variable have been taken into account by *Jiang* and *Ebner* [11.34]. Applying various approaches, including simulations, they then discuss the wetting of the substrate by an oil-rich or a water-rich phase beneath a bulk water-rich or oil-rich phase. Such a wetting film can be also destroyed, depending on the coating of the substrate by a layer of surfactant molecules which effectively alters the net substrate-adsorbate potential; this phenomenon is well known to chemists.

11.1.3 Adsorbates and Two-Dimensional Systems

As stated above, the Ising Hamiltonian (11.1) may be interpreted as modelling the *multilayer adsorption* on a substrate, assuming short-range adatom–adatom and substrate–adatom interactions. A long-range potential, arising from the substrate, has been mimiced by a magnetic field which falls off with increasing distance from the surface.

Interesting interfacial problems may also happen in *submonolayer adsorbates*, i.e. in the realm of 2D physics. Concretely, let us consider chemisorbed adsorbates, describable by lattice gas models of the form

$$\mathscr{H} = -(\varepsilon + \mu) \sum_{i=1}^{N} c_i - \sum_{i \neq j} \varphi_{ij} c_i c_j - \sum_{i \neq j \neq k} \varphi_t c_i c_j c_k - \cdots, \tag{11.3}$$

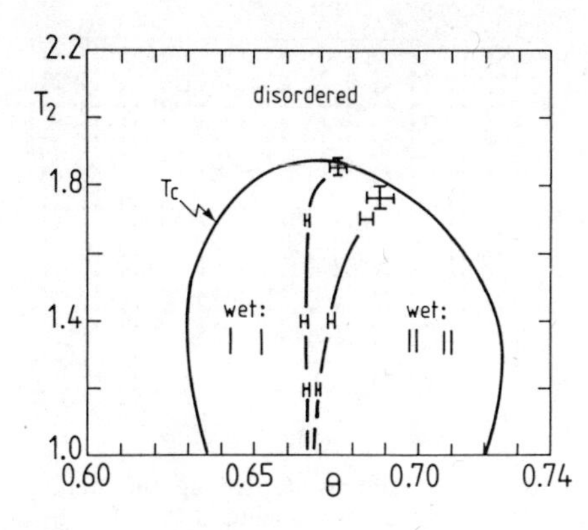

Fig. 11.6. Phase diagram of the lattice gas model of H/Fe (110) as a function of coverage Θ and temperature T_2, showing the two wetting lines in the ordered (3 × 1) phase. From [11.37]

where the sums run over all preferred adsorption sites, i, j, k, imposed by the substrate; $c_i(=0, 1)$ is the occupation variable. φ_{ij} denotes pair interactions between the adatoms; in addition, multi-body forces $(\varphi_t, \ldots)$ may be present. The binding energy gained, if an atom is adsorbed, is ε, and μ is the chemical potential. As has been mentioned before, the equivalent Ising version of the lattice gas model is readily obtained by substituting $S_i = 2c_i - 1$, yielding then a 2D Ising model in a field.

In the framework of such 2D lattice gas or Ising models, many interfacial phenomena have been identified, some of which correspond to those discussed for the 3D systems. A novel type of wetting occurs at domain boundaries of complex ordered adsorbate phases.

In complex ordered systems, several physically distinct, but equivalent domains, say $A, B, C, \ldots$, may coexist. Wetting occurs if the domain boundary or interface between two of them becomes, upon variation of external parameters such as temperature or pressure, unstable with respect to the intrusion of a third domain, i.e. $A{:}C \rightarrow A{:}B{:}C$. Perhaps the simplest realisation is in (3 × 1) phases, where, on a rectangular adsorption lattice, two successive rows, which are predominantly occupied by adatoms, will be followed by a predominantly empty row. Concrete examples are the adsorbates O/Pd (110) and H/Fe (110). In both cases, Hamiltonians of form (11.3) have been proposed and shown to reproduce quite well the experimental phase diagrams in the coverage-temperature plane [11.35, 36]. For these models, MC simulations have been performed to study the wetting transitions [11.37, 38]. In both cases, two different wetting transition lines are located in the ordered (3 × 1) phase, at which the domain boundaries become unstable and decompose, see Fig. 11.6. The wetting mechanisms can conveniently be analysed in terms of two types of domain boundaries or walls, "light walls" and "heavy walls".

A pictorial illustration is given in Fig. 11.7, where typical MC equilibrium arrangements of adatoms in the dry region of the phase diagram (**a**), and in the wet region (**b**) at lower coverage are shown. The different types of domains, A, B, C, are defined by shifting the position of the empty row; in an obvious

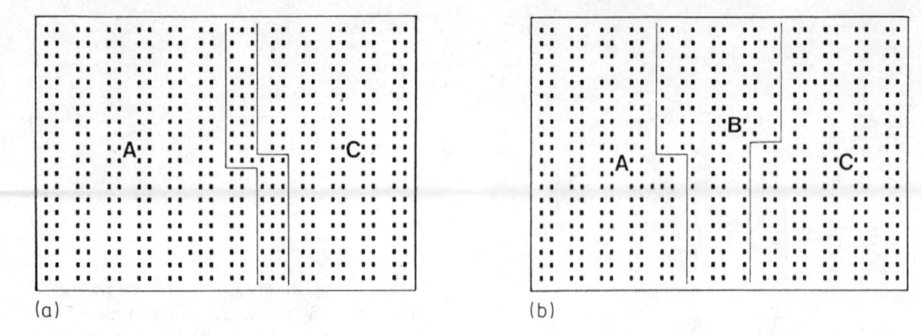

Fig. 11.7 a, b. Typical MC equilibrium arrangements of the adatoms illustrating the wetting transition at the domain boundary of the (3×1) phase in the lattice gas model of O/Pd (110) upon decreasing the coverage. From [11.38]

notation, A corresponds to $(++-)$, B to $(+-+)$ and C to $(-++)$. The boundary conditions are chosen such that the adatoms in the first three rows are fixed in state A and those in the last three rows are fixed in state C. The A:C domain boundary consists of three successive rows of adatoms, a heavy wall "$-+++-$" (the superlight wall "$+--+$", which one might naively expect, is unstable). Upon lowering the coverage, adatoms will preferably be desorbed at the domain boundary, leading to the decomposition of the heavy wall into two light walls "$-+-$", separated by a slab of domain B. The thickness of this slab is expected to diverge, in the thermodynamic limit, on approach to the wetting transition. Symbolically, the transition may be written as

$$A \| C \rightarrow A|B|C. \tag{11.4}$$

In general, any domain boundary is either a heavy or a light wall, which will split into two walls of the other type upon varying coverage and/or temperature, separated by the intervening third phase. As seen from Fig. 11.6, only in a narrow region of the ordered (3×1) phase both types of wall will remain stable.

Similar wetting instabilities at domain boundaries have been observed in lattice gas and Ising models describing $(p \times 1)$ superstructures. In particular, the ANNNI model in a field (for a special choice of its parameters, it is the Ising analogue of the lattice gas model of O/Pd (110)) displays a rich variety of phenomena for $p = 2, 3$, and 4, as has been elaborated by *Rujan* et al. [11.39]. In the (2×1) case, the disordered phase wets the interface between the two degenerate low-temperature structures $(+-)$ and $(-+)$, at the bulk first-order transition. In the (3×1) phase, there are always two kinds of walls, light and heavy ones, with their actual structures being determined by the neighbouring ground-states of the (3×1) phase. In the (4×1) case, there are three physically distinct types of walls, corresponding to the possible phase shifts between the domains. The wetting instability has been studied in detail [11.39, 40]; the transition seems to be of first order. A MC study on similar wetting phenomena in the (4×1) phase of an Ising model with competing multispin interactions, the "$(2+4)$" Ising model, has been undertaken by *Ceva* and *Riera* [11.41].

In complex ordered adsorbates of other types, interesting local structures may occur at the domain boundaries. For example, *Sadiq* and *Binder* [11.42] considered a lattice gas model on a square lattice, with repulsive interactions between nearest and next-nearest neighbours, which exhibits four-fold degenerate (2×1) superstructures. Several types of walls are possible which locally lower or increase the density of adatoms, but no wetting takes place.

The interfacial phenomena discussed for the 3D models appear again in the context of lattice gas models for adsorbates, albeit in a modified form, demonstrating then the effect of the reduced dimensionality. Confined geometries may also play an important role because the substrate of a real adsorbate system is of course finite, consisting of several terraces. At the terrace edges, pinning fields and altered interactions are present, in analogy to the ingredients of Hamiltonian (11.1).

To study the resulting finite-size interfacial effects in a very transparent way, adsorbates on stepped surfaces have been considered in simulations by *Albano* et al. [11.43–46]. The regularly spaced steps are supposed to be completely straight (at higher temperatures, there is the possibility of step roughening [11.47]). Assuming extremely short-range interactions, one can restrict the analysis to one terrace of width L, perpendicular to the step direction, and length M, $M \gg L$. The lattice gas model is then defined on $L \times M$ preferred adsorption sites, subject to free boundary conditions perpendicular to the step direction and, say, periodic boundary conditions parallel to it. The adatoms interact only with their nearest neighbours. In the Ising representation of this lattice gas model, there will be, in addition to the bulk field, H, also boundary fields H_1 and H_L, acting on the spins in the boundary rows.

For H_1 different from H_L, the adatoms will be preferentially adsorbed at one of the terrace boundaries, introducing an interface between a high-coverage and a low-coverage region. A wetting transition may be identified between a phase where the interface is pinned to the boundary and a phase where it is unbound. In contrast to the 3D case, the depinning proceeds smoothly (and not row by row), reflecting the fact that the roughening transition temperature is zero in the 2D situation. The wetting transition temperatures have been estimated for attractive interactions between the adatoms, vanishing bulk field, H, and boundary fields of equal strength, but opposite sign, $H_1 = -H_L$ [11.43–46]. The MC estimates are in good agreement with the exact solution by *Abraham*, valid in the thermodynamic limit, for the 2D nearest-neighbour Ising model [11.48].

The exact approach also describes asymptotic properties of droplets on a wall in the 2D nearest neighbour Ising model [11.48]. There, all boundary spins on a square lattice (of linear dimensions K_1 and K_2) are assumed to be positive, except for L adjacent ones, say, in the middle of the lower boundary. Then a droplet of negative magnetisation is nucleated on that wall, with the base length

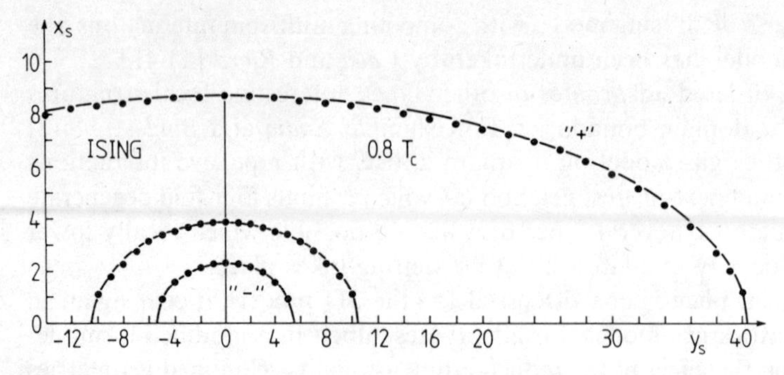

Fig. 11.8. Ising droplets on a wall with three different base lengths at $T = 0.8T_c$. From [11.43]

being L. From the exact solution for the semi-infinite model, $K_1, K_2 \to \infty$ and L finite, at temperatures below T_c, the droplet is known to take on an elliptical shape for large base length L [11.48],

$$x_s^2 + a_1 y_s^2 = a_2, \tag{11.5}$$

where the x- and y-axes are perpendicular and parallel to the wall, see Fig. 11.8. The minor radius, x_m, diverges with the square root of the base length, $x_m = \sqrt{a_2} \sim L^{1/2}$ as $L \to \infty$ [11.48]. Simulations, at $T = 0.8T_c$, show that this asymptotic behaviour is approximated closely only for rather large values of L, although the shape of the droplets is nearly elliptical already for quite small base lengths, [11.49], see Fig. 11.8. An explanation can be given by relating the shape of the droplet to a random walk [11.50]. Then one obtains the identity $x_m = b\sqrt{L}/\sqrt{2}$, where b is the mean step length of the random walk, and corrections to the asymptotic behaviour, $x_m \sim L^{1/2}$, may be attributed to a L-dependent step length. Indeed, the step length perpendicular to the wall will be reduced for elongated droplets, as depicted in Fig. 11.8, since fluctuations towards the wall will be suppressed, just because of the closeness of the wall. At T_c, the droplet is expected to be semicircular, with radius $L/2$, for $L \to \infty$, as follows exactly from, e.g., conformal invariance [11.51].

The equilibrium shapes of Ising droplets around a nucleus in the center of the system have also been studied numerically and analytically [11.49, 51]. The centre spin is fixed to be negative, while the boundary spins of $N \times N$ lattices are fixed to be positive. At T_c, a circular droplet of negative magnetisation is formed, and its radius, R, is found to diverge, for large system sizes, as

$$R \propto N^x, \quad x = \frac{\beta}{\nu\eta}, \tag{11.6}$$

where β, ν and η are the canonical critical exponents; putting in their exactly known values, one finds $x = 1/2$.

11.2 Interfaces in Multistate Models

Let us now consider lattice models where the variable, S_i, defined on site i, can be in at least three different states, say, $\alpha, \beta, \gamma, \ldots$ Standard examples are the q-state Potts models with Hamiltonians of the form

$$\mathscr{H} = -\sum_{i,j} J_{ij}\delta_{s_i s_j}, \quad S_i = 0, 1, 2, \ldots, q-1, \tag{11.7}$$

q-state clock models, and spin-1 Ising models ($S_i = 0, \pm 1$) such as the Blume–Emery–Griffiths model.

Interfaces may be introduced by suitable boundary conditions, to separate for example, an α-rich region from a β-rich region. Sometimes, there will be a preference to then adsorb the other states, $\gamma, \ldots$, at that interface, or the γ-rich phase will even wet the $\alpha{:}\beta$ interface by decomposing it, similarly to the wetting at domain boundaries. An illustration of this "interfacial adsorption" effect is given in Fig. 11.9 for a spin-1 Ising model, the 2D Blume–Capel model. Related critical interfacial behaviour may occur at or below the bulk phase transition, which can then be characterised by new critical exponents.

The early work on interfacial adsorption or wetting phenomena in multistate models has been reviewed elsewhere [11.52, 3]. One of the remaining open problems has been the critical behaviour of the interfacial adsorption in the 2D Potts model at $q > 4$, having a bulk transition of first order. Simulations suggested that the thickness of the interfacial adsorption layer, l, diverges on approach to the bulk transition temperature in the form of a power law, $l \propto \tau^a$,

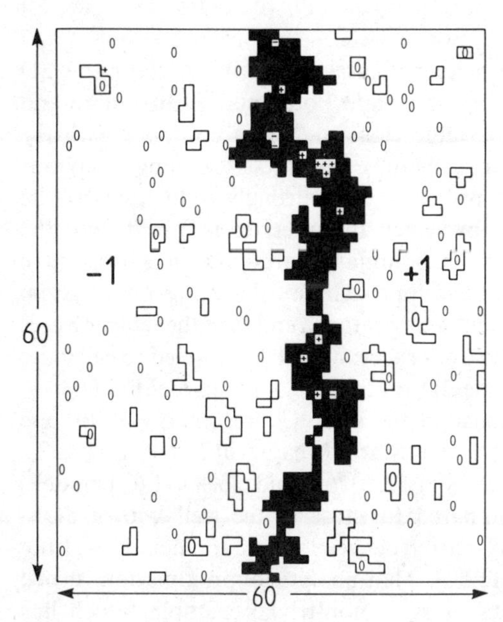

Fig. 11.9. Typical MC equilibrium configuration close to the tricitical point of the 2D Blume–Capel model. The non-boundary states (0) adsorbed at the interface are shown blackened. From [11.52]

with a being close to one [11.52]. Further MC work for a lattice model on lipid membranes by *Mouritsen* and *Zuckermann* gave supportive data [11.53]. However, analytic arguments [11.54] indicate that in the simulations, the asymptotic critical regime has not yet been reached, in which the critical exponent a is suggested to be 1/3, as for wetting in the 2D Blume–Capel model or in other 2D models with a bulk transition of first order [11.39, 55].

The distinction between wetting and depinning (the terms are commonly used synonymously in the description of Ising interfaces, as had also been done in the previous section) has been emphasised in a study of the 2D, nearest neighbour ferromagnetic three-state Potts model and its generalisation, the three-state chiral clock (CC_3) model [11.56] (for a definition of the CC_3 model, see Sect. 11.4). The variables on the top and the bottom boundaries are fixed in different states, say 0 and 2, respectively. In addition, one assumes reduced couplings between the first (or top) row and the second row, compared to the ones in the rest of the system. Then the 0:2 interface tends to be pinned to the top boundary. Tuning the parameters of the models suitably, four different scenarios are possible: (i) the 1-rich region wets the 0:2 interface, $0{:}2 \rightarrow 0{:}1{:}2$, with the 0-rich region being bound to the top; (ii) wetting occurs, and the 0-rich region depins from its boundary; (iii) depinning takes place, but the 0:2 interface remains intact, and (iv) the non-wet 0:2 interface stays near the top boundary. Usually, critical wetting and depinning are decoupled. However, for a special choice of the parameters, the critical fluctuations coincide, and a novel multicritical point is encountered [11.56].

The interfacial adsorption contributes to the interface free energy, σ, in multistate models. Quite a few papers have dealt with the problem of determining σ accurately in simulations of ferromagnetic Potts models with nearest neighbour interactions [11.57–60]. For instance, the ratio $U = \sigma\xi_b^{d-1}/k_B T$ (which has been mentioned before in the discussion of critical wetting in 3D Ising models) has been computed for three-state Potts models on square and triangular lattices, yielding, at T_c, a value close to 1, in accordance with the conjecture that $U = 1$ [11.58]. Of course, thereby the finite-size scaling behaviour of σ has to be taken into account properly. It is strongly influenced by the choice of the boundary conditions, by which the interface is forced onto the system. For example, systems with fixed boundary states and on a torus have been considered [11.57, 59]. In a different approach for calculating σ, the system is partitioned into two halves at different temperatures, and then the temperatures are adjusted adiabatically. The method has successfully been applied to calculate σ near the first-order transition of the 2D seven-state Potts model [11.60].

From the temperature dependence of the interface free energy, $\sigma(T)$, one may determine the bulk transition temperature, because $\sigma(T) = 0$ at $T \geqq T_c$. Simulational applications include Potts [11.52, 61, 62] and clock [11.63] models. In 2D systems, results may be compared to those of the well known SOS-approximation [11.64], which may easily be generalised to include wetting phenomena in multistate models [11.52]. That analytic approximation turned out to give rather reliable estimates of T_c; a non-trivial example, which has been checked numerically, is the three-state Potts models with ferromagnetic and antiferromagnetic bonds [11.61]. Certainly, the interface free energy can also be calculated without difficulty for higher-dimensional models by using the MC method, see e.g., the study on the 3D q-state antiferromagnetic Potts model with $3 \leqq q \leqq 6$ [11.62], to estimate the bulk transition temperature.

It might be tempting to analyse many of the interfacial properties discussed in the section on Ising models for Potts models too, but not much simulational work has been performed. For example, MC investigations on wetting effects in the 3D nearest neighbour three-state Potts model with surface fields and modified surface interactions are missing, presumably mainly because of the lack of experimental applications. A noticeable exception is the simulation on multilayer adsorption with long-range substrate-adsorbate interactions in a Potts lattice gas [11.65]. Again, a sequence of layering transitions of first order is found. The MC studies on 2D Ising droplets have been extended to the Potts case by *Selke* [11.49]. For $q = 3$, it has been concluded that interfacial adsorption modifies the critical behaviour quantitatively, but not qualitatively. For instance, the size-dependence of the critical radius, R, of the droplet in the centre of the system is $R \propto N^x$ with $x \approx 1/2$, in agreement with a scaling argument which yields $x = \beta/\nu\eta$, see (11.6), from which x is obtained to be exactly 1/2.

Interfacial adsorption is expected to also play an interesting role in a Potts-type model describing molecular monolayers adsorbed on a surface; a continuation of the MC survey may clarify the situation [11.66].

11.3 Dynamical Aspects

So far, equilibrium properties of interfaces have been discussed. The dynamics of interfaces can conveniently be studied in simulations, too. In particular, the growth of interfaces and wetting layers as well as the effect of wetting on domain growth in the bulk have been investigated extensively.

11.3.1 Growth of Wetting Layers and Interfaces

Mon et al. [11.67, 68] considered the dynamical behaviour of the nearest-neighbour 3D Ising model, Hamiltonian (11.1), with $L \times L \times D$ spins in the wet part of the phase diagram, where the "$-$" region is no longer pinned to the surface, and at temperatures above T_R. They then monitored the temporal evolution of the magnetisation profile during a MC run, starting from an initial configuration where all spins are in the "$+$" state. Standard one-spin-flip Glauber kinetics with nonconserved order parameter where used. Typical results are depicted in Fig. 11.10, showing the spreading of the wet "$-$" region with time t into the bulk. For the growth of its thickness, l, two distinct dynamical regimes are identified with a crossover from a fluctuation-dominated logarithmic time dependence, $l \propto \ln t$, for large size L and small time t to quasi-1D diffusive

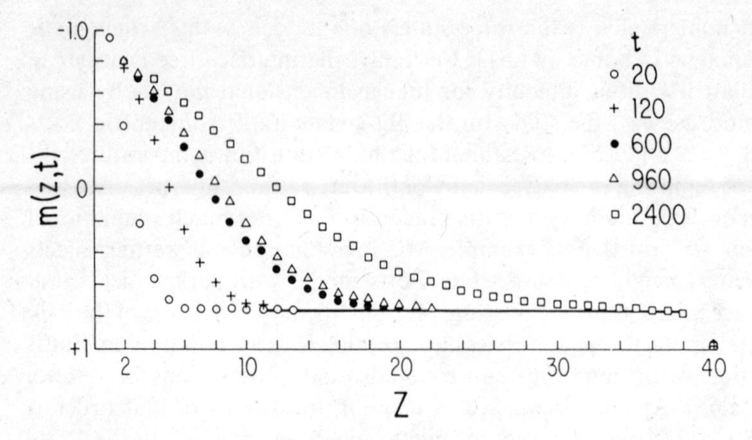

Fig. 11.10. Time-evolution of the magnetisation profile, showing the growth of the wetting layer in the 3D nearest-neighbour Ising model (11.1) at $J/k_B T = 0.25$, $J_s/J = 0.5$, $H = 0$ and $H_1/J = -5.0$ for a $8 \times 8 \times 40$ lattice. From [11.67]

growth for small L and large t, $l \propto (t/L^2)^{1/2}$. The time scale for the onset of the latter behaviour varies with L as $t \approx (L \ln L)^2$. The logarithmic time dependence, describing the growth at late stages in the thermodynamic limit, $L \to \infty$, follows from a scaling hypothesis [11.69]. Finite-size effects have been included in a scaling theory on static and dynamical critical wetting properties [11.70].

The simulations on the growth laws of the wet region turned out to be computationally very demanding [11.67, 68], and, due to the need of much further computer time, no study of the other possible situations in the complicated phase diagram of the model, including layering, prewetting, first-order and tricritical wetting, has yet been performed. Pertinent approximate analytic descriptions for some of these possibilities have been put forward, based on time-dependent Ginzburg–Landau equations [11.71].

For 2D discrete systems the scaling hypothesis by *Lipowsky* gives an asymptotic growth law for the thickness of the wetting layer, $l \propto t^{1/4}$, assuming the thermodynamic limit, short-range forces and kinetics with a nonconserved order parameter [11.69]. The prediction has been confirmed in simulations (a) on wetting in the CC_3 model [11.72, 73] and (b) in the (4×1) (or $\langle 2 \rangle$) phase of the ANNNI model [11.73], and (c) on interface growth in the nearest neighbour Ising model [11.74].

(a) The CC_3 model mimics the (3×1) phase of an adsorbate, see e.g., [11.75, 76]. As discussed above, the interface between two domains, say 0:2, may be decomposed by the intervening third domain at and above the wetting transition. To monitor the decomposition process and to record the temporal evolution of the width of the wetting layer "1", one prepares the system above the wetting transition in a non-wet initial configuration, 0:2. Care is needed in the analysis of the MC data, especially in order to take into account correctly the crossover, at large times, to the quasi-1D diffusive behaviour, which sets in because of the finite size of the system. Of course, the analysis becomes more reliable for larger sizes [11.72, 73].

(b) A similar type of wetting at domain boundaries takes place in the (4×1) phase of the ANNNI model. Indeed, in the simulations of *Ala-Nissila* et al. [11.73], the growth of the wetting layer is consistent with the predicted power law with the exponent 1/4.

(c) The same growth law is expected to govern the interface growth in the 2D Ising model with antiperiodic boundary conditions, connecting say, the top and the bottom of the system. The width, w, of the resulting interface may be defined by [11.74, 77]

$$w^2 = \langle z^2 \rangle - \langle z \rangle^2, \tag{11.8}$$

where $\langle z^n \rangle$ denotes the thermal average of the nth moment of the gradient of the magnetisation profile perpendicular to the interface. At the start of the simulation, performed at temperatures $T_c > T > T_R = 0$, all spins are in the same state. The increase of w with time, t, may then be described by a power law and an effective exponent, $x_{\mathrm{eff}} = \partial \ln w / \partial \ln t$. *Stauffer* and *Landau* [11.74] show that x_{eff} approaches the value 1/4 from above at sufficiently large times. At even later times, finite size effects will supposedly drive the system again into a quasi-1D diffusive behaviour.

First results on the interface dynamics in the 2D Ising model of submonolayer adsorbates on stepped surfaces [11.43–46] have been reported by *Binder* [11.78]; additional complications arise because the finite terraces do not support true long-range order in the adsorbate. At this point, a MC study on the smoothing process of initially rough steps on stepped surfaces may be mentioned [11.79]. The MC data seem to be compatible with the suggestion that the nucleation theory of Lifshitz and Slyozov holds in this situation, yielding a power law for the growth of the smooth region with an exponent 1/3 [11.80].

11.3.2 Domain Growth

Interfacial properties have an important effect in systems far from equilibrium which undergo a dynamical process of (bulk) ordering from the disordered phase into the ordered one. This process can be generated, for example, by instantaneously quenching the system from the high-temperature disordered phase below its order–disorder transition line. The consequent ordering proceeds through formation of locally ordered domains, whose average linear size, r, grows in time, at late stages, usually, in the form of a power law,

$$r \propto t^n, \tag{11.9}$$

where the exponent n depends, among others, on the ground-state degeneracy of the system and the conservation laws of the dynamics. Much analytical and numerical work has been done to identify dynamical universality classes, see e.g. the review by *Gunton* [11.81].

Here, attention is drawn to recent simulations which study the possible impact of domain boundary wetting on the ordering processes in models which mimic submonolayer adsorbates of type (3×1) [11.82, 83] or (4×1) [11.84–87].

Houlrik and *Knak Jensen* [11.82] monitored the domain growth following a quench in the 2D CC_3 model, using the standard one-spin-flip MC updating algorithm with nonconserved order parameter. For quenches to the ground state, two qualitatively different regimes can be distinguished. In case of a dry ground-state, the growth proceeds algebraically, as in (11.9), with an exponent, n, being consistent with 1/2. In case of a wet ground-state, the growth is inhibited after fairly short times, due to a large number of non-removable "vertices", i.e., points where three domains meet. However, in quenches to non-zero temperatures the ordering process is no longer completely stopped by the vertices, and the domain growth is algebraic with n being well compatible with a value 1/2, as for the dry case. It is concluded that both cases are in the same dynamical universitality class, together with the three-state Potts model (for a discussion of simulations of domain-growth kinetics in q-state Potts models, see also [11.88]).

The conclusion is supported by a study of *Ala-Nissila* and *Gunton* on the lattice gas model for O/Pd (110) [11.83]. In its (3×1) phase, universal domain growth is suggested with $r \propto t^{1/2}$. The result is explained by a phenomenological theory, which couples the curvature driven motion of the interface to the vertices. Actually, the theory also describes the remarkably abrupt change in the anisotropy of growth, as one crosses the wetting line: in the dry part of the phase diagram, the domains grow more rapidly in the direction of the competing interactions between the adatoms, whereas just the opposite occurs in the wet part, where the wetting process dissolves certain vertex–antivertex configurations, introducing thereby a new growth mode [11.83].

Analogous features have been observed in the (4×1) (or $\langle 2 \rangle$) phase of the ANNNI model, using Glauber kinetics [11.84–86]. The asymptotic growth law seems to be identical in the dry and the wet case, $r \propto t^{1/2}$. Again, the wetting transition is signalled by a change in the spatial anisotropy of the domain growth, due to the disappearance of vertex–antivertex configurations in the wet phase. Simulational results on the ordering process in the (4×1) phase of the ANNNI model with Kawasaki spin-exchange dynamics, i.e. with conserved magnetization, are less conclusive [11.87]. A convincing power-law fit to the data could not be made; presumably larger system sizes and larger times had to be considered. Nevertheless, by analysing MC configurations, the dynamical role of various types of domain walls and vertices can be identified also for this kind of kinetics [11.87].

11.4 Spatially Modulated Structures

In this section, recent advances on simulations of discrete lattice models with short-range competing interactions will be discussed. These interactions may lead to spatially modulated structures, as has been reviewed, for example, in [11.4] and [11.89].

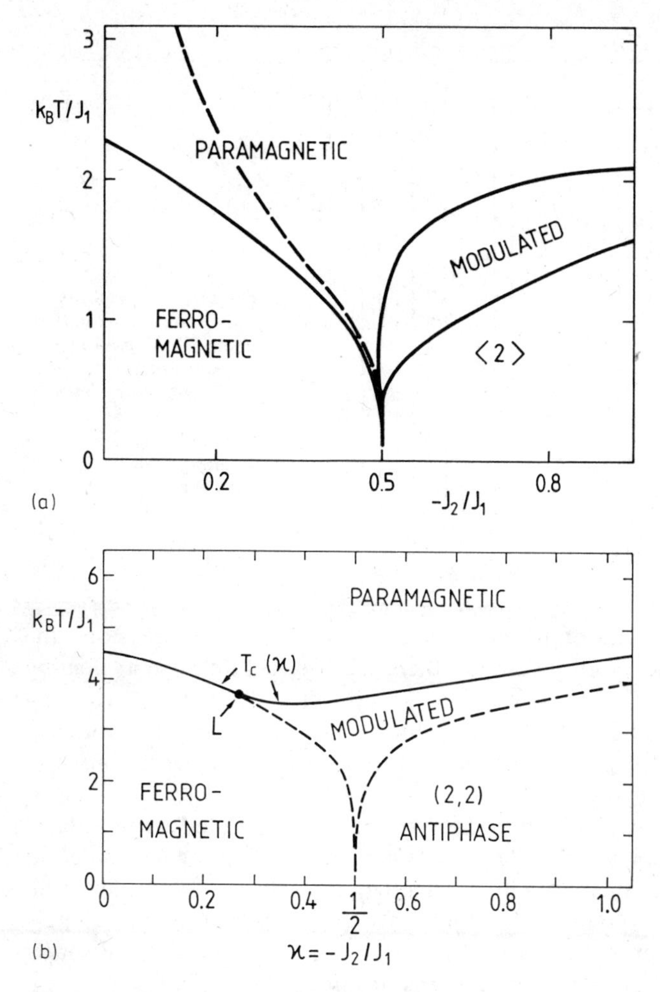

Fig. 11.11. Phase diagrams of the (**a**) 2D and (**b**) 3D ANNNI model with $J_0 = J_1$. In the 2D case, apart from the phase boundaries, the disorder line (dashed curve) is shown [11.4]

A prototype model is the ANNNI model, a spin-$\frac{1}{2}$ Ising model with ferromagnetic couplings, $J_1 > 0$, between spins in adjacent layers and competing antiferromagnetic couplings, $J_2 < 0$, between spins in next-nearest layers, augmented by ferromagnetic nearest-neighbour interactions, $J_0 > 0$, in the layers (of course, $J_1 < 0$ implies also competition between different structures, and the results for $J_1 > 0$ can be easily transcribed to this case by simply reversing the spins in each second layer). The spatially modulated structures of the model were first visualised in simulations more than ten years ago [11.90].

The phase diagrams of the ANNNI model in two and three dimensions are depicted in Fig. 11.11. In three dimensions, the long-range ordered spatially modulated phase displays many interesting features like sequences of high-order

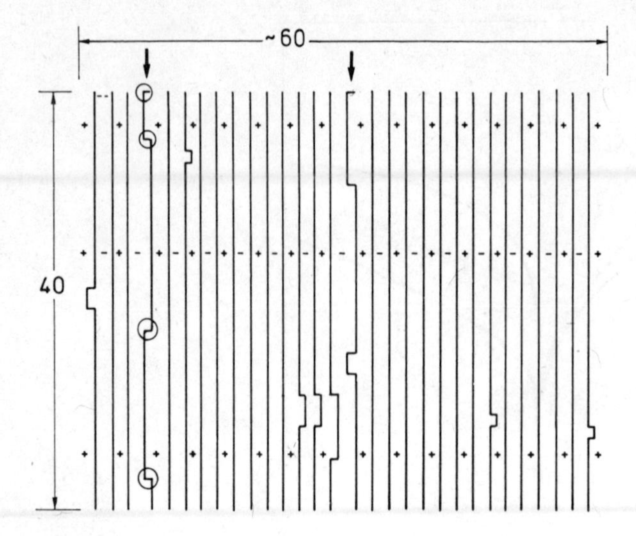

Fig. 11.12. Typical MC equilibrium configuration of the 2D ANNNI model in the floating incommensurate phase slightly above the $\langle 2 \rangle$ boundary. The 3-walls separating the "$+ + - -$" regions are marked by arrows. From [11.4]

commensurate phases and branching processes [11.4]. In two dimensions, the enhanced thermal fluctuations wash out these subtle structures, and a "floating incommensurate phase" is believed to be formed with algebraically decaying spin correlation functions. In both dimensions, typically spin configurations in the spatially modulated phase near the transition to the $\langle 2 \rangle$ phase consist of regions of the low-temperature phase "$+ + - -$", separated by $(+ + +)$ or $(- - -)$ walls, see Fig. 11.12. These arrays of interfaces are locked in at fixed positions in the 3D case, whereas they are depinned in the floating incommensurate phase. The walls interact with each other through pair and multi-wall forces, leading to the complex structures in the model [11.4]. (Interestingly enough, some of the properties of the 2D ANNNI model, such as the disorder line, may be mimiced, e.g., by a simple interacting line model, as has been shown in a simulation by *Müser* and *Rys* [11.91].)

The phase diagrams of Fig. 11.11 are rather well established. Nevertheless, exact results are scarce, and further clarifications of some of the details are useful. In particular, the question on the extent of the floating incommensurate phase has not been completely settled. Therefore, numerical methods which identify this phase are of interest. In addition to standard finite-size analyses and a phenomenological renormalisation group MC method [11.4], two interesting approaches have been proposed and/or applied recently [11.92, 93].

Barber and *Derrida* [11.92] used a dynamical MC method, in which they compared the evolution of two different initial spin configurations submitted to the same thermal noise. The floating phase seems to have a characteristic time behaviour. The phase diagram has been scanned by four sweeps, and results suggest a somewhat larger extent of the floating phase than found previously. A further experimentatation with, and analysis of, the method would be worthwhile.—In a MC study by *Arizmendi* et al. [11.93], the floating and disordered phases are distinguished by the different finite-size dependences of the correlation lengths. Additional MC data of the spin correlation function have been obtained on the Delft special-purpose Ising computer [11.94].

The 3D ANNNI model displays a special multicritical point, the Lifshitz point, at which the disordered, the ferromagnetic and the spatially modulated phases meet. To estimate its critical exponents very accurately, the proper choice of the scaling direction seems to be relevant [11.4, 95]. Because there are two correlation lengths with different critical exponents, anisotropic finite-size scaling analyses are appropriate in the vicinity of that point, as has been emphasized by *Binder* and *Wang* [11.96]. Of course, that sort of analysis would also be useful in determining the critical exponents for other cases, such as for a tricritical Lifshitz point (which has been demonstrated to exist in an Ising model on the hcp lattice by a simulation [11.97]) or for the thickness of a wetting layer (for which it has already been done in a study of the 2D Blume–Capel model [11.55]).

Motivated by the fascinating properties of the ANNNI model, several discrete lattice models with competing interactions have been introduced and studied by means of numerous techniques, including simulations, as reviewed in [11.4]. Only a few results will be sketched here.

To investigate the stability of the main features of the ANNNI model upon small perturbations, one may allow, for example, for interactions between more distant layers. The first step is to extend the range of interactions up to third-nearest layers. The MC study by *Hassold* et al. [11.98] on the 2D version of the model shows that the main features are indeed preserved, such as the floating incommensurate phase and the disorder line, which separates in the disordered part of the phase diagram regions with commensurate and incommensurate correlations.

Another variant of the ANNNI model is the biaxial next-nearest-neighbour Ising (or BNNNI) model. For instance, on a square lattice, the competing couplings, J_1 and J_2, would now act along both axes. Its phase diagram closely resembles that of the ANNNI model, Fig. 11.11, with a chessboard structure of type (4×4) replacing the $\langle 2 \rangle$ phase of type (4×1). It remains an open question whether there is also an intermediate floating incommensurate phase, as was suggested in an early simulation by *Selke* and *Fisher* [11.99]. However, *Landau* and *Binder* [11.100] concluded from a finite-size analysis of their MC data of the specific heat that the model has a single first-order transition between the ordered (4×4) and the disordered phases. In a subsequent simulation, *Velgakis* and *Oitmaa* [11.101], in agreement with [11.99], noticed modulated structures at temperatures slightly above the boundary of the chessboard phase. Due to finite-size limitations, the wavenumbers characterising these structures change discontinuously with temperature. In typical MC equilibrium spin configurations, one sees regions with a nearly perfect commensurate (4×4) structure (running preferentially in a staircase-like fashion diagonally through the system), separated by $(+ + +)$ or $(- - -)$ walls [11.102, 103], similar to the situation in the uniaxial case [11.4, 99, 104] shown in Fig. 11.12. (In [11.104], the nucleation processes of these walls are illustrated.) However, the existence of such walls does not prove a floating incommensurate phase; walls could be present in a disordered incommensurate phase as well. In the latter case, at the melting transition of the chessboard phase, the free energy of wall crossing is expected to be negative. A calculation of that quantity would help to clarify the issue.

Some details of the phase diagram also need to be clarified for the "(2 + 4)" Ising model, in which ferromagnetic couplings, $J_1 > 0$, between nearest neighbours compete with an antiferromagnetic four-spin interaction term

$$-J_2 S_{i\alpha} S_{i+1\alpha} S_{i+2\alpha} S_{i+3\alpha}, \qquad (11.10)$$

connecting up to third-nearest layers; $S_{i\alpha}$ denotes the spin at site α in layer i. As in Fig. 11.11, there are two main low-temperature commensurate phases, the ferromagnetic phase, at $-J_2/J_1 < \frac{1}{2}$, and a ferrimagnetic one "+ + + −", at $-J_2/J_1 > \frac{1}{2}$. In contrast to Fig. 11.11, mean-field theory suggests the absence of additionally spatially modulated phases [11.105]. On the other hand, in simulations of the 2D version of the model (which may describe lipid bilayers), modulated magnetisation patterns have been observed by *Scott* [11.106], indicating at least local incommensurate correlations. Further work is required to unambiguously determine the phase diagrams. As mentioned before, the (2 + 4) Ising model displays a wetting transition line in its (4 × 1) phase [11.41].

Most of the features of the ANNNI model also show up in the CC_3 model [11.4]. The complex spatially modulated structures stem from a chiral interaction term between spins in adjacent layers of the lattice

$$-J\cos\left(\frac{2\pi}{3}(S_i - S_j + \Delta)\right), \quad S_i = 0, 1, 2, \qquad (11.11)$$

where $J > 0$ and Δ, the chirality parameter, is a real number. In the layers, the usual ferromagnetic interactions of Potts type, see (11.7), couple nearest-neighbour spins.

The phase diagrams of the 2D and 3D versions of the CC_3 model have been widely studied [11.4]. Recent simulations on the 3D model have added knowledge on the patterns in the modulated phase [11.107] and on the transition between the modulated and disordered phases [11.108]. In agreement with renormalisation-group calculations, that transition is found to belong to the XY universality class (which amounts to a transition of Kosterlitz–Thouless type in two dimensions) [11.108]. The 2D version of the CC_3 model has been investigated using MC renormalisation-group techniques [11.82]. The phase boundaries and the wetting transition line have been determined accurately, in accordance with earlier MC results [11.4, 52]. Simulations on related 2D models have been undertaken for Baxter's hard hexagon model by computing its structure factor through the melting of the (3 × 1) phase, yielding information of experimental interest for interpreting diffraction data [11.109], as well as for a variant of the CC_3 model describing the ripple phase in lipid bilayers [11.110].

The lattice model of microemulsions proposed by *Widom* [11.29] leads to a complicated phase diagram with an abundance of spatially modulated phases. In its Ising transcription, ferromagnetic couplings, $J > 0$, between nearest neighbours compete with antiferromagnetic couplings between geometric and axial next-nearest neighbours, $2M < 0$ and $M < 0$, respectively. The competing interactions are now isotropic. Mean-field theory has provided guidance in studying the phase diagram [11.30]; its predictions have been checked numerically for the 3D and 2D version of the model [11.31–33].

In particular, *Jan* and *Stauffer* [11.31] and *Dawson* et al. [11.32] retrieved many of the results of mean-field theory in three dimensions. However, there are a few noticeable differences. For instance, the transition between the ferromagnetic (two-phase equilibrium of oil and water) and the paramagnetic (homogeneous solution) phases is confirmed to be of second order, but the transition between the complex modulated (e.g., lamellar) and the disordered phases is also of second order in mean-field theory, whereas in the simulations it turned out to be of first order. Furthermore, the simulations show pronounced shifts in the location of some phase boundaries, as compared to mean-field theory. In two dimensions, simulations by *Stauffer* and *Jan* [11.33] indicate that one of the main phases at low temperatures, predicted by mean-field theory and, indeed, present in three dimensions, now becomes unstable.

To mimic microemulsions in a porous medium, the site-diluted variant of the lattice model of Widom has been studied in simulations. In particular, the lowering of the ferromagnetic transition temperature with increasing dilution has been computed in the 2D [11.111] and the 3D [11.112] case. In the study on the 2D model, the ferromagnetic phase boundary of the BNNNI model has also been considered [11.111]. However, no attempt has been made to determine the effect of the dilution on the type of the singularities occurring at the transition. Extensive simulations on this very subtle aspect have been performed recently for the 2D nearest neighbour Ising model, settling a controversy between conflicting renormalisation group theories [11.113, 114].

11.5 Conclusions

Monte Carlo simulations have been very useful in studying numerous interfacial problems, including properties of freely fluctuating interfaces, interfaces in contact with a wall or surface, and arrays of interacting interfaces. Results on discrete lattice models with short-range forces can be compared to predictions of analytic theories and to possible experimental applications, especially for adsorbates, alloys, and microemulsions. In simulating these interfacial problems one can make use of some of the advantages of the Monte Carlo method: a wide range of time and spatial scales is easily accessible; complicated and subtle structures can be readily identified and their temporal evolution can be monitored. For example, the method may be used, like a scanning tunneling microscope, to visualise wetting processes at domain boundaries of submonolayer adsorbates (which has not yet been achieved experimentally).

A note of caution should be added. Because of the pronounced interface fluctuations, usually long runs and/or large systems are needed to obtain accurate thermal averages. This problem will have to be taken into account in future simulational work dealing more extensively with the effect of long-range forces, which are expected to play an important role in many experimental systems. Of course, very fast Monte Carlo algorithms and a great deal of computer time will be necessary to obtain reliable data.

Acknowledgements. I thank many colleagues working in this area who informed me about their research. I apologise to those persons whose contributions I have mentioned inadequately or overlooked.

References

11.1 B. Widom: Surface tension of fluids, in *Phase Transitions and Critical Phenomena*, ed. by C. Domb, M.S. Green, Vol. 2 (Academic, New York 1972) p. 79

11.2 D. Jasnow: Rep. Prog. Phys. **47**, 1059 (1984)

11.3 S. Dietrich: Wetting phenomena, in *Phase Transitions and Critical Phenomena*, ed. by C. Domb, J.L. Lebowitz, Vol. 12 (Academic, New York 1988) p. 1

11.4 W. Selke: Phys. Rep. **170**, 213 (1988)

11.5 R. Pandit, M. Wortis: Phys. Rev. B **25**, 3226 (1982); R. Pandit, M. Schick, M. Wortis: Phys. Rev. B **25**, 5112 (1982)

11.6 H. Nakanishi, M.E. Fisher: Phys. Rev. Lett. **49**, 1565 (1982)

11.7 K. Binder, D.P. Landau, D.M. Kroll: J. Magn. Magn. Mater. **54–57**, 669 (1986)

11.8 K. Binder, D.P. Landau, D.M. Kroll: Phys. Rev. Lett. **56**, 2272 (1986)

11.9 K. Binder, D.P. Landau: Phys. Rev. B **37**, 1745 (1988)

11.10 K. Binder, D.P. Landau, S. Wansleben: Phys. Rev. B **40**, 6971 (1989)

11.11 E. Bürkner, D. Stauffer: Z. Phys. B **55**, 241 (1983)

11.12 K.K. Mon, S. Wansleben, D.P. Landau, K. Binder: Phys. Rev. Lett. **60**, 708 (1988)

11.13 K.K. Mon, S. Wansleben, D.P. Landau, K. Binder: Phys. Rev. B **39**, 7089 (1989)

11.14 K.K. Mon, D.P. Landau, D. Stauffer: Phys. Rev. B **42**, 545 (1990)

11.15 G. Gompper, D.M. Kroll, R. Lipowsky: Phys. Rev. B **42**, 961 (1990)

11.16 G. Gompper, D.M. Kroll: Phys. Rev. B **37**, 3821 (1988)

11.17 C. Ebner: Phys. Rev. A **23**, 1925 (1981)

11.18 I.M. Kim, D.P. Landau: Surface Sci. **110**, 415 (1981)

11.19 M.J. de Oliveira, R.B. Griffiths: Surface Sci. **71**, 687 (1978)

11.20 A.K. Sen, C. Ebner: Phys. Rev. B **33**, 5076 (1986)

11.21 A. Patrykiejew, D.P. Landau, K. Binder: Surface Sci. **238**, 317 (1990)

11.22 D. Nicolaides, R. Evans: Phys. Rev. B **39**, 9336 (1989)

11.23 V. Privman: Phys. Rev. Lett. **61**, 183 (1988)

11.24 K.K. Mon: Phys. Rev. Lett. **57**, 866 (1986); 1963 (1986)

11.25 D.M. Kroll, G. Gompper: Phys. Rev. B **36**, 7078 (1987)

11.26 G. Gompper, D.M. Kroll: Phys. Rev. B **38**, 459 (1988)

11.27 K. Binder, W. Kinzel, W. Selke: J. Magn. Magn. Mater. **31–34**, 1445 (1983)

11.28 W. Helbing, B. Dünweg, K. Binder, D.P. Landau: Z. Phys. B **80**, 401 (1990)

11.29 B. Widom: J. Chem. Phys. **84**, 6943 (1986)

11.30 K.A. Dawson, M.D. Lipkin, B. Widom: J. Chem. Phys. **88**, 5149 (1988)

11.31 N. Jan, D. Stauffer: J. Phys. France **49**, 623 (1988)

11.32 K.A. Dawson, B.L. Walker, A. Berera: Physica A **165**, 320 (1990)

11.33 D. Stauffer, N. Jan: J. Chem. Phys. **87**, 6210 (1987)

11.34 Y. Jiang, C. Ebner: Phys. Rev. A **37**, 2091 (1988)

11.35 G. Ertl, J. Küppers: Surface Sci. **21**, 61 (1970)

11.36 W. Selke, W. Kinzel, K. Binder: Surface Sci. **125**, 74 (1983)

11.37 I. Sega, W. Selke, K. Binder: Surface Sci. **154**, 331 (1985)

11.38 W. Selke: Ber. Bunsenges., Phys. Chem. **90**, 232 (1986)
11.39 P. Rujan, W. Selke, G. Uimin: Z. Phys. B **65**, 235 (1986)
11.40 T. Ala-Nissila, J. Amar, J.D. Gunton: J. Phys. A **19**, L41 (1986)
11.41 H. Ceva, J.A. Riera: Phys. Rev. B **38**, 4705 (1988)
11.42 A. Sadiq, K. Binder: J. Stat. Phys. **35**, 517 (1984)
11.43 E.V. Albano, K. Binder, D.W. Heermann, W. Paul: Surface Sci. **223**, 151 (1989)
11.44 E.V. Albano, K. Binder, D.W. Heermann, W. Paul: Z. Phys. B **77**, 445 (1989)
11.45 E.V. Albano, K. Binder, D.W. Heermann, W. Paul: J. Chem. Phys. **91**, 3700 (1989)
11.46 E.V. Albano, K. Binder, D.W. Heermann, W. Paul: J. Stat. Phys. **61**, 161 (1990)
11.47 B. Salanon, F. Fabre, J. Lapujoulade, W. Selke: Phys. Rev. B **38**, 7385 (1988)
11.48 D.B. Abraham: Phys. Rev. Lett. **44**, 1165 (1980)
11.49 W. Selke: J. Stat. Phys. **56**, 609 (1989)
11.50 M.E. Fisher: J. Stat. Phys. **34**, 667 (1984)
11.51 T.W. Burkhardt, W. Selke, T. Xue: J. Phys. A **22**, L1129 (1989)
11.52 W. Selke: Interfacial adsorption in multi-state models, in *Static Critical Phenomena in Inhomogeneous Systems*, ed. by A. Pekalski, J. Sznajd; Lecture Notes in Physics, Vol. 206 (Springer, Berlin, Heidelberg 1984) p. 191
11.53 O.G. Mouritsen, M.J. Zuckermann: Phys. Rev. Lett. **58**, 389 (1987)
11.54 J. Bricmont, J.L. Lebowitz: J. Stat. Phys. **46**, 1015 (1987)
11.55 W. Selke, D.A. Huse, D.M. Kroll: J. Phys. A **17**, 3019 (1984); W. Selke, J. Yeomans: J. Phys. A **16**, 2789 (1983)
11.56 W. Pesch, W. Selke: Z. Phys. B **69**, 295 (1987)
11.57 K.K. Mon, D. Jasnow: Phys. Rev. A **30**, 670 (1984)
11.58 D. Jasnow, K.K. Mon, M. Ferer: Phys. Rev. B **33**, 3349 (1986)
11.59 H. Park, M. den Nijs: Phys. Rev. B **38**, 565 (1988)
11.60 J. Potvin, C. Rebbi: Phys. Rev. Lett. **62**, 3062 (1989)
11.61 G. Sun, Y. Ueno, Y. Ozeki: J. Phys. Soc. Jpn. **57**, 156 (1988)
11.62 Y. Ueno, G. Sun, I. Ono: J. Phys. Soc. Jpn. **58**, 1162 (1989); I. Ono, A. Yamagata: J. Magn. Magn. Mater. **90–91**, 309 (1990)
11.63 A. Yamagata, I. Ono: J. Phys. A **24**, 265 (1991)
11.64 E. Müller-Hartmann, J. Zittartz: Z. Phys. B **27**, 261 (1977)
11.65 C. Ebner: Phys. Rev. B **28**, 2890 (1983)
11.66 M.P. Allen, K. Armitstead: J. Phys. A **22**, 3011 (1989)
11.67 K.K. Mon, K. Binder, D.P. Landau: Phys. Rev. B **35**, 3683 (1987); **42**, 994 (1990)
11.68 K.K. Mon, K. Binder, D.P. Landau: J. Appl. Phys. **61**, 4409 (1987)
11.69 R. Lipowsky: J. Phys. A **18**, L585 (1985)
11.70 D.M. Kroll, G. Gompper: Phys. Rev. B **39**, 433 (1989)
11.71 I. Schmidt, K. Binder: Z. Phys. B **67**, 369 (1987)
11.72 M. Grant, K. Kaski, K. Kankaala: J. Phys. A **20**, L571 (1987)
11.73 T. Ala-Nissila, K. Kankaala, K. Kaski: J. Phys. A **22**, 2629 (1989)
11.74 D. Stauffer, D.P. Landau: Phys. Rev. B **39**, 9650 (1989)
11.75 M.E. Fisher: Faraday Trans. **82**, 1569 (1986)
11.76 Y.-N. Yang, E.D. Williams, R.L. Park, N.C. Bartelt, T.L. Einstein: Phys. Rev. Lett. **64**, 2410 (1990)
11.77 R.H. Swendsen: Phys. Rev. B **15**, 542 (1977)
11.78 K. Binder: Growth kinetics of wetting layers at surfaces, in *Kinetics of Ordering and Growth at surfaces*, ed. by M. Lagally (Plenum, New York 1990), p. 31
11.79 W. Selke: J. Phys. C **20**, L455 (1987)
11.80 J. Villain: Europhys. Lett. **2**, 531 (1986)
11.81 J.D. Gunton: J. Stat. Phys. **34**, 1019 (1984)
11.82 J.M. Houlrik, J.S. Knak Jensen: Phys. Rev. B **34**, 7828 (1986)
11.83 T. Ala-Nissila, J.D. Gunton: Phys. Rev. B **38**, 11418 (1988)
11.84 K. Kaski, T. Ala-Nissila, J.D. Gunton: Phys. Rev. B **31**, 310 (1985)
11.85 T. Ala-Nissila, J.D. Gunton: J. Phys. C **20**, L387 (1987)
11.86 T. Ala-Nissila, J.D. Gunton, K. Kaski: Phys. Rev. B **37**, 179 (1988)
11.87 T. Ala-Nissila, J.D. Gunton, K. Kaski: Phys. Rev. B **33**, 7583 (1986)
11.88 G.S. Grest, M.P. Anderson, D.J. Srolovitz: Phys. Rev. B **38**, 4752 (1988)
11.89 J. Yeomans: The theory and application of axial Ising models, in *Solid State Physics*, Vol. 41, ed. by H. Ehrenreich, D. Turnbull (Academic, New York 1988) p. 151
11.90 W. Selke, M.E. Fisher: Phys. Rev. B **20**, 257 (1979)
11.91 H.E. Müser, F.S. Rys: Z. Phys. B **76**, 107 (1989)
11.92 M.N. Barber, B. Derrida: J. Stat. Phys. **51**, 877 (1988); D. Hansel, G. Meunier, A. Verga: J. Stat. Phys. **61**, 329 (1990)

11.93 C.M. Arizmendi, A.H. Rizzo, L.N. Epele, C.A. Garcia Canal: Z. Phys. B **83**, 273 (1991)
11.94 G.J. Smeenk: Diploma Thesis, Technische Hogeschool Delft (1985); A. Compagner: Private communication
11.95 K. Kaski, W. Selke: Phys. Rev. B **31**, 3128 (1985); Z. Mo, J. Ferer: Phys. Rev. B **43**, 10890 (1991)
11.96 K. Binder, J.-S. Wang: J. Stat. Phys. **55**, 87 (1989)
11.97 E. Domany, J.E. Gubernatis: Phys. Rev. B **32**, 3354 (1985)
11.98 G.N. Hassold, J.F. Dreitlein, P.D. Beale, J.F. Scott: Phys. Rev. B **33**, 3581 (1986)
11.99 W. Selke, M.E. Fisher: Z. Phys. B **40**, 71 (1980); W. Selke: Z. Phys. B **43**, 335 (1981)
11.100 D.P. Landau, K. Binder: Phys. Rev. B **31**, 5946 (1985)
11.101 M.J. Velgakis, J. Oitmaa: J. Phys. A **21**, 547 (1988)
11.102 J. Oitmaa: private communication (1987)
11.103 M. Aydin, M.C. Yalabik: J. Phys. A **22**, 3981 (1989)
11.104 P. Rujan, W. Selke, G.V. Uimin: Z. Phys. B **53**, 221 (1983)
11.105 P.J. Jensen, K.A. Penson, K.H. Bennemann: Phys. Rev. A **40**, 1681 (1989)
11.106 H.L. Scott: Phys. Rev. A **37**, 263 (1988)
11.107 W.S. McCullough, H.L. Scott: J. Phys. A **22**, 4463 (1989)
11.108 M. Siegert, H.U. Everts: J. Phys. A **22**, L783 (1989)
11.109 N.C. Bartelt, T.L. Einstein, L.D. Roelofs: Phys. Rev. B **35**, 4812 (1987)
11.110 W.S. McCullough, H.L. Scott: Phys. Rev. Lett. **65**, 931 (1990)
11.111 M.J. Velgakis: Physica A **159**, 167 (1989)
11.112 D. Stauffer, J.S. Ho, M. Sahimi: J. Chem. Phys. **94**, 1385 (1991)
11.113 V.B. Andreichenko, Vl.S. Dotsenko, W. Selke, J.-S. Wang: Nucl. Phys. B **344**, 531 (1990)
11.114 J.-S. Wang, W. Selke, Vl.S. Dotsenko, V.B. Andreichenko: Europhys. Lett. **11**, 301 (1990); Physica A **164**, 221 (1990)
11.115 D.B. Abraham, J. Heiniö, K. Kask: Preprint (1991)
11.116 S.S. Manna, H.J. Herrmann, D.P. Landau: J. Stat. Phys. (1991) in print
11.117 P.J. Upton: Phys. Rev. B **44** (1991) in print; W. Selke: Physica A (1991) in print; W. Selke, N.M. Surakic, P.J. Upton: To be published
11.118 F. Karsch, A. Patkos: Nuclear Physics B **350** (1991) 563
11.119 R. Schuster, J.V. Barth, G. Ertl, R.J. Behm: Preprint (1991)
11.120 J.E. Finn, P.A. Manson: Phys. Rev. A **39** (1989) 6402; E. Velasco, P. Tarazona: Phys. Rev. A **42** (1990) 2454

Note Added in Proof: There are quite a few intriguing "post-deadline" results of simulations on the topic of this chapter. Only a few of them can be mentioned here. In particular, the spreading of droplets and the kinetics of the precursor film has been monitored for SOS models [11.115]; the dynamics of an Ising droplet in a magnetic field has been studied [11.116]. The possibility of a non-thermodynamic singularity caused by an interface meandering between defect lines or walls has been investigated analytically and numerically in 2D SOS and Ising models [11.117]. Wetting properties of the three-state Potts model in the 3D case have been found to be compatible with mean-field estimates [11.118]. A domain wall structure similar to that shown in Fig. 11.12 has been observed by scanning tunneling microscopy for the adsorbate system K/Cu(110) and reproduced in simulating an appropriate lattice gas model [11.119].—Attention is also drawn to the MC simulation of the prewetting line for a solid–fluid interface model and a related recent density-functional calculation [11.120].

12. Spin Glasses, Orientational Glasses and Random Field Systems

Allan P. Young, Joseph D. Reger and Kurt Binder

With 11 Figures

In this chapter, we draw attention to Monte Carlo simulations of various kinds of disordered solids, where, due to the preparation of the sample, structural disorder is frozen in ("quenched disorder"). Typically, these systems are produced by random dilution of different atomic species. This structural disorder introduces disorder in the effective interactions responsible for the ordering of the system; it may be of the "random bond"-type or of the "random field"-type. If this disorder is very strong, it may give rise to qualitatively new types of ordering phenomena such as spin glasses [12.1–4] in the case of magnetic systems and, most recently, vortex glasses [12.5] in the case of superconductors, etc. For many such problems analytical methods can only give very restricted information, and hence simulations are very important.

We start by reviewing recent work on spin glasses (Sect. 12.1) and discuss an interesting generalisation, the Potts glass, in Sect. 12.2. The latter model can be viewed as a model of an anisotropic quadrupolar glass (or orientational glass, respectively). Isotropic orientational glasses are then treated in Sect. 12.3, focusing attention on "random bond" disorder throughout. Sect. 12.4 summarises work on the random field Ising problem, while Sect. 12.5 contains a discussion as well as an outlook to other related problems not treated in the present article, such as simulations of structural glasses and their glass transitions. In this chapter, we do not attempt to give a complete survey, but restrict ourselves to a description of the main directions.

12.1 Spin Glasses

Spin glasses are magnetic systems which are disordered and "frustrated", by which one means that the interactions between the spins are in conflict with each other, so no single spin configuration can simultaneously minimise all the terms in the Hamiltonian. This section will concentrate on the insights into the spin glass problem that have come from Monte Carlo simulations. For background and many additional references the reader is referred to the reviews [12.1, 2]. Simulations of spin glass systems are more difficult than those of pure systems for several reasons. First for all, because one has a disordered system, one must do a configurational average over the disorder, in addition to the usual thermal average. Although intensive properties [12.6] of disordered systems are, in general, expected to be "self-averaging", which means that any one large system gives the same result, with probability unity, as an average over all systems with the same probability distribution for the disorder, the sizes that one can simulate in practice are much too small for self-averaging to hold, and one typically needs to average over several hundred different samples. Secondly, as also seen in experiment, relaxation times become very large as one lowers the temperature, so it is non-trivial to ascertain whether the system being simulated has reached equilibrium. Relaxation times also increase with system size, so the sizes one can study are rather small. As a result, finite size scaling [12.3] has played an important role in these studies.

12.1.1 The Spin Glass Transition

An important question which has received a lot of attention, is whether a sharp transition can occur in spin glass systems at finite temperature. To see how a spin glass transition may arise it is helpful to make an analogy with ferromagnetism. At a second-order phase transition there is a correlation length which diverges. For a ferromagnet, this is simply the distance over which the spins have a net parallel alignment. Since correlations decay exponentially, to within a good approximation, we can write

$$\langle S_i S_j \rangle \propto \exp(-R_{ij}/\xi_F), \tag{12.1}$$

which defines the ferromagnetic correlation length ξ_F. Here, R_{ij} is the distance between lattice points i and j. For convenience we are assuming Ising spins, where $S_i = \pm 1$, but this is not essential. As the ferromagnetic transition temperature, T_c, is approached from above, ξ_{SG} diverges with a power of $t = (T - T_c)/T_c$, i.e. $\xi_F \propto t^{-\nu}$.

In a spin glass there is no net tendency towards ferromagnetism or antiferromagnetism so ξ_F stays short range. Nonetheless, if one looks at all pairs of spins a given relative distance apart, they will have substantial correlation, but some pairs will be ferromagnetically correlated and others will have an antiferromagnetic correlation, so the *average* is close to zero. This cancellation coming from random signs will not occur if we square the correlation function, so we define the spin-glass correlation length, ξ_{SG} by

$$[\langle S_i S_j \rangle^2]_{av} \propto \exp(-R_{ij}/\xi_{SG}), \tag{12.2}$$

where the symbol $[\cdots]_{av}$ denotes an average over different bond configurations. It is also useful to discuss the spin-glass or Edwards–Anderson susceptibility, obtained by summing the squared correlation function over all distances, i.e.

$$\chi_{SG} = \sum_j [\langle S_i S_j \rangle^2]_{av}. \tag{12.3}$$

A spin glass transition is characterised by the divergence of ξ_{SG} and χ_{SG} as the temperature is lowered, i.e.

$$\xi_{SG} \propto t^{-\nu}, \tag{12.4}$$

$$\chi_{SG} \propto t^{-\gamma},$$

without a divergence of ξ_F. χ_{SG} diverges in the same way [12.1] as the nonlinear susceptibility χ_{nl}, defined by

$$m = \chi_F h - \chi_{nl} h^3 \cdots, \tag{12.5}$$

which can be measured experimentally. Here, m is the magnetization and h is the field.

12.1.2 The Edwards Anderson Model

In order to study this problem theoretically, it is useful to have a relatively simple model which incorporates the essential physical ingredients of randomness and frustration. The most commonly studied model is that of *Edwards* and *Anderson* [12.4] in which the Hamiltonian for Ising spins is

$$\mathscr{H} = -\sum_{\langle i,j \rangle} J_{ij} S_i S_j, \tag{12.6}$$

where the spins lie on a regular lattice and the interactions, assumed here to be nearest neighbour only, are independent random variables with a probability distribution $P(J_{ij})$. We will consider the "pure spin-glass case" where the mean of the distribution is zero, and will also set the standard deviation to unity to define the temperature scale. Two distributions are especially popular, the $\pm J$ distribution in which

$$P(J_{ij}) = \tfrac{1}{2}[\delta(J_{ij} - 1) + \delta(J_{ij} + 1)], \tag{12.7a}$$

and the Gaussian distribution where

$$P(J_{ij}) = \frac{1}{\sqrt{2\pi}} e^{-J_{ij}^2/2}. \tag{12.7b}$$

Models with vector spins rather than Ising spins have also been studied. For example the isotropic Heisenberg spin glass is obtained from (12.6) by replacing $S_i S_j$ by the scalar product $\boldsymbol{S}_i \cdot \boldsymbol{S}_j$.

12.1.3 Phase Transitions

The first simulations of the $d = 3$ Ising spin glass which were detailed enough to seriously test for the existence of a transition [12.7–10], were done on rather special high speed machines. The simulations of *Ogielski* and *Morgenstern* [12.7] and *Ogielski* [12.8] used a processor specially built by *Condon* and *Ogielski* [12.11] to perform simulations on Ising systems, while the simultaneous work of *Bhatt* and *Young* [12.9, 10] used a commercially available massively parallel processor, known as the Distributed Array Processor.

Bhatt and *Young* [12.9, 10] performed a finite size scaling analysis on the moments of the order parameter q. Two identical independent copies of the system were simulated for each set of bonds, and one quantity calculated was the instantaneous mutual overlap between them, which is defined by

$$Q(t) = \frac{1}{N} \sum_i S_i^{(1)}(t_0 + t) S_i^{(2)}(t_0 + t), \tag{12.8}$$

where t_0 is the equilibration time. The order parameter distribution, $P(q)$, is then

$$P(q) = \frac{1}{t_m} \sum_{t=1}^{t_m} [\delta(q - Q(t))]_{\text{av}}, \tag{12.9}$$

where the time average over t_m steps in (12.9) is equivalent to the thermal average denoted by $\langle \cdots \rangle_T$. It is possible to show [12.9, 10] whether or not the system has reached equilibrium by seeing if results for $P(q)$ computed in two different ways agree with each other. In a symmetry broken state, the *Edwards–Anderson* [12.14] order parameter $[\langle S_i \rangle_T^2]_{\text{av}}$, is just the first moment of $P(q)$. Following *Binder* [12.12], the dimensionless combinations of moments,

$$g = \frac{1}{2}\left(3 - \frac{\langle q^4 \rangle}{\langle q^2 \rangle^2}\right), \tag{12.10}$$

has the finite size scaling form

$$g = \bar{g}(L^{1/\nu}(T - T_c)), \tag{12.11}$$

where $\bar{g}$ is a scaling function, and $\langle \cdots \rangle$ indicates an average over the distribution $P(q)$. It follows that g is independent of size at T_c, so T_c can be located by finding the intersection of curves for g against T for different sizes.

This works well in $d = 4$, as shown in Fig. 12.1. The curves intersect at about $T = 1.75$, and splay out again at lower tempratures, which is an indication of spin glass ordering. The data for $d = 3$, which is a little less clear cut, is shown in Fig. 12.2. While the data at higher temperatures show the expected behavior and the curves come together at about $T = 1.2$, which is therefore the estimate of T_c, the curves for different sizes do not splay out again below T_c. This may be due to corrections to scaling. We shall see below that $d = 3$ is close to the lower critical dimension, d_l, the dimension below which the $T_c = 0$, and it is expected that corrections to finite size scaling are particularly severe in this limit. From scaling fits Bhatt and Young obtained $\gamma = 3.0 \pm 0.9$.

In huge simulations, *Ogielski* [12.7, 8] investigated the correlation length and relaxation times for large systems at little above T_c, where there should not be any finite size effects. From these results and a finite size scaling analysis of additional data obtained near T_c, he obtained the rather accurate results [12.8] $T_c = 1.175 \pm 0.025$, $\nu = 1.3 \pm 0.1$ and $\gamma = 2.9 \pm 0.3$.

The existence of a finite transition temperature was also found using domain wall renormalisation group (DWRG) techniques [12.13]. Here one computes the change in the ground state energy, E_{def}, upon changing boundary conditions, e.g. from periodic to antiperiodic, in one direction. For a spin glass $[E_{\text{def}}]_{\text{av}}$ is zero, but $[E_{\text{def}}^2]_{\text{av}}^{1/2}$ and $[|E_{\text{def}}|]$ are nonzero. If the energy change increases with size then one argues that there is order at low but finite temperature, whereas

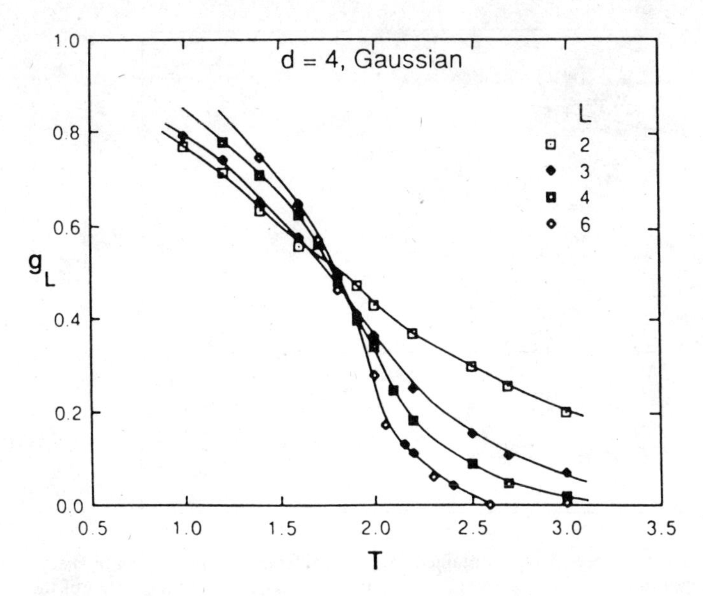

Fig. 12.1. Results for g against T in four dimensions for the Ising spin glass with a Gaussian distribution. T_c is estimated to be 1.75 ± 0.05 from the intersection of the curves. The curves for different sizes splay out below T_c showing that the spin glass order parameter is nonzero in this region. The lines are just guides to the eye. From [12.10]

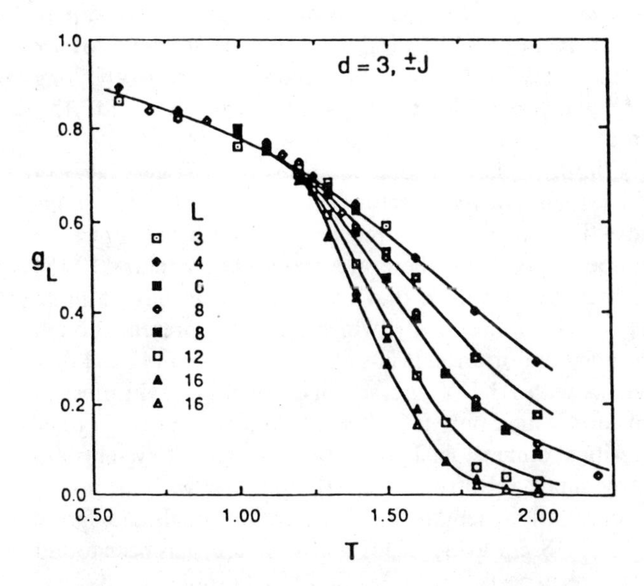

Fig. 12.2. Plot of g against T in three dimensions for the $\pm J$ distribution. The lines are just guides to the eye. The data for the open diamonds $L=8$, and the filled triangles $L=16$ are from *Ogielski* (private communication). The remaining data are from [12.10]

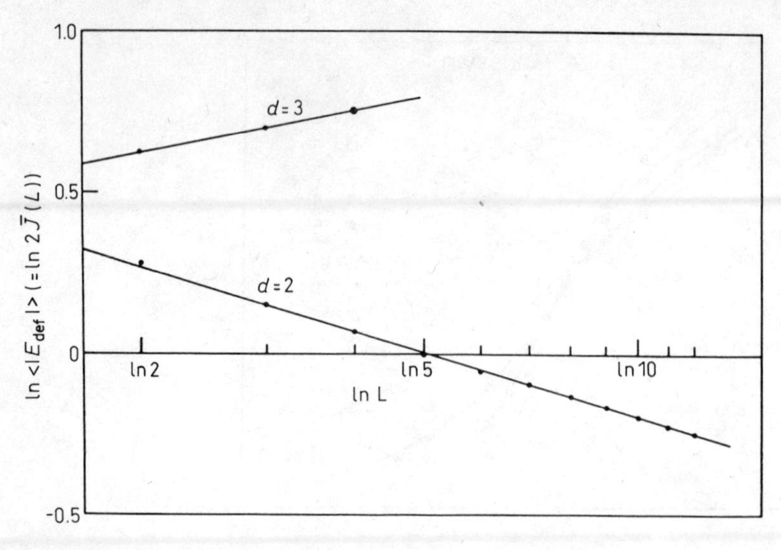

Fig. 12.3. Logarithm of the "defect energy", obtained from the difference in ground state energy when the boundary conditions are changed in one direction, as a function of the logarithm of the linear dimension, L, for the Ising spin glass in two and three dimensions. If the defect energy increases with L, as is the case for $d = 3$, then one expects T_c to be finite, whereas if it decreases with L, as in $d = 2$, then there is only a transition at $T = 0$. From [12.13]

if the energy decreases with size there is no order because, for any finite T, there will be some length scale where the energy change is of order $k_B T$, so the system will be disordered on this scale. In Fig. 12.3 we show results of *Bray* and *Moore* [12.13] which indicate order in 3 but not in 2 dimensions. Subsequently, long series expansions [12.14], which considerably expanded earlier work [12.15], also found a finite T_c in $d = 3$.

Perhaps surprisingly, both Monte Carlo simulations [12.16] and DWRG [12.17] find that $T_c = 0$ for an *isotropic* vector spin glass with short range interactions in $d = 2$ and 3. The lower critical dimension, above which T_c would be finite, seems to be about 4. By now there are many experiments [12.18] which show rather convincingly, however, that real systems do have a finite transition temperature [12.19], so anisotropy, which is always present though often small, must be of crucial importance. It has also been argued [12.20] that isotropic vector spin glasses with RKKY interactions, which occur in most of the experimental systems and which only fall off as (distance)$^{-3}$, are at, rather than below, their lower critical dimension. This means that RKKY systems are extremely sensitive to any small anisotropy [12.20], and that without anisotropy, the spin glass correlation length and susceptibility should grow exponentially as $T \to 0$ [12.21]. Some evidence for this behaviour has been found from simulations [12.22]. One expects [12.20, 23] that all 3D spin glass systems with anisotropy should have a phase transition in the universality class of the short range Ising spin glass model. Experimental values of critical exponents

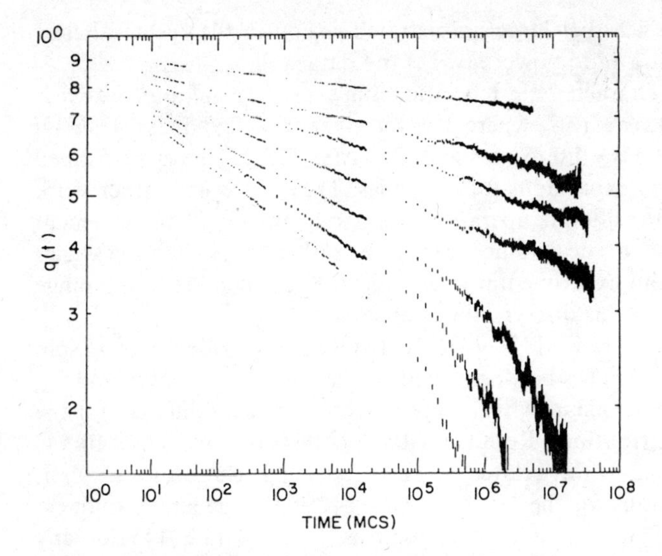

Fig. 12.4. Dynamic correlation function $q(t) = [\langle S_i(0)S_i(t)\rangle_T]_{av}$, for temperatures at and around T_c. Data are shown with error bars for temperatures $T = 1.30, 1.25, 1.10, 1.00, 0.90, 0.70$ (from bottom to top). The lattice size is 32^3. From [12.8]

vary quite a bit, but γ, for example, is generally found to be in the range 2–3.5, and is therefore in rough agreement with the simulations on Ising systems. Quite recently though, *Geschwind* et al. [12.24] have developed a new method for analysing experimental results from which they find that previous estimates of γ (and the dynamic exponent z discussed below) were too low. They find a best fit for γ of about 4.5 which is rather above the Monte Carlo values, but it is possible that critical regime had not been reached in the simulations, in which case the quoted exponents are really effective exponents, only valid over a certain range of reduced temperature and/or system size.

The Monte Carlo technique can also give valuable information on dynamics. The most detailed simulation of spin glass dynamics is that of *Ogielski* [12.8]. He finds empirically that the auto-correlation function $q(t)$, defined by

$$q(t) = \langle S_i(t_0)S_i(t_0 + t)\rangle, \tag{12.12}$$

fits the following functional form:

$$q(t) \propto t^{-x} \exp(-\omega t^{\beta}). \tag{12.13}$$

Some of his results for $q(t)$ are shown in Fig. 12.4. Above the transition temperature of the pure ferromagnet, $T_c^{pure} \approx 4.5$, he finds $\beta = 1$ so the decay is exponential. However, in the region $T_c < T < T_c^{pure}$, β decreases below unity giving rise to "stretched exponential" behaviour, showing that there are long time tails in the fluctuations. Such long time tails in this "Griffiths [12.25] phase" are expected on general grounds [12.26] because there is a non-zero probability of having a rather large unfrustrated region in the system. Below T_c, Ogielski finds that a pure power law fits the data well.

Close to T_c, the characteristic relaxation time, τ, of an infinite system is expected to diverge as $\tau \propto t^{-z\nu}$, where z is the dynamical exponent. *Ogielski* [12.8] finds that $z\nu = 7.0 \pm 1.0$. *Reger* and *Zippelius* [12.27] have performed high temperature series expansions for dynamics. They agree with other work in predicting a finite transition temperature and obtain $z\nu = 6 \pm 2$, in agreement with Ogielski. These theoretical values seem to be rather lower than the values obtained for $z\nu$ in recent experimental work [12.24, 28], which are in the range 10–11.5. The reason for this discrepancy is unclear.

Several studies have shown [12.29–31, 10, 14] that $T_c = 0$ for the Ising spin glass in 2 dimensions, so d_l lies between 2 and 3. The correlation length and χ_{SG} diverge at $T = 0$ with exponents which appear to be somewhat different for the Gaussian and $\pm J$ distributions. Results for the $\pm J$ distribution [12.29, 10, 14] find γ in the range 4–5.3, with ν related to γ by the scaling relation $\gamma = (2 - \eta)\nu$, with $\eta \approx 0.2$. Calculations for the Gaussian case [12.30, 31] generally quote ν, with values ranging from about 3 to as high as 4.4 ± 0.3 [12.31]. For any continuous distribution, the ground state is non-degenerate so $\eta = 0$ in $d = 2$, and hence the corresponding value for γ for the Gaussian distribution from [12.31] is $\gamma = 8.8 \pm 0.6$, much higher than estimates for the $\pm J$ distribution. Presumably, the large ground state degeneracy in the $\pm J$ case puts it into a different universality class from models with a continuous distribution, if the transition is at $T = 0$. When $T_c > 0$, however, one does not expect the ground state degeneracy to play an important role in the critical region, so the $\pm J$ and continuous distributions should be in the same universality class.

These results for $d = 2$, have been confirmed by experiments [12.32] on a quasi 2D spin glass which give $\gamma = 4.5 \pm 0.2$, in good agreement with the simulations on the $\pm J$ model. Crossover between two and three dimensional behaviour on changing the film thickness has also been observed [12.33] in *CuMn* layers alternating with insulating *Si* layers in a multi-layer structure. The experimental results appears to be in agreement with theoretical predictions [12.34].

12.1.4 The-Low Temperature State

In this section we discuss the *equilibrium* state below T_c. Although it is hard to probe the equilibrium state by experiments, which are always out of equilibrium to some extent, there are important theoretical issues which need to be understood. Two possible models have been discussed. The first assumes that short range systems are similar to the infinite range model of *Sherrington* and *Kirkpatrick* [12.35] (SK), which appears to have been solved by *Parisi* [12.36]. Below T_c, any given realisation of the SK model has many thermodynamic states [12.37], unrelated by symmetry, which differ in (total) free energy from the ground state only by an amount of order unity. This leads to a non-trivial distribution, $P(q)$, below T_c, with a peak at finite q and a tail with a finite weight

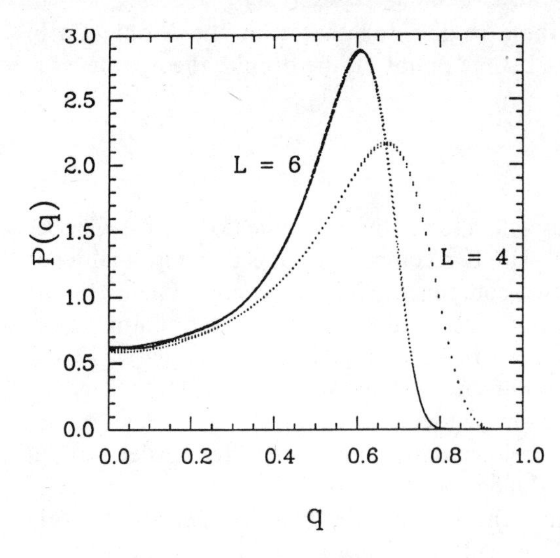

Fig. 12.5. A plot of $P(q)$ for $T = 1.2$ for the 4-d Ising spin glass with Gaussian bonds. For comparison, the transition temperature is at about $T = 1.8$. Data is shown for two sizes, $L = 6$ and $L = 4$. For each size there are two sets of points representing data calculated by the two methods, which are expected to agree with each other only when the system has reached equilibrium. Since they agree with each other so well that the difference is barely resolvable on the figure, the system is evidently well equilibrated. Notice that the weight in the tail does not depend significantly on size. From [12.41]

at $q = 0$. Such a distribution has indeed been observed by simulations on the SK model [12.38]. The alternative picture is the "droplet model" of *Fisher* and *Huse* [12.39], which is equivalent to the one parameter scaling picture of *Bray* and *Moore* [12.40]. There, the energy to turn over a domain of linear size L is postulated to vary as L^{θ} where θ is an exponent characterising the zero temperature fixed point. As a result, there is only a single thermodynamic state, apart from those related to it by global symmetry, and $P(q)$ for an infinite system is just a single delta function. For a finite system, $P(q)$ can have finite weight down to $q = 0$, but the weight vanishes like $L^{-\theta}$, as the linear size of the lattice, L, goes to infinity.

Careful Monte Carlo simulations of the $d = 4$ Ising spin glass below T_c have been carried out by *Reger* et al. [12.41]. Because of large sample-to-sample variations and long relaxation times this required a massive computational effort. The simulations have been performed on a parallel transputer array containing 40 processors. For sizes $2 < L < 6$, the weight in the tail decreases very little with L, and the results look qualitatively like the SK model. Results for one temperature are shown in Fig. 12.5. In parallel work, *Caracciolo* et al. [12.42a] have looked at the 3D spin glass in a magnetic field. From a variety of measurements they also argue that the system behaves in a manner similar to the SK model, though it is been argued [12.42b] that the simulations were not performed at sufficiently low temperature to probe the ordered phase.

Some analytical results for $P(q)$ have also been obtained recently. It is possible to perturb away from the SK limit by considering each spin to be connected to a *finite* number of others, z, chosen at random [12.43]. The first terms in the expansion of $P(q)$ in powers of $1/z$ have recently been found [12.44] for this model, as well as the expansion in powers of $1/d$ for a hypercubic d-dimensional lattice. Surprisingly, $P(q = 0)$ is found to increase to first order in both these models. While these results, together with the Monte Carlo simulations [12.41, 42a], may cast some doubt on the droplet theory, the issue is far from settled.

12.1.5 The Vortex Glass

The order parameter in a superconductor is the wavefunction of the Cooper pairs, which is a complex number with two components. It is therefore analogous to the order parameter in a two-component (XY) magnetic system. In the Meissner phase, below the field H_{c_1}, the order parameter is uniform and is therefore analogous to a ferromagnet. In a pure type II superconductor [12.45], the Abrikosov flux lattice phase occurs for fields between H_{c_1} and H_{c_2}. Quenched disorder destroys the flux lattice on long length scales [12.46], just as random fields, discussed in Sect. 12.4. below, destroy ferromagnetic order with a continuous symmetry below 4 dimensions [12.47]. What happens on scales larger than the distance over which the flux lattice is disordered is rather poorly understood. According to the conventional "flux creep" picture [12.48], no radically new physics comes into play, the flux tubes are pinned by the disorder but can move individually under the presence of a current, so there is dissipation. Consequently, type II superconductors above H_{c_1} are not strictly superconductors but have finite resistance. For low T_c materials at least, this is so small as to be unobservable in practice for $H < H_{c_2}$, but this may not be the case for high T_c compound, where fluctuation effects are larger. Since the field frustrates the system, the necessary ingredients, randomness and frustration, for a spin glass are both present. We will therefore discuss the possibility that disordered type II superconductors in a field greater than H_{c_1}, may instead enter a spin-glass-like state, called the "vortex glass" [12.5] which would have truly zero resistance for vanishingly small current. *Hertz* [12.49] seems to have been the first to notice that a (random) vector potential can induce frustration much like random bonds. The analogy between spin glass models and random superconductors was further developed by *Shih* et al. [12.50] in their theory of granular superconductors. It is only recently [12.5, 51], however, that the vortex glass has been proposed as a distinct phase for bulk superconductors such as the high T_c materials. Let us simplify the discussion by choosing, in the spirit of *Edwards* and *Anderson* [12.4], the simplest model with the correct features. Firstly, we discretise space and consider a simple cubic lattice, on each site of which there is a classical vector of unit length and phase ϕ_i. We further assume nearest-neighbour couplings J_{ij}, which are all equal to unity, and a random vector potential, A_{ij}, between nearest neighbours, so the Hamiltonian is

$$\mathscr{H} = -\sum_{\langle i,j \rangle} \cos(\phi_i - \phi_j - A_{ij}). \tag{17.14}$$

If the A_{ij} are constrained to take the values 0 or π then (12.14) is just the Edwards and Anderson XY spin glass with $\pm J$ distribution. However, as noted above, this does not appear to have a transition in $d = 3$. A more realistic model for a superconductor is to have the A's uniformly distributed between 0 and 2π [12.52]. This model is called the "gauge glass". Both the gauge glass and the XY spin glass are invariant under the global rotation $\phi_i \to -\phi_j + \text{const.}$ However, the "reflection" $\phi_i \to -\phi_i$ is *not* a symmetry of the gauge glass because the magnetic field breaks time-reversal invariance, though it is symmetry of the XY spin glass. This difference means that the two models may be in different universality classes and may also have different lower critical dimensions [12.52]. *Huse* and *Seung* [12.52] have performed extensive Monte Carlo simulations of the gauge glass model in three dimensions and found that its behaviour is closer to that of the Ising spin glass, which does appear to have a finite T_c, than to the XY spin glass, which does not. More work is in progress to establish whether or not a vortex glass phase occurs in three dimensions.

12.2 Potts Glasses

12.2.1 Introduction to Potts Glasses

Since understanding of Ising spin glasses [12.1] has greatly improved over the past years, glass models with more general features are now receiving more attention. Certain experimental systems, such as dilute ortho-hydrogen [12.53], do not have the spin inversion symmetry of the Ising Hamiltonian (12.6), which is invariant under the transformation $S_i \to -S_i$, for all i, since $S_i = \pm 1$. Systems lacking this symmetry are better described by a Potts model [12.54, 55] which is a generalisation of the Ising model to p states. The Hamiltonian of the Potts model is conventionally written as

$$\mathscr{H} = -\sum_{\langle i,j\rangle} J_{ij}\delta_{n_i n_j}, \tag{12.15}$$

where the Potts states are labelled by an integer variable $n_i = 1, 2, 3, \ldots, p$. Clearly, for $p = 2$ the inversion symmetry is restored and (12.15) can be rewritten in the form of the Ising Hamiltonian (12.6).

It is more convenient, however, to use the so called "simplex" representation of Potts spins. One introduces the spin variables $\boldsymbol{S}_i = \boldsymbol{u}^{(n_i)}$, where $\boldsymbol{u}^{(k)}$, for $k = 1, 2, \ldots, p$ form a set of basis vectors embedded in a $(p-1)$-dimensional space to reflect the correct symmetry of the Potts Hamiltonian. The basis vectors satisfy $\boldsymbol{u}^{(i)} \cdot \boldsymbol{u}^{(k)} = (\delta_{ik}p - 1)/(p-1)$ and $\sum_{i=1}^{p} \boldsymbol{u}^{(i)} = 0$. In this representation the Hamiltonian is given by

$$\mathscr{H} = -\sum_{\langle i,j\rangle} J'_{ij}\boldsymbol{S}_i \cdot \boldsymbol{S}_j. \tag{12.16}$$

This form is more convenient to use since it is identical to that of vector spin glasses. We have to bear in mind, however, that there is no inversion symmetry because the vector $-\boldsymbol{u}^{(k)}$ does not occur in the set (except for $p = 2$). Note also that $J'_{ij} = J_{ij}(p-1)/p$. We work in units in which $[J_{ij}^2]_{\text{av}} = 1$.

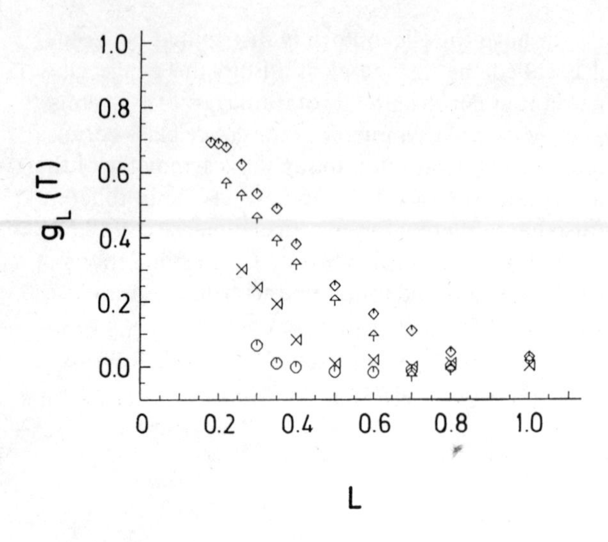

Fig. 12.6. Reduced cumulant $g_L(T)$ as a function of temperature T for different sizes of the $d = 3$ three-state Potts glass with Gaussian random bonds. The curves should intersect at the critical point T_c. From Ref. [12.73]

The effect of random dilution on both ferromagnetic and antiferromagnetic Potts models has been studied and the approximate phase diagrams are known for a range of values of p [12.56–58]. Our main interest here is in models with an equal amount of ferromagnetic and antiferromagnetic bonds. The latter lead to frustration and in sufficiently high dimensions, to a Potts glass phase below the critical temperature T_c. Note that although the frustration decreases, the degeneracy associated with antiferromagnetic couplings increases with p.

The Potts glass model has been suggested as a first step towards understanding the structural glass transition [12.59, 60] and it may be applicable to strongly anisotropic orientational glasses, where the anisotropy confines the quadrupole moment to one of p discrete directions [12.61].

We review here briefly the rather rich mean-field theory of Potts glasses and then focus on results obtained for more realistic short range models.

12.2.2 Mean-Field Theory

Although early investigations of the mean-field theory of the Potts glass using the replica method failed to find a self-consistent stable low temperature phase [12.62–64], *Parisi*'s replica symmetry breaking (RSB) scheme [12.65] yields a stable glass phase for all p. The first level of Parisi's RSB hierarchy is sufficient to stabilise the solution [12.65] just below T_c. In this region the solution with an infinite sequence of RSB levels [12.65, 66] found for models with spin inversion symmetry, does not exist. The solution with one level RSB becomes unstable at a second transition temperature, T_2 ($T_2 < T_c$), at which an infinite sequence of RSB levels appears. While the second transition is continuous for all values of p, the first transition from the disordered state becomes discontinuous for $p > 4$. Note that for the non-random infinite range Potts model, the transition is discontinuous for all $p > 2$. Interestingly, the discontinuous transition in the glass model for $p > 4$ is not a first-order one in the strict sense, since there is no latent heat, though T_c is not a critical point either, since the glass susceptibility does not diverge at T_c. It can be argued [12.67] that the absence of latent heat originates in the exponentially large number of local minima at T_c. In a sense, the system is not much more ordered after freezing into these states than in the high temperature phases.

There exist intriguing analogies between the dynamical theory of the Potts glass and the mode-coupling approach to the glass transition [12.68]. The source of these is possibly to be found in the identical structure of non-linearities in the corresponding equations [12.60]. It is therefore hoped that the study of Potts-glass models will also lead to a better understanding of structural glasses.

12.2.3 The Critical Dimensions

Mean-field theory, as discussed above, is expected to describe the correct physics above $d = 6$, which is the upper critical dimension, d_u, of the Potts glass model. The first step on the way to lower dimensions may be an ε-expansion around d_u. The renormalisation group analysis [12.69] in $6 - \varepsilon$ dimensions predicts a fluctuation-driven first-order transition for $2 < p < 4$, in marked contrast to the mean-field results.

Apart from the *upper* critical dimension d_u, the *lower* critical dimension d_l is of central interest in the study of short-range models. The critical temperature decreases with decreasing dimension and is zero for $d < d_l$. One therefore expects for the spin glass correlation length

$$\xi_{SG} \propto \begin{cases} (T - T_c)^{-\nu} & d > d_l, \\ \exp(CT^{-\sigma}) & d = d_l, \\ T^{-\nu} & d < d_l. \end{cases} \tag{12.17}$$

Note that $\nu \to \infty$ as $d \to d_l$ from either direction and so at d_l the correlation length diverges exponentially. We have written (12.17) assuming that $T_c = 0$ for $d = d_l$, though a non-zero T_c in this case cannot be ruled out a priori.

For $T_c = 0$ there is just one independent static critical exponent, provided the ground state is not extensively degenerate. In that case there is an additional relation $2 - \eta = d$, so that $\gamma = d\nu$. The exponent σ is equal to 1 in uniform systems (such as the Ising model for which $d_l = 1$). According to *McMillan's* scaling theory [12.21], however, one should have $\sigma = 2$ for random-bond systems, like Potts glasses.

12.2.4 The Short-Range Potts Model

We now discuss which of the possibilities in (12.17) is realised in short-range models.

(a) Phenomenological $T=0$ Scaling. The Migdal-Kadanoff renormalisation group analysis does not lead to any conventional spin-glass behaviour in the $\pm J$ Potts glass [12.70]. This may well be an artifact due to the combination of high degeneracy introduced by the antiferromagnetic bonds and the structure of the hierarchical lattice.

Phenomenological scaling ideas, which have proved to be a powerful technique for short-range Ising spin glasses, have been applied recently to short range Potts glasses [12.71]. From the interfacial energy of a domain wall, introduced arbitrarily, the authors compute an effective coupling $\delta E(L)$ on length-scale L. Equilibrium properties are determined from the behaviour of the exponent y, characterizing the scaled dependence of $\delta E(L)$ on L at $T=0$, i.e. $\delta E(L) \propto L^y$. Below d_l, y is negative and related to ν by $|y|=\nu$, while y is positive for $d>d_l$.

For $d=2$, *Banavar* and *Cieplak* [12.71] used sizes $3 \leqq L \leqq 10$ and found y to be clearly negative. For $d=3$, they were constrained to sizes $L=2$ and 3 and found $y=-0.11$ for the $\pm J$ distribution and $y=0.10$ for the Gaussian distribution. While one interpretation of this might be that the values of d_l are different for the two distributions, we feel that, given the very small sizes, the results are consistent with the more plausible hypothesis that the critical dimensions are equal and the common value is close to 3.

(b) Monte Carlo Simulations. The early Monte Carlo study [12.72] of the 3-state Potts glass with Gaussian interactions was unable to determine whether a phase transition occurred in three dimensions, though an upper bound on the critical temperature, $T_c < 0.23$, was obtained. In addition to static quantities, the analogue of the Edwards–Anderson order parameter, $q(t)$, was also computed and found to follow a Kohlrausch law in a certain range of temperature. Recently, a more detailed study was carried out [12.73] in which finite size scaling [12.3] was applied to the Monte Carlo data obtained for sizes $L \leqq 8$ on a simple cubic lattice with a Gaussian distribution of bonds.

The order parameter q was defined as a rotationally invariant measure of the overlap between two replicas of the same system. The authors computed the first moments of $P(q)$ and use the "renormalised coupling" [12.10, 12, 74] g, considered in (12.10, 11). As discussed for the Ising spin glass, the curves of $g(T,L)$ for different L are expected to intersect at T_c, as seen from the finite size scaling form in (12.11). From Fig. 12.6 one can exclude any transition for $T>0.15$, which is 18% of the mean-field value of $T_c^{MF}=\sqrt{z/p}$, which has the value 0.8165 for coordination number $z=6$ and $p=3$. It is unlikely however, that T_c is so small and yet finite. Further support for a $T=0$ transition comes from a slight modification of standard finite size scaling. Consider the moments

$$\langle q^n \rangle = L^{-n(d-2+\eta)/2} f_n(L^{1/\nu}(T-T_c)). \tag{12.18}$$

and note that the susceptibility is simply related to the second moment, $\chi_{SG}=N\langle q^2 \rangle$. Since it is not known a priori whether the system is below, exactly

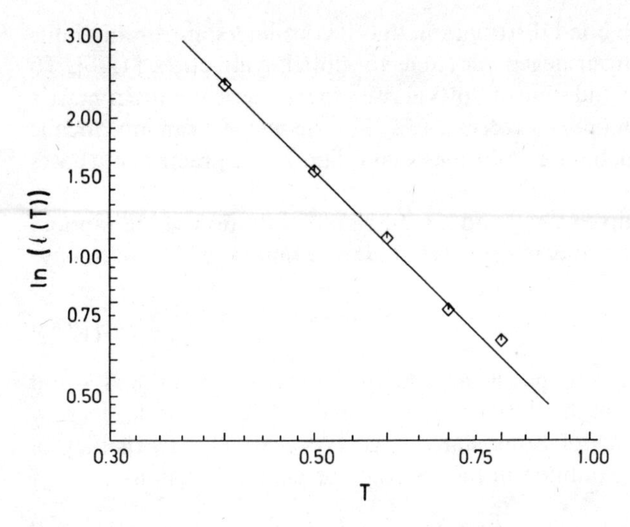

Fig. 12.7. Log–log plot of ln l versus temperature for the $d=3$ three-state Potts glass with Gaussian random bonds. $l(T)$ is proportional to the correlation length $\zeta_{SG}(T)$, see [12.73]. The straight line has a slope of -1.97, close to the value of -2 expected for a system at its lower critical dimension

at, or above the lower critical dimension, the authors use a scaling analysis that does not enforce a particular asymptotic form of the correlation length ($\xi_{SG} \propto |T-T_c|^{-\nu}$ is implicit in (12.18)), but obtain it from the data themselves. Using the more general scaling form

$$\chi_{SG} = L^{2-\eta}\tilde{\chi}\left(\frac{L}{\xi_{SG}(T)}\right) \tag{12.19}$$

in which the value of $\xi_{SG}(T)$ at each temperature is taken as an adjustable parameter, and the relation

$$\chi_{SG} \propto \xi_{SG}^{2-\eta}, \tag{12.20}$$

good scaling is found and the results are consistent with the system being exactly at its lower critical dimension, with

$$\xi_{SG} \propto \exp(CT^{-1.97}) \quad \text{and} \quad \chi_{SG} \propto \xi_{SG}^{1.5}. \tag{12.21}$$

The agreement with McMillan's prediction, $\sigma = -2$ for the exponent of ξ is impressive, see Fig. 12.7.

(c) Transfer Matrix Calculations. From the results of the Monte Carlo simulations [12.73] one can conclude by comparing (12.20) and (12.21) that the exponent η describing the decay of the spatial correlations at the critical point is $\eta \sim 0.5$ in $d=3$ for the 3-state Potts model. By contrast, for a system with a unique ground state, the correlations do not fall off with distance at $T=0$ and so [12.75] $2-\eta=d$. This is certainly the case for Ising spin glasses, where one can easily show for any continuous bond distribution, that the ground state for any finite lattice is unique apart from degeneracy due to global symmetries [12.1]. To better understand the ground state of Potts glasses, the numerical transfer matrix method [12.76] has been applied recently [12.77]. The method can in principle be used in any dimension, but for Potts glasses it is limited for practical purposes to $d=2$.

Using system sizes up to 9×9 and a Gaussian distribution of the random bonds, the ground state entropy per spin is found to be finite [12.77] and to have the values

$$S_0^{p=3}=0.0127, \quad S_0^{p=4}=0.1318, \tag{12.22}$$

i.e. the ground state is macroscopically degenerate in the two-dimensional 3- and 4-state Potts glass, just as in the Potts antiferromagnet [12.78a]. The degeneracy increases with the number of Potts states p, as expected. The degeneracy of the ground state is also exhibited in the spactial correlation function

$$G(R,T)=[\langle \boldsymbol{S}_i \cdot \boldsymbol{S}_j \rangle_T^2]_{\text{av}}, \quad R=|\boldsymbol{R}_i-\boldsymbol{R}_j|. \tag{12.23}$$

Fitting this to the expected power law behaviour,

$$G(R,T=0) \propto R^{-\eta}, \tag{12.24}$$

one finds

$$\eta=0.18 \pm 0.03 \tag{12.25}$$

for $d=2$, $p=3$, showing that the correlations do indeed decrease with distance. One can conclude that the 2D Potts glass with nearest-neighbour Gaussian couplings has a macroscopically degenerate ground state. This result is in contrast with Ising, XY or Heisenberg spin glasses with continuous bond distributions and is a distinctive property of the Potts model.

(d) High-Temperature Series Expansions. High temperature series have been developed [12.78b] for the 3-state Potts glass with $\pm J$ interactions on hypercubic lattices. There are 11 nontrivial terms (order 22) available for $d>4$ and 12 terms (order 24) for $d \leqq 4$. For $d \geqq 6$ the analysis of the Edwards–Anderson susceptibility yields an exponent that is consistent with mean-field theory. In three dimensions there is excellent agreement at higher temperatures with Monte Carlo data, but at low temperatures the series suffers from convergence problems, so more terms or a more sophisticated analysis are needed.

12.3 Orientational Glasses

12.3.1 Introduction to Orientational Glasses

While in spin glasses [12.1] the elementary "single-particle" degree of freedom that is considered is a dipole moment, in "orientational glasses" or "quadrupolar glasses" it is a quadrupole moment. Just as (magnetic) spin glasses are obtained in the laboratory by random dilution of magnets which have competing ferromagnetic and antiferromagnetic interactions with nonmagnetic ions (a prototype example is the ferromagnet EuS diluted with Sr [12.1]), orientational glasses result from random dilution of molecular crystals with atoms which have no quadrupole moments, e.g. N_2 diluted with Ar, KCN diluted with KBr [12.79]. In the following we shall mostly assume that the dominant effect of this randomness is disorder and competition in the pairwise interaction among quadrupoles, thus emphasising the analogy with spin glasses. The assumption is nontrivial (and certainly also not strictly true as far as real materials are concerned), since the disorder affects the effective single-site potential acting on the quadrupole moments. As a consequence, "random-field" disorder is also possible, and has been invoked in approximate analytical theories attempting [12.80] to explain glassy behaviour in quadrupolar glasses. We disregard such "random-field" disorder here, but will come back briefly to more realistic models of quadrupolar glasses at the end of this section. At this point, we discuss models of the random-bond type, i.e. suitable generalisations of the models of (12.6, 16), replacing the component S_i^u of an m-component spin vector $\boldsymbol{S}_i$ by the component $f_i^{\mu\nu}$ of the electric quadrupole moment tensor.

$$f_i^{\mu\nu} = \int \rho(\boldsymbol{x})\left(x_\mu x_\nu - \frac{\delta_{\mu\nu}}{m} \sum_{\lambda=1}^{m} x_\lambda^2 \right) d\boldsymbol{x}, \tag{12.26}$$

where $\rho(\boldsymbol{x})$ is the charge distribution of the molecule associated with the lattice site i in d-dimensional space $\boldsymbol{x}$, and $\{\mu, \nu\} = (1, 2, \ldots, m)$. We shall be mainly interested in $m = 3$ since planar ($m = 2$) isotropic orientational glasses are equivalent [12.81] to the XY model spin glass $\mathscr{H} = -\sum_{\langle i,j \rangle} J_{ij} \cos(\theta_i - \theta_j)$. Now the generalisation of this is

$$\mathscr{H}_{\text{quadrupolar}} = -\sum_{\langle i,j \rangle} \sum_{\mu\nu} J_{ij}^{\mu\nu} f_i^{\mu\nu} f_j^{\nu\mu}. \tag{12.27}$$

In the following, we shall focus on a simple specialised case: considering only axially symmetric charge distributions, the quadrupole moment tensor can be written in terms of the components $\{S_i^\mu\}$ of a unit vector aligned along the preferred axis,

$$f_i^{\mu\nu} = \left(S_i^\mu S_i^\nu - \frac{1}{m} \delta^{\mu\nu} \right) f, \tag{12.28}$$

with f describing the strength of the quadrupole moment. Henceforth our units are chosen such that $f = 1$. Secondly, we assume an isotropic interaction, i.e. $J_{ij}^{\mu\nu} = J_{ij}$ independent of μ, ν. Then (12.27) becomes

$$\mathscr{H}_{\text{isotropic quadrupolar}} = -\sum_{\langle i,j \rangle} J_{ij} \left[(\boldsymbol{S}_i \cdot \boldsymbol{S}_j)^2 - \frac{1}{m} \right]. \tag{12.29}$$

At this point, we note that some anisotropic versions of (12.27) are believed to be related to models of the previous section: for the 2D model ($d=2$, $m=2$) with cubic anisotropy, there are two preferred orientations of the moment (e.g. the x-axis and the y-axis of a square lattice) and hence the model should belong to the corresponding universality class of an Ising spin glass (Sect. 12.1). For the case $d=3$, $m=3$ with cubic anisotropy again preferring the three lattice directions, the model should belong to the class of the three-state Potts glass (Sect. 12.2). Numerical work on the latter model [12.61] is compatible with this conjecture.

Although systems with a fully isotropic interaction of the form (12.29) hardly occur in nature, understanding their behaviour as a limiting case is of fundamental interest. Thus a number of studies have been devoted to this problem [12.82–85], as reviewed in Sect. 12.3.2. In the Sect. 12.3.4 more realistic models of a different type will be briefly mentioned.

12.3.2 Static and Dynamic Properties of the Isotropic Orientational Glass ($m=3$) in Two and Three Dimensions

As in the previous sections, we restrict attention to a symmetric distribution of bonds, $P(J_{ij})=P(-J_{ij})$, such as the Gaussian distribution in (12.7b). While for a spin glass, the spin pair correlation of ferromagnetic type is trivial [12.1],

$$g_F(\mathbf{r})=[\langle \mathbf{S}_i \cdot \mathbf{S}_j \rangle_T]_{av}=\delta_{ij}, \quad \mathbf{r}=\mathbf{r}_i-\mathbf{r}_j, \tag{12.30}$$

and only the glass-like correlation defined in (12.2) exhibits interesting behaviour, for orientational glasses this is no longer true, because there is a single state where two quadrupoles are oriented in parallel, while the state where the two quadrupole moments are oriented perpendicularly to each other is infinitely degenerate. Aa a consequence, correlations of both ferromagnetic type and of

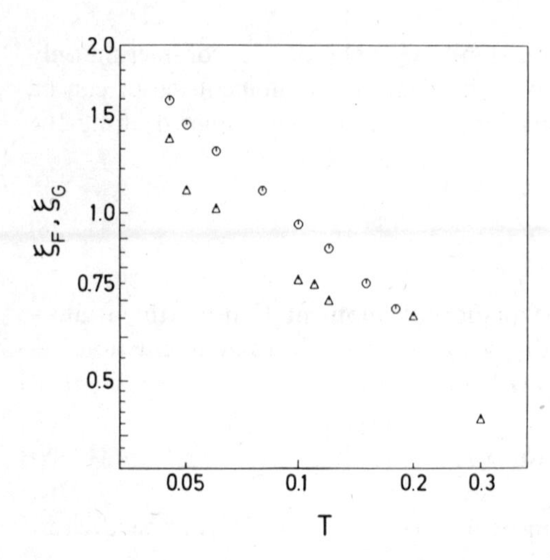

Fig. 12.8. Log–log plot of the two correlation lengths $\xi_F(T)$, triangles, and $\xi_G(T)$, octogons, of the 2D isotropic three-component quadrupolar glass versus temperature (measured in units of the width of the Gaussian distribution (12.7b)). From [12.83]

glass-like type need to be considered;

$$g_F(\mathbf{r}) = \left[\left\langle \sum_{\mu\nu} f_i^{\mu\nu} f_j^{\nu\mu} \right\rangle_T\right]_{\mathrm{av}} = \left[\left\langle (\mathbf{S}_i \cdot \mathbf{S}_j)^2 - \frac{1}{m} \right\rangle_T\right]_{\mathrm{av}} \propto \exp\left[\frac{-r}{\xi_F(T)}\right], \quad (12.31)$$

$$g_G(\mathbf{r}) = \left[\left\langle \sum_{\mu\nu} f_i^{\mu\nu} f_j^{\nu\mu} \right\rangle_T^2\right]_{\mathrm{av}} = \left[\left\langle (\mathbf{S}_i \cdot \mathbf{S}_j)^2 - \frac{1}{m} \right\rangle_T^2\right]_{\mathrm{av}} \propto \exp\left[\frac{-r}{\xi_G(T)}\right]. \quad (12.32)$$

Figure 12.8 gives an example for $d=2$, where a phase transition occurs at zero temperature only. Both correlation lengths $\xi_F(T)$, $\xi_G(T)$ increase as T is lowered, but it is clear that ξ_G is larger than ξ_F at the temperatures of interest [12.83]. These data were obtained from 20×20 lattices averaging typically of the order of 10^2 samples. Although even at the lowest temperatures the correlation lengths are very small, the pronounced critical slowing down of this model prevented studies at lower temperatures.

Defining a glass "susceptibility" χ_G similar to that of previous section,

$$\chi_G = 1 + \sum_{\mathbf{R} \neq 0} g_G(\mathbf{R}), \quad (12.33)$$

one can again attempt to study the behaviour at the zero temperature transition,

$$\xi_G \propto T^{-\nu}, \quad \chi_G \propto T^{-\gamma}, \quad T \to 0, \quad (12.34)$$

see Fig. 12.9. From these results, the estimates

$$\gamma(d=2) \cong 1.35, \quad \nu(d=2) \cong 0.63, \quad \gamma(d=3) \cong 2.7, \quad \nu(d=3) \cong 1.02, \quad (12.35)$$

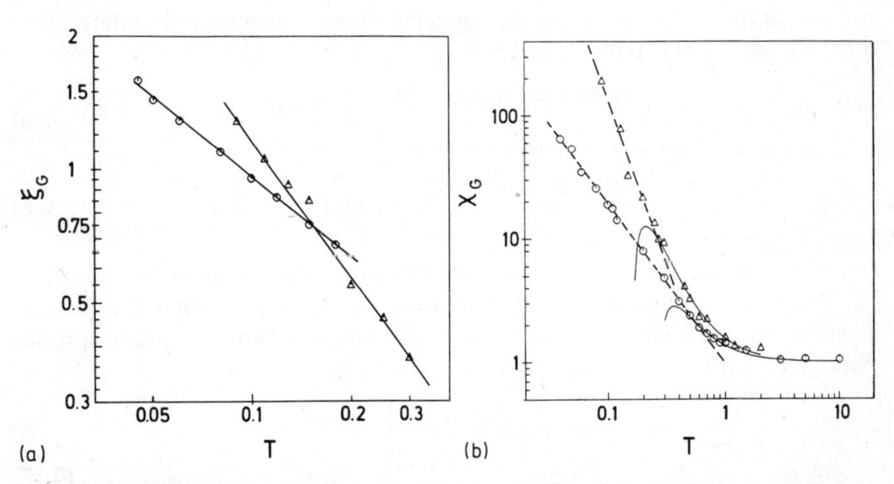

Fig. 12.9. Log–log plot (**a**) of the glass correlation length $\xi_G^{(T)}$ and (**b**) the glass susceptibility χ_G, (12.33) of the isotropic three-component orientational glass with Gaussian bonds plotted versus temperature. Octagons refer to $d=2$, triangles to $d=3$. The full (**a**) and broken (**b**) straight lines illustrate the estimate of the exponents ν, γ (12.35), from these data, assuming that the asymptotic regime where the power laws (12.34) hold, has actually been reached. The full curves in (**b**) represent the high-temperature series expansion (12.36) of *Carmesin* and *Ohno* [12.84]. From [12.83]

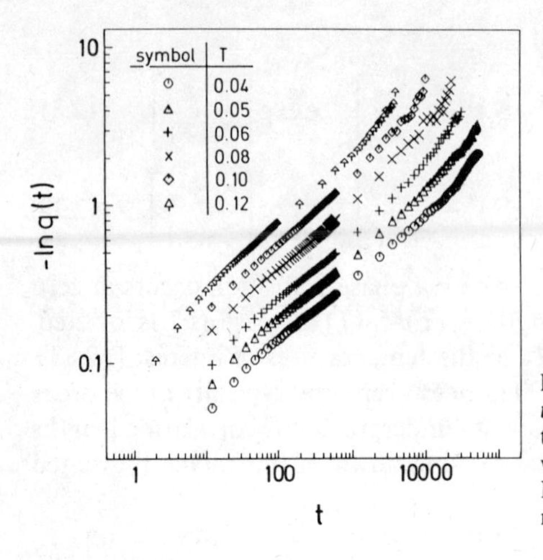

Fig. 12.10. Log–log plot of $-[\ln q(t)]$ vs time t, for the autocorrelation function $q(t)$ of the three-component isotropic orientational glass, with Gaussian bonds on the square lattice. Different symbols stand for different temperatures as indicated. From [12.83]

have been deduced [12.83]. *Very* roughly, these numbers are consistent with the relation $d\nu = \gamma$ [12.75] which should hold for phase transitions at $T = 0$ with a nondegenerate ground state. Considering the experience with the $d = 2$ Ising spin glass, where exponents extracted from a similar temperature range [12.76] clearly are systematically too low [12.29, 31], it is possible that the estimates quoted in (12.35) are only "effective" exponents for the temperature range considered, and that the true asymptotic exponents are somewhat higher. Nevertheless, the conclusion of [12.82, 83], that there is no phase transition at nonzero temperature seems quite plausible. Note that at high temperatures the Monte Carlo data are in good agreement with the corresponding high temperature series expansion [12.84],

$$\chi_G = 1 + \frac{16}{(45T^2)} + \frac{0.001548}{T^4} - \frac{0.00021739}{T^6} + O(T^{-8}), \quad d = 2, \tag{12.36a}$$

$$\chi_G = 1 + \frac{8}{(15T^2)} + \frac{0.02128}{T^4} - \frac{0.00097667}{T^6} + O(T^{-8}), \quad d = 3, \tag{12.36b}$$

but the series are too short to estimate the critical behaviour of the model.

Next we draw attention to the time autocorrelation function of the quadrupole moments, i.e. the analog of the time-dependent Edwards–Anderson order parameter $q(t)$ [12.1] of Ising spin glasses (12.12),

$$q(t) = \frac{\left[\frac{1}{N}\sum_i \left\langle \{\mathbf{S}_i(t')\cdot\mathbf{S}_i(t'+t)\}^2 - \frac{1}{m}\right\rangle_T\right]_{\mathrm{av}}}{\left(1 - \frac{1}{m}\right)} \tag{12.37}$$

where we have chosen a normalisation such that $q(0) = 1$, while $q(t \to \infty) \to 0$ for $T \geqq T_c$. As an example, Fig. 12.10 presents data for $d = 2$, in a representation where the Kohlrausch–Williams–Watts law [12.86],

$$q(t) = A\exp\left\{-\left[\frac{t}{\tau(T)}\right]^{y(T)}\right\}, \tag{12.38}$$

would reduce to a straight line for an amplitude A being equal to 1 (trying values of $A \neq 1$ did not improve the straight line fit in Fig. 12.10 significantly). Note that the data have considerably less precision than the corresponding data for Ising $\pm J$ spin glasses (Fig. 12.4) [12.8], which is inevitable, since the latter model allows much faster programs invoking multispin-coding techniques [12.87], transition probabilities can be tabulated etc., while the present model, where both continuous spins and continuous bond distributions occur, reached only a speed of about 1×10^6 attempted updates per second on supercomputers such as Fujitsu VP 100 used for these calculations. Therefore a significant test of the empirical "law" (12.38) could *not* be performed, but rather it has been used as a practically convenient fitting function to estimate an effective autocorrelation time $\tau_A(T)$, defined by $\tau_A(T) = \int_0^\infty q(t)dt$, and a corresponding parameter $y(T)$ characterising the width of the relaxation time spectrum. The results obtained from such a fit are collected in Fig. 12.11. It is seen that lowering the temperature decreases the effective "Kohlrausch exponent", y, significantly, reaching a value of about 0.3 in the critical region (near $T = 0$). In this respect, the behaviour is similar to Ising spin glasses [12.8], although there T_c is nonzero.

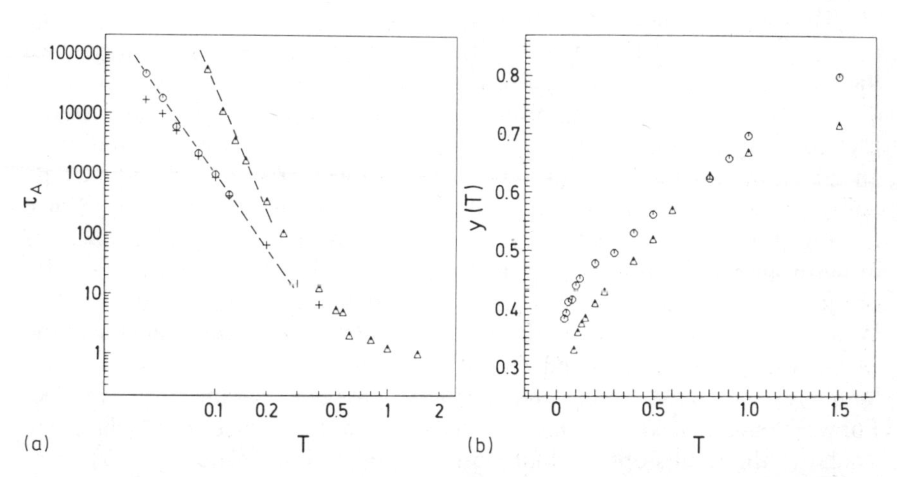

Fig. 12.11. (**a**) Log–log plot of the effective autocorrelaton time $\tau_A(T)$ of isotropic orientational glasses versus temperature. Triangles refer to $d = 3$, octagons to $d = 2$, the broken straight lines indicate an estimate of the dynamic exponent in the assumed power law $\tau_A(T) \sim T^{-z\nu}$. Note that $\tau_A(T)$ was found by fitting data, such as that shown in Fig. 12.10, to (12.38) and integrating (12.38) which gives as $\tau_A = (A\tau/y)\Gamma(1/y)$. In addition, the crosses denote data for $\tau_A(T)$ obtained from a direct numerical integration of the data for $q(t)$, which were cut off at times for which statistical errors became appreciable. (**b**) Effective exponent $y(T)$ plotted vs temperature. Triangles refer to $d = 3$, octagons to $d = 2$. From [12.83]

If one assumes a power law divergence of the relaxation time at the zero-temperature transition, $\tau_A \sim T^{-z\nu}$, one would obtain $z\nu(d=2) \sim 4.3$ and $z\nu(d=3) \sim 6.8$ [12.83]. However it has been observed [12.83] that, in $d=3$, a fit to an Arrhenius-like behaviour, $\tau_A(T) \propto \exp(\Delta E/T)$, where ΔE is some effective energy barrier with $5 \leq \Delta E \leq 8$, is also consistent with the data at low temperatures. Clearly more work is required to resolve the situation, but this will require either a much faster algorithm or much faster computers (or both).

12.3.3 More Realistic Models

The model (12.29) studied in the previous section is an interesting limiting case, but certainly does not correspond to any real material, where effects due to anisotropy, "random fields" (i.e. terms in the Hamiltonian coupling linear to the local quadrupole moment) and asymmetry in the bond distribution should be considered. In fact, it is also desirable to consider site-diluted models rather than the random-bond models discussed throughout this chapter, in order to be able to compare the dependence of various physical quantities on the concentration of the diluent.

As for spin glasses [12.1], only very modest attempts have been made to tackle this rather ambitious program. Results available so far [12.88] are encouraging, but of a completely qualitative character: for the most part the data of [12.88] have used workstations. In our view, a major effort using supercomputers or special-purpose computers is essential before quantitatively reliable results can be obtained.

Thus we only give a brief description of the two models of [12.88]: One model is a site-diluted triangular lattice, with rigid rotors at the lattice sites; the other model starts with rigid rotors at random positions in the continuum (the positions being fixed according to a "random car parking" algorithm). Electrostatic-type quadrupole–quadrupole interactions are used, amounting to an anisotropic interaction of the type of (12.27), together with random-field type terms $[\boldsymbol{S}_i \cdot (\boldsymbol{r}_i - \boldsymbol{r}_j)]^2$ where $\boldsymbol{r}_i \rightarrow \boldsymbol{r}_j$ is the distance between quadrupoles i, j. Since the strength of this interaction decreases only with the inverse fifth power of distance, an enormous computational effort would be required for a full study of such a model. Also other possible interactions, such as anisotropic van der Waals forces, may contribute [12.88], and thus the ratio of the strengths of the "random-bond"-like terms and the "random-field"-like terms may vary. Thus, for nonzero random fields, one finds a nonzero glass order parameter, as expected. For weak random fields the relaxation of the glass order parameter is qualitatively similar to the results obtained for isotropic orientational glasses [12.83].

There has even been an attempt to bypass the study of simplified models by Monte Carlo techniques, and instead to proceed directly to a simulation of chemically realistic models of orientational glasses with molecular dynamics (MD) techniques [12.89, 90]. The first of these studies [12.89] tries to model $(KBr)_{0.27}(KCN)_{0.73}$, simulating a cell containing 512 ions, out of which 70 randomly distributed Br^- ions occur on the anion sublattice. Only two different samples were examined. MD runs at each temperature were carried out over 3500 time steps of 2 fs, the first 500 of which were discarded for "equilibration". Starting from the high-temperature cubic rotator phase, the effect of isobarically cooling each sample was examined, and various orientational order parameters of the CN^- ions were examined. The results show encouraging qualitative similarity to experimental data. The same statement applies to the second study [12.90], which considers the $Ar_{1-x}(N_2)_x$ system, where as well as the temperature dependence even the concentration dependence was studied, for a box containing 256 molecules which are simulated for up to 40 000 time steps of 3×10^{-14} s each. In this case too, only two samples were studied.

Although these results are clearly interesting, in our view they should be considered as preliminary: the linear dimensions of the systems studied are extremely small, the physical time scale accessible is certainly too short for an adequate equilibration, and an appropriate sample average is clearly needed. Due to these various limitations, these investigations should be considered more as feasibility studies. Another problem that will hamper the direct comparison with experiment is the imprecise knowledge of interatomic potentials.

12.4 The Random-Field Ising Model

While it soon was realized that ferromagnets exposed to arbitrarily weak random fields are unstable with respect to domain formation for dimensions $d \leqq d_l = 4$ in the isotropic case [12.91] (XY and Heisenberg model) and for $d \leqq d_l = 2$ in the Ising case [12.91–96], the behaviour in the $d = 3$ random-field Ising model (RFIM) has long been controversial [12.96, 97], since some experiments and some theories suggested an analogous domain instability for $d = 3$. Although this controversy has been settled by rigorous arguments [12.95] establishing the existence of a spontaneous magnetisation for weak enough random fields at nonzero temperatures, the phase transition from this ferromagnetic phase to the disordered phase at finite temperature has remained poorly understood [12.96–110]. This problem has been addressed by recent simulations [12.100, 101, 106], as have the dynamical properties [12.111–116] of the low temperature phase.

The Hamiltonian of the RFIM is

$$\mathscr{H} = -J \sum_{\langle i,j \rangle} S_i S_j - \sum_i H_i S_i, \quad [H_i H_j]_{\mathrm{av}} = H_{\mathrm{R}}^2 \delta_{ij}. \tag{12.39}$$

Given the result that, for small enough T and small enough random field strength H_{R}, the model (12.39) has a ferromagnetically ordered phase, the first question to be answered is whether the system disorders via a first-order transition or a second-order transition when the temperature is raised. Mean-field theory [12.117] predicts a second-order transition (at least if H_{R}/J

is smaller than some threshold value corresponding to a tricritical point), but the upper critical dimension d_u where mean field theory should be qualitatively valid is $d_u = 6$, and hence one should not rely on mean field theory in $d = 3$. While early Monte Carlo work [12.118] considered systems too small to address this issue, large scale simulations of (12.39) at the Distributed Array Processor (DAP) [12.100], using linear dimensions $L = 64$ and 200 000 MCS/spin-site, suggested a weak first-order transition for $H_R = J$. In contrast, simulations of the diluted nearest-neighbour Ising antiferromagnetic in a uniform field H (which qualitatively can be mapped to (12.39) [12.119]) performed at the Ising model special purpose processor of AT&T Bell Laboratories [12.101], suggest a second-order transition. This model has been studied at a dilution of 30% of the lattice sites, at a field $H/T = 1.5$, at lattice sizes $L = 16, 32$ and 64. This simulation [12.101] used up to 10^6 MCS/spin. It is not clear, however, whether the different conclusions of [12.100, 101] reflect a physical difference of the models considered, or merely incomplete equilibrium and/or insufficient sampling over random-field configurations: note also that the experimental judgements on the order of the transition have been similarly controversial [12.97].

If a second-order transition occurs, it would be very interesting since it involves *three independent exponents* $\eta, \bar{\eta}$ and ν, unlike ordinary critical phenomena where only two exponents occur. This problem can be understood by considering the wave-vector-dependent susceptibilities $\left[S_k \equiv N^{-1} \sum_i S_i \exp(\mathrm{i}\boldsymbol{k} \cdot \boldsymbol{R}_i) \right]$,

$$\chi(\boldsymbol{k}) = N[\langle S_k S_{-k} \rangle_T - \langle S_k \rangle_T \langle S_{-k} \rangle_T]_{\mathrm{av}}, \tag{12.40}$$

$$\chi^{\mathrm{dis}}(\boldsymbol{k}) = N[\langle S_k \rangle_T \langle S_{-k} \rangle_T]_{\mathrm{av}}, \tag{12.41}$$

where the superscript "dis" stands for "disconnected". Now the critical behaviour of these quantities is [12.96]

$$\chi(\boldsymbol{k}) = \xi^{2-\eta} \tilde{\chi}(k\xi), \quad \xi \propto \left(\frac{T}{T_c(H_R)} - 1 \right)^{-\nu}, \tag{12.42}$$

$$\chi^{\mathrm{dis}}(\boldsymbol{k}) = \xi^{4-\bar{\eta}} \tilde{\chi}^{\mathrm{dis}}(k\xi), \tag{12.43}$$

where ξ is the correlation length, $\tilde{\chi}$ and $\tilde{\chi}^{\mathrm{dis}}$ being suitable scaling functions. *Villain* [12.98] and *Fisher* [12.107] proposed a modified hyperscaling relation,

$$2 - \alpha = (d - 2 + \bar{\eta} - \eta)\nu, \tag{12.44}$$

where α is the specific heat exponent (remember the standard form of hyperscaling is $2 - \alpha = d\nu$). Using (12.44) and the standard thermodynamic scaling relation, all critical exponents can be found from $\eta, \bar{\eta}$ and ν. The most reliable estimates of ν and $\bar{\eta}$ have been obtained by *Ogielski* [12.102] using a sophisticated optimisation algorithm to determine exact ground states. These give [12.102]

$$\nu \cong 1.0, \quad \bar{\eta} \cong 1.1, \tag{12.45a}$$

and the Monte Carlo estimate [12.101] of η is

$$\eta = 0.5 \pm 0.1. \tag{12.45b}$$

These values are consistent with, but more accurate than, domain wall renormalisation group calculations. Note that there is a rigorous inequality [12.104],

$$\bar{\eta} \leq 2\eta, \tag{12.46}$$

and that additional arguments have been given [12.105] supporting the equality, $\bar{\eta} = 2\eta$. There is clearly some numerical evidence for the equality, as seen in (12.45a, b). Unfortunately these theoretical values, when inserted into (12.44) give $\alpha \cong 0.5$, whereas, experimentally [12.97] a logarithmic specific heat divergence is found, corresponding to $\alpha = 0$. The cause of this discrepancy is unclear at present.

The fact that this problem is very difficult to treat by simulations becomes evident when we recall the prediction [12.98, 107] of an "activated critical dynamics", i.e. a relaxation time exhibiting an exponential divergence,

$$\tau \propto \exp\left[\mathrm{const}\left(\frac{H_{\mathrm{R}}}{T}\right)\xi^{2+\eta-\bar{\eta}}\right]. \tag{12.47}$$

Only qualitative evidence in favour of (12.47) has been found from simulations [12.101], though recent experimental work provides somewhat stronger support [12.97].

Another interesting feature is the occurrence of nonequilibrium domains, when one quenches the RFIM from $T > T_c$ to $T < T_c$. Both the initial growth rate of these domains has been studied [12.113, 114] and the predicted logarithmic growth laws at intermediate stages after the quench have been checked [12.112, 115, 116]. In view of the obvious numerical difficulties in studying such problems, we do not go into detail here.

These brief discussion clearly shows that the understanding of the RFIM is far from complete, and despite considerable effort, additional large-scale simulations of that model would be very desirable. A much simpler problem, on the other hand, is a random *surface* field acting at the surface of a semi-infinite magnet, which is found not to change the critical behaviour [12.121].

12.5 Concluding Remarks and Outlook

Monte Carlo simulations have played an important role in establishing, after a long period of controversy, that there is a finite-temperature transition for the Ising spin glass. For dynamics, practically all the information that we have, has come from simulations. The critical exponents agree only roughly with experiments and more work needs to be done to sort out the discrepancy. Crossover effects may be complicating the situation in experiments and the simulations may not be fully in the asymptotic scaling region. The situation is less satisfactory for the other models discussed in this article: To date there is much less known about Potts glasses and orientational glasses than about spin glasses. Some of the remaining unsolved questions concern: the expected existence of a finite-temperature phase transition in the Potts glass for $d=4$, the nature of this transition for higher values of p, and, of course, the properties of the low temperature phase. A particularly intriguing problem is the critical behaviour of the relaxation time, both for the Potts and orientational glasses, and for the random field systems. Finally we note that spin glass ideas continue to play a useful role in other areas of physics, even in such an apparently distant field as the macroscopic properties of high temperature superconductors.

Another intriguing question is the relation between the glass transition in spin glasses and orientational galsses and the fluid–glass transition [12.122]. Numerous attempts have been made to contribute to the understanding of this "ordinary" glass transition by simulations (e.g. [12.123–141]). For the most part, this work concerns dynamical phenomena on comparatively short time scales, and hence the connection to the work reviewed here is not obvious.

Acknowledgements. A.P.Y. would like to thank M.P.A. Fisher for educating him about vortex glasses and R.N. Bhatt for a stimulating collaboration on spin glasses. K.B. thanks H.-O. Carmesin and D. Hammes for a fruitful collaboration on various models for orientational glasses. We thank M. Scheucher for the collaboration on the Potts-glass model. This work was supported in part by NSF grant DMR 87-21673, and by the Deutsche Forschungsgemeinschaft (DFG), grant SFB 262/D1.

References

12.1 K. Binder, A.P. Young: Rev. Mod. Phys. **58**, 801 (1986)
12.2 J.L. van Hemmen, I. Morgenstern (eds). Heidelberg Colloquium on Glassy Dynamics, (Springer, Berlin, Heidelberg 1987)
12.3 V. Privman (ed.): Finite Size Scaling and Numerical Simulation of Statistical Systems (World Scientific, Singapore 1990); M.N. Barber: In *Phase Transitions and Critical Phenomena*, by C. Domb, J. Lebowitz (Academic, New York 1983) Vol. 8, p. 146
12.4 S.F. Edwards, P.W. Anderson: J. Phys. F. **5**, 965 (1975)
12.5 M.P.A. Fisher: Phys. Rev. Lett. **62**, 1415 (1989)
12.6 Examples are the free energy per spin, and the correlation between a pair of spins a given relative distance apart, averaged over all such pairs in the system
12.7 A.T. Ogielski, I. Morgenstern: Phys. Rev. Lett. **54**, 928 (1985)
12.8 A.T. Ogielski: Phys. Rev. B **32**, 7384 (1985)
12.9 R.N. Bhatt, A.P. Young: Phys. Rev. Lett. **54**, 924 (1985)
12.10 R.N. Bhatt, A.P. Young: Phys. Rev. B **37** 5606 (1988)
12.11 J.H Condon, A.T. Ogielski: Rev. Sci. Instrum. **56**, 1691 (1985)
12.12 K. Binder: Z. Phys. B **43**, 119 (1981)
12.13 A.J. Bray, M.A. Moore; J. Phys. C **17**, L463 (1984) W.L. McMillan: Phys. Rev. B **30**, 476 (1984); A.J. Bray, M.A. Moore: Phys. Rev. B **31**, 631 (1985) W.L. McMillan: Phys. Rev. B **31**, 340 (1985)
12.14 R.R.P. Singh, S. Chakravarty: Phys. Rev. Lett. **57**, 245 (1986); R.R.P. Singh, M.E. Fisher: J. Appl. Phys. **63**, 3994 (1988)
12.15 R. Fisch, A.B. Harris: Phys. Rev. Lett. **38**, 785 (1977)

12.16 J.A. Olive, A.P. Young, D. Sherrington: Phys. Rev. B **34**, 6341 (1986); S. Jain, A.P. Young: J. Phys. C **19**, 3913 (1986)
12.17 J.R. Banavar, M. Cieplak: Phys. Rev. Lett. **48**, 832 (1982); J.R. Banavar, M. Cieplak: Phys. Rev. B **26**, 2662 (1982); W.L. McMillan: Phys. Rev. B **31**, 342 (1985); B.W. Morris, S.G. Colborne, M.A. Moore, A.J. Bray, J. Canisius: J. Phys. C **19**, 1157 (1986)
12.18 A.P. Malozemoff, Y. Imry, B. Barbara: J. Appl. Phys. **52**, 7672 (1982); R. Omari, J.J. Prejean, J. Souletie: J. de Phys. **44**, 1079 (1983); P. Beauvillain, C. Chappert, J.P. Renard: J. de Phys. Lett. **45**, L665 (1984)
H. Bouchiat: J. Phys. (Paris) **47**, 71 (1986); L.P. Levy, A.T. Ogielski: Phys. Rev. Lett. **57**, 3288 (1986); L.P. Levy: Phys. Rev. B **38**, 4963 (1988)
12.19 Note that these probe the system at much longer space and time scales than the simulations
12.20 A.J. Bray, M.A. Moore, A.P. Young: Phys. Rev. Lett. **56**, 2641 (1986)
12.21 W.L. McMillan: J. Phys. C **17**, 3179 (1984)
12.22 J.D. Reger, A.P. Young: Phys. Rev. B **37**, 5493 (1988)
12.23 A.J. Bray, M.A. Moore: J. Phys. C **15**, 3897 (1982)
12.24 S. Geschwind, D.A. Huse, G.E. Devlin: Phys Rev. B **41**, 2650 (1990); S. Geschwind, D.A. Huse, G.E. Devlin: Phys. Rev. B **41**, 4854 (1990)
12.25 R.B. Griffiths: Phys. Rev. Lett. **23**, 17 (1969)
12.26 M. Randeria, J.P. Sethna, R.G. Palmer: Phys. Rev. Lett. **58**, 765 (1985)
12.27 J.D. Reger, A. Zippelius: Phys. Rev. Lett. **57**, 3225 (1986)
12.28 K. Gunnarson, P. Svendlingh, P. Norblad, L. Lundgren, H. Aruga, A. Ito: Phys. Rev. Lett. **61**, 754 (1988)
12.29 A.P. Young: Phys. Rev. Lett. **50**, 917 (1983); W.L. McMillan: Phys. Rev. B **28**, 5216 (1983); R.H. Swendsen, J.-S. Wang, Phys. Rev. Lett. **57**, 2606 (1986); H.F. Cheung, W.L. McMillan: J. Phys. C **16**, 7027 (1983)
12.30 I. Morgenstern, K. Binder: Phys. Rev. Lett. **43**, 1615 (1979); A.J. Bray, M.A. Moore: J. Phys. C **17**, L463 (1984); W.L. McMillan: Phys. Rev. B **29**, 4026 (1984); B **30**, 476 (1984); H.F. Cheung, W.L. McMillan: J. Phys. C **16**, 7033 (1983)
12.31 D.A. Huse, I. Morgenstern: Phys. Rev. B **32**, 3032 (1985)
12.32 C. Dekker, A.F.M. Arts, H.W. de Wijn: J. Appl. Phys. **63**, 4334 (1988); Phys. Rev. B **38**, 8985 (1988); C. Dekker, A.F.M. Arts, H.W. de Wijn, A.J. van Duyneveldt, J.A. Mydosh: Phys. Rev. B. to be published
12.33 J.A. Cowen, G.C. Kenning, J.M. Slaughter: J. Appl. Phys. **61**, 4080 (1987)
12.34 D.S. Fisher, D.A. Huse: Phys. Rev. B **36**, 8937 (1987)
12.35 D. Sherrington, S. Kirkpatrick: Phys. Rev. Lett. **35**, 1972 (1975)
12.36 G. Parisi: Phys. Rev. Lett. **43**, 1754 (1979); J. Phys. A **13**, 1101 (1980)
12.37 G. Parisi: Phys. Rev. Lett. **50**, 1946 (1983); A.P. Young, A.J. Bray, M.A. Moore: J. Phys. C **17**, L149 (1984); M. Mézard, G. Parisi, N. Sourlas, G. Toulouse, M. Virasoro: Phys. Rev. Lett. **52**, 1156 (1984); J. Phys. (Paris) **45**, 843 (1984); M. Mézard, G. Parisi, M. Virasoro: *Spin Glass Theory and Beyond*, (World Scientific, Singapore 1987)
12.38 A.P. Young: Phys. Rev. Lett. **51**, 1206 (1983)
12.39 D.S. Fisher, D.A. Huse: Phys. Rev. Lett. **56**, 1601 (1986); Phys. Rev. B **36**, 8937, (1987)
12.40 A.J. Bray, M.A. Moore: in *Heidelberg Colloquium on Glassy Dynamics*, ed. by J.L. van Hemmen, I. Morgenstern, Lecture Notes in Physics, Vol. 275 (Springer, Berlin, Heidelberg 1987), p. 121; A.J. Bray: Comments Cond. Mat. Phys. **14**, 21 (1988)
12.41 J.D. Reger, R.N. Bhatt, A.P. Young: Phys. Rev. Lett. **64**, 1859 (1990)
12.42(a) S. Caracciolo, G. Parisi, S. Patarnello, N. Sourlas: Lett. **11**, 783 (1990)
12.42(b) D.S. Fisher, D.A. Huse: preprint
12.43 L. Viana, A.J. Bray: J. Phys. C **8**, 3037 (1985)
P.K. Lai, Y.Y. Goldschmidt: J. Phys. A **22**, 399 (1989)
12.44 A. Georges, M. Mézard, J. Yedidia: preprint
12.45 See e.g. A.L. Fetter, P.C. Hohenberg: In *Superconductivity*, ed. by R.D. Parks (Dekker, New York 1969)
12.46 A.I. Larkin, Yu. N. Ovchnikov: J. Low Temp. Phys. **34**, 409 (1979)
12.47 Y. Imry, S.-K. Ma: Phys. Rev. Lett. **35**, 1399 (1975)
12.48 P.W. Anderson, Y.B. Kim: Rev. Mod. Phys. **36**, 39 (1964)
12.49 J.A. Hertz: Phys. Rev. B **18**, 4875 (1978)
12.50 W.Y. Shih, C. Ebner, D. Stroud: Phys. Rev. B **30**, 134 (1984)
12.51 D.S. Fisher, M.P.A. Fisher, D.A. Huse: Phys. Rev. B **43**, 130 (1991)
12.52 D.A. Huse, H.S. Seung: Phys. Rev. B **42**, 1059 (1990)
12.53 N.S. Sullivan, M. Devoret, D. Esteve: Phys. Rev. B **30**, 4935 (1984)

12.54 R.B. Potts: Proc. Camb. Phil. Soc. **48**, 106 (1952)
12.55 F.Y. Wu: Rev. Mod. Phys. **54**, 235 (1982)
12.56 B.W. Southern, M.F. Thorpe: J. Phys. C **12**, 5351 (1979)
12.57 W. Kinzel, E. Domany: Phys. Rev. B **23**, 3421 (1981)
12.58 J. Adler, Y. Gefen, M. Schick, W.-H. Shih: J. Phys. A **20**, L227 (1987)
12.59 R. Kree, L.A. Turski, A. Zippelius: Phys. Rev. B **58**, 1656 (1987)
12.60 T.R. Kirkpatrick, D. Thirumalai: Phys. Rev. B **36**, 5388 (1987); **37**, 5342 (1988)
D. Thirumalai, T.R. Kirkpatrick: Phys. Rev. B **38**, 4881 (1988)
12.61 H.-O.. Carmesin: J. Phys. A **22**, 297 (1989)
12.62 D. Elderfield, D. Sherrington: J. Phys. C **16**, L497; L971; L1169 (1983) D. Elderfield: J. Phys. A **17**, L517 (1984)
12.63 H. Nishimori, M.J. Stephen: Phys. Rev. B **27**, 5644 (1983)
12.64 E.J.S. Lage, A. Erzan: J. Phys. C **16**, L873 (1983); E.J.S. Lage, J.M. da Silva: J. Phys. C **17**, L593 (1984)
12.65 D.J. Gross, I. Kanter, H. Sompolinsky: Phys. Rev. Lett. **55**, 304 (1985)
12.66 G. Cwilich, T.R. Kirkpatrick: J. Phys. A **22**, 4971 (1989)
12.67 T.R. Kirkpatrick, P.G. Wolynes: Phys. Rev. B **36**, 8552 (1987)
12.68 A. Leutheusser: Phys. Rev. A **29**, 2765 (1984); W. Götze: Z. Phys. B **60**, 195 (1985); U. Bengtzelius: Phys. Rev. A **33**, 3433 (1986)
12.69 Y.Y. Goldschmidt: Phys. Rev. B **31**, 4369 (1985); Nucl. Phys. B **295** [FS21], 409 (1988)
12.70 J.R. Banavar, A.J. Bray: Phys. Rev. B **38**, 2564 (1988)
12.71 J.R. Banavar, M. Cieplak: Phys. Rev. B **39**, 9633 (1989)
12.72 H.-O. Carmesin, K. Binder: J. Phys. A **21**, 4053 (1988)
12.73 M. Scheucher, J.D. Reger, K. Binder, A.P. Young: Phys. Rev. B **42**, 6881 (1990)
12.74 K. Binder: Ferroelectrics **43**, 73 (1987)
12.75 K. Binder: Z. Phys. B**48**, 319 (1982)
12.76 I. Morgenstern, K. Binder: Phys. Rev. B **22**, 288 (1980); I. Morgenstern: In Lecture Notes in Physics Vol. 192, ed by L. van Hemmen, I. Morgenstern (Springer, Berlin, Heidelberg 1983)
12.77 M. Scheucher, J.D. Reger, K. Binder, A.P. Young: Europhys Lett. **14**, 119 (1991)
12.78(a) J.-S. Wang, R.H. Swendsen, R. Kotecky: Phys. Rev. B **42**, 2465 (1990)
12.78(b) R.R.P. Singh to be published
12.79 For reviews of experiments and materials, see K. Knorr: Physica Scripta, **T19**, 531 (1987); A. Loidl; Ann. Rev. Phys. Chem. **40**, 29 (1989)
12.80 K.H. Michel: Phys. Rev. Lett. **57**, 2188 (1986); Phys. Rev. B **35**, 1414 (1987); Z. Phys. B **68**, 259 (1987)
12.81 H.-O. Carmesin: Phys. Lett. A **125**, 294 (1987)
12.82 H.-O. Carmesin, K. Binder: Europhys. Lett. **4**, 269 (1987); Z. Phys. B **68**, 375 (1987)
12.83 D. Hammes. H.-O. Carmesin, K. Binder: Z. Phys. B **76**, 115 (1989)
12.84 H.-O. Carmesin: Z. Phys. B **73**, 381 (1988); H.-O. Carmesin, K. Ohno: J. Magn. Mag. Mat., in press
12.85 K. Binder: Ferroelectrics, **104**, 3 (1990); J. Noncryst. Solids, **131–133**, 262 (1991)
12.86 R. Kohlrausch: Ann. Phys. (Leipzig) **12**, 393 (1847); G. Williams, D.C. Watts: Trans. Faraday Soc. **66**, 80 (1970)
12.87 S. Wansleben: Comp. Phys. Comm. **43**, 315 (1987); G. Bhanot, D. Duke, R. Salvador: Phys. Rev. B **33**, 7841 (1986); J. Stat. Phys. **44**, 985 (1986);
R. Zorn, H.J. Herrmann, C. Rebbi: Comp. Phys. Comm. **23**, 337 (1981)
12.88 P.J. Holdsworth, M.J.P. Gingras, B. Bergersen, E.P. Chan: J. Phys. Condensed Matter, in press. For an early approach along similar lines (fcc lattice) see M. Devoret, D. Esteve: J. Phys. C **16**, 1827 (1983)
12.89 T.J. Lewis, M.L. Klein: Phys. Rev. Lett. **57**, 2698 (1986); Phys. Rev. B **40**, 7080 (1989)
12.90 H. Klee, H.-O. Carmesin, K. Knorr: Phys. Rev. Lett. **61**, 1855 (1988)
12.91 Y. Imry, S.K. Ma: Phys. Rev. Lett. **35**, 1399 (1975)
12.92 J. Villain: J Phys. Lett. (Paris) **43**, L551 (1982)
12.93 G. Grinstein, S.K. Ma: Phys. Rev. Lett. **49**, 685 (1982); Phys. Rev. B **28**, 2588 (1983)
12.94 K. Binder: Z. Phys. B **50**, 343 (1983)
12.95 J.Z. Imbrie: Phys. Rev. Lett. **53**, 1747 (1984), Commun. Math. Phys. **98**, 145 (1985)
12.96 For reviews of the theory of the RFIM, see J. Villain: in *Scaling Phenomena in Disordered Systems*, R. Pynn, A. Skjeltorp, (eds.) p. 423 (Plenum, New York 1985); Y. Imry: J. Stat. Phys. **34**, 849 (1984); G. Grinstein: J. Appl. Phys. **55**, 2371 (1984); T. Nattermann, J. Villain: Phase Transition, **11**, 5 (1988)

12.97 For reviews of experiments on random–field systems, see R.J. Birgeneau, R.A. Cowley, G. Shirane, H. Yashizawa: J. Stat. Phys. **34**, 817 (1984); R.J. Birgeneau, Y. Shapira, G. Shirane, R.A. Cowley, H. Yoshizawa: Physica, **37 B&C**, 83 (1986); For more recent experiments, see D.P. Belanger, V. Jaccarino, A.R. King, R.M. Nicklow: Phys. Rev. Lett. **59**, 930 (1987); J. Phys. (Paris), **49C–8**, 1229 (1988); A.E. Nash, A.R. King, V. Jaccarino: UCSB preprint, Experimental verification of activated critical dynamics in the $d = 3$ random field Ising model
12.98 J. Villain: J. de Phys. **46**, 1843 (1985)
12.99 A.J. Bray, M.A. Moore: J. Phys. C **18**, L927 (1985)
12.100 A.P. Young, M. Nauenberg: Phys. Rev. Lett. **54**, 2429 (1985)
12.101 A.T. Ogielski, D.A. Huse: Phys. Rev. Lett. **56**, 1298 (1986)
12.102 A.T. Ogielski: Phys. Rev. Lett. **57**, 1252 (1986)
12.103 Note that the exponent $\bar{\eta}$, quoted in [12.102] is related to our $\bar{\eta}$ by $\bar{\eta} = \bar{\eta} - 2$
12.104 M. Schwartz, A. Soffer: Phys. Rev. Lett. **55**, 2499 (1985)
12.105 M. Schwartz: University of Tel Aviv preprint
12.106 J.L. Cambier, M. Nauenberg: Phys. Rev. B **34**, 7998 (1986)
12.107 D.S. Fisher: Phys. Rev. Lett. **56**, 416 (1986)
12.108 D.A. Huse: Phys. Rev. B **36**, 5383 (1987)
12.109 D.A. Huse, D.S. Fisher: Phys. Rev. B **35**, 6841 (1987)
12.110 T. Nattermann, J. Vilfan: Phys. Rev. Lett. **61**, 223 (1988)
12.111 D. Stauffer, C. Hartzstein, K. Binder, A. Aharony: Z. Phys. B **55**, 325 (1984)
12.112 D. Chowdhury, D. Stauffer: Z. Phys. B **60**, 249 (1984)
12.113 E.T. Gawlinski, K. Kaski, M. Grant, J.D. Gunton: Phys. Rev. Lett. **53**, 2266 (1966); E.T. Grawlinski, M. Grant, J.D. Gunton, K. Kaski: Phys. Rev. B **31**, 281 (1985)
12.114 D. Chowdhury, E.T. Gawlinski, J.D. Gunton: Computer Phys. Commun. to be published
12.115 S.R. Anderson: Phys. Rev. B **36**, 8435 (1987)
12.116 E. Pytte, J.F. Fernandez: Phys. Rev. B **31**, 616 (1985)
12.117 A. Aharony: Phys. Rev. B **8**, 3318 (1978)
12.118 D.P. Landau, H.H. Lee, W. Kao: J. Appl. Phys. **49**, 1356 (1978); E.B. Rasmussen, M.A. Novotny, D.P. Landau: J. Appl. Phys. **53**, 1925 (1983); D. Andelman, H. Orland, L. Wijewordhana: Phys. Rev. Lett. **52**, 145 (1984); L. Jacobs, M. Nauenberg: Physica **128A**, 529 (1984)
12.119 S. Fishman, A. Aharony: J. Phys. C **22**, L729 (1979)
12.120 H.-F. Chung: Phys. Rev. B **33**, 6191 (1986)
12.121 K.K. Mon, M.P. Nightingale: Phys. Rev. B **37**, 3815 (1988)
12.122 J. Jäckle: Rep. Prog. Phys. **49**, 171 (1986)
12.123 C.A. Angell, J.H.R. Clark, L.V. Woodcock: Adv. Chem. Phys. **48**, 379 (1981)
12.124 C.A. Angell: Ann. N.Y. Sci. **371**, 136 (1981)
12.125 F.F. Abraham: J. Chem. Phys. **72**, 359 (1980); **75**, 498 (1981)
12.126 J.R. Fox, H.C. Andersen: Ann. N.Y. Acad. Sci. **371**, 123 ((1981); J. Phys. Chem. **88**, 4019 (1984)
12.127 J.J. Ullo, S. Yip: Phys. Rev. Lett. **54**, 1509 (1985)
12.128 S. Nośe, F. Yonezawa: Solid State Comm. **56**, 1005 (1985); B. Bernu, Y. Hiwatari, P. Hansen: J. Phys. C **18**, L371 (1985)
12.129 F. Yonezawa, M. Kimura: J. Non-Cryst. Solids **61/62**, 761 (1984)
12.130 B. Bernu, J.P. Hansen, Y. Hiwatari, G. Pastore: Phys. Rev. A **36**, 4891 (1987); G. Pastore, B. Bernu, J.P. Hansen, Y. Hiwatari: Phys. Rev. A **38**, 454 (1988)
12.131 J.L. Barrat, J.P. Hansen, J. Totniji: J. Phys. C **21**, 4511 (1988)
12.132 J.L. Barrat, J.N. Roux, J.P. Hansen, M.L. Klein: Europhys. Lett. **7**, 707 (1988)
12.133 J.N. Roux, J.L. Barrat, J.P. Hansen: J. Phys. Cond. Matter **1**, 7171 (1989)
12.134 J.L. Barrat, J.N. Roux, J.P. Hansen: preprint (1990); G.F. Signorini, J.L. Barrot, M.L. Klein: J. Chem. Phys. **92**, 1294 (1990)
12.135 G. Wahnström: J. Non-Cryst. Solids (1991, in press)
12.136 J.L. Barrat: J. Non-Cryst. Solids (1991, in press)
12.137 N. Pistoor, K. Kremer: In *Dynamics of Disordered Materials* Springer Proc. Phys. Vol. 37 ed. by D. Richter, A.J. Dianoux, W. Petry, J. Teixeira (Springer, Berlin, Heidelberg 1989)
12.138 R. Dejean de la Batie, J.-L. Viovy, L. Monnerie: J. Chem. Phys. **81**, 567 (1984)
12.139 H.P. Wittmann, K. Kremer, K. Binder: to be published
12.140 D. Rigby, R.J. Roe: J. Chem. **87**, 7285 (1987)
12.141 G.H. Fredrickson, H.C. Andersen: J. Chem. Phys. **83**, 5822 (1985); H. Nakanishi, H. Takano, Phys. Lett. **A115**, 117 (1986); E. Leutheusser, H. De Raedt: Solid State Commun. **54**, 457 (1986); G.H. Fredrickson: Ann. N.Y. Sci. **484**, 185 (1986); J. Reiter: to be published

Subject Index